**Space Trajectories**

# Space Trajectories

Basic and Advanced Topics

*Max Cerf*
ArianeGroup, Les Mureaux, France

WILEY

Published by John Wiley & Sons, Inc., Hoboken, New Jersey.
Published simultaneously in Canada.



***Library of Congress Cataloging-in-Publication Data Applied for:***

Hardback ISBN: 9781394293797

Cover Design: Wiley
Cover Images: © Magnilion/Getty Images, © NEMES LASZLO/SCIENCE PHOTO LIBRARY/Getty Images

Set in 9.5/12.5pt STIXTwoText by Straive, Chennai, India

*To Mallow, who witnessed the completion of this book in March 2024.*

# Contents

**About the Author** *xv*
**Foreword** *xvii*
**Acknowledgments** *xix*
**Introduction** *xxi*

**Part I Free Orbital Motion** *1*

**1 Two-Body Problem** *3*
1.1 Introduction *3*
1.2 Keplerian Motion *3*
1.2.1 Dynamic Model *4*
1.2.2 Prime Integrals *5*
1.2.3 Orbit Shape *7*
1.3 Motion Time Law *12*
1.3.1 Elliptical Orbit *12*
1.3.2 Hyperbolic Orbit *14*
1.3.3 Parabolic Orbit *15*
1.3.4 Lagrange Coefficients *16*
1.3.5 Universal Variable *18*
1.4 Orbital Parameters *23*
1.4.1 Classical Orbital Parameters *23*
1.4.2 Relation to Position and Velocity *23*
1.4.3 Equinoctial Orbital Parameters *25*
1.4.4 Earth Orbits *25*
1.5 Conclusion *31*
1.5.1 The Key Points *31*
1.5.2 To Go Further *31*

**2 Perturbed Motion** *33*
2.1 Introduction *33*
2.2 Unperturbed Motion *34*
2.2.1 Keplerian Model *34*
2.2.2 Orbital Parameters *34*
2.2.3 Useful Frames *36*

2.3 Perturbed Motion *36*
2.3.1 Osculating Orbit *37*
2.3.2 Derivation Formula *37*
2.3.3 Gauss Equations *38*
2.3.4 Lagrange Equations *41*
2.3.5 Equinoctial Parameters *43*
2.3.6 Integration Methods *44*
2.4 Gravitational Perturbations *48*
2.4.1 Gravitational Potential *49*
2.4.2 Spherical Body *49*
2.4.3 Nonspherical Body *51*
2.4.4 First Zonal Term *54*
2.5 Other Perturbations *58*
2.5.1 Third-Body Attraction *58*
2.5.2 Atmospheric Friction *64*
2.5.3 Radiation Pressure *65*
2.6 Conclusion *65*
2.6.1 The Key Points *65*
2.6.2 To Go Further *66*

**3 Three-Body Problem** *67*
3.1 Introduction *67*
3.2 Circular-Restricted Three-Body Problem *68*
3.2.1 Motion Equations *68*
3.2.2 Accessible Region *69*
3.2.3 Lagrange Points *71*
3.3 Periodic Orbits *74*
3.3.1 Linearized Solution *74*
3.3.2 Periodic Solution *80*
3.3.3 Halo Orbits *83*
3.4 Transfers *86*
3.4.1 Linearization in the Orbit Vicinity *86*
3.4.2 Invariant Manifolds *88*
3.4.3 Transfer Strategies *89*
3.5 Conclusion *93*
3.5.1 The Key Points *93*
3.5.2 To Go Further *93*

**4 Orbit Determination** *95*
4.1 Introduction *95*
4.2 Measurements *96*
4.2.1 Observation System *96*
4.2.2 Measurements *100*
4.2.3 Directions of Stars *102*
4.3 Preliminary Orbit Estimation *104*
4.3.1 Position Measurements *104*
4.3.2 Direction Measurements *107*

4.4 Continuous Orbit Estimation *110*
4.4.1 Least Squares *111*
4.4.2 Differential Correction *113*
4.4.3 Kalman Filtering *116*
4.5 Conclusion *119*
4.5.1 The Key Points *119*
4.5.2 To Go Further *119*

**5 Collision Risks** *121*
5.1 Introduction *121*
5.2 Orbit Uncertainties *122*
5.2.1 Orbital Motion *122*
5.2.2 Ellipsoid of Uncertainty *124*
5.2.3 Gaussian Model *126*
5.3 Conjunction *129*
5.3.1 Numerical Simulation *129*
5.3.2 Orbit-to-Orbit Distance *129*
5.3.3 Trajectory-to-Orbit Distance *133*
5.3.4 Combined Covariance *134*
5.4 Risk of Collision *136*
5.4.1 Short Conjunction *137*
5.4.2 Probability of Collision *138*
5.4.3 Analytical Formula *140*
5.4.4 Maximum Probability *141*
5.4.5 Long Conjunction *143*
5.5 Conclusion *145*
5.5.1 The Key Points *145*
5.5.2 To Go Further *145*

**Part II Controlled Orbital Motion** *147*

**6 Impulsive Transfer** *149*
6.1 Introduction *149*
6.2 Orbit Target *150*
6.2.1 Problem Formulation *150*
6.2.2 Transfer in One Impulse *151*
6.2.3 Transfer in Two or Three Impulses *155*
6.3 Point Target *160*
6.3.1 Problem Formulation *160*
6.3.2 Minimum-Energy Transfer *161*
6.3.3 Minimum-Eccentricity Transfer *162*
6.3.4 Noncollinear Transfer *163*
6.3.5 Collinear Transfer *168*
6.4 Point and Time Target *171*
6.4.1 Lambert's Problem *171*
6.4.2 Lambert's Theorem *171*

6.4.3 Transfer Time Equation *175*
6.4.4 Universal Variable *181*
6.4.5 Solution Methods *183*
6.5 Conclusion *184*
6.5.1 The Key Points *184*
6.5.2 To Go Further *184*

**7 Orbital Rendezvous** *187*
7.1 Introduction *187*
7.2 Phasing and Transfer *188*
7.2.1 Orbital Model *188*
7.2.2 Phasing *189*
7.2.3 Transfer *190*
7.2.4 Visibility *197*
7.3 Target in Circular Orbit *198*
7.3.1 Hill–Clohessy–Wiltshire Equations *198*
7.3.2 Free Motion *200*
7.3.3 Maneuvers *204*
7.3.4 Approach Scenario *206*
7.4 Control Laws *206*
7.4.1 Optimum Control *206*
7.4.2 Specific Controls *210*
7.5 Conclusion *214*
7.5.1 The Key Points *214*
7.5.2 To Go Further *214*

**8 Optimal Thrust Level** *215*
8.1 Introduction *215*
8.2 Problem Formulation *216*
8.2.1 Optimal Control Problem *216*
8.2.2 Conditions for Optimality *217*
8.2.3 Property of the Velocity Costate *218*
8.3 Analytical Solution *221*
8.3.1 Direction of Thrust *221*
8.3.2 Costate Vector *225*
8.3.3 Injection Point and Direction *226*
8.3.4 Reduced Problem *228*
8.3.5 Performance Estimate *229*
8.4 Application *230*
8.4.1 Optimized Thrust Profile *230*
8.4.2 Fixed Thrust Level *232*
8.5 Conclusion *236*
8.5.1 The Key Points *236*
8.5.2 To Go Further *237*

**9 Low-Thrust Transfer** *239*
9.1 Introduction *239*
9.2 Problem Formulation *240*
9.2.1 Dynamics *240*
9.2.2 Optimal Control Problem *242*
9.2.3 Local Control Laws *245*
9.2.4 Edelbaum's Solution *249*
9.3 Transfers to the Geostationary Orbit *251*
9.3.1 Dynamic Model *251*
9.3.2 Optimal Control Problem *253*
9.3.3 Solution Method *255*
9.3.4 Application Cases *258*
9.4 Transfers Between Circular Orbits *262*
9.4.1 Dynamic Model *262*
9.4.2 Optimal Control Problem *263*
9.4.3 Form of Optimal Trajectories *265*
9.4.4 Solution Method *268*
9.4.5 Application Case *270*
9.5 Conclusion *273*
9.5.1 The Key Points *273*
9.5.2 To Go Further *273*
9.5.3 Authorizations *274*

**10 Space Debris Cleaning** *275*
10.1 Introduction *275*
10.2 Problem Formulation *276*
10.2.1 Debris Orbits *276*
10.2.2 Cleaning Program *277*
10.2.3 Optimization Problem *279*
10.3 Transfer Problem *280*
10.3.1 Generic Transfer Strategy *280*
10.3.2 High-Thrust Propulsion *281*
10.3.3 Low-Thrust Propulsion *282*
10.3.4 Reduced Formulation *284*
10.4 Solution Method *285*
10.4.1 Simulated Annealing *285*
10.4.2 Cost Function *287*
10.4.3 Solution Process *290*
10.5 Application Case *290*
10.5.1 High-Thrust Case *292*
10.5.2 Low-Thrust Case *294*
10.5.3 Comparison of High and Low-Thrust Solutions *296*
10.6 Conclusion *297*
10.6.1 The Key Points *297*
10.6.2 To Go Further *297*
10.6.3 Authorizations *297*

**Part III Launch and Reentry** *299*

**11 Launch into Orbit** *301*
11.1 Introduction *301*
11.2 Launcher Dynamics *302*
11.2.1 Force Models *302*
11.2.2 Motion Equations *304*
11.3 Launcher Configuration *307*
11.3.1 Propulsive ΔV *307*
11.3.2 Staging *308*
11.3.3 Preliminary Design *310*
11.3.4 Versatility *311*
11.4 Trajectory Optimization *312*
11.4.1 Constraints *312*
11.4.2 Risk Mitigation *315*
11.4.3 Optimal Control Problem *316*
11.4.4 Numerical Methods *318*
11.4.5 Trajectory Segmentation *320*
11.5 Guidance *322*
11.5.1 Implicit Guidance *323*
11.5.2 Explicit Guidance *324*
11.5.3 Flight Control *326*
11.6 Conclusion *327*
11.6.1 The Key Points *327*
11.6.2 To Go Further *327*

**12 Launcher Staging** *329*
12.1 Introduction *329*
12.2 Staging Problem *330*
12.2.1 Launcher Modeling *330*
12.2.2 Trajectory Modeling *332*
12.2.3 Global Problem *334*
12.3 Impulsive Method *335*
12.3.1 Propulsive Velocity Increment *335*
12.3.2 Staging Problem *336*
12.3.3 Staging Loops *337*
12.4 Coupled Method *339*
12.4.1 Control Law *339*
12.4.2 Iterations and Margins *340*
12.4.3 Versatile Configuration *341*
12.4.4 Application Case 1: Number of Boosters *342*
12.4.5 Application Case 2: Upper Stage Loading *345*
12.5 Propellant Reserve *348*
12.5.1 Performance and Reserve *348*
12.5.2 Iterative Method *349*
12.5.3 Embedded Method *350*
12.5.4 Application Case *351*

12.6 Conclusion *354*
12.6.1 The Key Points *354*
12.6.2 To Go Further *354*
12.6.3 Authorizations *354*

**13 Flat Earth Solutions** *355*
13.1 Introduction *355*
13.2 Ballistic Launch at Constant Acceleration *355*
13.2.1 Modeling *356*
13.2.2 Optimal Solution *357*
13.2.3 Application Case *358*
13.3 Injection into Orbit at Constant Acceleration *359*
13.3.1 Modeling *359*
13.3.2 Optimal Solution *359*
13.3.3 Application Case *361*
13.4 Vertical Launch at Variable Thrust *363*
13.4.1 Modeling *363*
13.4.2 Optimal Solution *364*
13.4.3 Application Case *366*
13.5 Injection into Orbit at Variable Thrust *367*
13.5.1 Modeling *367*
13.5.2 Optimal Solution *368*
13.5.3 Switching Function *370*
13.5.4 Numerical Solution *372*
13.6 Transition to the Round Earth *374*
13.6.1 Change of Coordinates *374*
13.6.2 Continuation Method *376*
13.6.3 Application Case *378*
13.7 Conclusion *381*
13.7.1 The Key Points *381*
13.7.2 To Go Further *381*

**14 Interplanetary Trajectory** *383*
14.1 Introduction *383*
14.2 Trajectory Modeling *384*
14.2.1 Patched Conics *384*
14.2.2 Sphere of Influence *385*
14.2.3 Heliocentric Phase *387*
14.2.4 Planetocentric Phase *390*
14.3 Escape Conditions *392*
14.3.1 Keplerian Motion *392*
14.3.2 Hyperbolic Orbit *393*
14.3.3 Escape Velocity *395*
14.3.4 Fly-by Maneuver *397*
14.4 Mission Scenario *399*
14.4.1 Launch Strategy *399*
14.4.2 Tisserand Criterion *401*

14.4.3 Scenario Optimization *405*
14.4.4 Correction Maneuvers *406*
14.5 Conclusion *409*
14.5.1 The Key Points *409*
14.5.2 To Go Further *410*

**15 Atmospheric Reentry** *411*
15.1 Introduction *411*
15.2 Motion Equations *411*
15.2.1 Reentry Missions and Vehicles *411*
15.2.2 Rotating Geocentric Frame *413*
15.2.3 Spherical Coordinates *414*
15.2.4 Forces Applied *416*
15.2.5 Differential Equations of Motion *419*
15.2.6 Trajectory Optimization *420*
15.3 Orbital Phase *421*
15.3.1 Deorbitation *422*
15.3.2 Ballistic Range *424*
15.3.3 Return Maneuver *425*
15.4 Atmospheric Phase *427*
15.4.1 Simplified Equations *427*
15.4.2 Reentry Corridor *428*
15.4.3 Flight Corridor *430*
15.4.4 Accessible Area *432*
15.4.5 Steep Reentry *435*
15.5 Conclusion *436*
15.5.1 The Key Points *436*
15.5.2 To Go Further *436*

**Short Bibliography** *437*
**Web links** *438*
**Index** *439*

## About the Author

Max Cerf has been an engineer at ArianeGroup since 1990, specializing in optimization of space trajectories and spacecraft. He is also a university professor and author of a book on optimization techniques. He is graduate of Ecole Centrale de Paris (1989), holds a PhD (2012) and a "Habilitation à Diriger des Recherches" (2019), and was appointed "Chevalier de l'Ordre National du Mérite" (2015).

# Foreword

The conquest of space is a major adventure in the history of mankind.

In the 17th century, Newton discovered the laws governing the motion of celestial bodies, thus founding classical mechanics, notably the theory of universal gravitation. Although mathematically simple, these laws conceal an immense wealth of possible dynamics. The Newtonian gravitational field possesses a host of remarkable properties, discovered over the centuries through the development of increasingly sophisticated mathematics, which are of great interest to space agencies. For example, the existence of periodic orbits, points of equilibrium, invariant manifolds, and other sometimes unexpected properties make possible space missions that are awe-inspiring.

In this remarkable book, in his inimitably clear style, Max Cerf reveals all the secrets of space dynamics and mission design. Senior engineer at ArianeGroup, expert in mission analysis and optimization, Max Cerf is a master of mission design and space vehicle control. In addition to being an accomplished and renowned engineer and researcher, he is also an exceptional teacher, as the reader will appreciate from the pleasant and fascinating reading of this book. From rocket launch and orbiting, through problems of atmospheric reentry, orbital transfer, and space debris clean-up, to interplanetary missions, Max Cerf provides all the keys you need to quickly master all aspects and implementation.

This book will appeal to students, researchers, and engineers looking for a global view of trajectory design problems in astrodynamics, and to acquire the main techniques, particularly in the fields of optimization and optimal control. After reading this book, all you want to do is take part in this amazing adventure!

*Emmanuel Trélat*
Sorbonne Université, Paris

## Acknowledgments

I express my gratitude to the people who have supported me in one way or another during my career and contributed to the content of this book, and more especially:

- *Lauren Poplawski for accepting and managing the book's publication by Wiley;*
- *Nandhini Karuppiah, Govindanagaraj Deenadayalu, and Isabella Proietti for their assistance throughout the editorial process;*
- *the many ArianeGroup colleagues with whom I worked over the years, with a special mention for Mathieu Verlet, Arnaud Ruiz, Stéphane Reynaud, and Hervé Gilibert;*
- *Christophe Sauty and Ana Gomez of Paris Observatory, who enabled me to complete the master's degree in astronomy in parallel with my daily work;*
- *Emmanuel Trélat for his laudatory foreword and our deep collaboration on various space trajectory problems;*
- *Thomas Haberkorn, Jean-Baptiste Caillau, and Hasnaa Zidani for our many fruitful exchanges.*

# Acknowledgements

# Introduction

Space trajectories is a field at the crossroads of mechanics and mathematics. Its development can be divided into two periods.

The first period followed the discovery of Kepler's laws and their demonstration by Newton who laid the foundations of classical mechanics. During this period spanning the 18th and the 19th centuries, research focused on celestial mechanics with the goal of predicting the motion of planets as precisely as possible.

The second period began with the invention of rocket propulsion at the beginning of the 20th century. The possibility of sending maneuvering vehicles in space stimulated the development of optimal control theory, which was originally mainly oriented toward space applications.

This book is aimed at students, researchers, and engineers interested in space applications and who wish to gain an overview of the field. It recalls the essential concepts of orbital mechanics and also presents advanced space applications involving trajectory optimization.

The content is divided into three parts, each of which composed of five chapters.

- The first part deals with free orbital motion with the following topics: Keplerian motion, perturbed motion, three-body problem, orbit determination, and collision risks in orbit.
- The second part deals with controlled orbital motion with the following topics: impulsive transfer, orbital rendezvous, thrust level optimization, low-thrust transfer, and space debris cleaning.
- The third part deals with ascent and reentry with the following topics: launch into orbit, launcher staging, analytical solutions in flat Earth, interplanetary mission, and atmospheric reentry.

These topics have been selected for their practical relevance to space applications. Each chapter covers the necessary concepts, so that it can be read independently of the rest of the book.

# Part I

# Free Orbital Motion

# 1

# Two-Body Problem

## 1.1 Introduction

The two-body problem is the fundamental model of orbital mechanics. It describes the motion of two bodies in pure gravitational interaction to the exclusion of all other forces. In the restricted two-body problem, the mass of one body is assumed to be negligible, while the other is considered as point-like. This modeling is appropriate to the motion of artificial satellites around a celestial body.

The two-body problem has an analytical solution called Keplerian motion. The trajectories are conics, satisfying properties of conservation of energy and angular momentum. The conic nature (circle, ellipse, parabola, or hyperbola) depends only on the initial conditions of position and velocity and the motion time law is determined by Kepler's equation. The Keplerian orbit of a satellite around the attracting body is represented geometrically by its orbital parameters, which are analytically related to the position and velocity.

The first section presents the dynamic model and derives the main prime integrals of the Keplerian motion, namely those of angular momentum, eccentricity vector, and energy. The conic shape is then determined from the initial conditions of position and velocity.

The second section deals with the time law of the motion. Position and time are linked by Kepler's equation, which is transcendental. Kepler's equation takes different forms depending on the conic nature. It can also be expressed in a universal form valid for any kind of orbit. The solution of Kepler's equation requires a numerical iterative method.

The third section defines the orbital parameters locating the orbit in space and establishes their relation to position and velocity. The classical orbital parameters have singularities for circular or equatorial orbits, which motivates the definition of the equinoctial parameters, devoid of such singularities. An overview of Earth's orbits is finally presented, with a focus on Earth's coordinate systems and on specific properties such as geo-synchronism or sun-synchronism.

## 1.2 Keplerian Motion

The two-body problem, or Kepler's problem, concerns the relative motion of two particles exerting between them the force of gravitational attraction. The resulting dynamics, called Keplerian motion, is presented in this section.

*Space Trajectories: Basic and Advanced Topics*, First Edition. Max Cerf.

### 1.2.1 Dynamic Model

The two-body problem is formulated in a Galilean (or inertial) reference frame.

Both bodies are considered as material points of masses $m_1$ and $m_2$ and positions $\vec{r}_1$ and $\vec{r}_2$, respectively. The gravitational force exerted by the second body on the first body is

$$\vec{F}_{2\to1} = G\frac{m_1 m_2}{r^3}\vec{r} \tag{1.1}$$

The constant G is the universal gravitational constant and $\vec{r} = \vec{r}_2 - \vec{r}_1$ denotes the position of body 2 relative to body 1. Newton's third law (action and reaction) implies that body 1 exerts an opposite force on body 2: $\vec{F}_{1\to2} = -\vec{F}_{2\to1}$.

---

**Universal Gravitational Constant**

The accurate knowledge of the value of G is of considerable importance, both for the computation of space trajectories and for the confirmation of theories such as general relativity.

The first reliable measurement of G was made in 1798 by Henry Cavendish, using a torsion pendulum. The possibility of on-board experiments carried out by satellites has greatly improved the accuracy. The reference value adopted in the international system is: $G \approx 6.67408 \cdot 10^{-11}\ m^3\ kg^{-1}\ s^{-2}$.

---

Newton's second law (also called fundamental relation of dynamics) applied to each body in the inertial frame yields the two differential equations

$$\begin{cases} \dfrac{d^2\vec{r}_1}{dt^2} = \dfrac{\vec{F}_{2\to1}}{m_1} = G\dfrac{m_2}{r^3}\vec{r} \\ \dfrac{d^2\vec{r}_2}{dt^2} = \dfrac{\vec{F}_{1\to2}}{m_2} = -G\dfrac{m_1}{r^3}\vec{r} \end{cases} \tag{1.2}$$

After subtracting the first equality from the second, we obtain the differential equation for the motion of body 2 relative to body 1.

$$\frac{d^2\vec{r}}{dt^2} = -\frac{\mu}{r^3}\vec{r} \tag{1.3}$$

where $\mu = G(m_1 + m_2)$ is called the gravitational constant of the two-body system.

In the restricted two-body problem, the mass of body 2 is negligible compared with that of body 1 (main attracting body). The constant $\mu = Gm_1$ is then called the gravitational constant of the main attracting body. This model is well suited to practical applications studying the motion of artificial satellites.

The values of $\mu$ for Earth, Moon, and Sun are (in $m^3/s^2$):

$$\mu_{Earth} \approx 3.986\,005 \cdot 10^{14}, \quad \mu_{Moon} \approx 4.902\,835 \cdot 10^{12}, \quad \mu_{Sun} \approx 1.327\,23 \cdot 10^{20}$$

The solution to the two-body problem is called **Keplerian motion**, after Johann Kepler, who first set out the laws of motion.

---

**Kepler's Laws**

The laws of planetary motion around the Sun were set out empirically by Johann Kepler (1571–1630) in *Astronomia Nova*, published in 1609.

The three laws of Kepler are as follows:

- the planets describe ellipses with the Sun as one of their foci;
- the radius vector sweeps equal areas in equal times;
- the square of the period of revolution is proportional to the cube of the semi-major axis.

Kepler had guessed these three laws from the collection of observations made by the Danish astronomer Tycho Brahé (1456–1601) on the motion of the planets.

Their demonstration came much later owing to Isaac Newton (1642–1727).

In *Philosophiae Naturalis Principia Mathematica*, published in 1687, Newton postulated the laws of motion (inertia, force, and reaction) and the law of gravitation, from which he rigorously established the solution to the two-body problem.

These fundamental results form the basis of orbital and celestial mechanics since the 18th century.

---

### 1.2.2 Prime Integrals

A prime integral of a dynamic system is a quantity that remains constant over time. Knowledge of prime integrals helps to delimit the system evolution, even when no analytical solution exists. The two-body problem admits several prime integrals, such as those of angular momentum, energy, and eccentricity vector.

#### 1.2.2.1 Angular Momentum

The angular momentum per unit mass is $\vec{h} = \vec{r} \wedge \vec{v}$, where $\vec{v} = \frac{d\vec{r}}{dt}$ is the velocity of body 2 relative to body 1. Calculating the derivative of $\vec{h}$ with the assumption of a central acceleration given by (1.3) yields

$$\frac{d\vec{h}}{dt} = \frac{d\vec{r}}{dt} \wedge \vec{v} + \vec{r} \wedge \frac{d\vec{v}}{dt} = \vec{v} \wedge \vec{v} + \vec{r} \wedge \left(-\frac{\mu}{r^3}\vec{r}\right) = \vec{0} \tag{1.4}$$

The angular momentum vector $\vec{h}$ is therefore constant throughout the motion.

If $\vec{r}$ and $\vec{v}$ are collinear, then $\vec{h} = \vec{0}$ and the motion takes place along a straight line. Such a trajectory is said to be degenerate. Unless otherwise specified, we assume in the sequel that $\vec{h} \neq \vec{0}$.

As the position and the velocity are constantly perpendicular to $\vec{h}$, the motion takes place in a plane called the orbital plane.

Taking body 1 as origin O and a reference axis Ox in this plane (Figure 1.1):

- the position of body 2 is defined by the polar angle $\theta$ and the radius vector r;
- the velocity of body 2 has radial and orthoradial components $\dot{r}$ and $r\dot{\theta}$ (where $\dot{x}$ denotes the time derivative of x).

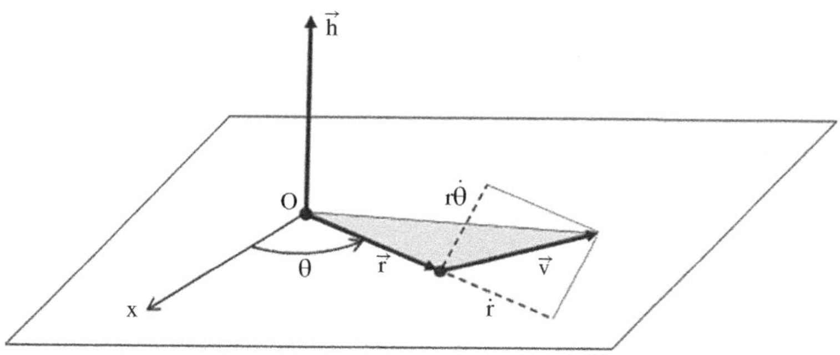


**Figure 1.1** Orbital plane.

The modulus of the angular momentum is

$$h = \|\vec{r} \wedge \vec{v}\| = r^2\dot{\theta} \tag{1.5}$$

In an infinitesimal time dt, the radius vector sweeps the area dS of the triangle with sides $\vec{r}$ and $\vec{v}dt$. This area is

$$dS = \frac{1}{2}\|\vec{r} \wedge \vec{v}dt\| = \frac{1}{2}hdt \tag{1.6}$$

The areal velocity (area swept per unit time) is therefore constant, just as the angular momentum. This establishes **Kepler's second law** or **law of areas.**

#### 1.2.2.2 Energy

The mechanical energy per unit mass noted w is the sum of the kinetic energy $\frac{v^2}{2}$ and the potential energy $-\frac{\mu}{r}$, from which the gravitational force $-\frac{\mu}{r^3}\vec{r}$ derives.

$$w = \frac{v^2}{2} - \frac{\mu}{r} \tag{1.7}$$

Taking the scalar product of each member of (1.3) with $\vec{v}$ and using $\vec{r} \cdot \vec{v} = r\dot{r}$ (Figure 1.1), we obtain

$$\frac{d\vec{v}}{dt} \cdot \vec{v} = -\frac{\mu}{r^3}\vec{r} \cdot \vec{v} = -\frac{\mu}{r^3}r\dot{r} = -\mu\frac{\dot{r}}{r^2} \Rightarrow \frac{d}{dt}\left(\frac{v^2}{2}\right) = \frac{d}{dt}\left(\frac{\mu}{r}\right) \tag{1.8}$$

Instantaneous variations in kinetic and potential energy cancel each other out. This statement is in fact true for any force deriving from a potential.

The mechanical energy $w = \frac{v^2}{2} - \frac{\mu}{r}$ is therefore constant throughout the motion.

The accessible domain depends on the energy sign. To escape the main attracting body, the energy must be zero or positive (kinetic energy ≥ potential energy).

In this case, the escape velocity $v_{esc}$ associated with zero energy ($w = 0$) and the residual velocity at infinity $v_\infty$ (for $r \to \infty$) are respectively given by

$$\begin{aligned} w &= \frac{v^2}{2} - \frac{\mu}{r} = 0 \quad \text{(for escape)} \quad \Rightarrow \quad v_{esc} = \sqrt{\frac{2\mu}{r}} \\ w &= \frac{v^2}{2} - \frac{\mu}{r} = \frac{v_\infty^2}{2} \quad (\text{for } r \to \infty) \quad \Rightarrow \quad v_\infty = \sqrt{2w} \end{aligned} \tag{1.9}$$

#### 1.2.2.3 Eccentricity Vector

The eccentricity vector $\vec{e}$ (also called Laplace vector) is defined as:

$$\vec{e} = \frac{\vec{v} \wedge \vec{h}}{\mu} - \frac{\vec{r}}{r} \tag{1.10}$$

To calculate the derivative of $\vec{e}$, we observe that the radial unit vector $\vec{u}_r = \frac{\vec{r}}{r}$, which lies in the orbital plane, rotates around the fixed axis $\vec{u}_h = \frac{\vec{h}}{h}$ with an angular velocity $\dot{\theta}$ (see Figure 1.1).

Its derivative is therefore orthoradial and it can be expressed as:

$$\frac{d\vec{u}_r}{dt} = \dot{\theta}\,\vec{u}_h \wedge \vec{u}_r \tag{1.11}$$

Using (1.5) to substitute $\dot{\theta}$, then (1.3) to make $\vec{v}$ appear, we obtain

$$\frac{d\vec{u}_r}{dt} = \frac{h}{r^2}\vec{u}_h \wedge \vec{u}_r = \vec{h} \wedge \frac{\vec{r}}{r^3} = -\frac{1}{\mu}\vec{h} \wedge \frac{d^2\vec{r}}{dt^2} = -\frac{1}{\mu}\vec{h} \wedge \frac{d\vec{v}}{dt} \tag{1.12}$$

By deriving equation (1.10), with $\frac{d\vec{h}}{dt} = \vec{0}$ from (1.4), and using (1.12), the terms cancel each other out and the derivative of $\vec{e}$ is zero.

The eccentricity vector $\vec{e} = \frac{\vec{v}\wedge\vec{h}}{\mu} - \frac{\vec{r}}{r}$ is therefore constant throughout the motion.

#### 1.2.2.4 Useful Relation Between (h, w, e)

By squaring each member of equation (1.10), we obtain the useful relation

$$e^2 = \frac{1}{\mu^2}(\vec{v}\wedge\vec{h})^2 - \frac{2}{\mu r}(\vec{v}\wedge\vec{h})\cdot\vec{r} + 1 = \frac{v^2h^2}{\mu^2} - \frac{2h^2}{\mu r} + 1 = \frac{2h^2}{\mu^2}w + 1 \tag{1.13}$$

This relation can be rewritten by defining two constants p and a from the angular momentum and from the energy respectively, with the convention $a = \infty$ if $w = 0$.

$$p \underset{\text{def}}{=} \frac{h^2}{\mu} \quad \text{and} \quad w = \frac{v^2}{2} - \frac{\mu}{r} \underset{\text{def}}{=} -\frac{\mu}{2a} \tag{1.14}$$

Substituting in (1.13) yields a relation between the constants p, a, and e.

$$p = a(1 - e^2) \tag{1.15}$$

### 1.2.3 Orbit Shape

This section establishes the orbit equation in the orbital plane, successively in polar and Cartesian coordinates. The frame origin O is at the center of mass of the attracting body and the axis Ox is directed along the eccentricity vector $\vec{e}$. In the case $\vec{e} = \vec{0}$, the axis Ox can be chosen arbitrarily. The axis Oy completes the frame in the orbital plane, which is orthogonal to the angular momentum $\vec{h}$ (Figure 1.1).

The scalar product with $\vec{r}$ of the two members of equation (1.10) yields

$$\vec{e}\cdot\vec{r} = \frac{(\vec{v}\wedge\vec{h})\cdot\vec{r}}{\mu} - r = \frac{(\vec{r}\wedge\vec{v})\cdot\vec{h}}{\mu} - r = \frac{h^2}{\mu} - r = p - r \tag{1.16}$$

with the constant p defined in (1.14).

On the other hand, we have $\vec{e}\cdot\vec{r} = er\cos\theta$. Substituting in (1.16), we obtain

$$r = \frac{p}{1 + e\cos\theta} \tag{1.17}$$

This equation in polar coordinates defines a conic of parameter p, eccentricity e and major axis parallel to $\vec{e}$. The angle θ from $\vec{e}$ to $\vec{r}$ is the **true anomaly**.

**Kepler's first law** regarding the trajectory shape (conic) is thus established.

The eccentricity value determines the nature of the conic.

- $e < 1$ yields an ellipse with $w < 0$ from (1.13). With a negative energy, the central attraction cannot be escaped and the orbit is closed.
- $e = 1$ yields a parabola with $w = 0$ from (1.13). With a null energy, the central attraction is escaped with a null residual velocity at infinity (1.9).
- $e > 1$ yields a hyperbola with $w > 0$ from (1.13). With a positive energy, the central attraction is escaped with a non-null residual velocity at infinity (1.9).

The polar coordinates (r, θ) are transformed into Cartesian coordinates (x, y) by

$$\begin{cases} x = r\cos\theta \\ y = r\sin\theta \end{cases}, \quad r^2 = x^2 + y^2 \tag{1.18}$$

Substituting $\vec{e} \cdot \vec{r} = er\cos\theta = ex$ in (1.16) yields: $r = p - ex$.
By squaring both members, we obtain the conic equation in Cartesian coordinates.

$$(1 - e^2)\,x^2 + 2pex + y^2 = p^2 \tag{1.19}$$

At this step, we need to distinguish the cases $e < 1$, $e > 1$ or $e = 1$, corresponding to an ellipse, a hyperbola and a parabola, respectively.

#### 1.2.3.1 Ellipse: $e < 1$ and $a > 0$ (because $w < 0$)

By substituting $p = a(1 - e^2)$ from (1.15), equation (1.19) becomes

$$\frac{(x + ae)^2}{a^2} + \frac{y^2}{a^2(1 - e^2)} = 1 \tag{1.20}$$

and defining $c = ae$ and $b = a\sqrt{1 - e^2}$, we get

$$\left(\frac{x + c}{a}\right)^2 + \left(\frac{y}{b}\right)^2 = 1 \tag{1.21}$$

This is the Cartesian equation of an ellipse of semi-major axis a along Ox and semi-minor axis b along Oy. The distance from the center to the focus is c.

Figure 1.2 shows the geometric elements of an elliptical orbit. The center C is located at (−c; 0), on the negative side of O. The focus O (attracting body) and the virtual focus O' are located on either side of the center C.

The extreme points called **apses** are the periapsis or pericenter P (closest point to the attracting body O) and the apoapsis or apocenter A (farthest point from the attracting body O). Their names depend on the celestial body: apogee and perigee (Earth), aphelion and perihelion (Sun), apolune and perilune (Moon)…

The distances from the focus to the pericenter P and apocenter A are

$$\begin{cases} OP = r_P = a(1 - e) \\ OA = r_A = a(1 + e) \end{cases} \tag{1.22}$$

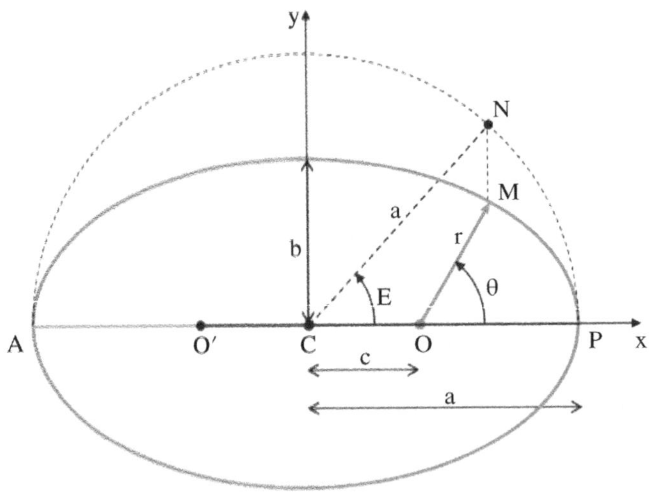


**Figure 1.2** Elliptical orbit.

The ellipse is deduced from the eccentric circle (of center C and radius a) by an affinity orthogonal to the major axis AP and of ratio b/a. The position is defined by the true anomaly $\theta$ or by the **eccentric anomaly** E, depicted in Figure 1.2.

Expressing the Cartesian coordinates with the true and eccentric anomalies

$$\begin{cases} x = a\cos E = r\cos\theta + c \\ y = b\sin E = r\sin\theta \end{cases} \quad \text{with } r = \frac{p}{1+e\cos\theta} \quad \text{and} \quad \begin{cases} p = a(1-e^2) \\ c = ae \\ b = a\sqrt{1-e^2} \end{cases} \tag{1.23}$$

we deduce relations between the anomalies $\theta$, E, and the radius vector r.

$$\begin{cases} \cos E = \dfrac{e+\cos\theta}{1+e\cos\theta}, \quad \sin E = \dfrac{\sqrt{1-e^2}\sin\theta}{1+e\cos\theta} \\ \cos\theta = \dfrac{\cos E - e}{1-e\cos E}, \quad \sin\theta = \dfrac{\sqrt{1-e^2}\sin E}{1-e\cos E} \quad \rightarrow \quad \tan\dfrac{\theta}{2} = \sqrt{\dfrac{1+e}{1-e}}\tan\dfrac{E}{2} \\ r = \dfrac{p}{1+e\cos\theta} = a(1-e\cos E) \end{cases} \tag{1.24}$$

In the degenerate case ($h = 0, e = 1$), the ellipse flattens and reduces to the segment from A to P, yielding a **rectilinear ellipse**. The focus O coincides then with the pericenter ($c = a$). The position and velocity along the straight line are given by

$$r = a(1 - e\cos E), \quad \dot{r} = \frac{\sqrt{\mu a}}{r}\sin E \tag{1.25}$$

#### 1.2.3.2 Hyperbola: e > 1 and a < 0 (because w > 0)

In a similar way to the elliptical case, substituting $p = a(1-e^2)$ in (1.19) yields

$$\frac{(x+ae)^2}{a^2} + \frac{y^2}{a^2(1-e^2)} = 1 \tag{1.26}$$

and defining $c = -ae$ and $b = -a\sqrt{e^2-1}$ (with $a < 0, e > 1$), we get

$$\left(\frac{x-c}{a}\right)^2 - \left(\frac{y}{b}\right)^2 = 1 \tag{1.27}$$

This is the Cartesian equation of a hyperbola of semi-major axis a along Ox and center-focus distance equal to c.

Figure 1.3 shows the geometric elements of a hyperbolic orbit. The center C is located at (c; 0), on the positive side of O. The focus O (attracting body) and the virtual focus O′ are located on either side of the center C. Only one hyperbola branch is drawn, corresponding to the actual trajectory around the attracting body.

The polar angle of the asymptote is found by taking the limit $r \to \infty$ in (1.17).

$$\cos\theta_\infty = -\frac{1}{e} \tag{1.28}$$

The distance from the focus O to the pericenter P is (recalling that $a < 0, e > 1$)

$$OP = r_P = a(1-e) \tag{1.29}$$

The Cartesian coordinates can be expressed with the **hyperbolic anomaly** H.

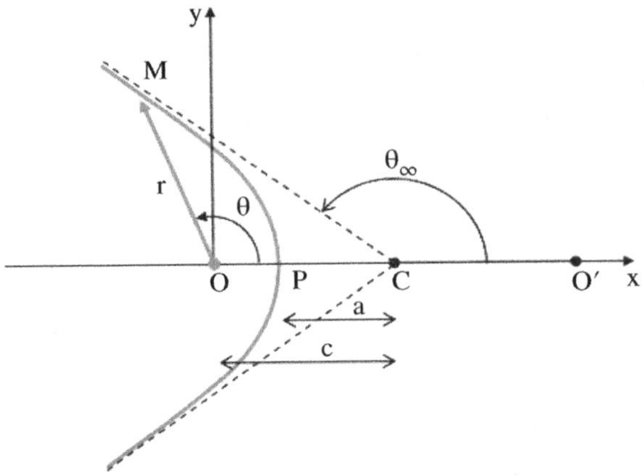


**Figure 1.3** Hyperbolic orbit.

Similarly to the elliptical case (1.23) and (1.24), angle E is replaced by angle H and the trigonometric functions are replaced by hyperbolic functions (with $a < 0$).

$$\begin{cases} x = -a\cosh H = r\cos\theta + c \\ y = \ \ b\sinh H = r\sin\theta \end{cases} \quad \text{with} \quad r = \frac{p}{1 + e\cos\theta} \quad \text{and} \quad \begin{cases} p = a(1 - e^2) \\ c = -ae \\ b = -a\sqrt{e^2 - 1} \end{cases} \tag{1.30}$$

We deduce relations between the anomalies θ, H, and the radius vector r.

$$\begin{cases} \cosh H = \dfrac{e + \cos\theta}{1 + e\cos\theta}, \quad \sinh H = \dfrac{\sqrt{e^2 - 1}\sin\theta}{1 + e\cos\theta} \\ \cos\theta = \dfrac{e\text{-}\cosh H}{e\cosh H - 1}, \quad \sin\theta = \dfrac{\sqrt{e^2 - 1}\sinh H}{e\cosh H - 1} \quad \rightarrow \quad \tan\dfrac{\theta}{2} = \sqrt{\dfrac{e+1}{e-1}}\tanh\dfrac{H}{2} \\ r = \dfrac{p}{1 + e\cos\theta} = a(1 - e\cosh H) \end{cases} \tag{1.31}$$

In the degenerate case ($h = 0, e = 1$), the hyperbola flattens and reduces to the half straight line ending at P, yielding a **rectilinear hyperbola**. The focus O coincides then with the pericenter P. The position and velocity along the straight line are given by

$$r = a(1 - e\cosh H), \quad \dot{r} = \frac{\sqrt{-\mu a}}{r}\sinh H \tag{1.32}$$

#### 1.2.3.3 Parabola: e = 1, w = 0

In that case, equation (1.19) yields

$$y^2 = 2p\left(\frac{1}{2}p - x\right) \tag{1.33}$$

This is the Cartesian equation of a parabola of axis Ox with vertex P at $\left(\frac{p}{2}; 0\right)$.
A parametric form is obtained by introducing the **parabolic anomaly** B.

$$B = \tan\frac{\theta}{2} \quad \Rightarrow \quad r = \frac{1}{2}p(1 + B^2) \tag{1.34}$$

In the degenerate case ($h = 0, e = 1$), the parabola reduces to the half straight line ending at P, yielding a **rectilinear parabola**. The focus O coincides then with the pericenter P. The position and velocity along the straight line are obtained by integrating the energy equation $w = \frac{\dot{r}^2}{2} - \frac{\mu}{r} = 0$.

Denoting $t_P$ the time at pericenter, we obtain

$$r = \frac{B^2}{2}, \quad \dot{r} = \frac{\sqrt{\mu}}{r}B \quad \text{with} \quad B^3 \underset{\text{def}}{=} 6\sqrt{\mu}(t - t_P) \tag{1.35}$$

Before ending this section, we derive some useful formulas for the velocity vector and its modulus.

#### 1.2.3.4 Velocity

The vector product with $\vec{h}$ of the two members of equation (1.10) yields

$$\vec{h} \wedge \vec{e} = \vec{h} \wedge \frac{\vec{v} \wedge \vec{h}}{\mu} - \vec{h} \wedge \frac{\vec{r}}{r} \tag{1.36}$$

Expanding the double vector product $\vec{h} \wedge (\vec{v} \wedge \vec{h}) = h^2\vec{v} - (\vec{h} \cdot \vec{v})\vec{h} = h^2\vec{v}$ yields the velocity vector as a function of $\vec{h}$, $\vec{e}$ and the radial unit vector $\vec{u}_r$.

$$\vec{v} = \frac{1}{p}\vec{h} \wedge (\vec{e} + \vec{u}_r) \quad \text{with} \quad \vec{u}_r = \frac{\vec{r}}{r} \tag{1.37}$$

On the other hand, the velocity modulus can be expressed from the energy (1.14).

$$v^2 = \mu\left(\frac{2}{r} - \frac{1}{a}\right) \tag{1.38}$$

This formula is valid for any conic with the sign of a defined by the energy (1.14):

- $a > 0$ for an ellipse ($w < 0$);
- $a = \infty$ for a parabola ($w = 0$);
- $a < 0$ for a hyperbola ($w > 0$).

From this formula, we deduce:

- the **circular velocity** $v_{sat}$ at radius r to stay on a circular orbit ($r = a = Cte$);
- the **escape velocity** $v_{esc}$ at radius r to escape to infinity ($w = 0$ and $a = \infty$).

$$v_{sat} = \sqrt{\frac{\mu}{r}}, \quad v_{esc} = \sqrt{\frac{2\mu}{r}} = \sqrt{2}v_{sat} \tag{1.39}$$

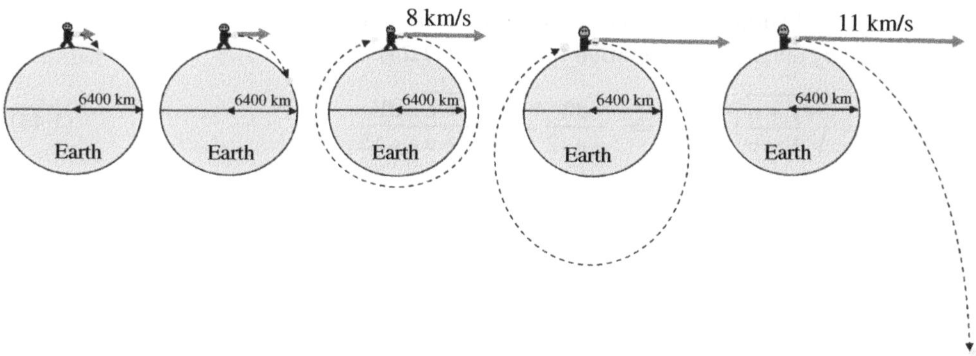


**Figure 1.4** Effect of the initial velocity.

At the Earth's surface, the circular velocity is 7.905 km/s (taking for the Earth's equatorial radius r = 6378.137 km). This is the strict minimum horizontal velocity required for a projectile to perfectly follow the surface of a hypothetical round Earth with no atmosphere.

Increasing the initial velocity allows the projectile to move away from the Earth's surface before returning to the point of departure. On reaching the escape velocity, whose value is 11.180 km/s at the Earth's surface, energy becomes zero, and the projectile goes to infinity.

Figure 1.4 illustrates the possible trajectories of a projectile launched horizontally from the Earth's surface, as a function of initial velocity.

## 1.3 Motion Time Law

The orbit equation takes different forms, depending on the conic nature. In this section, we deal with the position on the orbit as a function of time. The elliptical, hyperbolic, and parabolic cases are first separately considered, then the universal formulation is presented.

### 1.3.1 Elliptical Orbit

We first calculate the period of revolution on an elliptical orbit by the law of areas. The constant areal velocity is equal to the total area S of the ellipse divided by the period T. Integrating equation (1.6) yields

$$\frac{S}{T} = \frac{1}{2}h \quad \text{with} \quad \begin{cases} S = \pi ab = \pi a^2\sqrt{1-e^2} \\ h = \sqrt{\mu p} = \sqrt{\mu a(1-e^2)} \end{cases} \tag{1.40}$$

We obtain **Kepler's third law** linking the period and the semi-major axis.

$$\frac{T^2}{a^3} = \frac{4\pi^2}{\mu} \tag{1.41}$$

The mean angular velocity (called **mean motion**) noted n is defined by

$$n = \frac{2\pi}{T} = \sqrt{\frac{\mu}{a^3}} \tag{1.42}$$

To derive the motion time law, we start from (1.5) giving the angular momentum.

$$h\,dt = r^2 d\theta \tag{1.43}$$

The true anomaly differential $d\theta$ is then expressed as a function of the eccentric anomaly differential dE by deriving the equation for $\tan\frac{\theta}{2}$ in (1.24).

$$\tan\frac{\theta}{2} = \sqrt{\frac{1+e}{1-e}}\tan\frac{E}{2} \quad \Rightarrow \quad \frac{d\theta}{\cos^2\frac{\theta}{2}} = \sqrt{\frac{1+e}{1-e}}\frac{dE}{\cos^2\frac{E}{2}} \tag{1.44}$$

Using the relations (1.24), we have

$$\cos^2\frac{\theta}{2} = \frac{a(1-e)}{r}\cos^2\frac{E}{2} \tag{1.45}$$

which is substituted in (1.44) to give

$$r\,d\theta = b\,dE \tag{1.46}$$

Substituting $d\theta$ in (1.43) with $h = \sqrt{\mu p} = \sqrt{\mu a(1 - e^2)}$, we obtain the differential equation linking time and eccentric anomaly, with n defined by (1.42).

$$(1 - e \cos E)dE = ndt \tag{1.47}$$

Integrating from the time $t_p$ of passage at the pericenter (anomaly $E_p = 0$) and defining the **mean anomaly** by $M = n(t - t_p)$, we obtain **Kepler's equation**.

$$E - e \sin E = M \tag{1.48}$$

Kepler's equation has no analytical solution. Because of its practical importance, it has been the subject of considerable research since the 18th century and many numerical methods have been investigated. Solutions by series expansions were useful until the advent of computers. Nowadays, iterative methods are preferred like Newton's or Laguerre's methods presented hereafter.

#### 1.3.1.1 Newton's Method

Newton's method is applied to find a root of the function

$$f(E) = E - e \sin E - M \tag{1.49}$$

The time t is given defining the mean anomaly $M = n(t - t_p)$. Newton's iteration formula is

$$E_{k+1} = E_k - \frac{f(E_k)}{f'(E_k)} = E_k + \frac{M - (E_k - e \sin E_k)}{1 - e \cos E_k} \tag{1.50}$$

The starting point can be simply chosen as $E_0 = M$. A more accurate initial guess is based on the observation that $\begin{cases} f(M) = -e \sin M < 0 \\ f(M + e) = e[1 - \sin(M + e)] > 0 \end{cases}$. The root is therefore in the interval $[M, M + e]$ and it is estimated by linear interpolation as

$$E_0 = M + \frac{e \sin M}{1 + \sin M - \sin(M + e)} \tag{1.51}$$

Newton's method may lack robustness, especially for eccentricities close to 1.
This motivates the alternative method of Laguerre.

#### 1.3.1.2 Laguerre's Method

Laguerre's method aims at finding the root of polynomials. Its advantage is to be globally convergent (whereas Newton's method is sensitive to the starting point) with a cubic convergence rate, whatever the polynomial degree. Although Kepler's equation is transcendental, Laguerre's method can be customized to it with some specific assumptions. It leads to the following iteration formula:

$$E_{k+1} = E_k - \frac{5f(E_k)}{f'(E_k) \pm 2\sqrt{|4f'(E_k)^2 - 5f(E_k)f''(E_k)|}} \tag{1.52}$$

The denominator sign is chosen to minimize the difference $|E_{k+1} - E_k|$.
The starting point can be $E_0 = M$, or it can be defined as above by (1.51).

---

**Justification of Laguerre's Formula**

Reference: *An improved algorithm due to Laguerre for the solution of Kepler's equation, B. Conway, Celestial Mechanics 39 (1986)*

Let us consider a polynomial of degree n: $g(x) = (x - x_1)(x - x_2)\cdots(x - x_n)$ and suppose we are interested in finding the specific root $x_1$. By assuming (drastically) that all other roots are identical: $x_2 = x_3 = \cdots = x_n$, the function $g(x)$ is replaced by $G(x) = (x - x_1)(x - x_2)^{n-1}$. Let us now define the functions $G_1$ and $G_2$ as

$$\begin{cases} G_1(x) = \dfrac{d \ln G(x)}{dx} = \dfrac{G'(x)}{G(x)} = \dfrac{1}{x - x_1} + \dfrac{n-1}{x - x_2} \\ G_2(x) = -\dfrac{d^2 \ln G(x)}{dx^2} = \dfrac{G'(x)^2 - G(x)G''(x)}{G^2(x)} = \dfrac{1}{(x - x_1)^2} + \dfrac{n-1}{(x - x_2)^2} \end{cases}$$

If $x_0$ is an arbitrary estimate of the root $x_1$ we are looking for, we have

$$\begin{cases} G_1(x_0) = \dfrac{1}{x_0 - x_1} + \dfrac{n-1}{x_0 - x_2} \\ G_2(x_0) = \dfrac{1}{(x_0 - x_1)^2} + \dfrac{n-1}{(x_0 - x_2)^2} \end{cases}$$

Solving this system with two unknowns $x_1$ and $x_2$ yields for $x_1$

$$x_1 = x_0 - \frac{n}{G_1(x_0) \pm \sqrt{(n-1)\left[nG_2(x_0) - G_1(x_0)^2\right]}}$$

Assume that the estimate $x_0$ is already close to $x_1$; the denominator sign is chosen to deviate as little as possible from $x_0$. The iteration (1.52) is obtained by applying this formula to the function f (1.49), with $G_1$ and $G_2$ being the derivatives of ln f with the arbitrary choice $n = 5$, which works well in practice.

---

Extensive tests [R12] have shown that Laguerre's method properties are retained when applied to Kepler's equation. It is insensitive to the initial guess and the convergence requires on average less iterations than Newton's method. But its computational cost per iteration is higher, so that Newton's method remains competitive as long as eccentricity is not close to 1.

### 1.3.2 Hyperbolic Orbit

The time law equation is established in a similar way to the elliptical case, starting from (1.5). The true anomaly differential $d\theta$ is expressed as a function of the hyperbolic anomaly differential dH by deriving the equation for $\tan\frac{\theta}{2}$ in (1.31).

$$\tan\frac{\theta}{2} = \sqrt{\frac{e+1}{e-1}}\tanh\frac{H}{2} \Rightarrow \frac{d\theta}{\cos^2\dfrac{\theta}{2}} = \sqrt{\frac{e+1}{e-1}}\frac{dE}{\cosh^2\frac{E}{2}} \tag{1.53}$$

Using the relations (1.31), we have

$$\cos^2\frac{\theta}{2} = -\frac{a(e-1)}{r}\cosh^2\frac{H}{2} \tag{1.54}$$

which is substituted in (1.53) to give (with $b = -a\sqrt{e^2 - 1}$)

$$r\,d\theta = b\,dH \tag{1.55}$$

Substituting $d\theta$ in (1.5) with $h = \sqrt{\mu p} = \sqrt{\mu a(e^2 - 1)}$, we obtain the differential equation linking time and hyperbolic anomaly.

$$(e\cosh H - 1)dH = n dt \quad \text{with} \quad n \underset{\text{def}}{=} \sqrt{\frac{\mu}{(-a)^3}} \tag{1.56}$$

The mean motion n is defined in a similar way to the elliptical case, taking into account that $a < 0$ for a hyperbola.

Integrating from the time $t_P$ of passage at the pericenter (anomaly $H_P = 0$) and defining the **mean anomaly** by $M = n(t - t_P)$, we obtain **Kepler's equation** for a hyperbolic orbit.

$$e\sinh H - H = M \tag{1.57}$$

Numerical solution methods (Newton, Laguerre) apply in a similar way to the elliptical case, with the function (1.49) replaced by

$$f(E) = e\sinh H - H - M \tag{1.58}$$

### 1.3.3 Parabolic Orbit

Here again, the time law equation is established starting from (1.5). The true anomaly differential $d\theta$ is expressed as a function of the parabolic anomaly differential dB by deriving the equation for $\tan\frac{\theta}{2}$ in (1.34).

$$B = \tan\frac{\theta}{2} \quad \Rightarrow \quad dB = \frac{1}{2}(1 + B^2)d\theta \tag{1.59}$$

Using r expressed by (1.35), this can be written as:

$$r\, d\theta = p\, dB \tag{1.60}$$

Substituting $d\theta$ in equation (1.5), we obtain the differential equation linking time and parabolic anomaly.

$$(1 + B^2)dB = 2n\, dt \quad \text{with} \quad n \underset{\text{def}}{=} \sqrt{\frac{\mu}{p^3}} \tag{1.61}$$

The mean motion n is defined in a similar way to the elliptical case, using the parameter p instead of the semi-major axis a.

Integrating from the time $t_P$ of passage at the pericenter (anomaly $B_P = 0$), and defining the **mean anomaly** by $M = n(t - t_P)$, we obtain **Barker's equation** for a parabolic orbit.

$$B^3 + 3B = 6M \tag{1.62}$$

Unlike Kepler's equation, Barker's equation has an analytic solution.

By the change of variable $B = z - \frac{1}{z}$, equation (1.62) becomes

$$z^6 - 6Mz^3 - 1 = 0 \tag{1.63}$$

whose solutions are

$$z^3 = 3M \pm \sqrt{9M^2 + 1} \tag{1.64}$$

Since the product of the roots is $-1$, both solutions in (1.64) yield the same value of B, so that the unique solution of (1.62) is expressed as:

$$B = \tan\frac{\theta}{2} = \left(3M + \sqrt{9M^2 + 1}\right)^{\frac{1}{3}} - \left(3M + \sqrt{9M^2 + 1}\right)^{-\frac{1}{3}} \tag{1.65}$$

### 1.3.4 Lagrange Coefficients

The Keplerian trajectory is contained in the orbital plane defined by the initial conditions $(\vec{r}_0, \vec{v}_0)$ at a time $t_0$. It is therefore possible to express the position and velocity at any time t as a linear combination of the initial conditions.

$$\begin{cases} \vec{r}(t) = F(t)\,\vec{r}(t_0) + G(t)\,\vec{v}(t_0) \\ \vec{v}(t) = F_t(t)\vec{r}(t_0) + G_t(t)\vec{v}(t_0) \end{cases} \text{shortened to} \begin{cases} \vec{r} = F\,\vec{r}_0 + G\,\vec{v}_0 \\ \vec{v} = F_t\vec{r}_0 + G_t\vec{v}_0 \end{cases} \tag{1.66}$$

The functions F(t) and G(t) are called **Lagrange coefficients** and the functions $F_t(t)$ and $G_t(t)$ are their derivatives with respect to time.

To compute these functions, we use the perifocal frame $(\vec{u}_p, \vec{u}_n, \vec{u}_h)$ defined from the eccentricity and angular momentum vectors (see Figure 1.5).

- The unit vector $\vec{u}_p$ is directed toward the pericenter $\vec{e} = e\vec{u}_p$.
- The unit vector $\vec{u}_h$ is along the angular momentum $\vec{h} = h\vec{u}_h$.
- The unit vector $\vec{u}_n$ complete the trihedron.

In the perifocal frame, the position vector $\vec{r}$ is assessed by

$$\vec{r} = r\cos\theta\,\vec{u}_p + r\sin\theta\,\vec{u}_n \tag{1.67}$$

and the velocity vector given in (1.37) is assessed by

$$\vec{v} = \frac{1}{p}\vec{h} \wedge (\vec{e} + \vec{u}_r) = \frac{\mu}{h}(e\vec{u}_n + \vec{u}_\theta) = \frac{\mu}{h}[-\sin\theta\,\vec{u}_p + (e + \cos\theta)\,\vec{u}_n] \tag{1.68}$$

where $\vec{u}_r$ is the unit radial vector and $\vec{u}_\theta$ is the unit orthoradial vector (orthogonal to $\vec{u}_r$ in the orbital plane, directed along the motion).

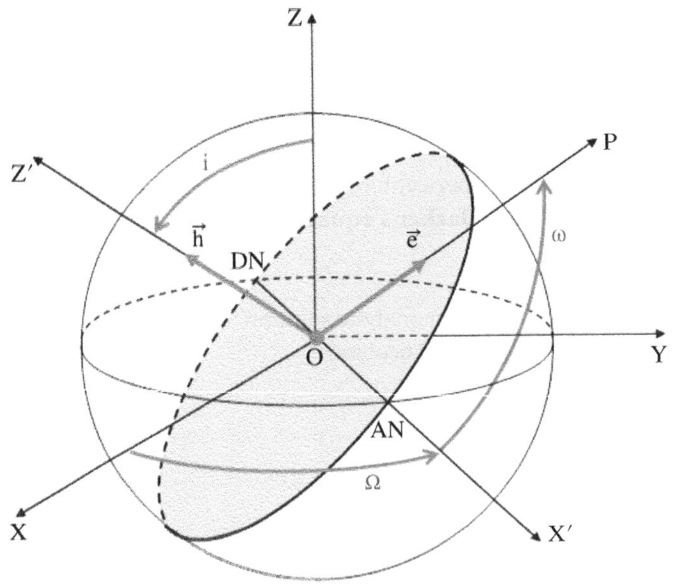


**Figure 1.5** Classical orbital parameters.

Solving (1.67) and (1.68) at the initial time for $\vec{u}_p$ and $\vec{u}_n$, we obtain

$$\begin{cases} \vec{u}_p = \dfrac{1}{p}(e + \cos\theta_0)\vec{r}_0 - \dfrac{1}{h}r_0 \sin\theta_0\vec{v}_0 \\ \vec{u}_n = \dfrac{1}{p}\sin\theta_0\vec{r}_0 \qquad + \dfrac{1}{h}r_0 \cos\theta_0\vec{v}_0 \end{cases} \tag{1.69}$$

Substituting these vectors $\vec{u}_p$ and $\vec{u}_n$ in (1.67) and (1.68) at time t yields

$$\begin{cases} \vec{r} = \dfrac{r}{p}(e\cos\theta + \cos\Delta\theta)\vec{r}_0 + \dfrac{rr_0}{h}\sin\Delta\theta\,\vec{v}_0 \\ \vec{v} = \dfrac{\mu}{hp}(e\sin\theta_0 - e\sin\theta - \sin\Delta\theta)\vec{r}_0 + \mu r_0(e\cos\theta_0 + \cos\Delta\theta)\,\vec{v}_0 \end{cases} \tag{1.70}$$

where $\Delta\theta = \theta - \theta_0$ denotes the true anomaly difference.

These formulas can be rearranged by introducing the variable $\sigma \underset{\text{def}}{=} \frac{\vec{r}\cdot\vec{v}}{\sqrt{\mu}}$.

Using (1.70), we have $\sigma \underset{\text{def}}{=} \frac{\vec{r}\cdot\vec{v}}{\sqrt{\mu}} = \frac{er\sin\theta}{\sqrt{p}}$, so that $e\sin\theta = \sqrt{p}\frac{\sigma}{r}$.

From the conic equation (1.17), we have also $r = \frac{p}{1+e\cos\theta} \Rightarrow e\cos\theta = \frac{p}{r} - 1$.

After substituting in (1.70) the terms in $e\sin\theta$ and $e\cos\theta$, with $h = \sqrt{\mu p}$, and identifying with (1.66), we get the following expressions of Lagrange coefficients as functions of $\sigma_0$, $r_0$, r, and $\Delta\theta$.

$$\begin{cases} F = 1 - \dfrac{r}{p}(1-\cos\Delta\theta), & F_t = \dfrac{\sqrt{\mu}}{pr_0}[\sigma_0(1-\cos\Delta\theta) - \sqrt{p}\sin\Delta\theta] \\ G = \dfrac{rr_0}{h}\sin\Delta\theta, & G_t = 1 + \dfrac{r_0}{p}(1-\cos\Delta\theta) \end{cases} \tag{1.71}$$

Another useful formula can be derived for $F_t$. For that purpose, we express the angular momentum from $(\vec{r}_0, \vec{v}_0)$ and from $(\vec{r}, \vec{v})$, using (1.66).

$$\begin{aligned} \vec{h} &= \vec{r}_0 \wedge \vec{v}_0 = \vec{r} \wedge \vec{v} = (F\,\vec{r}_0 + G\,\vec{v}_0) \wedge (F_t\,\vec{r}_0 + G_t\,\vec{v}_0) \\ &= (FG_t - GF_t)\,\vec{r}_0 \wedge \vec{v}_0 \end{aligned} \tag{1.72}$$

The coefficients F, G, $F_t$, and $G_t$ are therefore linked by

$$FG_t - GF_t = 1 \tag{1.73}$$

This allows to calculate $F_t$ by

$$F_t = \frac{FG_t - 1}{G} \tag{1.74}$$

Substituting F, G, and $G_t$ given by (1.71), we obtain for $F_t$

$$F_t = \sqrt{\frac{\mu}{p}}\frac{1-\cos\Delta\theta}{\sin\Delta\theta}\left(\frac{1-\cos\Delta\theta}{p} - \frac{1}{r_0} - \frac{1}{r}\right) \tag{1.75}$$

#### 1.3.4.1 Specific Formulas for Each Type of Conic

The formulas (1.70) depending on the true anomaly are valid for any type of conic.

They can be specialized to the conic type by using either the eccentric anomaly E for ellipses, the hyperbolic anomaly H for hyperbolas or the parabolic anomaly B for parabolas.

- For an ellipse ($e < 1, a > 0$), the position (1.67) and velocity (1.68) vectors are expressed with the eccentric anomaly using formulas (1.24).

$$\begin{cases} \vec{r} = a(\cos E - e)\,\vec{u}_p + \sqrt{ap}\sin E\,\vec{u}_n \\ \vec{v} = \dfrac{\sqrt{\mu}}{r}[-\sqrt{a}\sin E\,\vec{u}_p + \sqrt{p}\cos E\,\vec{u}_n] \end{cases} \tag{1.76}$$

By calculations similar to (1.69)–(1.71), we obtain

$$\begin{cases} F = 1 + \dfrac{a}{r_0}(\cos\Delta E - 1), & F_t = -\dfrac{\sqrt{\mu a}}{rr_0}\sin\Delta E \\ G = \dfrac{a\sigma_0}{\sqrt{\mu}}(1 - \cos\Delta E) + \sqrt{\dfrac{a}{\mu}}r_0\sin\Delta E, & G_t = 1 + \dfrac{a}{r}(\cos\Delta E - 1) \end{cases} \tag{1.77}$$

- For a hyperbola ($e > 1, a < 0$), the position (1.67) and velocity (1.68) vectors are expressed with the hyperbolic anomaly using formulas (1.31).

$$\begin{cases} \vec{r} = a(\cosh H - e)\,\vec{u}_p + \sqrt{-ap}\sinh H\,\vec{u}_n \\ \vec{v} = \dfrac{\sqrt{\mu}}{r}\left[-\sqrt{-a}\sinh H\,\vec{u}_p + \sqrt{p}\cosh H\,\vec{u}_n\right] \end{cases} \tag{1.78}$$

By calculations similar to (1.69)–(1.71), we obtain

$$\begin{cases} F = 1 + \dfrac{a}{r_0}(\cosh\Delta H - 1), & F_t = -\dfrac{\sqrt{-\mu a}}{rr_0}\sinh\Delta H \\ G = \dfrac{a\sigma_0}{\sqrt{\mu}}(1 - \cosh\Delta H) + \sqrt{\dfrac{-a}{\mu}}r_0\sinh\Delta H, & G_t = 1 + \dfrac{a}{r}(\cosh\Delta H - 1) \end{cases} \tag{1.79}$$

- For a parabola ($e = 1$), the position and velocity vectors are given by

$$\begin{cases} \vec{r} = \dfrac{p}{2}(1 - B^2)\,\vec{u}_p + pB\,\vec{u}_n \\ \vec{v} = \dfrac{h}{r}[-B\,\vec{u}_p + \vec{u}_n] \end{cases} \tag{1.80}$$

By calculations similar to (1.69)–(1.71), we obtain

$$\begin{cases} F = 1 - \dfrac{1}{2r_0}\Delta\sigma^2, & F_t = -\dfrac{\sqrt{\mu}}{rr_0}\Delta\sigma \\ G = \dfrac{r_0}{\sqrt{\mu}}\Delta\sigma + \dfrac{\sigma_0}{2\sqrt{\mu}}\Delta\sigma^2, & G_t = 1 - \dfrac{1}{2r}\Delta\sigma^2 \end{cases} \quad \text{with} \quad \sigma = \frac{\vec{r}.\vec{v}}{\sqrt{\mu}} = \sqrt{p}B \tag{1.81}$$

### 1.3.5 Universal Variable

The elliptical, hyperbolic, and parabolic cases can be merged using the universal variable formulation. This section follows the development presented in [R1].

Let us retrieve Kepler's equations (1.47), (1.56), and (1.61) for each kind of conic, with their expressions for n and r.

$$\begin{cases} (1 - e\cos E)dE = \sqrt{\dfrac{\mu}{a^3}}dt, & r = a(1 - e\cos E) \rightarrow rdE = \sqrt{\dfrac{\mu}{a}}dt \\ (e\cosh H - 1)dH = \sqrt{\dfrac{\mu}{(-a)^3}}dt, & r = a(1 - e\cosh H) \rightarrow rdH = \sqrt{\dfrac{\mu}{-a}}dt \\ (1 + B^2)dB = 2\sqrt{\dfrac{\mu}{p^3}}dt, & r = \dfrac{1}{2}p(1 + B^2) \rightarrow rdB = \sqrt{\dfrac{\mu}{p}}dt \end{cases} \tag{1.82}$$

The similarity of the three equations suggests defining a new variable $\chi$ by the differential equation

$$r\,d\chi = \sqrt{\mu}dt \tag{1.83}$$

The change of variable from t to $\chi$ is called Sundman's transformation, and $\chi$ is called the **universal variable** (or generalized anomaly) since it merges the three conic kinds. The goal is now to express Kepler's equation in variable $\chi$.

#### 1.3.5.1 Change of Variable from Time to Universal Variable

The first step consists in calculating the successive derivatives of the time t with respect to the variable $\chi$, starting from the definition (1.83).

$$\sqrt{\mu}\frac{dt}{d\chi} = r \Rightarrow \sqrt{\mu}\frac{d^2t}{d\chi^2} = \frac{dr}{d\chi} \tag{1.84}$$

From $r^2 = \vec{r}\cdot\vec{r}$, we have $r\frac{dr}{d\chi} = \vec{r}\cdot\frac{d\vec{r}}{d\chi} = \vec{r}\cdot\frac{d\vec{r}}{dt}\frac{dt}{d\chi} = \vec{r}\cdot\vec{v}\frac{r}{\sqrt{\mu}}$, so that

$$\frac{dr}{d\chi} = \frac{\vec{r}\cdot\vec{v}}{\sqrt{\mu}} \tag{1.85}$$

This expression was noted $\sigma$ in Section 1.3.4, just after (1.70). By substitution in (1.84), we have

$$\sqrt{\mu}\frac{d^2t}{d\chi^2} = \sigma \text{ with } \sigma \underset{\text{def}}{=} \frac{\vec{r}\cdot v}{\sqrt{\mu}} \tag{1.86}$$

Equation (1.86) is derived in turn.

$$\sqrt{\mu}\frac{d^2t}{d\chi^2} = \sigma \Rightarrow \sqrt{\mu}\frac{d^3t}{d\chi^3} = \frac{d\sigma}{d\chi} \tag{1.87}$$

From $\sigma = \frac{\vec{r}\cdot\vec{v}}{\sqrt{\mu}}$, we have

$$\sqrt{\mu}\frac{d\sigma}{d\chi} = \frac{d\vec{r}}{d\chi}\cdot\vec{v} + \vec{r}\cdot\frac{d\vec{v}}{d\chi} = \frac{dt}{d\chi}\left(\frac{d\vec{r}}{dt}\cdot\vec{v} + \vec{r}\cdot\frac{d\vec{v}}{dt}\right) = \frac{r}{\sqrt{\mu}}\left(v^2 - \mu\vec{r}\cdot\frac{\vec{r}}{r^3}\right) = \frac{r}{\sqrt{\mu}}\left(v^2 - \frac{\mu}{r}\right)$$

so that $\frac{d\sigma}{d\chi} = \frac{rv^2}{\mu} - 1$. Substituting in (1.87), we obtain

$$\sqrt{\mu}\frac{d^3t}{d\chi^3} = 1 - \alpha r \text{ with } \alpha \underset{\text{def}}{=} \frac{2}{r} - \frac{v^2}{\mu} \tag{1.88}$$

The constant $\alpha$ denotes the inverse of the semi-major axis, which is defined from the energy by $\frac{v^2}{2} - \frac{\mu}{r} = -\frac{\mu}{2a}$.

It is positive for an ellipse, negative for a hyperbola and null for a parabola.

Equation (1.88) is derived in turn, using (1.85).

$$\sqrt{\mu}\frac{d^3t}{d\chi^3} = 1 - \alpha r \Rightarrow \sqrt{\mu}\frac{d^4t}{d\chi^4} = -\alpha\frac{dr}{d\chi} = -\alpha\sigma \tag{1.89}$$

We observe from (1.86) and (1.89) that the derivatives of t satisfy the relation

$$\frac{d^4t}{d\chi^4} + \alpha\frac{d^2t}{d\chi^2} = 0 \tag{1.90}$$

The solution of this linear fourth-order differential equation depends on four integration constants noted $a_0$, $a_1$, $a_2$, $a_3$. This solution can be expressed by introducing "universal functions" as follows.

#### 1.3.5.2 Universal Functions

Let us consider the function $U_0(\chi, \alpha)$ defined by the series

$$U_0(\chi,\alpha) \underset{\text{def}}{=} 1 - \frac{\alpha\chi^2}{2!} + \frac{(\alpha\chi^2)^2}{4!} - \cdots + (--1)^k \frac{(\alpha\chi^2)^k}{(2k)!} + \cdots \tag{1.91}$$

and the functions $U_1(\chi, \alpha)$, $U_2(\chi, \alpha)$... defined by successive integrations

$$U_{n+1}(\chi,\alpha) \underset{\text{def}}{=} \int_0^{\chi} U_n(u,\alpha)\,du \tag{1.92}$$

Their expressions by series are

$$U_n(\chi,\alpha) = \chi^n \left[\frac{1}{n!} - \frac{\alpha\chi^2}{(n+2)!} + \frac{(\alpha\chi^2)^2}{(n+4)!} - \cdots + (-1)^k \frac{(\alpha\chi^2)^k}{(n+2k)!} + \cdots\right] \tag{1.93}$$

These functions all satisfy equation (1.90). Their values in $\chi = 0$ are

$$U_0(0,\alpha) = 1, \ \ U_1(0,\alpha) = U_2(0,\alpha) = \cdots = U_n(0,\alpha) = 0 \tag{1.94}$$

They are furthermore linearly independent.

Indeed, if $a_0U_0 + a_1U_1 + a_2U_2 + \cdots + a_nU_n = 0, \forall\chi$, then taking the value in $\chi = 0$ yields $a_0 = 0$. By deriving, we have $a_1U_0 + a_2U_1 + \cdots + a_nU_{n-1} = 0, \forall\chi$, and taking again the value in $\chi = 0$, we get $a_1 = 0$, and so on.

Since the functions $U_n(\chi, \alpha)$ are independent solutions of (1.90), the general solution of (1.90), which is of fourth order, can be searched under the form

$$\sqrt{\mu}(t - t_0) = a_0U_0(\chi,\alpha) + a_1U_1(\chi,\alpha) + a_2U_2(\chi,\alpha) + a_3U_3(\chi,\alpha) \tag{1.95}$$

To express $a_0$, $a_1$, $a_2$, $a_3$ in terms of initial conditions $(\vec{r}_0, \vec{v}_0)$ at $t_0$, we associate the value $\chi = 0$ to the initial time $t_0$, which leads in (1.95) to $a_0 = 0$. The other coefficients are obtained in three steps by successive derivations as follows.

Deriving (1.95) with $\sqrt{\mu}\frac{dt}{d\chi} = r$ from (1.84) yields

$$r = a_1U_0(\chi,\alpha) + a_2U_1(\chi,\alpha) + a_3U_2(\chi,\alpha) \tag{1.96}$$

Applying this relation at the initial time ($\chi = 0$) leads to $a_1 = r_0$.

Deriving (1.96) with $\frac{dr}{d\chi} = \sigma$ from (1.85) yields

$$\sigma = -\alpha r_0U_1(\chi,\alpha) + a_2U_0(\chi,\alpha) + a_3U_1(\chi,\alpha) \tag{1.97}$$

where we have also used $\frac{d}{d\chi}U_0(\chi,\alpha) = -\alpha U_1(\chi,\alpha)$, demonstrated from (1.93).

Applying this relation at the initial time ($\chi = 0$) leads to $a_2 = \sigma_0$.

Deriving (1.97) with $\frac{d\sigma}{d\chi} = \frac{rv^2}{\mu} - 1 = 1 - \alpha r$ obtained from (1.86) and (1.84) yields

$$1 - \alpha r = -\alpha r_0U_0(\chi,\alpha) - \alpha\sigma_0U_1(\chi,\alpha) + a_3U_0(\chi,\alpha) \tag{1.98}$$

Applying this relation at the initial time ($\chi = 0$) leads to $a_3 = 1$.

#### 1.3.5.3 Generalized Kepler's Equation

We can now substitute the four coefficients $a_0$, $a_1$, $a_2$, $a_3$ in (1.95) to obtain the **generalized Kepler's equation**.

$$\sqrt{\mu}(t - t_0) = r_0U_1(\chi,\alpha) + \sigma_0U_2(\chi,\alpha) + U_3(\chi,\alpha) \tag{1.99}$$

This equation must be solved using the functions $U_n(\chi, \alpha)$ corresponding to the conic type. Their explicit expressions are deduced from the series formula (1.93).

- For an ellipse ($\alpha > 0$)

$$\begin{cases} U_0(\chi,\alpha) = \cos(\sqrt{\alpha}\chi), \quad U_2(\chi,\alpha) = \dfrac{1-\cos(\sqrt{\alpha}\chi)}{\alpha} \\ U_1(\chi,\alpha) = \dfrac{\sin(\sqrt{\alpha}\chi)}{\sqrt{\alpha}}, \quad U_3(\chi,\alpha) = \dfrac{\sqrt{\alpha}\chi - \sin(\sqrt{\alpha}\chi)}{\alpha\sqrt{\alpha}} \end{cases} \tag{1.100}$$

- For a hyperbola ($\alpha < 0$)

$$\begin{cases} U_0(\chi,\alpha) = \cosh(\sqrt{-\alpha}\chi), \quad U_2(\chi,\alpha) = \dfrac{\cosh(\sqrt{-\alpha}\chi)-1}{-\alpha} \\ U_1(\chi,\alpha) = \dfrac{\sinh(\sqrt{-\alpha}\chi)}{\sqrt{-\alpha}}, \quad U_3(\chi,\alpha) = \dfrac{\sinh(\sqrt{-\alpha}\chi) - \sqrt{-\alpha}\chi}{-\alpha\sqrt{-\alpha}} \end{cases} \tag{1.101}$$

- For a parabola ($\alpha = 0$)

$$\begin{cases} U_0(\chi,\alpha) = 1, \quad U_2(\chi,\alpha) = \dfrac{\chi^2}{2} \\ U_1(\chi,\alpha) = \chi, \quad U_3(\chi,\alpha) = \dfrac{\chi^3}{6} \end{cases} \tag{1.102}$$

#### 1.3.5.4 Relation Between t, $\chi$, and $\sigma$

A useful formula can be derived by summing Kepler's equation (1.99) multiplied by $\alpha$ and (1.97), which gives $\sigma = \sigma_0 U_0(\chi,\alpha) + (1-\alpha r_0)U_1(\chi,\alpha)$.

$$\alpha\sqrt{\mu}(t-t_0) + \sigma = \sigma_0[U_0(\chi,\alpha) + \alpha U_2(\chi,\alpha)] + [U_1(\chi,\alpha) + \alpha U_3(\chi,\alpha)] \tag{1.103}$$

From the series expression (1.93) of $U_n(\chi,\alpha)$, we observe that

$$U_n(\chi,\alpha) + \alpha U_{n+2}(\chi,\alpha) = \frac{\chi^n}{n!} \tag{1.104}$$

which simplifies the terms in brackets in (1.103). We obtain an explicit relation between the variables t, $\chi$, and $\sigma$.

$$\chi = \alpha\sqrt{\mu}(t-t_0) + \sigma - \sigma_0 \tag{1.105}$$

#### 1.3.5.5 Lagrange Coefficients with Universal Variable

Solving equation (1.99) with the functions $U_n(\chi,\alpha)$ corresponding to the conic nature (sign of $\alpha$) yields the generalized anomaly $\chi$.

In order to calculate the corresponding position and velocity by (1.66), we need expressions of Lagrange coefficients as functions of $\chi$. For that purpose, we first calculate the derivatives of the position vector $\vec{r}$ with respect to $\chi$.

By using $\frac{dt}{d\chi} = \frac{r}{\sqrt{\mu}}$ and $\frac{d\vec{v}}{dt} = -\mu\frac{\vec{r}}{r^3}$, we obtain

$$\begin{cases} \dfrac{d\vec{r}}{d\chi} = \dfrac{d\vec{r}}{dt}\dfrac{dt}{d\chi} = \dfrac{r}{\sqrt{\mu}}\vec{v} \\ \dfrac{d^2\vec{r}}{d\chi^2} = \dfrac{1}{\sqrt{\mu}}\dfrac{dr}{d\chi}\vec{v} + \dfrac{r}{\sqrt{\mu}}\dfrac{d\vec{v}}{d\chi} = \dfrac{1}{\sqrt{\mu}}\sigma\vec{v} + \dfrac{r}{\sqrt{\mu}}\dfrac{d\vec{v}}{dt}\dfrac{dt}{d\chi} = \dfrac{\sigma}{\sqrt{\mu}}\vec{v} - \dfrac{1}{r}\vec{r} \\ \dfrac{d^3\vec{r}}{d\chi^3} = \dfrac{1}{\sqrt{\mu}}\dfrac{d\sigma}{d\chi}\vec{v} + \dfrac{\sigma}{\sqrt{\mu}}\dfrac{d\vec{v}}{d\chi} + \dfrac{1}{r^2}\dfrac{dr}{d\chi}\vec{r} - \dfrac{1}{r}\dfrac{d\vec{r}}{d\chi} \\ \quad = \dfrac{1}{\sqrt{\mu}}(1-\alpha r)\vec{v} + \dfrac{\sigma}{\sqrt{\mu}}\dfrac{r}{\sqrt{\mu}}\left(-\mu\dfrac{\vec{r}}{r^3}\right) + \dfrac{\sigma}{r^2}\vec{r} - \dfrac{1}{r}\dfrac{r}{\sqrt{\mu}}\vec{v} = \dfrac{-\alpha r}{\sqrt{\mu}}\vec{v} \end{cases} \tag{1.106}$$

so that

$$\frac{d^3\vec{r}}{d\chi^3} + \alpha\frac{d\vec{r}}{d\chi} = 0 \tag{1.107}$$

This linear third-order differential equation is similar to (1.90). Applying the same procedure as (1.91)–(1.94), the solution is searched under the form

$$\vec{r} = \vec{c}_0 U_0(\chi, \alpha) + \vec{c}_1 U_1(\chi, \alpha) + \vec{c}_2 U_2(\chi, \alpha) \tag{1.108}$$

with three constant vectors $\vec{c}_0, \vec{c}_1, \vec{c}_2$.

These vectors are determined by deriving (1.108) and identifying to (1.106) at the initial time ($\chi = 0$).

$$\begin{cases} \vec{r} = \vec{c}_0 U_0(\chi, \alpha) + \vec{c}_1 U_1(\chi, \alpha) + \vec{c}_2 U_2(\chi, \alpha) \\ \dfrac{d\vec{r}}{d\chi} = -\alpha\vec{c}_0 U_1(\chi, \alpha) + \vec{c}_1 U_0(\chi, \alpha) + \vec{c}_2 U_1(\chi, \alpha) = \dfrac{r}{\sqrt{\mu}}\vec{v} \\ \dfrac{d^2\vec{r}}{d\chi^2} = -\alpha\vec{c}_0 U_0(\chi, \alpha) - \alpha\vec{c}_1 U_1(\chi, \alpha) + \vec{c}_2 U_0(\chi, \alpha) = \dfrac{\sigma}{\sqrt{\mu}}\vec{v} - \dfrac{1}{r}\vec{r} \end{cases} \tag{1.109}$$

This yields

$$\vec{c}_0 = \vec{r}_0, \vec{c}_1 = \frac{r_0}{\sqrt{\mu}}\vec{v}_0, \vec{c}_2 = \left(\alpha - \frac{1}{r_0}\right)\vec{r}_0 + \frac{\sigma_0}{\sqrt{\mu}}\vec{v}_0 \tag{1.110}$$

Substituting these constants in (1.108), we obtain

$$\vec{r} = \left[U_0(\chi, \alpha) + \left(\alpha - \frac{1}{r_0}\right) U_2(\chi, \alpha)\right]\vec{r}_0 + \left[\frac{r_0}{\sqrt{\mu}}U_1(\chi, \alpha) + \frac{\sigma_0}{\sqrt{\mu}}U_2(\chi, \alpha)\right]\vec{v}_0 \tag{1.111}$$

By identification with (1.66), we find the Lagrange coefficients expressed with the generalized anomaly $\chi$, and also their time derivatives using $\frac{d\chi}{dt} = \frac{\sqrt{\mu}}{r}$.

$$\begin{cases} F = U_0(\chi, \alpha) + \left(\alpha - \dfrac{1}{r_0}\right) U_2(\chi, \alpha), & F_t = -\dfrac{\sqrt{\mu}}{r_0 r}U_1(\chi, \alpha) \\ G = \dfrac{r_0}{\sqrt{\mu}}U_1(\chi, \alpha) + \dfrac{\sigma_0}{\sqrt{\mu}}U_2(\chi, \alpha), & G_t = \dfrac{r_0}{r}U_0(\chi, \alpha) + \dfrac{\sigma_0}{r}U_1(\chi, \alpha) \end{cases} \tag{1.112}$$

These coefficients define the position $\vec{r}$ and velocity $\vec{v}$ at any time t from initial conditions $(\vec{r}_0, \vec{v}_0)$ at a time $t_0$, with the constants $\alpha = \frac{2}{r_0} - \frac{v_0^2}{\mu}, \sigma_0 = \frac{\vec{r}_0 \cdot \vec{v}_0}{\sqrt{\mu}}$ and with the generalized anomaly $\chi$ solution of Kepler's equation (1.99).

Comparing Lagrange coefficients in variable $\chi$ (1.112) with those obtained for each conic shape in (1.77), (1.79), and (1.81), we deduce the relations between $\chi$ and E, H, or B, with the convention $\chi = 0$ at time $t_0$.

$$\begin{cases} \chi = \sqrt{a}(E - E_0) & \text{(ellipse)} \\ \chi = \sqrt{a}(H - H_0) & \text{(hyperbola)} \\ \chi = \sqrt{p}(B - B_0) = \sigma - \sigma_0 & \text{(parabola)} \end{cases} \tag{1.113}$$

The universal variable formulation will be further studied in Section 6.4.4 dealing with Lambert's problem.

## 1.4 Orbital Parameters

Until now, we have studied the Keplerian motion in its orbital plane independently of the attracting body frame. This section introduces the orbital parameters, which position the Keplerian conic in space. Their relation to position and velocity is established and some specific properties of Earth orbits are presented, such as geo-synchronism and sun-synchronism.

### 1.4.1 Classical Orbital Parameters

The classical orbital parameters are defined in an inertial frame linked to the attracting body. The origin O is the body center of mass, which is a focus of the conical orbit. The Z axis is directed toward the North pole and the X and Y axes are in the equatorial plane. These axes are fixed (not rotating with the attracting body) and they are defined at a given reference epoch. The X axis corresponds to a reference meridian (usually passing by the vernal point or by Greenwich in the case of the Earth).

The conical Keplerian orbit is positioned in the reference frame OXYZ through the classical orbital parameters, which are the right ascension of the ascending node $\Omega$ (RAAN), the inclination i, the argument of the pericenter $\omega$, the semi-major axis a, and the eccentricity e. These five constants define the angular momentum $\vec{h}$ and the eccentricity vector $\vec{e}$ as follows (see Figure 1.5).

- The frame (OXYZ) is transformed into the frame (OX′Y′Z′) by a first rotation of axis Z and angle $\Omega$, and a second rotation of axis X′ and angle i. The X′ axis is the intersection of the equatorial and orbital planes.
  The **ascending node** AN, opposite the descending node DN, is the orbit point where the equator is crossed from South to North.
  The X′ axis is called the **line of nodes**. The Z′ axis defines the direction of the angular momentum $\vec{h}$, orthogonal to the orbital plane.
- The frame (OX′Y′Z′) is transformed into the frame (OPQW) by a rotation of axis Z′ and angle $\omega$. The axis OW is identical to OZ′ (orthogonal to the orbital plane). The axis OP is along the eccentricity vector $\vec{e}$, pointing toward the pericenter P.
- The orbit shape is defined by the values of the semi-major axis a and of the eccentricity e, from which the angular momentum modulus is determined.

The five orbital parameters $(a, e, i, \Omega, \omega)$ completely determine the Keplerian orbit in space. The current position on the orbit is defined by the true anomaly $\theta$, which varies with time according to Kepler's equation studied in Section 1.3.

### 1.4.2 Relation to Position and Velocity

In space applications, it is frequently needed to pass from orbital parameters to position and velocity, or conversely.

#### 1.4.2.1 Transition from Orbital Parameters to Position and Velocity

The transition from orbital parameters $(a, e, i, \Omega, \omega, \theta)$ to position and velocity in the frame (OXYZ) involves the orbit **perifocal frame** (OPQW), whose axes P, Q, and W are respectively oriented by the vectors $\vec{e}$, $\vec{n} = \vec{h} \wedge \vec{e}$, and $\vec{h}$.

The orbit lies in the (OPQ) plane, with the P and Q axes corresponding to the x and y axes in Figures 1.2 and 1.3.

The position and velocity components in (OPQW) are given by (1.67)–(1.68).

$$\vec{r}_{PQW} = \frac{p}{1+e\cos\theta}\begin{pmatrix}\cos\theta\\ \sin\theta\\ 0\end{pmatrix}, \quad \vec{v}_{PQW} = \sqrt{\frac{\mu}{p}}\begin{pmatrix}-\sin\theta\\ e+\cos\theta\\ 0\end{pmatrix} \tag{1.114}$$

The transition from (OXYZ) to (OPQW) is achieved by the three successive rotations Ω [Z], i [X′], ω [Z′] as explained earlier.

The change of coordinates from (OXYZ) to (OPQW) is obtained by multiplying the matrices of these rotations. By convention, the same letter (e.g. X) is used to designate the axis, the vector orienting the axis, or the coordinate along this axis.

$$\begin{pmatrix}P\\ Q\\ W\end{pmatrix} = \begin{pmatrix}\cos\omega & \sin\omega & 0\\ -\sin\omega & \cos\omega & 0\\ 0 & 0 & 1\end{pmatrix}\begin{pmatrix}1 & 0 & 0\\ 0 & \cos i & \sin i\\ 0 & -\sin i & \cos i\end{pmatrix}\begin{pmatrix}\cos\Omega & \sin\Omega & 0\\ -\sin\Omega & \cos\Omega & 0\\ 0 & 0 & 1\end{pmatrix}\begin{pmatrix}X\\ Y\\ Z\end{pmatrix} \tag{1.115}$$

The reverse change from (OPQW) to (OXYZ) is obtained by swapping the matrix product order and changing the angle signs.

$$\begin{pmatrix}X\\ Y\\ Z\end{pmatrix} = \begin{pmatrix}\cos\Omega & -\sin\Omega & 0\\ \sin\Omega & \cos\Omega & 0\\ 0 & 0 & 1\end{pmatrix}\begin{pmatrix}1 & 0 & 0\\ 0 & \cos i & -\sin i\\ 0 & \sin i & \cos i\end{pmatrix}\begin{pmatrix}\cos\omega & -\sin\omega & 0\\ \sin\omega & \cos\omega & 0\\ 0 & 0 & 1\end{pmatrix}\begin{pmatrix}P\\ Q\\ W\end{pmatrix} \tag{1.116}$$

Formulas (1.114) and (1.116) are combined to calculate the position and velocity components in the (OXYZ) reference frame from the six orbital parameters.

#### 1.4.2.2 Transition from Position and Velocity to Orbital Parameters

The transition from the position and velocity vectors $(\vec{r}, \vec{v})$ expressed in (OXYZ) to the orbital parameters is achieved by successively calculating:

- the energy $w = \frac{v^2}{2} - \frac{\mu}{r}$ and semi-major axis by $w = -\frac{\mu}{2a}$ (equation (1.14));
- the angular momentum $\vec{h} = \vec{r} \wedge \vec{v}$ and eccentricity by $h^2 = \mu p = \mu a(1 - e^2)$;
- the eccentricity vector $\vec{e} = \frac{\vec{v}\wedge\vec{h}}{\mu} - \frac{\vec{r}}{r}$ (equation (1.10)).

The inclination (comprised by convention between 0 and 180 deg) is obtained from $\vec{Z} \cdot \vec{h} = h\cos i$. The line of nodes is oriented along the vector $\vec{N} = \vec{Z} \wedge \vec{h}$ (OX′ axis in Figure 1.5), and the axis normal to the line of nodes in the orbital plane is oriented along $\vec{N}' = \vec{h} \wedge \vec{N}$.

The three vectors $\vec{N}$, $\vec{N}'$, and $\vec{e}$ are used to calculate:

- the right ascension of the ascending node Ω by

$$\begin{cases}\vec{X}\cdot\vec{N} = h\cos\Omega\ \sin i\\ \vec{Y}\cdot\vec{N} = h\sin\Omega\ \sin i\end{cases} \tag{1.117}$$

- the argument of the pericenter ω by

$$\begin{cases}\vec{e}\cdot\vec{N} = he\cos\omega\sin i\\ \vec{e}\cdot\vec{N}' = h^2 e\sin\omega\sin i\end{cases} \tag{1.118}$$

- the true anomaly $\theta$ by

$$\begin{cases} \vec{e} \cdot \vec{r} = er \cos\theta \\ \vec{e} \cdot \vec{v} = \sqrt{\frac{\mu}{p}} ev \sin\theta \end{cases} \tag{1.119}$$

With the above relations, the angles $\Omega$, $\omega$, and $\theta$ are determined unambiguously from their cosines and sines, provided that the vectors $\vec{N}$ and $\vec{e}$ are not null.

### 1.4.3 Equinoctial Orbital Parameters

The classical orbital parameters are indeterminate in two cases that occur very frequently in space applications:

- for an equatorial orbit, the line of nodes is not defined because $\vec{N} = \vec{0}$;
- for a circular orbit, the major axis is not defined because $\vec{e} = \vec{0}$.

In order to avoid such singularities, the equinoctial orbital parameters are based on the components of the vectors $\vec{e}$ and $\vec{h}$, instead of the angles i, $\Omega$, and $\omega$.

These parameters noted $(n, P_1, P_2, Q_1, Q_2, L)$ are related to $(a, e, i, \Omega, \omega, \theta)$ by

$$\begin{cases} P_1 = e \sin(\Omega + \omega), \quad Q_1 = \tan\frac{i}{2} \sin\Omega, \quad n = \sqrt{\frac{\mu}{a^3}} \\ P_2 = e \cos(\Omega + \omega), \quad Q_2 = \tan\frac{i}{2} \cos\Omega, \quad L = \Omega + \omega + M \end{cases} \tag{1.120}$$

The parameters $(P_1, P_2)$ define the eccentricity vector, which can be null in the case of a circular orbit, whereas the parameters $(Q_1, Q_2)$ define the angular momentum vector, which can be directed along the polar axis in the case of an equatorial orbit.

The position on orbit is defined either by the mean longitude $L = \Omega + \omega + M$, or equivalently by the true longitude $l = \Omega + \omega + \theta$.

Singularities remain for rectilinear orbits ($h = 0$) or for 180 deg inclination. These cases require specific formulations. Although other singularity-free orbital parameters exist, such as Delaunay or Poincaré elements, equinoctial parameters are the most frequently used for space applications.

### 1.4.4 Earth Orbits

Earth orbits are classified according to their altitude.

- Low Earth Orbits (LEO, at altitudes below 2000 km) are mainly used for observation of the Earth's surface and for telecommunications.
- Medium Earth Orbits (MEO, at altitudes about 20 000 km) are mainly used for navigation.
- High Earth orbits (HEO, at altitudes of 36 000 km and above) are mainly used for meteorology, television, or space observation.

This section presents the definition of the geocentric reference frame, the spherical and geodetic coordinates (relative to the Earth's shape), and the geosynchronous or sun-synchronous orbits.

#### 1.4.4.1 Geocentric Reference Frame

In the case of a satellite orbiting around the Earth, the origin O is the Earth's center of mass, OZ is the Earth's polar axis, and OXY is the Earth's equatorial plane. This geocentric frame (OXYZ) must be defined at a given epoch because the Earth is moving around the Sun and perturbations act on the Earth's axis of rotation.

The Earth's motion around the Sun defines the ecliptic plane. The orbit of the Earth is quasi-circular, with a radius of 149 597 871 km (= 1 astronomical unit) and a period of 365.25 days, giving an average angular rate $\omega_s = 0.986$ deg/day.

The Earth's equator is tilted from the ecliptic by the obliquity angle $\varepsilon = 23.45$ deg. The X axis is the intersection between the equator and the ecliptic. It is directed toward the vernal point $\gamma$, which corresponds to the spring equinox in the northern hemisphere (Figure 1.6).

Due to perturbations caused mainly by the Moon's attraction, the polar axis is not fixed in space. On average, it describes a cone (precession motion with a period of about 25 760 years), on which oscillations are superimposed (nutation motion with a period of about 18.6 years, due to the Moon orbit precession).

Strictly speaking, the geocentric reference frame (OXYZ) is not inertial, which means that Newton's second law is not valid in this frame. Nevertheless, it can be considered as quasi-inertial (i.e. sufficiently close to an inertial frame) to study the motion of Earth's satellites over usual durations of space applications (a few years at most).

---

**Galilean (or Inertial) Frame**

There is no known perfectly Galilean frame in the universe. In practice, we can define quasi-Galilean frames in which Newton's laws are verified with excellent accuracy. The choice of the origin and of the axes depends on the motion under study, in particular on the predominant forces and the propagation time.

---

A reference geocentric frame called J2000 has been defined on January 1, 2000 at 12 h. The geocentric frame at any time t is defined from J2000 by taking into account the Earth's motion on its mean orbit around the Sun (for the origin O) and the luni-solar precession

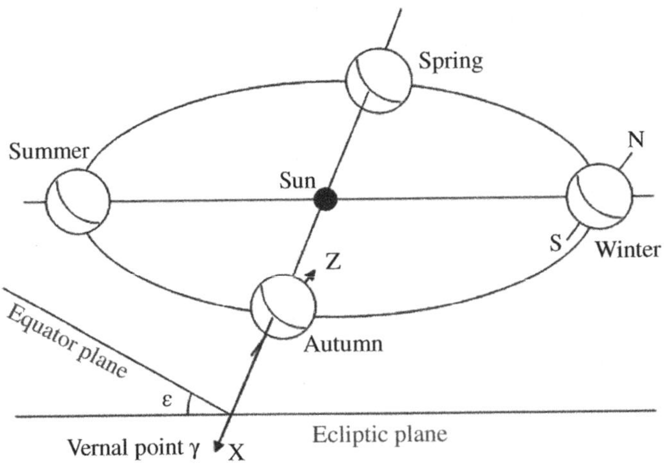


**Figure 1.6** Ecliptic and equator planes.

effect (for the axes X, Y, and Z). This results in the **mean equatorial frame** at time t. The true equatorial frame is obtained by applying furthermore the nutation effect.

Ephemerides such as those published by the Jet Propulsion Laboratory can be used to position the planets and their mean frames at a given date, relatively to J2000. A thorough description of the frames used in astrodynamics is presented in [R39].

### 1.4.4.2 Spherical Coordinates

For many space applications, it is convenient to express the position in spherical coordinates. The right ascension $\alpha$ is the angle between the reference meridian plane (OXZ) and the current meridian plane (OMZ). The declination $\delta$ is the angle between the equatorial plane (OXY) and the current position $\overrightarrow{OM}$.

Figure 1.7 shows an orbit of inclination i with ascending node in A. The angular distance from the node A to a point M of anomaly $\theta$ is the argument of latitude $\nu = \omega + \theta$. The meridian passing through M intersects the equator at L.

The transition from orbital parameters to spherical coordinates ($\alpha$,$\delta$) is expressed using spherical trigonometry relations in the triangles (ANM) and (AML).

$$\begin{cases} \sin\delta = \sin(\omega+\theta)\sin i \\ \cos\delta\sin(\alpha-\Omega) = \sin(\omega+\theta)\cos i \\ \cos\delta\cos(\alpha-\Omega) = \cos(\omega+\theta) \end{cases} \tag{1.121}$$

In particular, the right ascension depends on the anomaly by

$$\tan(\alpha-\Omega) = \tan(\omega+\theta)\cos i \tag{1.122}$$

Geographic coordinates (longitude $\lambda$, latitude $\varphi$) define the position in the rotating geocentric frame. Their definition is the same as right ascension and declination with the longitude measured from the current Greenwich meridian, which moves with the Earth's rotation. The angle between the vernal axis OX and Greenwich meridian at time t is the **sidereal time of Greenwich** (equivalent to the right ascension of Greenwich).

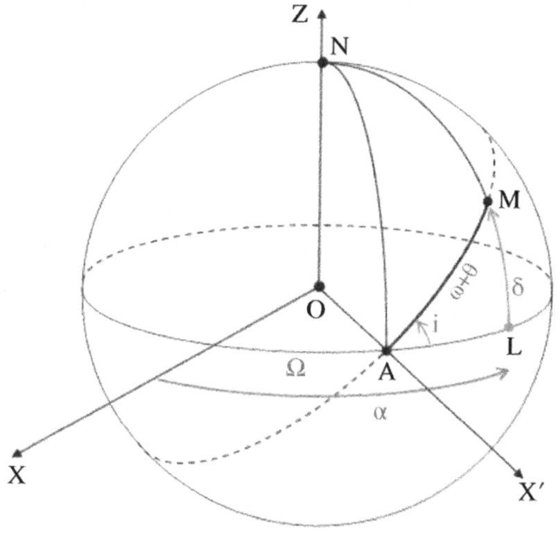


**Figure 1.7** Spherical coordinates.

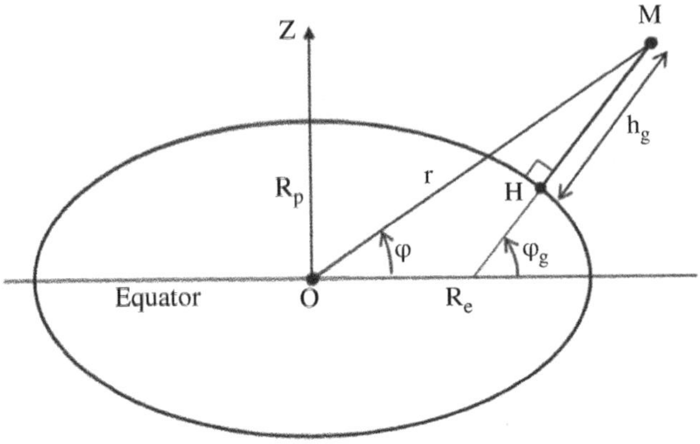


**Figure 1.8** Geodetic coordinates.

For a point of right ascension α and declination δ, the geographic coordinates are

$$\begin{cases} \lambda(t) = \alpha - \alpha_{Greenwich}(t) \\ \varphi(t) = \delta \end{cases} \tag{1.123}$$

#### 1.4.4.3 Earth Shape

In first approximation, the Earth's surface called the **geoid** is close to an ellipsoid of revolution with axis OZ, semi-major axis equal to the equatorial radius $R_e$, and semi-minor axis equal to the polar radius $R_p$.

The reference values of the WGS84 (World Geodetic System) model are:

- equatorial radius $R_e = 6\,378\,137$ m, polar radius $R_p = 6\,356\,752$ m;
- mass $M_g = 5.973\,33 \cdot 10^{24}$ kg, gravitational constant $\mu_g = 3.986\,004 \cdot 10^{14}$ m$^3$/s$^2$.

**Geodetic coordinates** (longitude $\lambda_g$, latitude $\varphi_g$, and altitude $h_g$) are defined relatively to the Earth's surface by considering the orthogonal projection on the ellipsoid, as shown in Figure 1.8.

The Cartesian, geographic, and geodetic coordinates are linked by

$$\begin{cases} x = r\cos\varphi\cos\lambda = (N + h_g)\cos\varphi_g\cos\lambda_g \\ y = r\cos\varphi\sin\lambda = (N + h_g)\cos\varphi_g\sin\lambda_g \\ z = r\sin\varphi = \left[N\left(1 - e_g^2\right) + h_g\right]\sin\varphi_g \end{cases} \quad \text{with } N = \frac{R_e}{\sqrt{1 - e_g^2\sin^2\varphi_g}} \tag{1.124}$$

where $e_g$ is the Earth surface eccentricity: $e_g = \frac{\sqrt{R_e^2 - R_p^2}}{R_e} = 0.081819$.

The Cartesian coordinates in the above formulas refer to the Earth's rotating frame based on the Greenwich meridian. Passing from Cartesian to geodetic coordinates is achieved, either by the following iterative algorithm, or by Bowring's formulas presented in Section 5.3.2.

$$\text{Initialization:} \left[\begin{array}{l} N^0 = R_e \\ h_g^0 = \sqrt{x^2 + y^2 + z^2} - \sqrt{R_e R_p} \\ \varphi_g^0 = \text{Arc tan}\left(\dfrac{z}{\sqrt{x^2 + y^2}}\dfrac{N^0 + h_g^0}{N^0\left(1 - e_g^2\right) + h_g^0}\right) \end{array}\right.$$

$$\text{Iterations:}\quad \begin{cases} N^{(k)} = \dfrac{R_e}{\sqrt{1 - e_g^2 \sin^2 \varphi_g^{(k-1)}}} \\[2ex] h_g^{(k)} = \dfrac{\sqrt{x^2 + y^2}}{\cos \varphi_g^{(k-1)}} - N^{(k)} \\[2ex] \varphi_g^{(k)} = \text{Arc tan}\left( \dfrac{z}{\sqrt{x^2 + y^2}} \dfrac{N^{(k)} + h_g^{(k)}}{N^{(k)}\left(1 - e_g^2\right) + h_g^{(k)}} \right) \end{cases} \tag{1.125}$$

#### 1.4.4.4 Geosynchronous Orbits

An Earth satellite is geosynchronous if its orbital period is equal to the Earth's period of rotation $T_g$. This period $T_g$ is defined in the inertial frame (OXYZ) and it differs from the solar day (24 h), which is the time interval between two passages of the Sun at the local meridian (i.e. between two consecutive midday hours).

To assess $T_g$, let us consider an observer on Earth, represented by the point on the Earth's surface in Figure 1.9. In 24 h, it rotates by an angle $360 + \alpha = 360.986$ deg in the inertial geocentric frame linked to the directions of fixed stars (like the vernal point). Its period of revolution $T_g$ in this inertial frame is therefore

$$T_g = 24 \times \frac{360}{360 + 0.986} = 23\,\text{h}\ 56\,\text{min}\ 04\,\text{s} \tag{1.126}$$

This period of rotation of the Earth in the inertial geocentric frame is called the **sidereal day**, as it corresponds to the time interval between two passages of a star at the meridian of a given location, as shown in Figure 1.9.

From the orbital period $T_g$, Kepler's third law (1.41) gives the semi-major axis equal to $a_g = 42\,164$ km. For the satellite to be **geostationary** (i.e. fixed relative to the Earth surface), its orbit must in addition have the same axis of rotation as the Earth (meaning the orbit is equatorial) and a constant angular velocity (meaning the orbit is circular, by Kepler's second law). The only geostationary orbit is therefore circular, equatorial at an altitude of 42 164–6378 = 35 786 km. This geostationary Earth orbit is widely used for television and meteorological applications [R7].

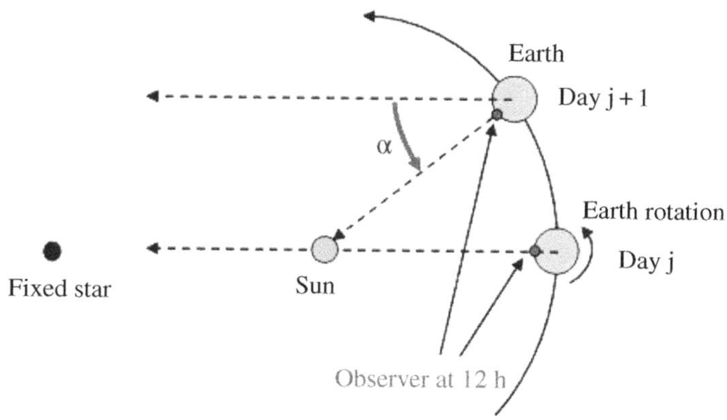


**Figure 1.9** Solar and sidereal day.

#### 1.4.4.5 Sun-Synchronous Orbit

An Earth satellite is sun-synchronous if its orbital plane maintains a constant angle with the direction of the Sun. This means that the lighting conditions on the satellite and in the areas it flies over are identical throughout the year.

In the geocentric frame, the Sun's orbit is close to a circle with a period of 365.25 days, inclination $\varepsilon = 23.45$ deg and ascending node on the OX axis ($\Omega = 0$ deg), which points to the vernal point. The passage at the ascending node corresponds to the vernal equinox (Figure 1.6).

Equation (1.122) yields the Sun's right ascension $\alpha_s$ as a function of time

$$\tan\alpha_s(t) = \tan(\omega_s t)\cos\varepsilon \tag{1.127}$$

where $\omega_s = 0.986$ deg/day is the (almost constant) Earth's mean motion and t is the time elapsed since the spring equinox.

A sun-synchronous orbit is characterized by the local time at its ascending node. The local time $h_{loc}$ is the difference between the Sun and local right ascensions, expressed in hours.

$$h_{loc} = \alpha - \alpha_s \tag{1.128}$$

Figure 1.10 shows a satellite in polar orbit with the ascending node marked by the point on the Earth's surface. The time elapsed between the left and right figure is three months. In the inertial frame OXYZ, the right ascension of the ascending node has increased by 90 deg between both figures. Such an orbit is sun-synchronous since its orbital plane keeps a constant angle (here 90 deg) with the Sun's direction. Its local time at ascending node (LTAN) is here 18 h, since the Sun is about to pass below the horizon (the Earth rotates in the direct sense).

For a perfect sun-synchronism, it would also be necessary to update the orbit inclination with the time, as the plane of the ecliptic does not coincide with the equator (obliquity of 23.45 deg). In practice, sun-synchronism is achieved by updating only the right ascension of the ascending node to follow the right ascension of the Sun. This requires an average precession rate equal to

$$\dot{\Omega} = \dot{\alpha}_s \approx \omega_s \tag{1.129}$$

The gravitational perturbation caused by the Earth's flattening creates a precession of the orbital plane. With the right choice of orbital parameters, the precession rate required for sun-synchronism can be obtained naturally. A location of a given latitude is then always flown over at the same local time. This natural precession effect is analyzed in Section 2.4.4.

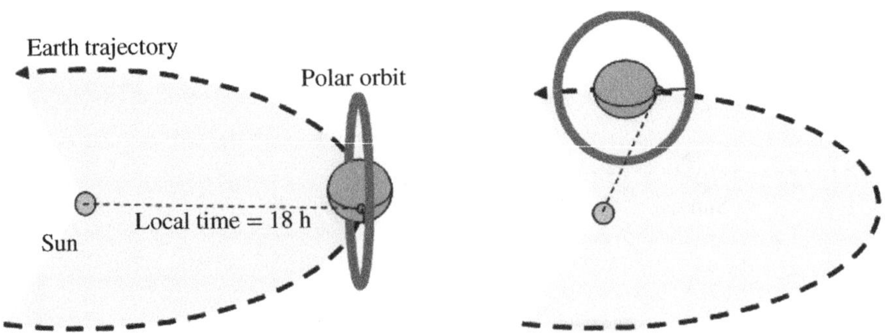


**Figure 1.10** Sun-synchronous orbit.

## 1.5 Conclusion

### 1.5.1 The Key Points

- A Keplerian trajectory is a conic (ellipse, parabola, or hyperbola), which can reduce to a straight line in degenerate cases.
- The main prime integrals of the Keplerian motion are those of energy, angular momentum, and eccentricity vector. The energy is negative for an ellipse, null for a parabola, and positive for a hyperbola.
- Kepler's equation is transcendental and must be solved by a numerical method to determine the position on the conic at a given time.
- The classical or equinoctial orbital parameters locate the orbit in an inertial frame linked to the attracting body.
- Geosynchronous orbits have a period equal to the Earth's rotation rate, sun-synchronous orbits keep a constant angle with respect to the Sun's direction.

### 1.5.2 To Go Further

- BATTIN R. – *An Introduction to the Mathematics and Methods of Astrodynamics*. AIAA (1999).

This book is one of the most exhaustive on the analysis and the properties of Keplerian motion. Chapters 3–6 present many useful results, always derived in a very elegant way using vector calculus and/or geometrical approaches. A particular emphasis is made on Kepler's equation. Chapter 4 presents the derivation of Kepler's equation, depending on the conic type, and establishes the universal formulation in a very clear manner. Chapter 5 presents numerous solution methods to Kepler's equation, either by series or by iterations.

- CHOBOTOV V. – *Orbital Mechanics Third Edition*. AIAA (2002).

This book is very pleasant to read and it presents the essentials of Keplerian motion in Chapters 3 and 4. Of particular interest is the extensive comparison of Newton's and Laguerre's method to solve Kepler's equation, presented in Chapter 4.

- ROY A. E. – *Orbital Motion*. Adam Hilger (1988).

This book is rather oriented toward celestial mechanics, from an astronomical point of view. Chapter 4 provides the basis of Keplerian motion, and presents the formulation adapted to degenerate conics (which is often not mentioned, although of practical interest for some applications).

- VALLADO D. – *Fundamentals of Astrodynamics and Applications*. Microcosm Press, Springer (2007).

This book is one of the best references in astrodynamics. Keplerian motion is presented in Chapter 1, whereas Chapter 2 is dedicated to Kepler's equation and the numerous methods to solve it.

# 2

# Perturbed Motion

## 2.1 Introduction

The Keplerian model provides a first trajectory approximation by considering only the attraction of the main body, assumed to be punctual. The trajectory is then a conic defined by its orbital parameters.

The forces ignored by the Keplerian model are called perturbations. They include the gravitational terms due to the Earth's mass distribution, the attraction of other bodies (Moon and Sun), the atmospheric friction, the radiation pressure due to the solar wind, forces due to the magnetic field, and tidal forces.

These forces are very weak compared to the central gravitational attraction and their effect can be analyzed as a progressive deviation of the satellite from its Keplerian orbit. The osculating orbit is defined as the Keplerian orbit that would be followed from a given time if all perturbations were to become zero.

The evolution of the osculating parameters under the effect of perturbations is described by Gauss or Lagrange equations. These differential equations can be solved by numerical methods, known as special perturbation methods, or with simplifications by analytical methods, known as general perturbation methods. The latter helps understanding the main effects of each type of perturbation.

After recalling the main results of Keplerian motion in the first section, the second section develops the differential equations followed by the orbital parameters. These equations are formulated either under Gauss form (valid for any kind of perturbative force), or under Lagrange form (valid only for perturbations deriving from a potential). The integration methods, numerical or semi-analytical, are then shortly discussed

The third section deals with gravitational perturbations through their expansion in spherical harmonics. This makes it possible to derive the precession effects of the orbital plane due to the Earth flattening, associated with the first zonal term.

The fourth section deals with other perturbations, such as third-body attraction, atmospheric drag, and radiation pressure. Simplified models allow us to analyze the main effects of the Sun and Moon on geostationary orbits and of the atmospheric drag on low Earth orbits.

*Space Trajectories: Basic and Advanced Topics*, First Edition. Max Cerf.

## 2.2 Unperturbed Motion

The unperturbed motion corresponds to the Keplerian model, which takes into account only the central gravitational field of the attracting body. This model described in Chapter 1 provides a first approximation of the actual motion. The properties useful for the perturbed motion analysis are recalled in this section.

### 2.2.1 Keplerian Model

The Keplerian model assumes two particles in $M_1$ and $M_2$ of respective masses $m_1$ and $m_2$ exerting between them the gravitational force of attraction of expression

$$\vec{F}_{2\to 1} = -\vec{F}_{1\to 2} = G\frac{m_1 m_2}{r^3}\vec{r} \quad \text{with } \vec{r} = \overrightarrow{M_1 M_2} \tag{2.1}$$

where $G \approx 6.674\,08 \cdot 10^{-11}\ m^3\ kg^{-1}\ s^{-2}$ is the universal gravitational constant.

In space applications, $M_1$ is the main body (Earth, for example) and $M_2$ is a satellite whose mass $m_2$ is negligible compared with $m_1$. The differential equations of the motion of $M_2$ relative to $M_1$ are

$$\begin{cases} \dot{\vec{r}} = \vec{v} \\ \dot{\vec{v}} = -\dfrac{\mu}{r^3}\vec{r} \underset{\text{noted}}{=} \vec{a}_c \end{cases} \tag{2.2}$$

where $\mu = G(m_1 + m_2) \approx Gm_1$ is the gravitational constant of the main body $M_1$, $\vec{v}$ is the velocity of $M_2$ relative to $M_1$, and $\vec{a}_c$ is the central acceleration due to $M_1$.

The Keplerian motion admits several prime integrals, in particular those of angular momentum $\vec{h}$, of energy w and of eccentricity vector $\vec{e}$, whose expressions as functions of position $\vec{r}$ and velocity $\vec{v}$ have been established in Section 1.2.2.

$$\vec{h} = \vec{r} \wedge \vec{v}, \quad w = \frac{v^2}{2} - \frac{\mu}{r}, \quad \vec{e} = \frac{\vec{v} \wedge \vec{h}}{\mu} - \frac{\vec{r}}{r} \tag{2.3}$$

These quantities are constant along the Keplerian motion. The vectors $\vec{h}$ and $\vec{e}$ are shown in Figure 2.1.

### 2.2.2 Orbital Parameters

A Keplerian trajectory is a conic represented by its orbital parameters in an inertial reference frame (OXYZ). In the case of an orbit around the Earth, O is the Earth's center of mass, OZ is the pole axis pointing north, and OXY is the equatorial plane with the OX axis pointing to the vernal point.

The classical orbital parameters, defined in Section 1.4.1, are the inclination i, the right ascension of the ascending node $\Omega$, the argument of the perigee $\omega$, the semi-major axis a and the eccentricity e. We recall their definition here.

The orbital plane (containing the conic) is defined from the equatorial plane OXY by a rotation of axis Z and angle $\Omega$, then a rotation of axis $X'$ and angle i, as shown in Figure 2.1. The ascending node A, opposite the descending node D, is the point where the equator is crossed from south to north. The $X'$ axis is the line of nodes. The major axis (oriented by the eccentricity vector $\vec{e}$ pointing toward the perigee) is obtained by a rotation of axis $Z'$

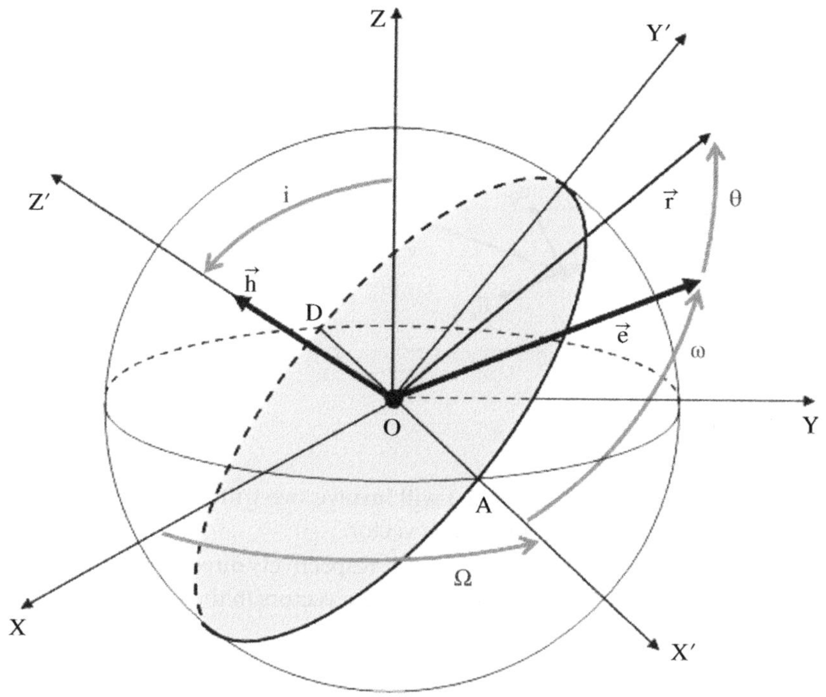


**Figure 2.1** Reference frame and orbital parameters.

and angle ω. The Z′ axis is oriented along the angular momentum $\vec{h}$. The orbit shape is determined by the semi-major axis a and the eccentricity e.

The five orbital parameters (a, e, i, Ω, and ω) are constants determining the Keplerian orbit. The position on the orbit is determined either from the perigee by the true anomaly θ, or from the ascending node by the latitude argument $\nu = \omega + \theta$.

The components of the position $\vec{r}$ in the frame (OXYZ) result from the three successive rotations of angles Ω, i, and ν.

$$\begin{aligned}\vec{r} = {} & r(\cos\Omega\cos\nu - \sin\Omega\sin\nu\cos i)\vec{X} \\ & + r(\sin\Omega\cos\nu + \cos\Omega\sin\nu\cos i)\vec{Y} \\ & + r(\sin\nu\sin i)\vec{Z}\end{aligned} \tag{2.4}$$

The perturbed motion analysis deals only with elliptical orbits and their evolution on long durations. For an ellipse, the true anomaly θ is related to the eccentric anomaly E by (1.24).

$$\cos E = \frac{e + \cos\theta}{1 + e\,\cos\theta}, \quad \sin E = \frac{\sqrt{1 - e^2}\sin\theta}{1 + e\,\cos\theta} \tag{2.5}$$

The eccentric anomaly E varies with time according to Kepler's equation (1.48).

$$E - e\,\sin E = M = n(t - t_P) \quad \text{with} \quad n = \sqrt{\frac{\mu}{a^3}} = \frac{2\pi}{T} \tag{2.6}$$

where M is the mean anomaly, $t_P$ is the time of passage at the perigee, n is the pulsation (or mean motion), and T is the period of revolution.

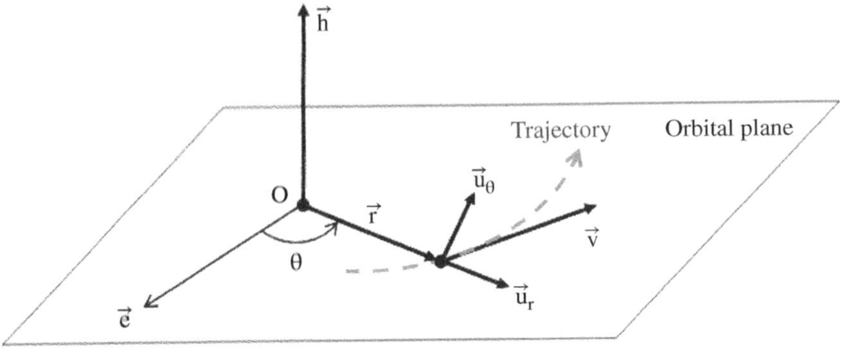


**Figure 2.2** Local orbital frame.

### 2.2.3 Useful Frames

The analysis of the perturbed motion in Section 2.3 will involve two trihedra, one based on the ascending node, and the other based on the radius vector.

The first trihedron is defined by the vectors $\vec{h}$, $\vec{N}$, and $\vec{N}'$ respectively directed along the Z′, X′, and Y′ axes shown in Figure 2.1. The components of these vectors in the reference frame (XYZ) are

$$\begin{cases} \vec{h} = \vec{r} \wedge \vec{v} = (\ \vec{X} \sin\Omega \sin i - \vec{Y} \cos\Omega \sin i + \vec{Z} \cos i)h \\ \vec{N} = \vec{Z} \wedge \vec{h} = (\ \vec{X} \cos\ \Omega + \vec{Y} \sin\Omega)h\ \sin i \\ \vec{N}' = \vec{h} \wedge \vec{N} = (-\vec{X} \sin\Omega\ \cos i + \vec{Y} \cos\Omega\ \cos i + \vec{Z}\ \sin i)h^2\ \sin i \end{cases} \tag{2.7}$$

The angle between $\vec{N}$ and $\vec{r}$ is $\nu = \omega + \theta$, which leads to the following relations that will be used later.

$$\begin{cases} \vec{N} \wedge \vec{r}\ = \vec{h}r\ \sin i\ \sin\nu \\ \vec{N}' \wedge \vec{r} = -h\vec{h}r\ \sin i\ \cos\nu \end{cases} \quad \text{with}\ \ \nu = \omega + \theta \tag{2.8}$$

The second trihedron is the local orbital frame noted $(\vec{u}_r, \vec{u}_\theta, \vec{u}_h)$ and shown in Figure 2.2. It is defined by the radius vector $\vec{r} = r\vec{u}_r$, the orthoradial direction $\vec{u}_\theta$ in the orbital plane and the angular momentum $\vec{h} = \vec{r} \wedge \vec{v} = h\vec{u}_h$.

Note: the vectors $(\vec{u}_r, \vec{u}_\theta, \vec{u}_h)$ are unitary, which is not the case for $(\vec{h}, \vec{N}, \vec{N}')$.

The velocity components in the local orbital frame are obtained by deriving the position vector: $\vec{r} = r\vec{u}_r \Rightarrow \vec{v} = \dot{\vec{r}} = \dot{r}\vec{u}_r + r\dot{\vec{u}}_r = \dot{r}\vec{u}_r + r\dot{\theta}\vec{u}_\theta$.

By using the Keplerian relations

$$r = \frac{p}{1 + e\cos\theta}, \quad p = \frac{h^2}{\mu} = a(1 - e^2), \quad h = r^2\dot{\theta} \tag{2.9}$$

we obtain the expression of the velocity vector $\vec{v}$

$$\vec{v} = \frac{\mu}{h}[e \sin\theta\, \vec{u}_r + (1 + e\cos\theta)\vec{u}_\theta] \tag{2.10}$$

## 2.3 Perturbed Motion

The perturbed motion is analyzed as a progressive deviation from the Keplerian trajectory. This section develops the differential equations followed by the orbital parameters.

**Celestial Mechanics and Perturbation Methods**

Celestial mechanics flourished in the 19th century. The aim was to explain and predict the motions of the planets in the solar system as accurately as possible. Limitations in computing capacity led to mathematical developments that were subsequently used in other branches of physics.

Joseph-Louis Lagrange (1736–1813), in *Mécanique Analytique* published in 1788, summarized his work on perturbation methods. This research was continued by numerous mathematicians and astronomers. It culminated with Le Verrier's discovery of Neptune in 1846, predicted from perturbations in the orbit of Uranus.

### 2.3.1 Osculating Orbit

The Keplerian model assumes that the only acting force is the central gravitational attraction (2.1). In the real world, many other forces must be taken into account. Any force other than central attraction creates a perturbing acceleration $\vec{a}_p$ that adds to the central acceleration $\vec{a}_c$. The equations of motion (2.2) become

$$\begin{cases} \dot{\vec{r}} = \vec{v} \\ \dot{\vec{v}} = \vec{a}_c + \vec{a}_p \end{cases} \tag{2.11}$$

If all perturbations cancel out from a time $t_0$, the motion becomes Keplerian.

The Keplerian conic followed from this time $t_0$ depends on the current position $\vec{r}_0$ and velocity $\vec{v}_0$. This conic is called the **osculating orbit** at $t_0$ and its orbital parameters are called the osculating parameters at $t_0$.

The actual motion from the current time $t_0$ to a later time $t_1$ is subject to perturbations and it will gradually deviate from the initial osculating orbit. The position $\vec{r}_1$ and velocity $\vec{v}_1$ at time $t_1$ define a new osculating orbit, which depends on the perturbations applied between $t_0$ and $t_1$.

### 2.3.2 Derivation Formula

A function $K(t, \vec{r}, \vec{v})$ of time, position, and velocity has the total time derivative

$$\frac{dK}{dt} = \frac{\partial K}{\partial t} + \frac{\partial K}{\partial \vec{r}} \cdot \dot{\vec{r}} + \frac{\partial K}{\partial \vec{v}} \cdot \dot{\vec{v}} = \frac{\partial K}{\partial t} + \frac{\partial K}{\partial \vec{r}} \cdot \vec{v} + \frac{\partial K}{\partial \vec{v}} \cdot (\vec{a}_c + \vec{a}_p) \tag{2.12}$$

This derivative is the sum of an unperturbed or Keplerian term (for $\vec{a}_p = \vec{0}$) and a term due to $\vec{a}_p$.

$$\frac{dK}{dt} = \left(\frac{dK}{dt}\right)_{\text{Kepler}} + \frac{\partial K}{\partial \vec{v}} \cdot \vec{a}_p \tag{2.13}$$

If the function K is constant along a Keplerian motion (like an orbital parameter or a prime integral), its derivative reduces to

$$\frac{dK}{dt} = \frac{\partial K}{\partial \vec{v}} \cdot \vec{a}_p \tag{2.14}$$

This formula extends to vector functions $\vec{K}$. In particular, it allows us to derive the angular momentum vector $\vec{h} = \vec{r} \wedge \vec{v}$.

$$\frac{d\vec{h}}{dt} = \frac{\partial \vec{h}}{\partial \vec{v}} \cdot \vec{a}_p = \vec{r} \wedge \vec{a}_p \tag{2.15}$$

From this, we can also assess the derivative of the angular momentum modulus.

Starting from $h^2 = \vec{h} \cdot \vec{h}$, we have $h\frac{dh}{dt} = \vec{h} \cdot \frac{d\vec{h}}{dt}$, and substituting $\frac{d\vec{h}}{dt}$ from (2.15), we get $h\frac{dh}{dt} = \vec{h} \cdot (\vec{r} \wedge \vec{a}_p) = (\vec{h} \wedge \vec{r}) \cdot \vec{a}_p = hr\vec{u}_\theta \cdot \vec{a}_p$, which yields

$$\frac{dh}{dt} = ra_\theta \tag{2.16}$$

### 2.3.3 Gauss Equations

Gauss equations express the variations of the orbital parameters (a, e, i, Ω, ω, θ) as a function of the perturbing acceleration $\vec{a}_p$ projected onto the local orbital frame: $\vec{a}_p = a_r\vec{u}_r + a_\theta\vec{u}_\theta + a_h\vec{u}_h$.

Since the orbital parameters are constant throughout the Keplerian motion, their derivatives can be assessed by formula (2.14).

#### 2.3.3.1 Variation of the Semi-major Axis

The semi-major axis a is defined from the energy.

$$w = \frac{v^2}{2} - \frac{m}{r} = -\frac{m}{2a} \quad \Rightarrow \quad \frac{1}{a} = -\frac{1}{m}\vec{v} \cdot \vec{v} + \frac{2}{r} \tag{2.17}$$

Applying (2.14), and substituting $\vec{v}$ given by (2.10), we obtain

$$\frac{da}{dt} = \frac{2a^2}{\mu}\vec{v} \cdot \vec{a}_p = \frac{2a^2}{h}[a_r e \cos\theta + a_\theta(1 + e\cos\theta)] \tag{2.18}$$

#### 2.3.3.2 Variation of the Eccentricity

The derivative of the eccentricity e is obtained from formula (2.9).

$$\frac{h^2}{\mu} = a(1 - e^2) \quad \Rightarrow \quad 2h\frac{dh}{dt} = \mu(1 - e^2)\frac{da}{dt} - 2\mu ae\frac{de}{dt} \tag{2.19}$$

Substituting the derivatives given by (2.16) and (2.18), we obtain

$$\frac{de}{dt} = \frac{p}{eh}[a_r e \sin\theta + a_\theta(1 + e\cos\theta)] - \frac{hr}{\mu ae}a_\theta \tag{2.20}$$

which can be expressed as:

$$\frac{de}{dt} = \frac{p}{h}\left[a_r \sin\theta + a_\theta\left(\cos\theta + \frac{e + \cos\theta}{1 + e\cos\theta}\right)\right] \tag{2.21}$$

#### 2.3.3.3 Variation of the Ascending Node Right Ascension and of the Inclination

The components of $\vec{h}$ in (OXYZ) are given by (2.7).

$$\begin{cases} \vec{X} \cdot \vec{h} = h \sin i \sin\Omega \\ \vec{Y} \cdot \vec{h} = -h \sin i \cos\Omega \\ \vec{Z} \cdot \vec{h} = h \cos i \end{cases} \tag{2.22}$$

Deriving these relations

$$\begin{cases} \vec{X} \cdot \frac{d\vec{h}}{dt} = \frac{dh}{dt}\sin i \sin\Omega + \frac{di}{dt}h \cos i \sin\Omega + \frac{d\Omega}{dt}h \sin i \cos\Omega \\ \vec{Y} \cdot \frac{d\vec{h}}{dt} = -\frac{dh}{dt}\sin i \cos\Omega - \frac{di}{dt}h \cos i \cos\Omega + \frac{d\Omega}{dt}h \sin i \sin\Omega \\ \vec{Z} \cdot \frac{d\vec{h}}{dt} = \frac{dh}{dt}\cos i - \frac{di}{dt}h \sin i \end{cases} \tag{2.23}$$

and solving by linear combinations, we obtain

$$\begin{cases} \dfrac{d\Omega}{dt} h \sin i = (\vec{X}\cos\Omega + \vec{Y}\sin\Omega) \cdot \dfrac{d\vec{h}}{dt} \\ \dfrac{di}{dt} h = (\vec{X}\cos i \sin\Omega - \vec{Y}\cos i \cos\Omega - \vec{Z}\sin i) \cdot \dfrac{d\vec{h}}{dt} \end{cases} \tag{2.24}$$

These equations can be written using the vectors $\vec{N}$ and $\vec{N}'$ defined in (2.7) and substituting the derivative of $\vec{h}$ given by (2.15).

$$\begin{cases} \dfrac{d\Omega}{dt} h \sin i = \dfrac{\vec{N}}{h \sin i} \cdot (\vec{r} \wedge \vec{a}_p) = \dfrac{\vec{N} \wedge \vec{r}}{h \sin i} \cdot \vec{a}_p \\ \dfrac{di}{dt} h = -\dfrac{\vec{N}'}{h^2 \sin i} \cdot (\vec{r} \wedge \vec{a}_p) = -\dfrac{\vec{N}' \wedge \vec{r}}{h^2 \sin i} \cdot \vec{a}_p \end{cases} \tag{2.25}$$

Using the vector products $\vec{N} \wedge \vec{r}$ and $\vec{N}' \wedge \vec{r}$ given by (2.8), we obtain

$$\begin{cases} \dfrac{d\Omega}{dt} = \dfrac{r \sin\nu}{h \sin i} a_h \\ \dfrac{di}{dt} = \dfrac{r \cos\nu}{h} a_h \end{cases} \quad \text{with} \quad \nu = \omega + \theta \tag{2.26}$$

#### 2.3.3.4 Variation of the True Anomaly

Equation (2.13) applied to the true anomaly θ gives

$$\frac{d\theta}{dt} = \left(\frac{d\theta}{dt}\right)_{\text{Kepler}} + \frac{\partial\theta}{\partial\vec{v}} \cdot \vec{a}_p \tag{2.27}$$

For the Keplerian motion, we have from (1.5): $h = r^2\left(\frac{d\theta}{dt}\right)_{\text{Kepler}}$.

To calculate the derivative $\frac{\partial\theta}{\partial\vec{v}}$, we start from (2.9) and (2.10) put into the form

$$\begin{cases} \mu r(1 + e\cos\theta) = h^2 \\ \mu r e \sin\theta = h\vec{r} \cdot \vec{v} \end{cases} \tag{2.28}$$

By deriving these relations with respect to $\vec{v}$ (i.e. with constant $\vec{r}$)

$$\begin{cases} \mu r \left(\dfrac{\partial e}{\partial\vec{v}} \cos\theta - \dfrac{\partial\theta}{\partial\vec{v}} e \sin\theta\right) = 2h\dfrac{\partial h}{\partial\vec{v}} \\ \mu r \left(\dfrac{\partial e}{\partial\vec{v}} \sin\theta + \dfrac{\partial\theta}{\partial\vec{v}} e \cos\theta\right) = \dfrac{\partial h}{\partial\vec{v}} \vec{r} \cdot \vec{v} + h\vec{r} \end{cases} \tag{2.29}$$

and combining the derivatives to eliminate $\frac{\partial e}{\partial\vec{v}}$, with $\vec{v}$ given by (2.10), we get

$$\mu r e \frac{\partial\theta}{\partial\vec{v}} = \left(\frac{\mu}{h} r e \cos\theta - 2h\right) \sin\theta \frac{\partial h}{\partial\vec{v}} + h\cos\theta\,\vec{r} \tag{2.30}$$

Using $r = \frac{p}{1+e\cos\theta}$ and $h^2 = \mu p$, (2.30) can be written as:

$$\frac{\partial\theta}{\partial\vec{v}} = \frac{1}{he}\left[\cos\theta(1 + e\cos\theta)\vec{r} - \cos\theta(2 + e\cos\theta)\frac{\partial h}{\partial\vec{v}}\right] \tag{2.31}$$

and substituted in (2.27), with the Keplerian derivative given by $\left(\frac{d\theta}{dt}\right)_{\text{Kepler}} = \frac{h}{r^2}$.

$$\frac{d\theta}{dt} = \frac{h}{r^2} + \frac{1}{he}\left[\cos\theta(1 + e\cos\theta)\vec{r} - \cos\theta(2 + e\sin\theta)\frac{\partial h}{\partial\vec{v}}\right] \cdot \vec{a}_p \tag{2.32}$$

To assess $\frac{\partial h}{\partial \vec{v}} \cdot \vec{a}_p$, we start from $h^2 = \vec{h} \cdot \vec{h}$ yielding $h\frac{\partial h}{\partial \vec{v}} = \vec{h} \cdot \frac{\partial \vec{h}}{\partial \vec{v}}$, and using (2.15) we have $\frac{\partial h}{\partial \vec{v}} = \vec{u}_h \cdot \frac{\partial \vec{h}}{\partial \vec{v}} = \vec{u}_h \cdot (\vec{r} \wedge \vec{a}_p)$.

By substituting into (2.32) with the acceleration $\vec{a}_p = a_r\vec{u}_r + a_\theta\vec{u}_\theta + a_h\vec{u}_h$ in the local orbital frame, we obtain the derivative of the true anomaly

$$\frac{d\theta}{dt} = \frac{h}{r^2} + \frac{r}{he}[\cos\theta(1 + e\cos\theta)a_r - \sin\theta(2 + e\cos\theta)a_\theta] \tag{2.33}$$

#### 2.3.3.5 Variation of the Eccentric Anomaly

Equation (2.13) applied to the eccentric anomaly E gives

$$\frac{dE}{dt} = \left(\frac{dE}{dt}\right)_{\text{Kepler}} + \frac{\partial E}{\partial \vec{v}} \cdot \vec{a}_p \tag{2.34}$$

The first term is assessed by deriving Kepler's equation (2.6).

$$E - e\sin E = M = n(t - t_P) \Rightarrow \left(\frac{dE}{dt}\right)_{\text{Kepler}} = \frac{n}{1 - e\cos E} = \frac{na}{r} \tag{2.35}$$

The derivative $\frac{\partial E}{\partial \vec{v}}$ is assessed by deriving (2.5): $\cos E = \frac{e + \cos\theta}{1 + e\cos\theta}$, which gives

$$\frac{\partial E}{\partial \vec{v}} = \frac{r}{a\sqrt{1 - e^2}}\left(\frac{\partial\theta}{\partial \vec{v}} - \frac{1}{1 - e^2}\sin\theta\frac{\partial e}{\partial \vec{v}}\right) \tag{2.36}$$

Substituting into (2.34) with the acceleration components in the local orbital frame, we obtain after calculation the derivative of the eccentric anomaly under the form

$$\frac{dE}{dt} = \frac{na}{r} + \frac{1}{nae}\left[(\cos\theta - e)a_r - \left(1 + \frac{r}{a}\right)\sin\theta\, a_\theta\right] \tag{2.37}$$

#### 2.3.3.6 Variation of the Mean Anomaly

Equation (2.13) applied to the mean anomaly M gives

$$\frac{dM}{dt} = \left(\frac{dM}{dt}\right)_{\text{Kepler}} + \frac{\partial M}{\partial \vec{v}} \cdot \vec{a}_p \tag{2.38}$$

The first term is obtained from the definition of M.

$$M = n(t - t_P) \Rightarrow \left(\frac{dM}{dt}\right)_{\text{Kepler}} = n \tag{2.39}$$

The derivative $\frac{\partial M}{\partial \vec{v}}$ is assessed by deriving Kepler's equation (2.6).

$$M = E - e\sin E \Rightarrow \frac{\partial M}{\partial \vec{v}} = \frac{\partial E}{\partial \vec{v}}(1 - e\cos E) - \frac{\partial e}{\partial \vec{v}}\sin E \tag{2.40}$$

The derivative $\frac{\partial e}{\partial \vec{v}}$ is assessed from the definition of $\vec{e}$ (2.3) and the derivative $\frac{\partial E}{\partial \vec{v}}$ is given by (2.36). After substitution and some tedious calculations (see [R1] or [R39]), we obtain the derivative of the mean anomaly under the form

$$\frac{dM}{dt} = n + \frac{b}{aeh}[(p\cos\theta - 2re)a_r - (p + r)\sin\theta\, a_\theta] \tag{2.41}$$

#### 2.3.3.7 Variation of the Perigee Argument

The angle $\nu = \omega + \theta$ is calculated from the vectors $\vec{N}$ and $\vec{r}$.

$$\begin{cases} \vec{N} \cdot \vec{r} = hr \sin i \cos \nu \\ \vec{N} = (\vec{X} \cos \Omega + \vec{Y} \sin \Omega) h \sin i \end{cases} \Rightarrow r \cos \nu = \vec{X} \cdot \vec{r} \cos \Omega + \vec{Y} \cdot \vec{r} \sin \Omega \tag{2.42}$$

Deriving with respect to $\vec{v}$

$$-r \sin \nu \frac{\partial \nu}{\partial \vec{v}} = -\vec{X} \cdot \vec{r} \sin \Omega \frac{\partial \Omega}{\partial \vec{v}} + \vec{Y} \cdot \vec{r} \cos \Omega \frac{\partial \Omega}{\partial \vec{v}} \tag{2.43}$$

then substituting $\vec{r}$ given by (2.4), and with the derivative of $\Omega$ (2.26), we obtain

$$\frac{\partial \nu}{\partial \vec{v}} = -\frac{\partial \Omega}{\partial \vec{v}} \cos i \Rightarrow \frac{d\nu}{dt} = -\frac{\partial \Omega}{\partial \vec{v}} \cdot \vec{a}_p \cos i = -\frac{r \cos \nu}{h \tan i} a_h \tag{2.44}$$

The derivative of $\omega$ is deduced by subtracting (2.32) to (2.44). It is given in (2.45), which recaps the orbital parameters variations under the form of Gauss equations.

#### 2.3.3.8 Gauss Equations

$$\begin{cases} \dfrac{da}{dt} = \dfrac{2a^2}{h}[a_r e \sin \theta + a_\theta(1 + e \cos \theta)] \\ \dfrac{de}{dt} = \dfrac{p}{h}\left[a_r \sin \theta + a_\theta \left(\cos \theta + \dfrac{e + \cos \theta}{1 + e \cos \theta}\right)\right] \\ \dfrac{di}{dt} = a_h \dfrac{r \cos \nu}{h} \\ \dfrac{d\Omega}{dt} = a_h \dfrac{r \sin \nu}{h \sin i} \\ \dfrac{d\omega}{dt} = \dfrac{r}{he}[-a_r \cos \theta(1 + e \cos \theta) + a_\theta \sin \theta(2 + e \cos \theta)] - a_h \dfrac{r \sin \nu}{h \tan i} \\ \dfrac{dM}{dt} = n + \dfrac{b}{aeh}[a_r(p \cos \theta - 2re) - a_\theta(p + r) \sin \theta] \end{cases} \tag{2.45}$$

### 2.3.4 Lagrange Equations

Lagrange equations express the variations of the orbital parameters in the case of a perturbation $\vec{a}_p$ deriving from a potential $U_p(\vec{r})$. The six orbital parameters considered for Lagrange equations are $(a, e, i, \Omega, \omega, \lambda = nt_p)$ and they are grouped in a vector noted $\vec{\alpha}$. The position and the velocity are functions of t and $\vec{\alpha}$, and the differential equations of motion (2.11) take the form

$$\begin{cases} \dfrac{d\vec{r}}{dt} = \dfrac{\partial \vec{r}}{\partial t} + \dfrac{\partial \vec{r}}{\partial \vec{\alpha}} \cdot \dfrac{d\vec{\alpha}}{dt} = \vec{v} \\ \dfrac{d\vec{v}}{dt} = \dfrac{\partial \vec{v}}{\partial t} + \dfrac{\partial \vec{v}}{\partial \vec{\alpha}} \cdot \dfrac{d\vec{\alpha}}{dt} = \vec{a}_c + \vec{a}_p \end{cases} \tag{2.46}$$

If $\vec{a}_p = \vec{0}$, the motion is Keplerian and the derivative of $\vec{\alpha}$ is null.

$$\frac{\partial \vec{r}}{\partial t} = \vec{v}, \quad \frac{\partial \vec{v}}{\partial t} = \vec{a}_c \tag{2.47}$$

This allows to reduce the equations of perturbed motion, which become

$$\frac{\partial \vec{r}}{\partial \vec{\alpha}} \cdot \frac{d\vec{\alpha}}{dt} = \vec{0}, \quad \frac{\partial \vec{v}}{\partial \vec{\alpha}} \cdot \frac{d\vec{\alpha}}{dt} = \vec{a}_P = \frac{\partial U_P}{\partial \vec{r}} \tag{2.48}$$

The system (2.48) of two vector equations can be written in a compact form by introducing the Lagrange matrix L.

$$L\frac{d\vec{\alpha}}{dt} = \frac{\partial U_P}{\partial \vec{\alpha}} \quad \text{with} \quad L = \left(\frac{\partial \vec{r}}{\partial \vec{\alpha}}\right)^t \frac{\partial \vec{v}}{\partial \vec{\alpha}} - \left(\frac{\partial \vec{v}}{\partial \vec{\alpha}}\right)^t \frac{\partial \vec{r}}{\partial \vec{\alpha}} \quad \text{and} \quad \frac{\partial U_P}{\partial \vec{\alpha}} = \frac{\partial U_P}{\partial \vec{r}}\frac{\partial \vec{r}}{\partial \vec{\alpha}} \tag{2.49}$$

The Lagrange matrix L is antisymmetric: $L = -L^t$.
Its elements called **Lagrange brackets** are defined by

$$L_{ij} \underset{\text{noted}}{=} [\alpha_i, \alpha_j] = \frac{\partial \vec{r}}{\partial \alpha_i} \cdot \frac{\partial \vec{v}}{\partial \alpha_j} - \frac{\partial \vec{r}}{\partial \alpha_j} \cdot \frac{\partial \vec{v}}{\partial \alpha_i} \tag{2.50}$$

We can also write the system (2.48) with the Poisson matrix defined as: $P = -L^{-1}$.

$$\frac{d\vec{\alpha}}{dt} = P^t \frac{\partial U_P}{\partial \vec{\alpha}} \quad \text{with} \quad P = L^{-t} = -L^{-1} = \frac{\partial \vec{\alpha}}{\partial \vec{r}}\left(\frac{\partial \vec{\alpha}}{\partial \vec{v}}\right)^t - \frac{\partial \vec{\alpha}}{\partial \vec{v}}\left(\frac{\partial \vec{\alpha}}{\partial \vec{r}}\right)^t \tag{2.51}$$

The elements of the Poisson matrix called **Poisson brackets** are defined by

$$P_{ij} \underset{\text{noted}}{=} (\alpha_i, \alpha_j) = \frac{\partial \alpha_i}{\partial \vec{r}} \cdot \frac{\partial \alpha_j}{\partial \vec{v}} - \frac{\partial \alpha_i}{\partial \vec{v}} \cdot \frac{\partial \alpha_j}{\partial \vec{r}} \tag{2.52}$$

The calculation of Lagrange or Poisson brackets is quite lengthy. It is based on the following relations between position, velocity, and orbital parameters.

$$\vec{r} = \frac{p}{1 + e\cos\theta} R \begin{pmatrix} \cos\theta \\ \sin\theta \\ 0 \end{pmatrix}, \quad \vec{v} = -\sqrt{\frac{\mu}{p}} R_{\Omega i \omega} \begin{pmatrix} -\sin\theta \\ e + \cos\theta \\ 0 \end{pmatrix}$$

$$\text{with } R_{\Omega i \omega} = \begin{pmatrix} \cos\Omega & -\sin\Omega & 0 \\ \sin\Omega & \cos\Omega & 0 \\ 0 & 0 & 1 \end{pmatrix} \begin{pmatrix} 1 & 0 & 0 \\ 0 & \cos i & -\sin i \\ 0 & \sin i & \cos i \end{pmatrix} \begin{pmatrix} \cos\omega & -\sin\omega & 0 \\ \sin\omega & \cos\omega & 0 \\ 0 & 0 & 1 \end{pmatrix} \tag{2.53}$$

This calculation is not detailed here (see [R1] or [R39]). It yields the variations of the orbital parameters under the form of Lagrange equations.

$$\begin{cases} \dfrac{da}{dt} = \dfrac{2}{na}\dfrac{\partial U_P}{\partial \lambda} \\ \dfrac{de}{dt} = -\dfrac{b}{na^3 e}\dfrac{\partial U_P}{\partial \omega} + \dfrac{b^2}{na^4 e}\dfrac{\partial U_P}{\partial \lambda} \\ \dfrac{di}{dt} = -\dfrac{1}{nab\sin i}\dfrac{\partial U_P}{\partial \Omega} + \dfrac{\cos i}{nab\sin i}\dfrac{\partial U_P}{\partial \omega} \\ \dfrac{d\Omega}{dt} = \dfrac{1}{nab\sin i}\dfrac{\partial U_P}{\partial i} \\ \dfrac{d\omega}{dt} = -\dfrac{\cos i}{nab\sin i}\dfrac{\partial U_P}{\partial i} + \dfrac{b}{na^3 e}\dfrac{\partial U_P}{\partial e} \\ \dfrac{d\lambda}{dt} = -\dfrac{2}{na}\dfrac{\partial U_P}{\partial a} - \dfrac{b^2}{na^4 e}\dfrac{\partial U_P}{\partial e}, \quad (\lambda = nt_P) \end{cases} \tag{2.54}$$

Lagrange equations are less general than Gauss equations, since they only apply to perturbations deriving from a potential. They are mainly used to analyze the effect of gravitational perturbations (Section 2.4).

Lagrange or Gauss equations consider the classical orbital parameters that are not defined for circular ($e = 0$) or equatorial ($i = 0$) orbits. These equations present singularities for such orbits, which are very frequent in space applications. The use of equinoctial parameters avoids these singularities.

### 2.3.5 Equinoctial Parameters

The equinoctial parameters have been defined in Section 1.4.3. Their relations to the classical parameters given in (1.120) are recalled here.

$$\begin{cases} P_1 = e\ \sin(\Omega + \omega), & Q_1 = \tan\dfrac{i}{2}\sin\Omega, \quad n = \sqrt{\dfrac{\mu}{a^3}} \\ P_2 = e\ \cos(\Omega + \omega), & Q_2 = \tan\dfrac{i}{2}\cos\Omega, \quad L = \Omega + \omega + M \end{cases} \tag{2.55}$$

The following angles are introduced:

- $\overline{\omega} = \Omega + \omega$ is called the perigee longitude;
- $L = \Omega + \omega + M$ is called the mean longitude;
- $l = \Omega + \omega + M$ is called the true longitude;
- $K = \Omega + \omega + E$ is called the eccentric longitude.

Kepler's equation in equinoctial parameters is derived by substituting angles E and M with K, L, and $\overline{\omega}$: $E - e\sin E = M \ \Rightarrow\ K - \overline{\omega} - e\sin(K - \overline{\omega}) = L - \overline{\omega}$, which can be written after expansion as:

$$K + P_1\ \cos K - P_2\ \sin K = L \tag{2.56}$$

For a given mean longitude L depending on time (like the mean anomaly), this transcendental equation in K has to be solved numerically to find the eccentric longitude K. We deduce then the eccentric anomaly E, the true anomaly $\theta$, the true longitude L, the radius vector r, the parameter p and the angular momentum h of the orbit. The calculation procedure is summarized as follows.

$$\begin{cases} L = K + P_1 \cos K - P_2 \sin K \ \rightarrow\ K \\ E = K - \overline{\omega} \ \rightarrow\ E, \theta \\ L = \overline{\omega} + \theta \\ r = a(1 - P_1\ \sin K - P_2 \cos K) \\ p = r(1 + P_1\ \sin L + P_2 \cos L) \ \rightarrow\ h = \sqrt{\mu p} \end{cases} \tag{2.57}$$

In the presence of perturbations, the equinoctial parameters follow the differential equations (2.58). These general equations are established directly from Gauss equations

([R1]) and they apply to any type of orbit and perturbation.

$$
\left\{
\begin{aligned}
\frac{da}{dt} &= \frac{2a^2}{h}\left[a_r(P_2 \sin L - P_1 \cos L) + a_\theta \frac{p}{r}\right] \\
\frac{dP_1}{dt} &= \frac{r}{h}\left[-a_r \frac{p}{r}\cos L + a_\theta\left(P_1 + \left(1+\frac{p}{r}\right)\sin L\right) - a_h P_2 (Q_1 \cos L - Q_2 \sin L)\right] \\
\frac{dP_2}{dt} &= \frac{r}{h}\left[a_r \frac{p}{r}\sin L + a_\theta\left(P_2 + \left(1+\frac{p}{r}\right)\cos L\right) + a_h P_1 (Q_1 \cos L - Q_2 \sin L)\right] \\
\frac{dQ_1}{dt} &= \frac{r}{2h} a_h \left(1 + Q_1^2 + Q_2^2\right)\sin L \\
\frac{dQ_2}{dt} &= \frac{r}{2h} a_h \left(1 + Q_1^2 + Q_2^2\right)\cos L \\
\frac{dl}{dt} &= n - \frac{r}{h}\left[a_r\left(\frac{a}{a+b}\frac{p}{r}(P_1 \sin L + P_2 \cos L) + \frac{2b}{a}\right)\right. \\
&\qquad \left. + a_\theta \frac{a}{a+b}\left(1+\frac{p}{r}\right)(P_1 \cos L - P_2 \sin L) + a_h (Q_1 \cos L - Q_2 \sin L)\right]
\end{aligned}
\right.
$$

$$
\text{with} \quad b = a\sqrt{1-e^2}, \quad e^2 = P_1^2 + P_2^2 \tag{2.58}
$$

### 2.3.6 Integration Methods

The main perturbations acting on an Earth satellite are by decreasing order of magnitude (in brackets the ratio between the perturbing acceleration $a_p$ and the central acceleration $\mu/r^2$):

- irregularities in the Earth's gravitational potential ($10^{-3}$ due to the flattening, less than $10^{-6}$ for the other terms);
- attraction of Moon and Sun ($10^{-7}$);
- atmospheric friction ($10^{-6}$ at an altitude of 200 km);
- radiation pressure ($10^{-8}$).

Other perturbations are much weaker and are generally neglected: tidal effects (which are usually taken into account by modifying the coefficients of the gravitational potential), Earth albedo (reflection of the solar flux by the Earth's surface), other planet attractions, relativistic effects due to the Earth motion, Lorentz force due to the Earth magnetic field ....

In the presence of perturbations, the trajectory follows equations of the form (2.11), (2.45), (2.54) or (2.58), which are equivalent formulations of the dynamics. These differential equations can be solved by special perturbation or general perturbation methods. For a detailed discussion of the various methods, see [R32] and [R39].

#### 2.3.6.1 Special Perturbations

Special perturbation methods numerically solve the equations of perturbed motion in the form (2.45), (2.54) or (2.58). These methods are applicable regardless of the perturbations considered. The trajectory obtained is specific to the initial conditions and forces considered, hence the name "special" perturbations. The numerical integration is performed using either Encke's or Cowell's method.

**Encke's method** (1791–1865) consists in integrating the position and velocity deviation from the osculating orbit. The integration begins at a time $t_0$, with the reference osculating orbit at that time, and continues until the deviation reaches a given tolerance. At this so-called rectification time $t_1$, a new reference osculating orbit is defined from the current position and velocity, and the integration of the deviation continues with respect to the new orbit.

**Cowell's method** (1870–1949) consists of directly integrating the differential equations of motion using a numerical method. Various methods can be used: single step (Runge–Kutta) or multiple steps (Adams–Bashford–Moulton), fixed or variable step, second order (Gauss-Jackson), .... These methods can be applied in the same way to formulations (2.11), (2.45), (2.54), or (2.58). The formulations (2.45), (2.54), and (2.58) with orbital parameters are more accurate and faster for small perturbations, as the second members are close to zero. Conversely, these formulations can become ill-conditioned for strong perturbations, in which case the formulation (2.11) with Cartesian coordinates is preferable.

One difficulty is to set the integration step in order to obtain the best compromise between computation time and accuracy. If the orbit is highly eccentric, the same time step produces a very different angular displacement at the pericenter or at the apocenter, according to the relations

$$hdt = r^2 d\theta = rbdE \tag{2.59}$$

The numerical error can be regularized by integrating as a function of anomaly rather than time. The generalized Sundman transformation: $dt = cr^k ds$ improves the integration efficiency in terms of time and accuracy. The standard choices are ($k = 1, c = b/h$) or ($k = 2, c = 1/h$) corresponding respectively to $s = E$ or $s = \theta$.

#### 2.3.6.2 General Perturbations

General perturbation methods determine an analytical solution to the perturbed motion. This is only possible in specific cases considering perturbations with a simple formulation. The aim is to analyze the overall effect of perturbations over time, hence the name "general" perturbations.

Under the effect of a perturbation p, the orbital parameters $\alpha$ follow the differential equations (2.45) or (2.54) of the form: $\dot{\alpha} = f(\alpha, p, \theta)$, where $\theta$ is the true anomaly.

If the perturbation is small compared to the central attraction, the evolution can be decomposed into a fast variation (over one period) due to the true anomaly, and a slow variation (on the scale of several revolutions). This slow variation includes secular terms (linear as a function of time) and periodic terms. The latter may derive from the periodicity of the perturbation (motion of the Moon or Sun) and/or of the periodicity of the orbital parameter itself (argument of the perigee, longitude of the ascending node).

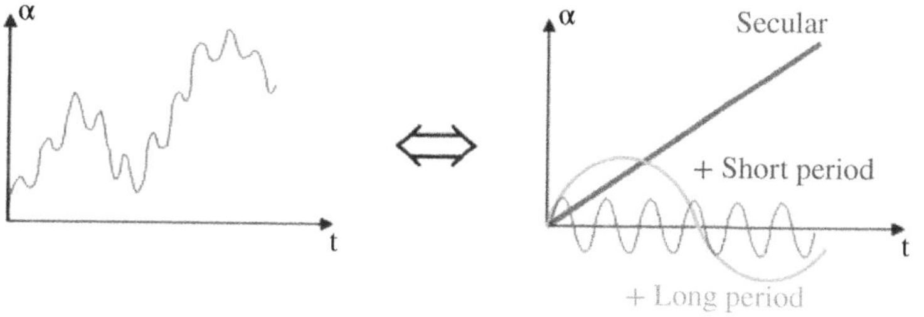


**Figure 2.3** Decomposition in secular, long-period, and short-period terms.

The general shape of the evolution is depicted in Figure 2.3, with its decomposition into secular, slow periodic, and fast periodic terms.

The averaging technique consists of assuming that the orbital parameters keep constant values $\overline{\alpha}$ over one revolution, and solving the equation for the evolution of the average values: $\dot{\overline{\alpha}} = f(\overline{\alpha}, \overline{p})$. The averaged equations can in some cases be solved analytically or, by default, integrated numerically. Such semi-analytical methods (analytical averaging over one revolution, followed by a numerical integration) make it possible to determine the evolution over very long periods. Among the most widely used are the DSST (Draper Semianalytical Satellite Theory) and SGP (Simplified General Perturbations) propagators.

#### 2.3.6.3 Averaging Method

We illustrate the averaging process in three simple cases of in-plane perturbations. The radial and orthoradial components of the perturbing acceleration are depicted in Figure 2.4. They project onto the orbit axes (x, y) by

$$\vec{a}_p = a_r\vec{u}_r + a_\theta\vec{u}_\theta = a_x\vec{u}_x + a_y\vec{u}_y \quad \text{with} \quad \begin{cases} a_r = \phantom{-}a_x\cos\theta + a_y\sin\theta \\ a_\theta = -a_x\sin\theta + a_x\cos\theta \end{cases} \tag{2.60}$$

- First case: constant inertial acceleration

This first case assumes that the components $(a_x, a_y)$ are constant, so that the acceleration $\vec{a}_p$ keeps a constant direction in the inertial frame (Cxy).

Substituting $(a_r, a_\theta)$ given by (2.60) in Gauss equations (2.45) for the semi-major axis and the eccentricity, and using $r = \frac{p}{1 + e\cos\theta}$, we have

$$\begin{cases} \dfrac{da}{dt} = \dfrac{2a^2}{h}[-a_x\sin\theta + a_y\cos\theta + a_y e] \\[2ex] \dfrac{de}{dt} = \dfrac{p}{h}\left[(-a_x\sin\theta + a_y\cos\theta)\dfrac{e+\cos\theta}{1+e\cos\theta} + a_y\right] \end{cases} \tag{2.61}$$

The averaging process consists of integrating these equations over one period.

$$\Delta a = \int_0^T \frac{da}{dt}dt, \quad \Delta e = \int_0^T \frac{de}{dt}dt \tag{2.62}$$

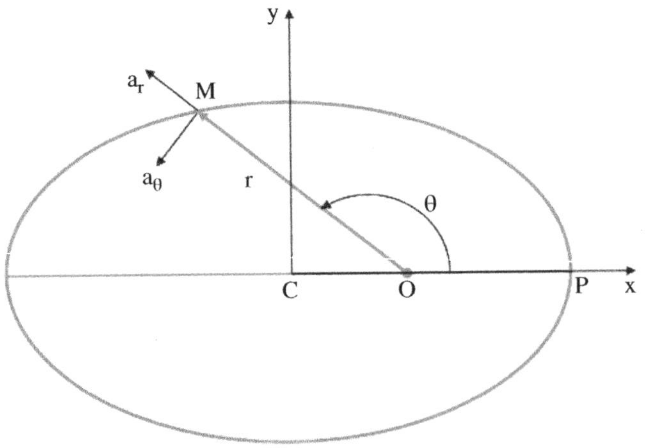


**Figure 2.4** Constant perturbing acceleration in the orbital plane.

In order to carry out the integration, we make a variable change from time to eccentric anomaly, using Kepler's equation (2.6)

$$E - e\ \sin E = M = n(t - t_P) \quad \Rightarrow \quad ndt = (1 - e\ \cos E)dE \tag{2.63}$$

and we pass from the true anomaly to the eccentric anomaly using (1.24).

The integrals (2.62) expressed in eccentric anomaly are

$$\begin{cases} \Delta a = \dfrac{2a^2}{nh}\displaystyle\int_0^{2\pi} [-a_x\sqrt{1-e^2}\sin E + a_y(1-e^2)\cos E]dE \\ \Delta e = \dfrac{p}{nh}\displaystyle\int_0^{2\pi} [-a_x\sqrt{1-e^2}\sin E\ \cos E + a_y(\cos^2 E - 2e\cos E + 1)]dE \end{cases} \tag{2.64}$$

The integration is carried out assuming that the eccentricity does not vary over one revolution (since we are interested in long-term effects). The result is

$$\begin{cases} \Delta a = 0 \\ \Delta e = 3\pi \dfrac{a^2\sqrt{1-e^2}}{\mu} a_y \end{cases} \tag{2.65}$$

A constant inertial acceleration has no secular effect on the semi-major axis, while the eccentricity is only affected by the $a_y$ component (along the minor axis).

An acceleration opposed to (resp. along) the velocity at the pericenter decreases (resp. increases) the eccentricity.

This constant inertial thrusting strategy can be used to reach a geostationary earth orbit ([GEO], circular at an altitude of 35 786 km) starting from an elliptical orbit having the same semi-major axis (super GTO, perigee at 500 km, apogee at 71 000 km).

- Second case: constant orthoradial acceleration

This second case assumes that $a_\theta$ is constant and $a_r = 0$. Gauss equations (2.45) for the semi-major axis and the eccentricity reduce to

$$\begin{cases} \dfrac{da}{dt} = \dfrac{2a^2}{h}(1 + e\cos\theta)a_\theta \\ \dfrac{de}{dt} = \dfrac{1}{h}[p\ \cos\theta + r(\cos\theta + e)]a_\theta \end{cases} \tag{2.66}$$

By variable change from time to eccentric anomaly and passing from the true anomaly to the eccentric anomaly, the integrals over one revolution are

$$\begin{cases} \Delta a = \dfrac{2a^2}{nh}\displaystyle\int_0^{2\pi} (1-e^2)\, a_\theta\, dE \\ \Delta e = \dfrac{p}{nh}\displaystyle\int_0^{2\pi} [2\ \cos E - e(1+\cos^2 E)]\, a_\theta\, dE \end{cases} \tag{2.67}$$

which yields

$$\begin{cases} \Delta a = \ 4\pi \dfrac{a^3\sqrt{1-e^2}}{\mu} a_\theta \\ \Delta e = -3\pi \dfrac{a^2 e\sqrt{1-e^2}}{\mu} a_\theta \end{cases} \tag{2.68}$$

For a circular orbit (e = 0), the orthoradial acceleration has no secular effect on the eccentricity and the average orbit remains circular. The semi-major axis variation per revolution is $\frac{T^2}{\pi} a_\theta$, where T is the average orbit period.

- Third case: constant tangential acceleration

This third case assumes that the acceleration remains oriented along the velocity.

Denoting $(a_t, a_n)$ the tangential and normal acceleration components and using the velocity expression (2.10), we have the relations

$$\begin{cases} a_r = \dfrac{h}{pv}[a_t e \sin\theta - a_n(1 + e\cos\theta)] \\ a_\theta = \dfrac{h}{pv}[a_t(1 + e\cos\theta) + a_n e \sin\theta] \end{cases} \tag{2.69}$$

Gauss equations (2.45) for the semi-major axis and the eccentricity take the form

$$\begin{cases} \dfrac{da}{dt} = \dfrac{2a^2v}{\mu} a_t \\ \dfrac{de}{dt} = \dfrac{1}{v}\left[2(\cos\theta + e)a_t - \dfrac{r}{a}\sin\theta\, a_n\right] \end{cases} \tag{2.70}$$

In the case of a constant tangential acceleration, we have $a_t$ constant and $a_n = 0$.

Similarly to the previous cases, we make a variable change from time to eccentric anomaly and pass from the true anomaly to the eccentric anomaly.

The integrals over one revolution are

$$\begin{cases} \Delta a = \dfrac{2a^3}{\mu} a_t \displaystyle\int_0^{2\pi} \sqrt{1 - e^2\cos^2 e}\, dE \\ \Delta e = \dfrac{2a^2(1 - e^2)}{\mu} a_t \displaystyle\int_0^{2\pi} \sqrt{\dfrac{1 - e\cos E}{1 + e\cos E}} \cos E\, dE \end{cases} \tag{2.71}$$

In the case of a circular orbit, these integrals yield

$$\Delta a = \frac{4\pi a^3}{\mu} a_t = \frac{T^2}{\pi} a_t, \quad \Delta e = 0 \tag{2.72}$$

and we unsurprisingly obtain the same results as (2.68).

## 2.4 Gravitational Perturbations

The gravitational force exerted by a point $M_2$ of mass $m_2$ on a point $M_1$ of mass $m_1$ is given by

$$\vec{F}_{2\to 1}(\vec{r}) = G\frac{m_1 m_2}{r^3}\vec{r} \quad \text{with} \quad \vec{r} = \overrightarrow{M_1 M_2} \tag{2.73}$$

For non-point bodies, the gravitational force is assessed by the integration of (2.73) over each body volume. The result generally differs from the one obtained with point bodies assumption and the difference is called gravitational perturbation.

### 2.4.1 Gravitational Potential

The calculation is eased by using the gravitational potential, instead of the acceleration. The gravitational potential created by $M_2$ in $M_1$ is defined as:

$$U_{2\to1} = G\frac{m_2}{r} = G\frac{m_2}{\sqrt{x^2+y^2+z^2}} \tag{2.74}$$

where (x, y, z) are the Cartesian coordinates of $\vec{r} = \overrightarrow{M_1M_2}$ in an arbitrary frame.

This potential is a function of the relative position $\vec{r}$. Its gradient is

$$\nabla U_{2\to1}(\vec{r}) = \frac{\partial U_{2\to1}}{\partial \vec{r}} = \begin{pmatrix} \frac{\partial U_{2\to1}}{\partial x} \\ \frac{\partial U_{2\to1}}{\partial y} \\ \frac{\partial U_{2\to1}}{\partial z} \end{pmatrix} = -\frac{Gm_2}{(\sqrt{x^2+y^2+z^2})^{\frac{3}{2}}}\begin{pmatrix} x \\ y \\ z \end{pmatrix} = -\frac{Gm_2}{r^3}\vec{r} \tag{2.75}$$

The gravitational acceleration in $M_1$ due to $M_2$ (2.73) is thus the gradient of the potential, with the opposite sign. This sign change is due to the convention chosen to define the potential in (2.74).

$$\vec{a}_{2\to1} = \frac{\vec{F}_{2\to1}}{m_1} = -\nabla U_{2\to1}(\vec{r}) \tag{2.76}$$

### 2.4.2 Spherical Body

The Earth (and most celestial bodies) can be considered in first approximation as homogeneous spheres. With this assumption, their gravitational field takes a very simple form, as demonstrated hereafter.

#### 2.4.2.1 Gravity Field of a Spherical Shell

Let us consider a spherical shell of center O, radius r, thickness dr, and uniform density ρ as shown in Figure 2.5, and try to calculate its gravitational potential in any point S located at a distance d from the center O. The point S may be either inside or outside the shell. The position of S with respect to O is defined by the vector $\vec{d} = \overrightarrow{OS}$.

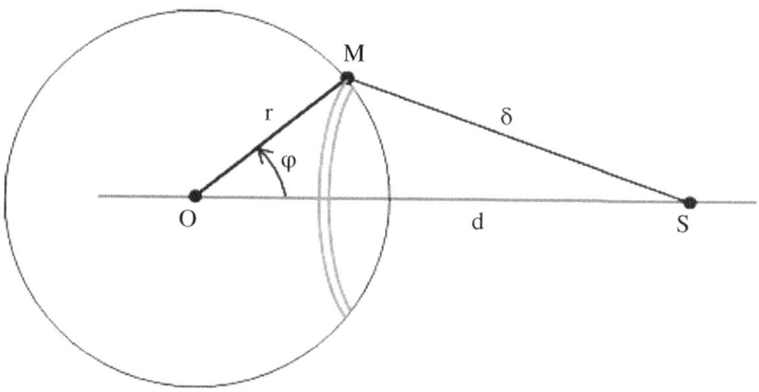


**Figure 2.5** Potential of a spherical shell.

Let us consider a ring on the shell of axis OS, apex half-angle φ, width rdφ and thickness dr. Its perimeter is $2\pi r \sin\varphi$ and its mass is $m_{ring} = 2\pi r^2 \rho \sin\varphi \, d\varphi \, dr$.

All ring points are at the same distance $\delta = \sqrt{r^2 + d^2 - 2rd\cos\varphi}$ from S and exert the same gravitational potential in S. The potential in S due to the ring is thus

$$U_{ring} = G\frac{m_{ring}}{\delta} = G\rho\frac{2\pi r^2 \sin\varphi \, d\varphi \, dr}{\sqrt{r^2 + d^2 - 2rd\cos\varphi}} \tag{2.77}$$

The potential in S due to the spherical shell is obtained by integration over the angle φ ranging from 0 to π.

$$\begin{aligned} U_{shell} &= 2\pi r^2 G\rho dr \int_{\varphi=0}^{\varphi=\pi} \frac{\sin\varphi \, d\varphi}{\sqrt{r^2 + d^2 - 2rd\cos\varphi}} \\ &= 2\pi G\rho\frac{r}{d}dr\left[\sqrt{r^2 + d^2 - 2rd\cos\varphi}\right]_{\varphi=0}^{\varphi=\pi} = 2\pi G\rho\frac{r}{d}dr[\delta(\varphi)]_{\varphi=0}^{\varphi=\pi} \end{aligned} \tag{2.78}$$

The value of the term in brackets depends whether the point S lies outside or inside the shell of radius r.

- If S lies outside the shell ($d \geq r$), then $\begin{cases} \delta(\varphi = 0) = d - r \\ \delta(\varphi = \pi) = d + r \end{cases}$ and we have

$$U_{shell}(d) = \frac{4\pi r^2}{d}G\rho dr = \frac{Gm_{shell}(r)}{d} \quad \text{if } d \geq r \tag{2.79}$$

since the shell mass is $m_{shell}(r) = 4\pi r^2 \rho dr$.

The shell potential (2.79) is that of a point mass in O having the shell mass. The gravity field of the shell in the point S of position $\vec{d}$ is then

$$\vec{g}_{shell}(\vec{d}) = \frac{Gm_{shell}(r)}{d^3}\vec{d} \quad \text{if } d \geq r \tag{2.80}$$

- If S lies inside the shell ($d < r$), then $\begin{cases} \delta(\varphi = 0) = r - d \\ \delta(\varphi = \pi) = r + d \end{cases}$ and we have

$$U_{shell}(d) = 4\pi r G\rho \, dr \quad \text{if } d < r \tag{2.81}$$

The shell potential (2.81) does not depend on d. The gravity field of the shell in the point S of position $\vec{d}$, which is the gradient of the potential, is then

$$\vec{g}_{shell}(\vec{d}) = \vec{0} \quad \text{if } d < r \tag{2.82}$$

#### 2.4.2.2 Gravity Field of a Sphere

Let us assess the gravity field of a sphere of radius R and center O in a point S.

This is done by summing the fields of concentric shells of radii from 0 to R. The result depends on whether the point S lies outside or inside the sphere of radius R.

If S lies outside the sphere, then formula (2.79) applies to all concentric shells of radii from 0 to R. The resulting gravity field in S of position $\vec{d} = \overrightarrow{OS}$ is

$$\vec{g}_{sphere}(\vec{d}) = \frac{Gm_{sphere}(R)}{d^3}\vec{d} \quad \text{if } d \geq R \tag{2.83}$$

where $m_{sphere}(R)$ is the total sphere mass (radius R).

The gravity field outside the sphere is that of a point mass located in its center, having the total sphere mass.

- If S lies inside the sphere, then formula (2.79) applies to shells from 0 to d, and formula (2.81) applies to shells of radii from d to R. We obtain

$$\vec{g}_{sphere}(\vec{d}) = \frac{Gm_{sphere}(d)}{d^3}\vec{d} \quad \text{if } d < R \tag{2.84}$$

where $m_{sphere}(d)$ is the mass of the inner sphere (radius d).

The gravity field inside the sphere is that of a point mass located in its center, but having only the mass of the inner sphere of radius d.

#### 2.4.2.3 Gravity Gradient

Another result that will be used later is the gradient of gravity in a position $\vec{r}$ due to a point mass O (or equivalently to a homogenous sphere of center O).

The gravity field created by O in M of Cartesian coordinates $\vec{r}(x, y, z)$ is

$$\vec{g}(\vec{r}) = -\frac{\mu}{r^3}\vec{r} = -\frac{\mu}{r^3}\begin{pmatrix} x \\ y \\ z \end{pmatrix} \quad \text{with } r = \sqrt{x^2 + y^2 + z^2} \quad \text{and } \mu = Gm_O \tag{2.85}$$

The gravity gradient is the matrix defined by $G = \nabla\vec{g} = \frac{\partial\vec{g}}{\partial\vec{r}}$.

By deriving the gravity (2.85) with respect to coordinates (x, y, z), we obtain

$$G = \frac{\mu}{r^5}\begin{pmatrix} 3x^2 - r^2 & 3xy & 3xz \\ 3xy & 3y^2 - r^2 & 3yz \\ 3xz & 3yz & 3z^2 - r^2 \end{pmatrix} = \frac{\mu}{r^5}(3\vec{r}\,\vec{r}^{\,t} - r^2 I_3) \tag{2.86}$$

where $I_3$ denotes the $3 \times 3$ identity matrix.

### 2.4.3 Nonspherical Body

The Earth is neither perfectly spherical, nor homogenous. Its gravitational field depends on its shape and mass distribution.

An Earth element of mass dm at point M of coordinates $(r_M, \lambda_M, \varphi_M)$ generates the gravitational potential dU at point S of coordinates $(r_S, \lambda_S, \varphi_S)$.

$$dU = G\frac{dm}{\delta} \quad \text{with } \delta = MS \tag{2.87}$$

The gravitational potential is the sum of the contributions of all mass elements.

Expressing the distance $\delta = MS$ as a function of the angle $\gamma = (\overrightarrow{OS}, \overrightarrow{OM})$ and of the distance ratio $s = r_M/r_S$ (Figure 2.6), the total potential at S is

$$U(S) = \frac{G}{r_S}\iiint_{Earth}\frac{dm}{\sqrt{1 - 2s\cos\gamma + s^2}} \tag{2.88}$$

#### 2.4.3.1 Spherical Harmonics

The integral (2.88) is developed using Legendre polynomials and functions.

We summarize hereafter the main definitions and useful results. For more details, see [R1], [R5], [R28], and [R39].

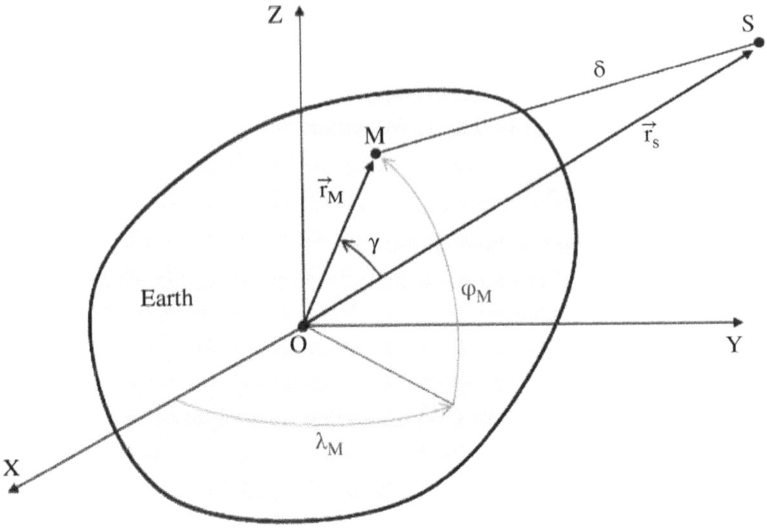


**Figure 2.6** Potential of a mass distribution.

Legendre polynomials $P_n(\alpha)$ are associated with the binomial expansion of the function $L(x, \alpha)$ called the generating function and defined as:

$$L(x, \alpha) = \frac{1}{\sqrt{1 - 2\alpha x + x^2}} = \sum_{n=0}^{\infty} P_n(\alpha) x^n \tag{2.89}$$

**Legendre polynomial** $P_n$ is of degree n and is calculated using Rodrigues' formula.

$$P_n(\alpha) = \frac{1}{2^n n!} \frac{d^n}{d\alpha^n} (\alpha^2 - 1)^n \tag{2.90}$$

**Legendre function** $P_{nm}$ of degree n and order m is linked to the m-th derivative of $P_n$.

$$P_{nm}(x) = (1 - x^2)^{m/2} \frac{d^m}{dx^m} P_n(x) \tag{2.91}$$

The polynomial of degree n is also the function of degree n and order 0: $P_n = P_{n0}$.

For the gravitational potential (2.88), the variable $\alpha = \cos\gamma$ is expressed as a function of the spherical coordinates of M and S: $\cos\gamma = \sin\varphi_M \sin\varphi_S + \cos\varphi_M \cos\varphi_S \cos(\lambda_M - \lambda_S)$.

With Legendre's addition (or decomposition) formula for polynomials

$$P_n(\cos\gamma) = P_n(\sin\varphi_M) P_n(\sin\varphi_S) + 2\sum_{m=1}^{n} \frac{(n-m)!}{(n+m)!} P_{nm}(\sin\varphi_M) P_{nm}(\sin\varphi_S) \cos(\lambda_M - \lambda_S) \tag{2.92}$$

the potential in S can be written as a sum of spherical harmonics

$$U(S) = \frac{\mu}{r_S} \left[ 1 - \sum_{n=2}^{\infty} J_n \left(\frac{R_E}{r_S}\right)^n P_n(\sin\varphi_S) + \sum_{n=2}^{\infty} \sum_{m=1}^{n} \left(\frac{R_E}{r_S}\right)^n P_{nm}(\sin\varphi_S)(C_{nm}\cos(m\lambda_S) + S_{nm}\sin(m\lambda_S)) \right] \tag{2.93}$$

This formulation separates the central term $\mu/r_S$, zonal terms depending only on the latitude $\varphi_S$ (with coefficients $J_n$), and terms depending both on the latitude and the longitude $\lambda_S$ (with coefficients $C_{nm}$ and $S_{nm}$).

The **zonal harmonics** (m = 0) associated with the coefficients $J_n$ correspond to a distribution of mass symmetrical with respect to the pole axis, and a division of the sphere along the parallels. The zonal harmonic of degree n cancels out for n latitudes.

The **sectoral harmonics** (m = n) correspond to a division of the sphere along the meridians. The sectoral harmonic of degree n cancels out for 2n longitudes.

The **tesseral harmonics** (m ≠ n) correspond to a division of the sphere surface into "squares." The (n,m) tesseral harmonic cancels out for (n−m) latitudes and 2m longitudes.

The geometry of the different spherical harmonics is depicted in Figure 2.7.

For a given mass distribution, the coefficients $J_n$, $C_{nm}$, and $S_{nm}$ are calculated by

$$C_{nm} = \frac{1}{R_e^n M_T} \iiint_{Terre} r_M^n \frac{(n-m)!}{(n+m)!} P_{nm}(\sin \varphi_M) \cos(m\lambda_M) dm \quad \text{and} \quad J_n \underset{def}{=} -C_{n,0}$$

$$S_{nm} = \frac{1}{R_e^n M_T} \iiint_{Terre} r_M^n \frac{(n-m)!}{(n+m)!} P_{nm}(\sin \varphi_M) \sin(m\lambda_M) dm \tag{2.94}$$

These coefficients are normalized by the Earth's equatorial radius $R_e$ and the Earth's mass $M_T$.

---

**Space Geodesy**

The observation of the motion of satellites enables an experimental determination of the potential coefficients and an estimation of the Earth's mass distribution. The first estimations of the $J_2$ coefficient and the Earth flattening were based on optical observations of the trajectory of Sputnik-2 in 1957.

Recent geodetic satellites use highly accurate accelerometers (up to $10^{-12}$ m/s$^2$ for the GOCE mission) to measure and compensate for non-gravitational perturbations. This allows us to assess the pure effect of the Earth's gravitational field on the satellite trajectory, and going back to the Earth's mass distribution.

Geodetic models combine these space-based measurements with ground measurements. The main reference models (in brackets, the degree of the model) are WGS84 (180), EGM96 (360), GRIM5 (120), and EIGEN6 (1440).

The main perturbation is due to the first zonal term, whose coefficient is estimated nowadays at $J_2 = 0.001\,082\,63$. This term is 1000 times weaker than the central term. The following potential terms ($J_3, J_4$) are of the order of $10^{-6}$.

---

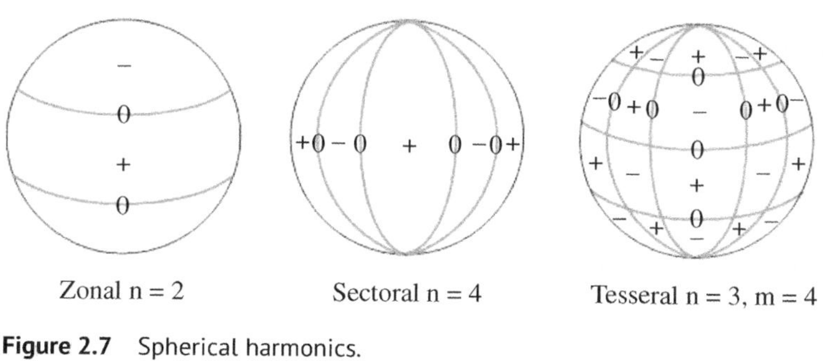


**Figure 2.7** Spherical harmonics.

#### 2.4.3.2 MacCullagh Approximation

With the first three Legendre polynomials: $P_0(\alpha) = 1, P_1(\alpha) = \alpha, P_2(\alpha) = \frac{3\alpha^2 - 1}{2}$, and using (2.89), the gravitational potential (2.88) can be approximated by

$$U(S) = \frac{G}{r_S} \sum_{n=0}^{\infty} \iiint_{\text{Earth}} \left(\frac{r_M}{r_S}\right)^n P_n(\cos\gamma)\, dm$$

$$\approx \frac{GM_T}{r_S} + \frac{G}{r_S^2} \iiint_{\text{Earth}} r_M \cos\gamma\, dm + \frac{G}{2r_S^3} \iiint_{\text{Earth}} r_M^2 (3\cos^2\gamma - 1)\, dm \tag{2.95}$$

By choosing the origin O at the center of mass of the attracting body, the second term cancels out. Furthermore, by observing that the moments of inertia A, B, C, and $I_S$ relative to the X, Y, Z, and OS axes are (Figure 2.6)

$$A = \iiint_{\text{Earth}} (y_M^2 + z_M^2)\, dm, \quad B = \iiint_{\text{Earth}} (x_M^2 + z_M^2)\, dm, \quad C = \iiint_{\text{Earth}} (x_M^2 + y_M^2)\, dm$$

$$I_S = \iiint_{\text{Earth}} (r_M \cos\gamma)^2 dm \quad \text{with} \quad r_M^2 = x_M^2 + y_M^2 + z_M^2 \tag{2.96}$$

we obtain the approximation of MacCullagh of the gravitational potential

$$U(S) \approx \frac{GM_T}{r_S} + \frac{G}{2r_S^3}(A + B + C - 3I_S) \tag{2.97}$$

### 2.4.4 First Zonal Term

The main perturbation is due to the Earth flattening. By considering the Earth as an ellipsoid of revolution, the potential (2.93) reduces to the first zonal term.

$$U(S) = \frac{\mu}{r_S} + U_p(S)$$

$$\text{with } U_p(S) = -\frac{\mu}{r_S} J_2 \left(\frac{R_e}{r_S}\right)^2 P_2(\sin\varphi_S) = \frac{\mu R_e^2 J_2}{2r_S^3}(1 - 3\sin^2\varphi_S) \tag{2.98}$$

The effects of this perturbation are analyzed from Lagrange equations (2.54). To eliminate short-period terms, we first assess the mean value of $U_p$ over a period. Choosing the mean anomaly M as integration variable and using the relations

$$r(1 + e\cos\theta) = p, \quad \cos\varphi = \sin(\omega + \theta)\sin i, \quad \frac{dM}{dt} = n, \quad h = r^2\frac{d\theta}{dt} \tag{2.99}$$

the mean value $U_p$ is expressed as:

$$\overline{U}_p = \frac{1}{2\pi} \int_{M=0}^{2\pi} U_p dM = \frac{\mu R_e^2 J_2 n}{4\pi hp} \int_{\theta=0}^{2\pi} (1 + e\cos\theta)[1 - 3\sin^2(\omega + \theta)\sin^2 i]d\theta \tag{2.100}$$

The terms in $\cos\theta$ and $\sin^2\theta$ have means 0 and $\pi$, respectively, so that

$$\overline{U}_p = \frac{\mu R_e^2 J_2}{4a^3(1 - e^2)^{3/2}}(2 - 3\sin^2 i) \text{ by using } \frac{n}{hp} = \frac{1}{a^3(1 - e^2)^{3/2}} \tag{2.101}$$

Deriving $U_p$ with respect to the orbital parameters and substituting into Lagrange equations (2.54), we get the mean parameter evolutions due to the $J_2$ term [R39].

$$\begin{cases} \frac{d\bar{a}}{dt} = 0, \ \frac{d\bar{e}}{dt} = 0, \ \frac{d\bar{i}}{dt} = 0 \\ \frac{d\bar{\Omega}}{dt} = -\frac{3J_2R_e^2\bar{n}}{2\bar{p}^2}\cos\bar{i}, \ \frac{d\bar{\omega}}{dt} = \frac{3J_2R_e^2\bar{n}}{4\bar{p}^2}(5\cos^2\bar{i} - 1) \\ \frac{d\bar{M}}{dt} = \bar{n} + \frac{3J_2R_e^2\bar{n}}{4\bar{p}^2}\sqrt{1-\bar{e}^2}(3\cos^2\bar{i} - 1) \end{cases} \tag{2.102}$$

The top bar above the orbital parameters reminds us that these equations are linked to their averaged evolutions over a period. The last equation for the mean anomaly $M = n(t - t_P)$ is obtained from Lagrange equation for the time of passage at pericenter $\lambda = nt_P$. We observe that:

- the semi-major axis, eccentricity, and inclination are constant on average;
- the derivatives $\dot{\bar{\Omega}}$, $\dot{\bar{\omega}}$, and $\dot{\bar{M}}$ are constant, since $\bar{a}, \bar{e}, \bar{i}$ are constant.

The angles $\Omega$ and $\omega$ therefore show secular variations called **nodal precession** for $\bar{\Omega}$ and **apsidal precession** for $\bar{\omega}$.s

These effects are physically explained by the equatorial bulge, which generates an additional attraction depicted by upward and downward arrows in Figure 2.8. This results in a torque $\vec{C}$ around the line of nodes AD, which simultaneously induces a rotation of the angular momentum $\vec{h}$ around the pole axis and a rotation of the orbit major axis (oriented by the vector $\vec{e}$) in the orbital plane.

This bulge also induces an acceleration in the orbital plane that changes the mean motion in the equation for M.

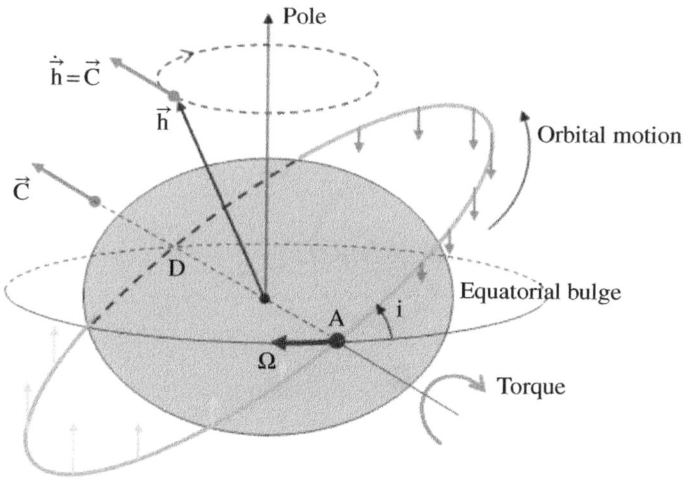


**Figure 2.8** Nodal precession due to the Earth flattening.

The formulas (2.102) for Ω, ω, and M are frequently used under the form

$$\begin{cases} \dot{\Omega} = -\frac{3}{2} J_2 n \left(\frac{R_e}{a}\right)^2 \frac{\cos i}{(1-e^2)^2}, \quad \dot{\omega} = \frac{3}{4} J_2 n \left(\frac{R_e}{a}\right)^2 \frac{5\cos^2 i - 1}{(1-e^2)^2} \\ \dot{M} = n + \frac{3}{4} J_2 n \left(\frac{R_e}{a}\right)^2 \frac{3\cos^2 i - 1}{(1-e^2)^{1.5}} \underset{\text{noted}}{=} n + \Delta n \end{cases} \tag{2.103}$$

where the top bars are omitted for conciseness.

We discuss here after some practical consequences of the perturbations due to the first zonal term.

#### 2.4.4.1 Nodal Precession

The nodal precession can be used to achieve **sun-synchronous** orbits. These orbits have the property of maintaining a constant angle between the orbital plane and the direction of the Sun. This means that the lighting conditions on the satellite and in the overflown areas are identical throughout the year, facilitating optical observation of the Earth's surface.

Sun-synchronism is achieved by choosing orbital parameters a, e, and i in (2.103) so that the nodal precession rate $\dot{\Omega}$ is equal to the apparent rotation rate of the Sun relative to the Earth (360 deg/year = 0.986 deg/day).

A very frequently used sun-synchronous orbit is a circular one at the altitude of 800 km with an inclination of 98.6 deg.

A sun-synchronous orbit is characterized by its local time at the ascending node (LTAN). Figure 2.9 shows a satellite in polar orbit with a local time of 18 h at the ascending node. The time difference between left and right is 3 months. In the inertial frame, the right ascension of the ascending node has increased by 90 deg.

#### 2.4.4.2 Apsidal Precession

The apsidal precession in (2.103) is canceled out at an inclination of 63.4 deg, called critical inclination. The Molnya orbit (40 000 km/500 km/63.4 deg) with a period of 12 h and the Tundra orbit (47 000 km/24 000 km/63.4 deg) with a period of 24 h are highly elliptical orbits designed for telecommunications. Their apogee located over Russia appears as nearly geostationary for 8 h. The choice of the critical inclination allows to keep the apogee location at the same place without the need for propelled correction maneuvers.

#### 2.4.4.3 Orbital Period

The Keplerian orbital period $T_K$ is given by (2.6): $T_K = \frac{2\pi}{n} = 2\pi\sqrt{\frac{a^3}{\mu}}$.

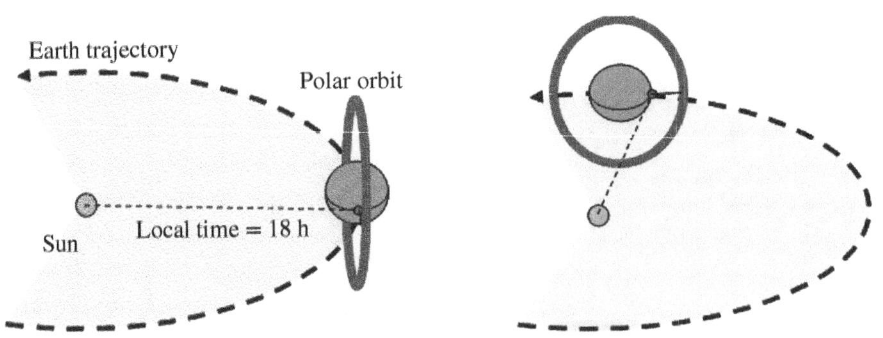


**Figure 2.9** Sun-synchronous orbit.

In the presence of perturbations, the orbital plane and the perigee location are no longer fixed and the orbital period can be defined in different ways.

- The anomalistic period $T_a$ is the time interval between two passages at the pericenter. It accounts only for the mean anomaly variation.

$$(n + \Delta n)T_a = 2\pi \tag{2.104}$$

- The nodal period $T_n$ is the time interval between two passages at the node. It accounts for the variations of the mean anomaly and the perigee argument.

$$(n + \Delta n + \dot{\omega})T_n = 2\pi \tag{2.105}$$

- The sidereal period $T_s$ accounts furthermore for the variation of the node in the reference inertial frame.

$$(n + \Delta n + \dot{\omega} + \dot{\Omega})T_s = 2\pi \tag{2.106}$$

The last equation is used for achieving an accurate geostationary orbit accounting for the $J_2$ perturbation. The geostationary orbit is circular ($e = 0$) and equatorial ($i = 0$), and the Earth's sidereal rotation rate is $T_g = 86\,164$ s = 23 h 56 min 04 s.

In order to be geostationary, the orbit sidereal period must be $T_s = T_g$. Substituting into (2.106) with $\dot{\Omega}$, $\dot{\omega}$, $\Delta n$ given by (2.103) for $e = 0$ and $i = 0$ yields

$$n\left[1 + 3J_2\left(\frac{R_e}{a}\right)^2\right] = \frac{2\pi}{T_g} \quad \text{with} \quad n = \sqrt{\frac{\mu}{a^3}} \tag{2.107}$$

Let us denote $\Delta a$ the semi-major axis deviation from the Keplerian value $a_K$, which is $a_K = \left[\mu\left(\frac{T_g}{2\pi}\right)^2\right]^{\frac{1}{3}} = 42\,164$ km

Substituting $a = a_K + \Delta a$ into (2.107) and expanding to first order, we obtain

$$\frac{\Delta a}{a_K} = 2J_2\left(\frac{R_e}{a_K}\right)^2 \tag{2.108}$$

The geostationary radius with $J_2$ is thus 2.09 km higher than the Keplerian value, which yields an altitude of 35 788 km (instead of 35 786 km).

---

**Earth Flattening**

The Earth flattening is the ratio $f = \frac{R_e - R_p}{R_e}$, where $R_e$ and $R_p$ are the equatorial and polar radius, respectively. The flattening is due to the rotation around the polar axis at an angular rate $\Omega_e$. On the other hand, it perturbates the Earth's gravity field through the first zonal term $J_2$.

The Earth figure is estimated by assuming that the Earth is a rotating fluid and that the gravitational and centrifugal forces cancel each other out at the surface.

The total potential (gravitational + centrifugal) at a point of coordinates $(r, \varphi)$ is

$$U(r,\varphi) = \frac{\mu}{r} - \frac{1}{2}\frac{\mu}{r}J_2\left(\frac{R_e}{r}\right)^2(3\sin^2\varphi - 1) + \frac{1}{2}\Omega_e^2 r^2\cos^2\varphi$$

The Earth's surface is an equipotential of equation $U(r, \varphi) = U(R_e, 0)$.

We assume that the Earth's surface deviates little from a circle of radius $R_e$, with an equation of the form $r(\varphi) = R_e + \delta r(\varphi)$.

Substituting into the equipotential equation and expanding to first order yields

$$\frac{\delta r(\varphi)}{R_e}\left[-\frac{\mu}{R_e}+\frac{3\mu}{2R_e}J_2(3\sin^2\varphi-1)+\Omega_e^2R_e^2\cos^2\varphi\right]=\left(\frac{3\mu}{2R_e}J_2+\frac{1}{2}\Omega_e^2R_e^2\right)\sin^2\varphi$$

Keeping only the central term in the first bracket, we obtain a surface equation of the form $r(\varphi)=R_e\left[1-\left(\frac{3}{2}J_2+\frac{1}{2}\frac{\Omega_e^2R_e^3}{\mu}\right)\sin^2\varphi\right]$.

The polar radius is then $R_p=r\left(\varphi=\frac{\pi}{2}\right)=R_e\left[1-\left(\frac{3}{2}J_2+\frac{1}{2}\frac{\Omega_e^2R_e^3}{\mu}\right)\right]$.

This leads to the **formula of Clairaut** (1743) linking the flattening f, the $J_2$ term and the angular rate $\Omega_e$: $f=\frac{R_e-R_p}{R_e}=\frac{3}{2}J_2+\frac{1}{2}\frac{\Omega_e^2R_e^3}{\mu}$.

With the values $\begin{cases}\mu=3.986\,005\cdot10^{14}\ \mathrm{m^3s^{-2}}\\ J_2=1.082\,63\cdot10^{-3}\\ \Omega_e=7.292\,115\cdot10^{-5}\ \mathrm{rad\,s^{-1}}\end{cases}$ and $R_e=6\,378\,137$ m, we obtain a flattening $f=\frac{1}{298}$ and a polar radius $R_p=6\,356\,734$ m.

#### 2.4.4.4 Gravity Acceleration Including First Zonal Term

Many simulation programs involve numerically integrating the motion differential equations under the form (2.11). For that purpose, it is necessary to express the acceleration components in Cartesian coordinates.

The gravitational potential accounting for the first zonal term is given in spherical coordinates by (2.98) reproduced here.

$$U(r,\varphi)=\frac{\mu}{r}+\frac{\mu R_e^2J_2}{2r^3}(1-3\sin^2\varphi) \tag{2.109}$$

The spherical coordinates depend on the Cartesian coordinates through

$$r=\sqrt{x^2+y^2+z^2},\quad \sin\varphi=\frac{z}{r} \tag{2.110}$$

Deriving the potential (2.109) with respect to (x, y, z), we obtain the Cartesian components of the gravity acceleration including the first zonal term.

$$\vec{a}=-\frac{\partial U}{\partial \vec{r}}\Rightarrow\begin{cases}a_x=-\frac{\mu x}{r^3}\left[1+\frac{3}{2}R_e^2J_2\frac{1}{r^2}\left(1-5\frac{z^2}{r^2}\right)\right]\\ a_y=-\frac{\mu y}{r^3}\left[1+\frac{3}{2}R_e^2J_2\frac{1}{r^2}\left(1-5\frac{z^2}{r^2}\right)\right]\\ a_z=-\frac{\mu z}{r^3}\left[1+\frac{3}{2}R_e^2J_2\frac{1}{r^2}\left(3-5\frac{z^2}{r^2}\right)\right]\end{cases} \tag{2.111}$$

## 2.5 Other Perturbations

### 2.5.1 Third-Body Attraction

Figure 2.10 shows three bodies (Earth T, Moon L and satellite S) in gravitational interaction. In an inertial frame of origin O, the satellite and Earth positions are defined by $\vec{\rho}_S=\overrightarrow{OS}$

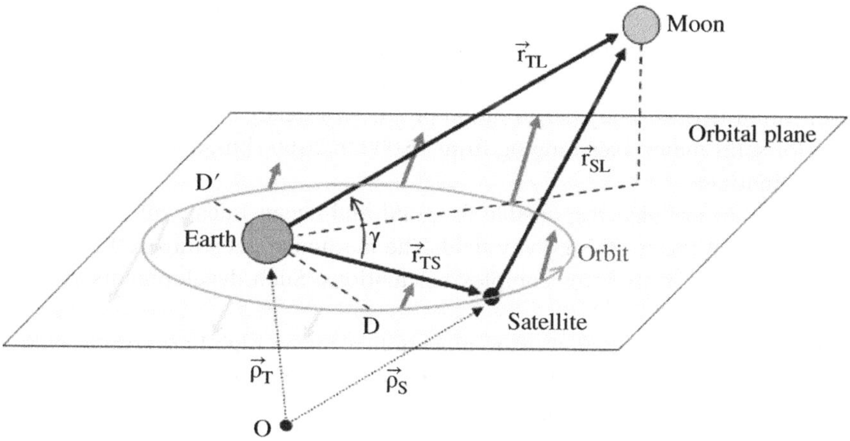


**Figure 2.10** Perturbation due to the Moon attraction.

and $\vec{\rho}_T = \overrightarrow{OT}$ respectively. Their equations of motion under the mutual gravitational attractions are

$$\begin{cases} \ddot{\vec{\rho}}_S = \dfrac{Gm_T}{r_{ST}^3}\vec{r}_{ST} + \dfrac{Gm_L}{r_{SL}^3}\vec{r}_{SL} \\ \ddot{\vec{\rho}}_T = \dfrac{Gm_S}{r_{TS}^3}\vec{r}_{TS} + \dfrac{Gm_L}{r_{TL}^3}\vec{r}_{TL} \end{cases} \tag{2.112}$$

By subtraction, the satellite's position relative to the Earth $\vec{r}_{TS} = \vec{\rho}_S - \vec{\rho}_T$ satisfies

$$\ddot{\vec{r}}_{TS} = \ddot{\vec{\rho}}_S - \ddot{\vec{\rho}}_T = -\frac{G(m_T + m_S)}{r_{TS}^3}\vec{r}_{TS} + Gm_L\left(\frac{\vec{r}_{SL}}{r_{SL}^3} - \frac{\vec{r}_{TL}}{r_{TL}^3}\right) \tag{2.113}$$

The first term is the attraction of the Earth alone (unperturbed motion). The second term shows that the perturbation arises from the *difference in the attractions* exerted by the Moon on the Earth and on the satellite. When the satellite is closer to the Moon, it is attracted more than the Earth, and the perturbation corresponds in Figure 2.10 to the arrows pointing upward toward the Moon. Conversely, the arrows pointing downward correspond to a repulsive effect with respect to the Earth. The resulting torque around the DD′ line induces a precession motion, similarly to the action of the equatorial bulge depicted in Figure 2.8.

The Moon perturbation can be expressed as the derivative with respect to the satellite position of the potential $U_p = Gm_L\left(\frac{1}{r_{SL}} - \frac{\vec{r}_{TS}\cdot\vec{r}_{TL}}{r_{TL}^3}\right)$.

Using the relation $r_{SL}^2 = r_{TS}^2 + r_{TL}^2 - 2r_{TL}r_{TS}\cos\gamma$ and the expansion with Legendre polynomials (2.89), the perturbation potential takes the form

$$U_p = \frac{Gm_L}{r_{TL}}\left[1 + \sum_{n=2}^{\infty}\left(\frac{r_{TS}}{r_{TL}}\right)^n P_n(\cos\gamma)\right] \tag{2.114}$$

This development allows the use of Lagrange equations (2.54), by expressing $U_p$ as a function of the orbital parameters of the satellite and of the perturbing body (called **third body**). In the case of an Earth artificial satellite, the main perturbations are due to the Moon and the Sun. Their own motion relative to the Earth induces secular and periodic effects on the satellite orbit.

In some cases, resonances may appear between the third-body perturbation and the Earth's gravitational perturbations. For example, if the apsidal precession rate $\dot{\omega}$ (2.103) matches the Sun's apparent motion at a rate of 0.986 deg/day, then the orbit axis rotates in synchronism with the Sun, and the third-body perturbation produces a steady effect over time. This resonance occurs for semi-major axes ranging from 15 000 to 20 000 km, depending on the eccentricity and inclination.

The Moon and the Sun are also responsible for solid and ocean tides, which generate themselves perturbation of the Earth's gravity field. The mathematical study of third-body perturbations is feasible, but it involves complex calculations. Such developments can be found in [R8], [R10], and [R42].

Since the influences of the Moon and Sun are most noticeable on high Earth orbits, especially on the geostationary orbit (GEO), it is interesting to develop a specific model in order to assess the main effects on circular orbits.

#### 2.5.1.1 Third-Body Effects on Circular Orbits

Let us consider a satellite initially in circular orbit around the Earth and a third attracting body (Moon or Sun). The satellite and third-body positions in the inertial geocentric frame are noted $\vec{r}$ and $\vec{r}_p$, their angular momentum are noted $\vec{h}$ and $\vec{h}_p$, and the corresponding unit vectors are noted $\vec{u}_r$, $\vec{u}_p$, $\vec{u}_h$ and $\vec{u}_{hp}$.

$$\vec{r} = r\vec{u}_r, \quad \vec{r}_p = r_p\vec{u}_p, \quad \vec{h} = h\vec{u}_h, \quad \vec{h}_p = h_p\vec{u}_{hp} \tag{2.115}$$

Equation (2.113) defines the satellite's motion relative to the Earth under the form

$$\ddot{\vec{r}} = \vec{a}_c + \vec{a}_p \quad \text{with } \vec{a}_c = -\frac{Gm_T}{r^3}\vec{r}, \quad \vec{a}_p = Gm_P\left(\frac{\vec{r}_p - \vec{r}}{\|\vec{r}_p - \vec{r}\|^3} - \frac{\vec{r}_p}{r_p^3}\right) \tag{2.116}$$

where the satellite mass is neglected ($m_S \ll m_T$) and the third-body mass is $m_P$.

With this equation similar to (2.11), we can apply (2.15), giving the variation of the angular momentum.

$$\dot{\vec{h}} = \vec{r} \wedge \vec{a}_p \tag{2.117}$$

At this stage, we make some simplifying assumptions on the third-body motion

- The third body is far from the satellite ($r \ll r_p$).

  This allows the first-order expansion $\frac{1}{\|\vec{r}_p - \vec{r}\|^3} \approx \frac{1}{r_p^3}\left(1 + 3\frac{r}{r_p}\vec{u}_r \cdot \vec{u}_p\right)$

- The third body is in circular Keplerian motion around the Earth.

  Its mean motion is then $n_p = \sqrt{\frac{G(m_T+m_P)}{r_p^3}} \Rightarrow \frac{Gm_P}{r_p^3} = n_p^2\frac{m_P}{m_T+m_P}$.

With these assumptions, the perturbing acceleration (2.116) takes the form

$$\vec{a}_p = 3\gamma_P n_p^2 r\,(\vec{u}_r \cdot \vec{u}_p)\vec{u}_p \quad \text{with } \gamma_P \underset{\text{def}}{=} \frac{m_P}{m_T + m_P} \tag{2.118}$$

where $\gamma_P$ is the mass ratio between the perturbing body and the two-body system, and the angular momentum equation (2.117) becomes

$$\dot{\vec{h}} = 3\gamma_P n_p^2 r^2(\vec{u}_r \cdot \vec{u}_p)(\vec{u}_r \wedge \vec{u}_p) \tag{2.119}$$

Our goal is to assess the evolution of the averaged angular momentum. For that purpose, we establish a preliminary result.

**Calculation of the Integral $\int_0^T(\vec{u}\cdot\vec{d})(\vec{u}\wedge\vec{d})\,dt$**

Let us consider a fixed orthonormal trihedron $(\vec{i},\vec{j},\vec{k})$, a unit vector $\vec{u}$ in uniform rotation at period T in the plane $(\vec{i},\vec{j})$, any fixed unit vector $\vec{d}$ in the plane $(\vec{i},\vec{k})$ and the angles $\lambda = (\vec{i},\vec{u})$ and $\alpha = (\vec{i},\vec{d})$. The vectors $\vec{u}$ and $\vec{d}$ are expressed by

$$\begin{cases} \vec{u} = \vec{i}\cos\lambda(t) + \vec{j}\sin\lambda(t) \\ \vec{d} = \vec{i}\cos\alpha + \vec{k}\sin\alpha \end{cases} \quad \text{with } \lambda(t) = \frac{2\pi}{T}t, \quad \alpha = \mathrm{Cte}$$

Let us calculate the integral $I = \int_0^T(\vec{u}\cdot\vec{d})(\vec{u}\wedge\vec{d})\,dt$ with these vectors.

$$I = \int_0^T \cos\alpha\cos\lambda(t)[-\vec{j}\sin\alpha\cos\lambda(t) - \vec{k}\cos\alpha\sin\lambda(t)]dt = -\vec{j}\sin\alpha\cos\alpha\,\frac{T}{2}$$

By using $\begin{cases} \vec{d}\wedge\vec{k} = -\vec{j}\cos\alpha \\ \vec{d}\cdot\vec{k} = \cos\alpha \end{cases}$, we obtain $I = \int_0^T(\vec{u}\cdot\vec{d})(\vec{u}\wedge\vec{d})\,dt = \frac{T}{2}(\vec{d}\cdot\vec{k})(\vec{d}\wedge\vec{k})$.

This formula is applied to integrate the angular momentum equation (2.119). The integration is made in two steps, with appropriate assumptions.

- First integration step: over the satellite orbital period

  This integration step is carried out with the following assumptions.
  - The satellite period T is much smaller than the third body period $T_p$, so that the unit vector $\vec{u}_p$ can be considered as constant.
  - The perturbation during one orbital period is small, so that the satellite orbit remains circular of radius r. The averaged angular momentum is noted $\vec{h}_m$.

The integral formula is applied to (2.119) with $\vec{u} = \vec{u}_r, \vec{d} = \vec{u}_p, \vec{k} = \vec{u}_{hm}$. This yields the variation rate of the averaged angular momentum over one satellite period.

$$\dot{\vec{h}}_m \underset{def}{=} \frac{1}{T}\int_0^T \dot{\vec{h}}\,dt = \frac{3\gamma_p n_p^2 r^2}{T}\int_0^T (\vec{u}_r\cdot\vec{u}_p)(\vec{u}_r\wedge\vec{u}_p)\,dt = \frac{3\gamma_p n_p^2 r^2}{2}(\vec{u}_p\cdot\vec{u}_{hm})(\vec{u}_p\wedge\vec{u}_{hm}) \tag{2.120}$$

This variation rate is orthogonal to $\vec{u}_{hm}$. The modulus of the averaged angular momentum is then constant worth $h_m = r^2 n$, where n is the satellite mean motion. By substituting $\vec{h}_m = r^2 n\vec{u}_{hm}$ in (2.120), we obtain

$$\dot{\vec{u}}_{hm} = \frac{3\gamma_p n_p^2}{2n}(\vec{u}_p\cdot\vec{u}_{hm})(\vec{u}_p\wedge\vec{u}_{hm}) \tag{2.121}$$

- Second integration step: over the third-body orbital period

  This integration step is carried out with the following assumptions.
  - The mean motion n of the satellite is constant over the third-body period.
  - The doubly averaged angular momentum direction is noted $\vec{u}_{hmm}$. Its evolution is defined from (2.121) integrated over the third-body period.

$$\dot{\vec{u}}_{hmm} \underset{def}{=} \frac{1}{T_p}\int_0^{T_p} \dot{\vec{u}}_{hm}\,dt = \frac{3\gamma_p n_p^2}{2nT_p}\int_0^{T_p} (\vec{u}_p\cdot\vec{u}_{hm})(\vec{u}_p\wedge\vec{u}_{hm})\,dt \tag{2.122}$$

– The perturbation during the third-body orbital period is small, so that the direction $\vec{u}_{hm}$ in the integral (2.122) can be considered as constant.

The integral formula is applied to (2.122), with $\vec{u} = \vec{u}_p, \vec{d} = \vec{u}_{hm}, \vec{k} = \vec{u}_{hp}$. We obtain

$$\dot{\vec{u}}_{hmm} = \frac{3\gamma_p n_p^2}{4n}(\vec{u}_{hmm} \cdot \vec{u}_{hp})(\vec{u}_{hmm} \wedge \vec{u}_{hp}) \tag{2.123}$$

The formula (2.123) gives the evolution of the averaged satellite orbital plane due to the third-body perturbation. To study the long-term evolution, the first-zonal-term effect must also be taken into account, as below.

For simplicity, we note $\vec{u}_h$ the averaged direction in the sequel. The components of $\vec{u}_h$ are given by (2.22) and from (2.102)–(2.103), the zonal term only changes the mean ascending node longitude. The semi-major axis, the eccentricity (null for a circular orbit), and the inclination are constant on average. We have thus

$$\vec{u}_h = \begin{pmatrix} \sin i \sin \Omega \\ -\sin i \cos \Omega \\ \cos i \end{pmatrix} \quad \text{with} \quad \dot{\Omega} = -\frac{3}{2} J_2 n \left(\frac{R_e}{a}\right)^2 \cos i \tag{2.124}$$

With the inclination being constant, the time derivative of $\vec{u}_h$ is then

$$\dot{\vec{u}}_h = \dot{\Omega} \sin i \begin{pmatrix} \cos \Omega \\ \sin \Omega \\ 0 \end{pmatrix} = -\dot{\Omega}\,(\vec{u}_h \wedge \vec{u}_z) = \frac{3}{2} J_2 n \left(\frac{R_e}{a}\right)^2 (\vec{u}_h \cdot \vec{u}_z)(\vec{u}_h \wedge \vec{u}_z) \tag{2.125}$$

where $\vec{u}_z$ is the polar axis of the geocentric frame.

Gathering formula (2.123) applied to both the Moon (index L) and the Sun (index S), and formula (2.125) accounting for the zonal term, we obtain

$$\dot{\vec{u}}_h = \frac{3}{4} n \vec{u}_h \wedge \left[\frac{\gamma_L n_L^2}{n^2}(\vec{u}_h \cdot \vec{u}_{hL})\,\vec{u}_{hL} + \frac{\gamma_S n_S^2}{n^2}(\vec{u}_h \cdot \vec{u}_{hS})\,\vec{u}_{hS} + 2J_2 \left(\frac{R_e}{a}\right)^2 (\vec{u}_h \cdot \vec{u}_z)\,\vec{u}_z\right] \tag{2.126}$$

This equation must be integrated numerically, accounting for the Moon precession motion, in order to assess accurately the variation of the satellite orbital plane.

It is nevertheless possible to estimate the main trends as follows.

#### 2.5.1.2 Stable Solution

Let us consider a single third body and search for a stable solution noted $\vec{u}_{he}$.

We have to solve the equation $\dot{\vec{u}}_{he} = \vec{0}$, which from (2.126) takes the form

$$\begin{gathered} \vec{u}_{he} \wedge [A(\vec{u}_{he} \cdot \vec{u}_{hp})\,\vec{u}_{hp} + B(\vec{u}_{he} \cdot \vec{u}_z)\,\vec{u}_z] = \vec{0} \\ \text{with } A = \frac{\gamma_p n_p^2}{n^2}, \quad B = 2J_2 \left(\frac{R_e}{a}\right)^2 \end{gathered} \tag{2.127}$$

The vector $\vec{u}_{hp}$ and $\vec{u}_z$ are noncollinear. For the vector product to be null, the vector in brackets must be collinear with $\vec{u}_{he}$.

$$A(\vec{u}_{he} \cdot \vec{u}_{hp})\,\vec{u}_{hp} + B(\vec{u}_{he} \cdot \vec{u}_z)\,\vec{u}_z = C\vec{u}_{he} \tag{2.128}$$

The vector $\vec{u}_{he}$ is thus in the plane $(\vec{u}_z, \vec{u}_{hp})$. This implies from (2.124) that the ascending node longitudes $\Omega_e$ and $\Omega_p$ are equal, and they can arbitrarily be set to zero: $\Omega_e = \Omega_p = 0$. By passing to the components

$$\vec{u}_{hp} = \begin{pmatrix} 0 \\ -\sin i_p \\ \cos i_p \end{pmatrix}, \quad \vec{u}_{he} = \begin{pmatrix} 0 \\ -\sin i_e \\ \cos i_e \end{pmatrix}, \quad \vec{u}_{he} \cdot \vec{u}_{hp} = \cos(i_p - i_e) \tag{2.129}$$

and substituting into the vector equation (2.127), we obtain from the first component

$$A \sin 2(i_p - i_e) - B \sin 2i_e = 0 \tag{2.130}$$

whose solution is

$$\tan 2i_e = \frac{A \sin 2i_p}{A \cos 2i_p + B} \quad \text{with } A = \frac{\gamma_p n_p^2}{n^2}, \quad B = 2J_2\left(\frac{R_e}{a}\right)^2 \tag{2.131}$$

### 2.5.1.3 Motion Around the Stable Solution

Let us now retrieve equation (2.126) with a single third body by assuming that the initial direction $\vec{u}_h$ is close to the stable one $\vec{u}_{he}$.

$$\dot{\vec{u}}_h = \frac{3}{4} n \vec{u}_h \wedge [\, A(\vec{u}_h \cdot \vec{u}_{hp})\, \vec{u}_{hp} + B(\vec{u}_h \cdot \vec{u}_z)\, \vec{u}_z] \tag{2.132}$$

Substituting $\vec{u}_h \approx \vec{u}_{he}$ into the scalar products into brackets and using (2.128) yields

$$\dot{\vec{u}}_h \approx \frac{3}{4} n \vec{u}_h \wedge [\, A(\vec{u}_{he} \cdot \vec{u}_{hp})\, \vec{u}_{hp} + B(\vec{u}_{he} \cdot \vec{u}_z)\, \vec{u}_z] = \frac{3}{4} nC\, \vec{u}_h \wedge \vec{u}_{he} \tag{2.133}$$

The variation rate of $\vec{u}_h$ is then almost orthogonal to the fixed vector $\vec{u}_{he}$ and the motion is close to a cone around the stable direction $\vec{u}_{he}$. Starting from the initial conditions ($\Omega = 0$, $i = 0$), the angle between $\vec{u}_{he}$ and $\vec{u}_h$ will remain close to $i_e$, and the rate $\dot{u}_h$ is constant. Its value is assessed from (2.132) taking $\vec{u}_h = \vec{u}_z$, which corresponds to the initial conditions ($\Omega = 0, i = 0$).

$$\dot{\vec{u}}_{h0} = \frac{3}{4} nA(\vec{u}_z \cdot \vec{u}_{hp})(\vec{u}_z \wedge \vec{u}_{hp}) \quad \Rightarrow \quad \dot{u}_{h0} = \frac{3}{8} nA \sin 2i_p \tag{2.134}$$

We can estimate the time $T_e$ to achieve a full revolution along the cone of axis $\vec{u}_{he}$ and half apex angle $i_e$. Introducing the satellite orbital period $T = 2\pi/n$, we get

$$T_e = \frac{2\pi \sin i_e}{\dot{u}_{h0}} = \frac{8T \sin i_e}{3A \sin 2i_p} \tag{2.135}$$

### 2.5.1.4 Numerical Application to the GEO

For the numerical application to a GEO, we assume that the Moon and Sun are in the same direction: $\vec{u}_{hL} = \vec{u}_{hS} \underset{\text{noted}}{=} \vec{u}_{hp}$.

This allows applying the aforementioned formulas by taking $A = \frac{\gamma_p n_p^2}{n^2} = \frac{\gamma_L n_L^2 + \gamma_S n_S^2}{n^2}$. The numerical values are

$$\begin{cases} T_L = 27.32 \text{ days}, \quad T_S = 365.25 \text{ days}, \quad T_{GEO} = 86\,164 \text{ s} = 0.997 \text{ days}, \quad i_p = 23.45 \text{ deg} \\ \gamma_L = \dfrac{1}{82.3}, \quad \gamma_L = 1, \quad J_2 = 1.082 \times 10^{-3}, \quad R_e = 6378.137 \text{ km}, \quad a_{GEO} = 42\,164 \text{ km} \end{cases}$$

$$\rightarrow \; A = 2.38 \times 10^{-5}, \quad B = 4.95 \times 10^{-5}$$

Equation (2.131) yields the stable inclination: $i_e = 7.36$ deg.

Equation (2.135) yields the revolution period: $T_e = 53.6$ years.

Equation (2.134) yields the mean rotation rate $\dot{u}_h = 0.86$ deg/year. The yearly correction for the station keeping in GEO is then $\Delta V \approx V_{GEO}\dot{u}_h = 46$ m/s.

### 2.5.2 Atmospheric Friction

Even at high altitudes, the residual atmosphere generates an aerodynamic force, which depends on the atmosphere density, on the satellite velocity, and on its shape and orientation in the flow. The velocity relative to the atmosphere depends on the Earth's rotation and on the wind. The vehicle shape and attitude define the drag $C_x$ and lift $C_z$ coefficients and its apparent surface $S_a$ in the flow.

Since all these variables are difficult to model, we generally consider a spherical shape with zero lift and with drag of order $C_x \approx 2.2$. Furthermore, the relative velocity is approximated by the absolute velocity (i.e. the velocity in the inertial reference frame), which does not depend on the position or the wind direction.

With these simplifying assumptions, the drag acceleration is expressed as

$$\vec{a}_p = -\frac{1}{2}\rho v^2 \frac{S_a C_x}{m} \frac{\vec{v}}{v} \tag{2.136}$$

The atmospheric density ρ depends mainly on the altitude, but also on many others parameters: geographical location, date (season), time (day/night), solar activity, magnetic field, etc. Accurate assessments are difficult, as physical phenomena are still poorly understood, especially at high altitudes. Increasingly comprehensive atmospheric models have been developed since the 1960s, such as USSA76, JACCHIA77, CIRA90, GRAM99, and NRLMSIS00.

For simplified evaluations, the exponential atmosphere model explicitly gives the density ρ as a function of the geodetic altitude z.

$$\rho(z) = \rho_0 e^{-\frac{z}{H}} \tag{2.137}$$

In this model, H is the scale factor (≈ 7100 m at the ground level) and $\rho_0$ is the density at the ground (1.293 kg/m$^3$).

The drag force is nonconservative and it reduces the satellite energy. For an elliptical orbit, the braking is maximal at the perigee leading to an apogee decrease and a progressive circularization of the orbit. Once the orbit is circular, the braking applies constantly and the trajectory follows a downward spiral.

The altitude loss per revolution is calculated with Gauss equation (2.45). For a circular orbit of radius a, we have $e = 0$, $h = av$. Integrating the semi-major axis equation over one period T (2.6) yields

$$\Delta a = \frac{2a}{v} a_p T = -2\pi\rho B a^2 \quad \text{with } B = \frac{SC_x}{m} \tag{2.138}$$

This equation is used to estimate the fallout time depending on the satellite **ballistic coefficient** B (Figure 2.11). A low Earth orbit whose perigee is below 200 km is considered as unstable, since atmospheric friction causes the satellite to fall within a few days. Beyond 500 km, the effect becomes negligible compared with other perturbations. These estimates can be highly inaccurate due to the poor knowledge of the upper atmosphere parameters.

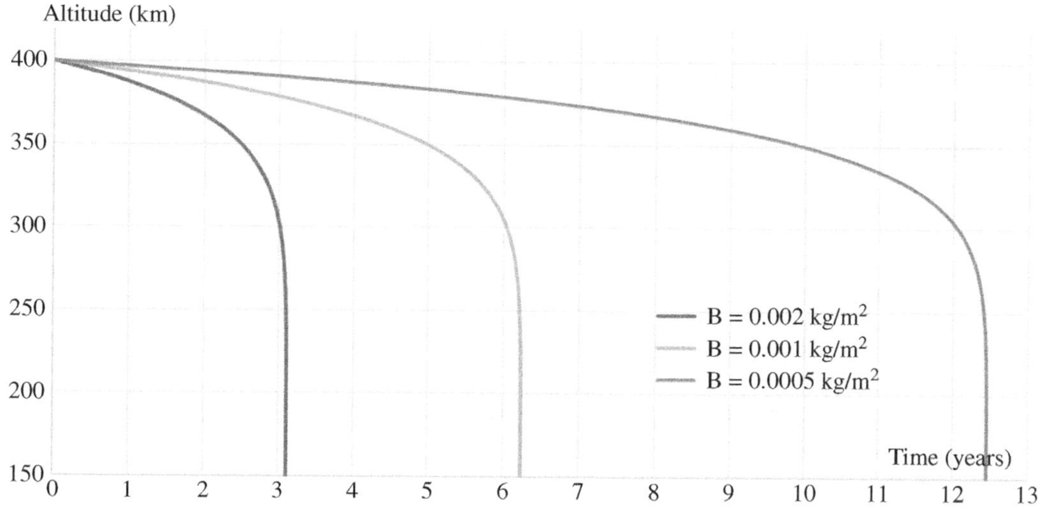


**Figure 2.11** Fallout duration from a circular orbit at 400 km.

### 2.5.3 Radiation Pressure

The radiation pressure force is due to the impact of photons emitted by the Sun on the satellite's surface. For a flat surface, its expression is

$$\vec{F}_r = -p_r \varepsilon_r S \cos\alpha \, \vec{n} \tag{2.139}$$

S is the surface exposed to the Sun, $\vec{n}$ is the normal to the surface, $\alpha$ is the angle between the normal and the Sun direction, $\varepsilon_r$ is the surface reflexivity (ranging from 0 to 2), and $p_r$ is the radiation pressure.

An accurate estimation of these parameters is difficult, as the shape, orientation, and surface properties of the satellite must be taken into account. In addition, the radiation pressure depends on the solar activity, which follows a periodic cycle of about 11 years, but is highly unpredictable from one cycle to the next.

For a simplified assessment, we consider the average solar flux $\Phi$ at the Earth level, which ranges from 1300 to 1500 W/m$^2$. The radiation pressure is then approximatively given by $p_r = \Phi/c \approx 4.6 \cdot 10^{-6}$ Pa, where c is the celerity of light.

The concept of solar sail exploits this radiation force to progressively change the satellite orbit by controlling the sail orientation with respect to the Sun. This concept could be used for interplanetary missions in the inner solar system.

## 2.6 Conclusion

### 2.6.1 The Key Points

- Perturbed motion is analyzed as a progressive deviation from the Keplerian trajectory, in terms of variation of the osculating orbital parameters.
- Gauss equations give the variation of the osculating parameters as a function of any perturbing acceleration expressed in the local orbital frame.

- Lagrange equations give the variation of the osculating parameters only for accelerations deriving from a potential, such as gravitational or third-body perturbation.
- Special perturbations consist of integrating numerically the perturbed motion equations, whereas general perturbations consist of forming semi-analytical solutions based on averaging.
- The main gravitational perturbation is due to the Earth flattening and it induces a nodal and an apsidal precession motion.

### 2.6.2 To Go Further

- BATTIN R. – *An Introduction to the Mathematics and Methods of Astrodynamics.* AIAA (1999).

As for Keplerian motion, this book derives in Chapter 10 the equation of perturbed motion in a very elegant way using vector calculus.

- CHAO C.-C. G. – *Applied Orbit Perturbation and Maintenance.* Aerospace Press (2005).

This book fully dedicated to perturbed motion is a reference for the analysis of Earth satellite orbits. It presents many useful results from semi-analytical developments. The effect of each type of perturbation (gravitational potential, third body, drag, solar pressure radiation) on each type of orbit (LEO, MEO, GEO) is studied separately in a very clear way.

- CHOBOTOV V. – *Orbital Mechanics Third Edition.* AIAA (2002).

As for Keplerian motion, this book presents in a very clear way the essentials of perturbations in Chapters 8–10.

- VALLADO D. – *Fundamentals of Astrodynamics and Applications.* Microcosm Press, Springer (2007).

Chapter 8 presents the special perturbation techniques with many details on numerical aspects. Chapter 9 presents the general perturbation techniques with the main semi-analytical methods useful for practical applications.

# 3

# Three-Body Problem

## 3.1 Introduction

The three-body problem concerns the motion of three material points in gravitational interaction. Unlike the two-body problem, there is no analytical solution. The problem is said to be restricted when one body has a negligible mass, which is the case of a spacecraft compared to celestial bodies. The two main bodies then follow a Keplerian motion, almost circular in the case of the Sun–Earth or Earth–Moon systems.

Because of its practical importance, the restricted circular three-body problem has given rise to many studies since the demonstrations of the collinear equilibrium points by Euler in 1767, then of the triangular equilibrium points by Lagrange in 1772, followed by Poincaré's work on the existence of periodic orbits. The discovery in 1906 of hundreds of Trojan asteroids in the Sun–Jupiter system confirmed the relevance of mathematical models.

Halo orbits located around the collinear points named $L_1$ and $L_2$ were discovered by Farquhar in 1966. These orbits are especially interesting for space missions because they maintain a constant relative geometry with respect to the two bodies. In the Sun–Earth system, the $L_1$ and $L_2$ points are located 1.5 million kilometers from Earth. They offer a constant thermal environment, suitable for observing the Sun in $L_1$ (ISEE, SOHO, and LISA missions) or for observing the space in $L_2$ (MAP, Gaia, Herschell-Planck, and JWST missions). In the Earth–Moon system, these points are located 60 000 km from the Moon. They are intended for the establishment of a permanent space station, as had been envisaged by Clarke as early as 1950.

Halo orbits are associated with families of trajectories, called invariant manifolds, naturally arriving to or departing from the orbit. These gravitational currents intrinsic to the three-body problem dynamics open up the prospect of economic space transfers to the Moon or beyond.

The first section of this chapter presents the modeling of the restricted circular three-body problem with the determination of Lagrange points.

The second section studies the stability in the vicinity of Lagrange points through linearization of the motion equations and the search for periodic orbits, which exist owing to nonlinear effects.

The third section develops the calculation of manifolds for Halo orbits and presents an overview of low-cost transfer strategies using the manifold properties.

*Space Trajectories: Basic and Advanced Topics*, First Edition. Max Cerf.

## 3.2 Circular-Restricted Three-Body Problem

This section presents the modeling of the circular-restricted three-body problem (CR3BP), with the motion equations in the rotating frame, the accessible region depending on the initial conditions, and the location of the equilibrium points.

### 3.2.1 Motion Equations

The three-body problem considers three material points $C_1$, $C_2$, and M of respective masses $m_1$, $m_2$, and m in gravitational interaction. When $C_1$ and $C_2$ are celestial bodies and M is a space vehicle, the mass of M is negligible and does not influence the motion of $C_1$ and $C_2$. In this case, the problem is said **restricted** and the celestial bodies $C_1$ and $C_2$, called primary bodies, follow a Keplerian motion due to their mutual attraction.

The CR3BP assumes furthermore that the body's motion is circular, which is quite close to the systems studied in practice. The angular velocity $\Omega$ of $C_1$ and $C_2$ around their center of mass O is then constant, given by Kepler's third law.

$$\Omega = \sqrt{\frac{G(m_1 + m_2)}{d^3}} \tag{3.1}$$

In this formula, $G = 6.674\,08 \cdot 10^{-11}\ \mathrm{m^3\,kg^{-1}\,s^{-2}}$ is the universal gravitational constant and $d = C_1C_2$ is the constant distance between the two bodies.

The system $(C_1, C_2)$ is assumed to be isolated and its center of mass O is taken as the origin of an inertial frame noted RG. Newton's second law in RG gives

$$m \left.\frac{d^2\vec{r}}{dt^2}\right|_{RG} = -Gmm_1 \frac{\vec{r}_1}{r_1^3} - Gmm_2 \frac{\vec{r}_2}{r_2^3} \tag{3.2}$$

where $\vec{r}_1$, $\vec{r}_2$, and $\vec{r}$ denote the position of M relative to $C_1$, $C_2$, and O, respectively.

To study the motion of M, we place ourselves in a frame rotating with the bodies $C_1$ and $C_2$. This synodic frame, called **Cross frame** and denoted by RC, is depicted in Figure 3.1. The body $C_1$ is supposed to be the most massive. The origin of RC is in the center of mass O, the z-axis is the axis of rotation of the system $(C_1, C_2)$, the x-axis is along $(C_1, C_2)$, and the y-axis completes the trihedron.

Taking the derivative in the rotating frame RC, instead of RG, adds into (3.2) the Coriolis and drive accelerations.

$$\left.\frac{d^2\vec{r}}{dt^2}\right|_{RC} + 2\vec{\Omega} \wedge \left.\frac{d\vec{r}}{dt}\right|_{RC} + \vec{\Omega} \wedge (\vec{\Omega} \wedge \vec{r}) = -\mu_1 \frac{\vec{r}_1}{r_1^3} - \mu_2 \frac{\vec{r}_2}{r_2^3} \tag{3.3}$$

where $\mu_1 = Gm_1$ and $\mu_2 = Gm_2$ are the gravitational constants of the attracting bodies $C_1$ and $C_2$ respectively.

Noting (x,y,z) the coordinates of M in RC, we obtain the differential system

$$\begin{cases} \ddot{x} - 2\Omega\dot{y} - \Omega^2 x = -\mu_1 \dfrac{x + d_1}{r_1^3} - \mu_2 \dfrac{x - d_2}{r_2^3} \\ \ddot{y} + 2\Omega\dot{x} - \Omega^2 y = -\left(\dfrac{\mu_1}{r_1^3} + \dfrac{\mu_2}{r_2^3}\right) y \\ \ddot{z} = -\left(\dfrac{\mu_1}{r_1^3} + \dfrac{\mu_2}{r_2^3}\right) z \end{cases} \tag{3.4}$$

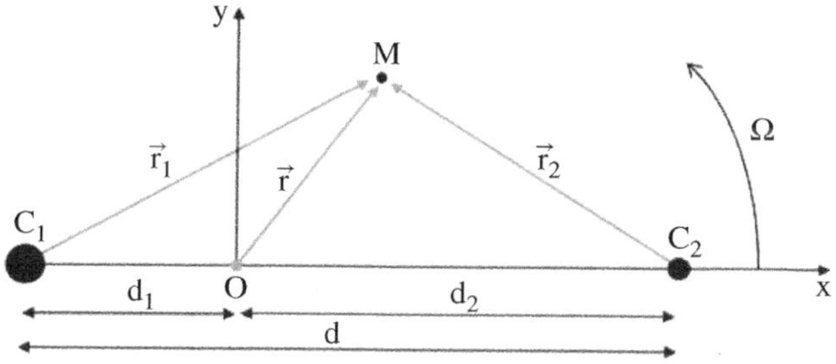


**Figure 3.1** Cross frame for the three-body problem.

Both bodies are fixed in RC and have coordinates $C_1(-d_1, 0, 0)$ and $C_2(d_2, 0, 0)$. Their respective distances $d_1$, $d_2$ to the center of mass O are given by

$$\begin{cases} d_1 + d_2 = d \\ d_1 m_1 = d_2 m_2 \end{cases} \Rightarrow \frac{d_1}{d} = \frac{m_2}{m_1 + m_2} \quad \textit{and} \quad \frac{d_2}{d} = \frac{m_1}{m_1 + m_2} \tag{3.5}$$

These equations are simplified by taking d as the distance unit and $\Omega^{-1}$ as the time unit. With these units, the relation (3.1) becomes $G(m_1 + m_2) = 1$.

The distances $d_1$, $d_2$ and the gravitational constants $\mu_1$, $\mu_2$ then depend only on the mass ratio $\mu = \frac{m_2}{m_1+m_2}$, which is comprised between 0 and 0.5 (since $C_1$ is by assumption the heaviest body).

$$\begin{cases} d_1 = \mu \\ d_2 = 1 - \mu \end{cases} \quad \text{and} \quad \begin{cases} \mu_1 = 1 - \mu \\ \mu_2 = \mu \end{cases} \tag{3.6}$$

For simplicity, we keep the notation (x,y,z) for the coordinates expressed in the normalized units. Substituting (3.6) into (3.4), we obtain dimensionless equations that depend only on the mass ratio $\mu$.

$$\begin{cases} \ddot{x} - 2\dot{y} - x = -(1-\mu)\dfrac{x+\mu}{r_1^3} - \mu\dfrac{x+\mu-1}{r_2^3} \\[2ex] \ddot{y} + 2\dot{x} - y = -\left(\dfrac{1-\mu}{r_1^3} + \dfrac{\mu}{r_2^3}\right) y \\[2ex] \ddot{z} = -\left(\dfrac{1-\mu}{r_1^3} + \dfrac{\mu}{r_2^3}\right) z \end{cases} \tag{3.7}$$

The normalized distances of M to the attracting bodies $C_1$ and $C_2$ are given by

$$\begin{cases} r_1^2 = (x+\mu)^2 + y^2 + z^2 \\ r_2^2 = (x+\mu-1)^2 + y^2 + z^2 \end{cases} \tag{3.8}$$

### 3.2.2 Accessible Region

The motion equations (3.3) can be formulated by defining a gravitational potential $U_g$ and a centrifugal potential $U_c$ from which the accelerations derive.

$$\begin{cases} U_g = \dfrac{\mu_1}{r_1} + \dfrac{\mu_2}{r_2} \Rightarrow \vec{a}_g = \dfrac{\partial U_g}{\partial \vec{r}} = -\mu_1 \dfrac{\vec{r}_1}{r_1^3} - \mu_2 \dfrac{\vec{r}_2}{r_2^3} \\[2ex] U_c = \dfrac{1}{2}(\vec{\Omega} \wedge \vec{r})^2 \Rightarrow \vec{a}_c = \dfrac{\partial U_c}{\partial \vec{r}} = \vec{\Omega} \wedge (\vec{\Omega} \wedge \vec{r}) \end{cases} \tag{3.9}$$

The total potential U is defined to within one additive constant as:

$$U = U_g + U_c = \frac{\mu_1}{r_1} + \frac{\mu_2}{r_2} + \frac{1}{2}(\vec{\Omega} \wedge \vec{r})^2 \tag{3.10}$$

The velocity and acceleration of M in the rotating frame RC are noted $\vec{v}$ and $\vec{a}$.
With the potential U, the motion equations (3.3) take the form

$$\vec{a} + 2\vec{\Omega} \wedge \vec{v} = \frac{\partial U}{\partial \vec{r}} \quad \text{with} \quad \vec{v} = \left.\frac{d\vec{r}}{dt}\right|_{RC}, \; \vec{a} = \left.\frac{d^2\vec{r}}{dt^2}\right|_{RC} = \left.\frac{d\vec{v}}{dt}\right|_{RC} \tag{3.11}$$

Taking the scalar product of both members with $\vec{v}$, the Coriolis term disappears.

$$\vec{a} \cdot \vec{v} = \frac{\partial U}{\partial \vec{r}} \cdot \vec{v} \quad \Rightarrow \quad \frac{d\vec{v}}{dt} \cdot \vec{v} = \frac{\partial U}{\partial \vec{r}} \cdot \frac{d\vec{r}}{dt} \quad \Rightarrow \quad \frac{1}{2}\frac{d(\vec{v} \cdot \vec{v})}{dt} = \frac{dU}{dt} \tag{3.12}$$

We obtain a prime integral of the motion called **Jacobi integral**.

$$2U - v^2 = K \quad \text{with} \quad K = Cte \tag{3.13}$$

The value of the Jacobi constant noted K depends on the initial conditions.

The accessible region, called **Hill region**, is bounded by the zero-velocity surface. It is defined by $2U \geq K$, which yields in dimensionless variables

$$2\frac{1-\mu}{r_1} + 2\frac{\mu}{r_2} + x^2 + y^2 \geq K \quad \text{with} \quad \begin{cases} r_1^2 = (x+\mu)^2 + y^2 + z^2 \\ r_2^2 = (x+\mu-1)^2 + y^2 + z^2 \end{cases} \tag{3.14}$$

Since the coordinates y and z appear squared in this inequation, the Hill region admits the planes of symmetry Oxy and Oxz, and the axis of symmetry Ox.

We can also observe that the region outside the cylinder of axis Oz and of equation $x^2 + y^2 \geq K$ is always accessible. To study the inner zone of this cylinder, we express $x^2 + y^2$ in terms of $r_1, r_2, z$.

$$\begin{cases} r_1^2 = (x+\mu)^2 + y^2 + z^2 \\ r_2^2 = (x+\mu-1)^2 + y^2 + z^2 \end{cases} \Rightarrow \; x^2 + y^2 = (1-\mu)r_1^2 + \mu r_2^2 - \mu(1-\mu) - z^2 \tag{3.15}$$

Substituting into (3.14), the Hill region is defined by the inequality

$$(1-\mu)\left(r_1^2 + \frac{2}{r_1}\right) + \mu\left(r_2^2 + \frac{2}{r_2}\right) \geq K + \mu(1-\mu) + z^2 \tag{3.16}$$

We observe that small values of $r_1, r_2$ are allowed, provided that z is small enough because (3.8) implies $z \leq r_1$ and $z \leq r_2$. This suggests examining the accessible region in the plane Oxy $(z = 0)$.

Figure 3.2 shows the shape of the potential energy $E_p = -U$ in the plane Oxy, whereas Figure 3.3 shows some level lines of the potential energy in that plane.

On these figures, the value of μ is set to 0.2 in order to accentuate the influence of the least massive body $C_2$. We observe two wells associated with the bodies $C_1$ and $C_2$ and the edge of the outer cylinder. There are three saddle points on the axis Ox marked $L_1$, $L_2$, and $L_3$ (in red), and two maxima symmetrical with respect to the axis Ox marked $L_4$ and $L_5$ (in green). These five equilibrium points $L_1$ to $L_5$ are called **libration** or **Lagrange points**. They are studied in Section 3.2.3.

The first member of (3.16) is a function of $(r_1, r_2)$ reaching a minimum value worth 3 for $r_1 = r_2 = 1$. This minimum potential (maximum potential energy) corresponds to the points $L_4$ and $L_5$ in Figure 3.2. If $K \leq K(L_4) = 3 - \mu(1-\mu)$, the inequality (3.16) is satisfied in all points and the whole Oxy plan is accessible. The value $K = K(L_4)$ in Figure 3.2 corresponds to a cross section by a horizontal plane passing through points $L_4$ and $L_5$. Increasing the value of K lowers this plane and reduces the accessible region defined by $2E_p \leq -K$ (Figure 3.3).

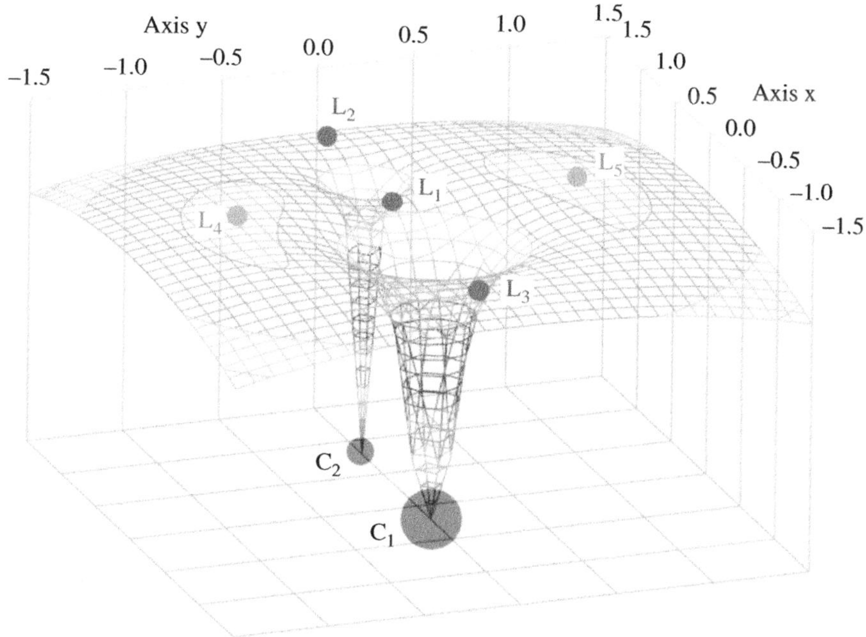


**Figure 3.2** Potential energy in the plane Oxy and Lagrange points ($\mu = 0.2$).

The reduction of the accessible region occurs in the following order:

- the orange lighter areas around $L_4$ and $L_5$ are eliminated first;
- the areas around the saddle points $L_3$, $L_2$, and $L_1$ are then eliminated in that order, which justifies the numbering of the points;
- for large values of K, only the inner central zones around the two bodies and the outer cylindrical zone remain.

Figure 3.3 shows level lines of the accessible region in the plane Oxy for different values of K. The positions of the points $L_1$ to $L_5$ and their associated K values are calculated in the following section.

### 3.2.3 Lagrange Points

The equilibrium points of the system (3.7) correspond to zero velocity and zero acceleration. The coordinates of these points satisfy the dimensionless system

$$\begin{cases} x - (1-\mu)\dfrac{x+\mu}{r_1^3} - \mu\dfrac{x+\mu-1}{r_2^3} = 0 \\ y - \left(\dfrac{1-\mu}{r_1^3} + \dfrac{\mu}{r_2^3}\right) y = 0 \\ \left(\dfrac{1-\mu}{r_1^3} + \dfrac{\mu}{r_2^3}\right) z = 0 \end{cases} \tag{3.17}$$

The last equation gives $z = 0$, because $0 < \mu \leq 0.5$. The equilibrium points are thus necessarily in the Oxy plan. The second equation gives two possible cases.

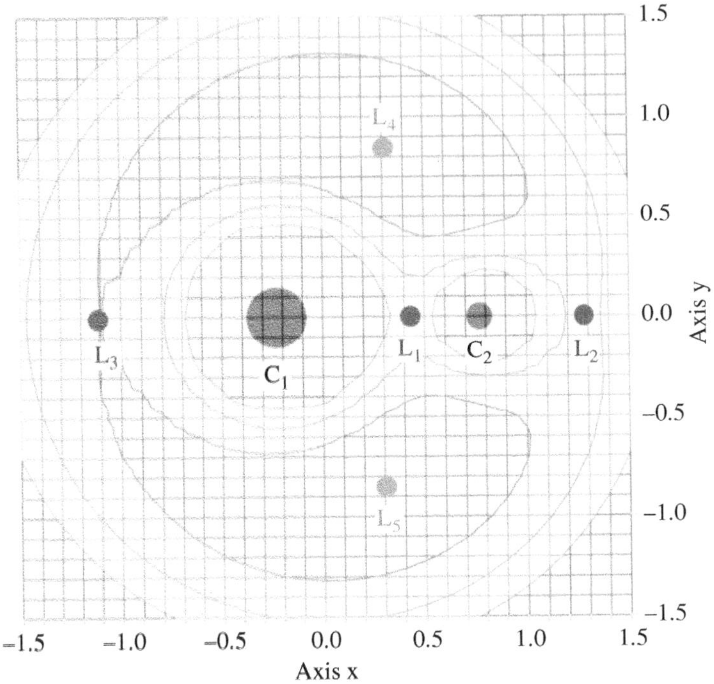


**Figure 3.3** Accessible region in the plane Oxy and Lagrange points ($\mu = 0.2$).

#### 3.2.3.1 First Case: $y \neq 0 \Rightarrow \frac{1-\mu}{r_1^3} + \frac{\mu}{r_2^3} = 1$

This relation eliminates x from the first equation and gives $r_1 = r_2 = 1$. We obtain two points symmetrical with respect to the axis Ox (in green in Figure 3.3), each forming an equilateral triangle with the bodies $C_1$ and $C_2$.

The coordinates of these two **triangular points** $L_4$ and $L_5$ are:

$$x = \frac{1}{2} - \mu, \quad y = \pm\frac{\sqrt{3}}{2}, \quad z = 0$$

The constant of Jacobi in these points is: $K(L_4) \underset{\text{noted}}{=} K_4 = 3 - \mu(1-\mu)$.

#### 3.2.3.2 Second Case: $y = 0$

Such equilibrium points are called **collinear** since they are located on the axis Ox.

Their abscissa satisfies the first equation (3.17) reproduced hereafter.

$$x - (1-\mu)\frac{x+\mu}{r_1^3} - \mu\frac{x+\mu-1}{r_2^3} = 0 \quad \text{with} \quad \begin{cases} r_1 = |x+\mu| \\ r_2 = |x+\mu-1| \end{cases} \tag{3.18}$$

Figure 3.4 shows the distances between the points O, $C_1$, and $C_2$, and the equilibrium point (noted $L_e$) on the axis Ox.

Equation (3.18) cannot be solved analytically. In order to find the number of solutions and their approximate value, we need to study the function in the first member of (3.18).

This function is not defined in $r_1 = 0$ and $r_2 = 0$.

On the other hand, its derivative, which is $1 + \frac{2(1-\mu)}{r_1^3} + \frac{2\mu}{r_2^3}$, is always positive, so that the function increases monotonically from $-\infty$ to $+\infty$ in each of the three intervals $x < -\mu$,

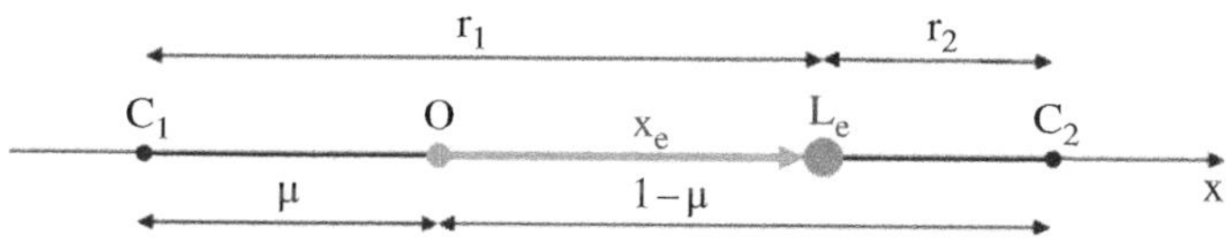


**Figure 3.4** Positions of the collinear equilibrium points.

$-\mu < x < 1-\mu$, and $1-\mu < x$. There is therefore a single root in each interval. This root is not algebraic, but it is possible to find an approximate value, as well as the associated value of K.

We describe here the solution method without the detailed calculation that is presented in [R27], [R31], [R35], [R41] and [R43].

- For the point $L_3$ to the left of $C_1$ ($x < -\mu$), we have the following relations: $r_2 = r_1 + 1$ and $x = -\mu - r_1$

  Equation (3.18) expressed in variable $r_1$ takes the form

$$(1 - r_1^3)(1 + r_1)^2 = Q_L r_1^3 (3 + 3r_1 + r_1^2) \quad \text{with } Q_L \underset{\text{def}}{=} \frac{\mu}{1-\mu} \tag{3.19}$$

  For $Q_L$ small, this equation yields $r_1 \approx 1$. This suggests an approximate solution of the form $r_1 = 1 + a_1 Q_L + a_2 Q_L^2 + \cdots$. After calculations, the result is

$$r_1 \approx 1 - \frac{7}{12} Q_L + \frac{7}{12} Q_L^2 - \frac{13223}{20736} Q_L^3 \quad \rightarrow \quad K_3 \approx 3 + \mu \tag{3.20}$$

- For the point $L_2$ to the right of $C_2$ ($1-\mu < x$), we have the following relations: $r_1 = 1 + r_2$ and $x = 1 - \mu + r_2$

  Equation (3.18) expressed in variable $r_2$ takes the form

$$r_2^3 \left(1 + r_2 + \frac{1}{3} r_2^2\right) = R_L^3 (1 + r_2)^2 (1 - r_2^3) \quad \text{with } R_L^3 \underset{\text{def}}{=} \frac{\mu}{3(1-\mu)} \tag{3.21}$$

  For $r_2$ small, this equation yields $r_2 \approx R_L$. This suggests an approximate solution of the form $r_2 = R_L (1 + a_1 R_L + a_2 R_L^2 + \cdots)$. After calculations, the result is

$$r_2 \approx R_L + \frac{1}{3} R_L^2 - \frac{1}{9} R_L^3 - \frac{31}{81} R_L^4 + \cdots \quad \rightarrow \quad K_2 \approx 3 + (9\mu)^{2/3} - 14\frac{\mu}{3} \tag{3.22}$$

- For the point $L_1$ between $C_1$ and $C_2$ ($-\mu < x < 1-\mu$), we have the following relations: $r_1 = 1 - r_2$ and $x = 1 - \mu - r_2$.

  Equation (3.18) expressed in variable $r_2$ takes the form

$$r_2^3 \left(1 - r_2 + \frac{1}{3} r_2^2\right) = R_L^3 (1 - r_2)^2 (1 - r_2^3) \quad \text{with } R_L^3 \underset{\text{def}}{=} \frac{\mu}{3(1-\mu)} \tag{3.23}$$

  For $r_2$ small, this equation yields $r_2 \approx R_L$. This suggests an approximate solution of the form $r_2 = R_L (1 + a_1 R_L + a_2 R_L^2 + \cdots)$. After calculations, the result is

$$r_2 \approx R_L - \frac{1}{3} R_L^2 - \frac{1}{9} R_L^3 - \frac{23}{81} R_L^4 + \cdots \quad \rightarrow \quad K_1 \approx 3 + (9\mu)^{2/3} - 10\frac{\mu}{3} \tag{3.24}$$

The three formulas (3.20), (3.22), and (3.24) give approximate values of the abscissae for the three points $L_3$, $L_2$, and $L_1$, respectively. For greater precision, equation (3.18) must be solved numerically.

When the mass ratio $\mu$ is small, the points $L_1$ and $L_2$ are close to the body $C_2$ and almost symmetrical at the distance $R_L$ defined by (3.21). This distance is comparable to the radius

$R_H$ of the Hill sphere and the radius $R_S$ of the sphere of influence, both based on perturbing accelerations in the vicinity of the body $C_2$.

These radii will be defined in Section 14.2.2 in the context of interplanetary missions. They depend on the mass ratio μ and on the distance d between the two bodies. Their respective values are assessed by the following formulas.

$$R_L = d\left(\frac{\mu}{3(1-\mu)}\right)^{1/3}, \quad R_H = d\left(\frac{\mu}{3}\right)^{1/3}, \quad R_S = d\left(\frac{\mu}{1-\mu}\right)^{2/5} \tag{3.25}$$

The positions of $L_1$ and $L_2$ for the Sun–Earth and Earth–Moon systems are given as follows. The respective spheres of influence have radii of 930 000 and 66 100 km.

| | Distance (km) | Period (days) | Parameter μ | Distance $C_2L_1$ (km) | Distance $C_2L_2$ (km) |
|---|---|---|---|---|---|
| Sun–Earth | 149 597 871 | 365.2564 | $3.0404 \cdot 10^{-6}$ | 1 497 621 | 1 507 683 |
| Earth–Moon | 384 399 | 27.3217 | $1.2151 \cdot 10^{-2}$ | 58 019 | 64 515 |

## 3.3 Periodic Orbits

The equilibrium points, in particular $L_1$ and $L_2$, have properties favorable to space missions. This section studies the motion in the vicinity of these points and presents the solution method to determine periodic orbits.

### 3.3.1 Linearized Solution

The first step in studying the dynamics in the vicinity of equilibrium points is to linearize the equations of motion (3.7). These equations are of the form

$$\begin{cases} \ddot{x} - 2\dot{y} = \nabla_x U \underset{\text{noted}}{=} U_x \\ \ddot{y} + 2\dot{x} = \nabla_y U \underset{\text{noted}}{=} U_y \\ \ddot{z} = \nabla_z U \underset{\text{noted}}{=} U_z \end{cases} \tag{3.26}$$

where the dimensionless potential U is defined by

$$U = U_g + U_c = \frac{1-\mu}{r_1} + \frac{\mu}{r_2} + \frac{x^2+y^2}{2} \quad \text{with} \quad \begin{cases} r_1^2 = (x+\mu)^2 + y^2 + z^2 \\ r_2^2 = (x+\mu-1)^2 + y^2 + z^2 \end{cases} \tag{3.27}$$

The gradient of the potential U at an equilibrium point $L_e(x_e, y_e, z_e = 0)$ is zero: $U_{xe} = U_{ye} = U_{ze} = 0$. By noting $(\delta x, \delta y, \delta z)$ the position deviations from $L_e$ and expanding U to order 2, we obtain the linearized equations of motion in the vicinity of $L_e$, also called **variational equations** of the system (3.26).

$$\begin{cases} \delta\ddot{x} - 2\delta\dot{y} = U_{xx}\delta x + U_{xy}\delta y + U_{xz}\delta z \\ \delta\ddot{y} + 2\delta\dot{x} = U_{yx}\delta x + U_{yy}\delta y + U_{yz}\delta z \\ \delta\ddot{z} = U_{zx}\delta x + U_{zy}\delta y + U_{zz}\delta z \end{cases} \quad \text{with} \quad \begin{cases} \delta x = x - x_e \\ \delta y = y - y_e \\ \delta z = z - z_e \end{cases} \tag{3.28}$$

The second derivatives of U are given after calculations by

$$\begin{cases} U_{xx} = 1 - c_2 + 3\left[(1-\mu)\frac{(x_e+\mu)^2}{r_1^5} + \mu\frac{(x_e+\mu-1)^2}{r_2^5}\right], & U_{xz} = -z_e\frac{\partial c_2}{\partial x} \\ U_{xy} = \qquad 3\left[(1-\mu)\frac{x_e+\mu}{r_1^5} + \mu\frac{x_e+\mu-1}{r_2^5}\right] y_e, & U_{yz} = -z_e\frac{\partial c_2}{\partial y} \\ U_{yy} = 1 - c_2 + 3\left[\frac{1-\mu}{r_1^5} + \frac{\mu}{r_2^5}\right] y_e^2, & U_{zz} = -z_e\frac{\partial c_2}{\partial z} - c_2 \end{cases} \quad (3.29)$$

where $c_2 = \frac{1-\mu}{r_1^3} + \frac{\mu}{r_2^3}$ (this notation will be retrieved in Section 3.3.2).

In the point $L_e$ ($z_e = 0$), the derivatives $U_{xz}$, $U_{yz}$ are zero. The third equation (3.28) in z is thus decoupled and it yields a purely oscillatory motion with the pulsation $\omega_z = \sqrt{c_2}$.

$$\delta\ddot{z} = -c_2\delta z \quad \Rightarrow \quad \delta z = A_z\cos(\omega_z t + \varphi_z) \quad (3.30)$$

The first two equations (3.28) in (x, y) take the form of a first-order linear system.

$$\dot{X} = FX \quad \text{with } X = \begin{pmatrix} \delta x \\ \delta y \\ \delta\dot{x} \\ \delta\dot{y} \end{pmatrix}, \quad F = \begin{pmatrix} 0 & 0 & 1 & 0 \\ 0 & 0 & 0 & 1 \\ U_{xx} & U_{xy} & 0 & 2 \\ U_{xy} & U_{yy} & -2 & 0 \end{pmatrix} \quad (3.31)$$

The general solution of this linear system is expressed as

$$\delta x = \sum_{k=1}^{4}\alpha_k e^{\lambda_k t}, \quad \delta y = \sum_{k=1}^{4}\beta_k e^{\lambda_k t} \quad (3.32)$$

where $(\lambda_k)_{k=1,4}$ are the eigenvalues of F.

These eigenvalues are the solutions of the characteristic equation

$$\lambda^4 + (4 - U_{xx} - U_{yy})\lambda^2 + U_{xx}U_{yy} - U_{xy}^2 = 0 \quad (3.33)$$

The coefficients $\alpha_k$, $\beta_k$ are determined by writing (3.32) at the initial time $t = 0$ and solving for given initial conditions.

$$\delta x_0 = \sum_{k=1}^{4}\alpha_k, \quad \delta\dot{x}_0 = \sum_{k=1}^{4}\alpha_k\lambda_k, \quad \delta y_0 = \sum_{k=1}^{4}\beta_k, \quad \delta\dot{y}_0 = \sum_{k=1}^{4}\beta_k\lambda_k \quad (3.34)$$

These coefficients $\alpha_k$, $\beta_k$ are also linked by the relations obtained by substituting the solution (3.32) into (3.28). After calculations, the result is

$$\left(\lambda_k^2 - U_{xx}\right)\alpha_k = (2\lambda_k + U_{xy})\beta_k, \quad k = 1 \text{ to } 4 \quad (3.35)$$

Equations (3.32)–(3.35) determine the linearized motion in the vicinity of the equilibrium point, from given initial conditions. The solution will be stable if all eigenvalues are negative or zero.

In order to study the stability further, we consider successively the triangular points and the collinear points, which do not have the same characteristic equation, and therefore, do not have the same properties.

#### 3.3.1.1 Triangular Points

The coordinates of the points $L_4$ and $L_5$ are: $x_e = \frac{1}{2} - \mu$, $y_e = \pm\frac{\sqrt{3}}{2}$, $z_e = 0$.

In these points, the second derivatives of the potential (3.29) take the values

$$U_{xx} = \frac{3}{4}, \quad U_{xy} = \pm\frac{3\sqrt{3}}{4}(1 - 2\mu), \quad U_{yy} = \frac{9}{4}, \quad U_{zz} = -1 \text{ and } c_2 = 1 \tag{3.36}$$

The motion in z (3.30) has the same pulsation as the two-body system: $\omega_z = 1$.

In order to analyze the motion in the plane Oxy, we have to solve its characteristic equation (3.33), which has the following roots

$$\lambda^2 = \frac{-1 \pm \sqrt{1 - 27\mu(1 - \mu)}}{2} \tag{3.37}$$

For the motion in the plane Oxy to be stable, all eigenvalues must be negative or zero, which yields the condition

$$1 - 27\mu(1 - \mu) \geq 0 \quad \Rightarrow \quad \mu \leq \frac{1}{2} - \frac{1}{2}\sqrt{\frac{23}{27}} \approx 0.0385 \tag{3.38}$$

This condition is verified in most practical cases, especially for the Sun–Earth and the Earth–Moon systems.

The eigenvalues (3.37) are then pure imaginary $\lambda = \pm i\omega_p$ and the motion defined by (3.32) has two harmonic components of pulsations $\omega_{p1}$, $\omega_{p2}$.

For $\mu$ small, these two pulsations in the plane are close to

$$\omega_{p1} \approx \sqrt{\frac{27\mu}{4}}, \quad \omega_{p2} \approx \sqrt{1 - \frac{27\mu}{4}} \approx 1 \tag{3.39}$$

The first pulsation is small and gives a long-period motion. This motion occurs along an ellipse whose axes are in a ratio $\sqrt{3\mu}$ and whose major axis makes with Ox the angle $\theta$ given by $\tan\theta = \sqrt{3}(1 - 2\mu)$.

The second pulsation is larger and close to that of the two-body system. It gives a fast motion called libration along an ellipse whose axes are in a ratio 1/2.

Both ellipses (slow and fast motion) are followed in the retrograde direction with respect to the two-body rotating frame. The superposition of these two motions yields an epicycloidal trajectory around the equilibrium point. A detailed analysis of these solutions is presented in [R27].

Figure 3.5 shows a trajectory in the vicinity of a triangular point, starting from the initial normalized conditions: $\delta x_0 = \delta y_0 = 10^{-5}$, $\delta\dot{x}_0 = \delta\dot{y}_0 = 0$.

The mass ratio is set to $\mu = 0.01$. The trajectory is plotted over two periods of the epicenter, with the first period in solid line and the second period in dotted line. This example is retrieved from [R27].

The epicycloid solution obtained by linearization is valid only in the vicinity of $L_4$ or $L_5$. Taking into account the full dynamics (3.26), it is possible to construct numerically orbits of larger amplitude around $L_4$ or $L_5$. The semi-major axis of the epicenter grows and it turns into a circular curve around the body $C_1$. In the same time, "tadpole" orbits appear around $L_4$ and around $L_5$, then "horseshoe" orbits going from the $L_4$ region to the $L_5$ region. Several hundred asteroids known as Trojans follow this kind of orbit in the Sun–Jupiter system.

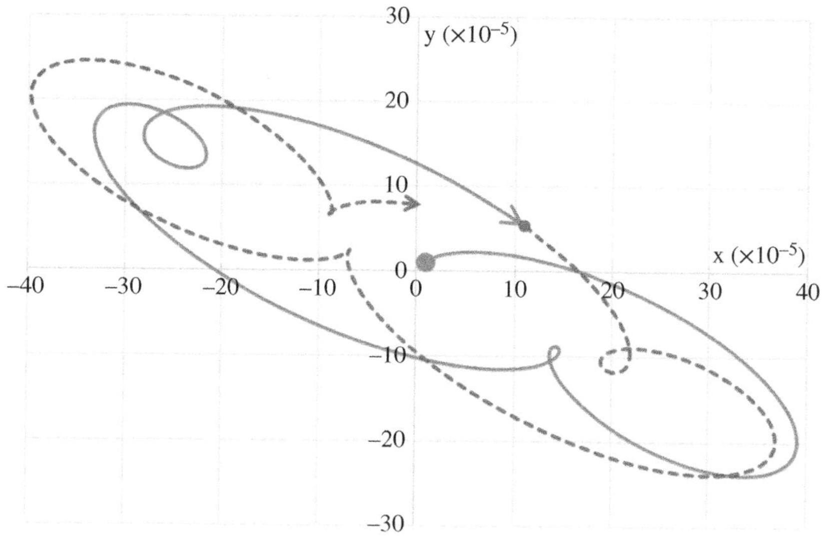


**Figure 3.5** Trajectory in the vicinity of a triangular point $L_4$ or $L_5$ ($\mu = 0.01$).

### 3.3.1.2 Collinear Points

The coordinates of the points $L_1$, $L_2$, and $L_3$ are of the form $(x_e, 0, 0)$. Their distances to the bodies $C_1$ and $C_2$ are $r_1 = |x_e + \mu|$, $r_2 = |x_e + \mu - 1|$, which gives for the second derivatives in (3.29)

$$\begin{cases} U_{xx} = 1 + 2c_2, & U_{xy} = 0 \\ U_{yy} = 1 - c_2, & U_{zz} = -c_2 \end{cases} \quad \text{with } c_2 = \frac{1-\mu}{r_1^3} + \frac{\mu}{r_2^3} \tag{3.40}$$

Equation (3.18) allows us to express $c_2$ as a function of $x_e$ and $r_2$.

$$\frac{1-\mu}{r_1^3} = \frac{x_e}{x_e+\mu} - \frac{\mu}{r_2^3}\left(1 - \frac{1}{x_e+\mu}\right) \quad \rightarrow \quad c_2 = \frac{\mu + x_e r_2^3}{(\mu + x_e) r_2^3} \tag{3.41}$$

By substituting $x_e$ with $r_2$ (see Figure 3.4), we get the expression of $c_2$, depending on the point considered $L_1$, $L_2$, or $L_3$.

- In $L_1$ and $L_3$ : $x_e = 1 - \mu - r_2 \quad \rightarrow \quad c_2 = 1 + \mu \frac{1+r_2+r_2^2}{r_2^3}$
- In $L_2$ : $x_e = 1 - \mu + r_2 \quad \rightarrow \quad c_2 = 1 + \mu \frac{(1-r_2)(1+r_2+r_2^2)}{(1+r_2)r_2^3}$

For the point $L_2$, it can be seen from (3.21) that $r_2 < 1$ (else the second member would be negative). We deduce from the aforementioned formulas that $c_2 > 1$ in each of the points $L_1$, $L_2$, and $L_3$. The pulsation of the motion in z (3.30) is therefore higher than that of the two-body system: $\omega_z = \sqrt{c_2} > 1$.

In order to analyze the motion in the plane Oxy, we have to solve its characteristic equation (3.33), which has the following roots:

$$\lambda^2 = \frac{c_2 - 2 \pm \sqrt{c_2(9c_2 - 8)}}{2} \tag{3.42}$$

The root product is equal to $1 + c_2 - 2c_2^2$, which is negative for $c_2 > 1$. We get therefore two real roots $\lambda = \pm\sigma$ and two pure imaginary roots $\lambda = \pm i\omega_p$.

The motion in the plane Oxy is the superposition of a stable component associated with the negative real eigenvalue $-\sigma$, an unstable component associated with the positive real eigenvalue $+\sigma$, and an oscillatory component at the pulsation $\omega_p$, which differs from the pulsation $\omega_z$ along z.

We can now express the general solution (3.32) of the linearized motion as:

$$\begin{cases} \delta x = A_p \cos(\omega_p t + \varphi_p) + A_s e^{-\sigma t} + A_i e^{\sigma t} \\ \delta y = -k_p A_p \sin(\omega_p t + \varphi_p) + k_\sigma (A_s e^{-\sigma t} - A_i e^{\sigma t}) \\ \delta z = A_z \cos(\omega_z t + \varphi_z) \end{cases} \tag{3.43}$$

In these equations, $A_p$, $A_s$, and $A_i$ denote the amplitudes of the oscillatory, stable, and unstable components in x, respectively. The amplitudes in y are those in x multiplied by coefficients $k_p$, $k_\sigma$, which depend on the eigenvalues.

These coefficients $k_p$, $k_\sigma$ are assessed from (3.35), which determines $\alpha_k$, $\beta_k$ introduced in the general solution (3.32). This yields

$$k_p = \frac{1 + 2c_2 + \omega_p^2}{2\omega_p}, \quad k_\sigma = \frac{1 + 2c_2 - \sigma^2}{2\sigma} \tag{3.44}$$

For $\mu$ small, the approximate values of $r_2$ and $c_2$ in the points $L_1$ and $L_2$ are given by formulas (3.22), (3.24), and (3.41).

$$r_2^3 \approx \frac{\mu}{3(1-\mu)} \quad \rightarrow \quad c_2 \approx 1 + \frac{\mu}{r_2^3} \approx 4 - 3\mu \tag{3.45}$$

The pulsations $\omega_z = \sqrt{c_2}$ and $\omega_p$ are assessed by (3.42) and the coefficient $k_p$ is assessed by (3.44).

$$\omega_z \approx 2, \quad \omega_p \approx \sqrt{2\sqrt{7} - 1} \approx 2.7016, \quad k_p \approx 3.208 \tag{3.46}$$

The pulsation in z is about twice that of the two-body system. The oscillatory motion in the plane Oxy is an ellipse whose axes are in a ratio of about 3.2. This ellipse is followed clockwise with a pulsation greater than that in z.

Let us now turn our attention to periodic solutions, and search for appropriate initial conditions. The initial conditions in x and y can be expressed from (3.43)

$$\begin{cases} \delta x_0 = A_p \cos\varphi_p + A_s + A_i \\ \delta y_0 = -k_p A_p \sin\varphi_p + k_\sigma (A_s - A_i) \\ \delta \dot{x}_0 = -\omega_p A_p \sin\varphi_p - \sigma (A_s - A_i) \\ \delta \dot{y}_0 = -\omega_p k_p A_p \cos\varphi_p - \sigma k_\sigma (A_s + A_i) \end{cases} \tag{3.47}$$

In order to achieve a purely oscillatory motion, the amplitudes $A_s$ and $A_i$ must vanish. This determines the initial velocity as a function of the initial position.

$$A_s = A_i = 0 \quad \Rightarrow \quad \delta \dot{x}_0 = \frac{\omega_p}{k_p} \delta y_0, \quad \delta \dot{y}_0 = -\omega_p k_p \delta x_0 \tag{3.48}$$

The difference of the pulsations in plane $\omega_p$ and out of plane $\omega_z$ produce an unclosed trajectory called **Lissajous orbit**. The stable and unstable components in the plane Oxy

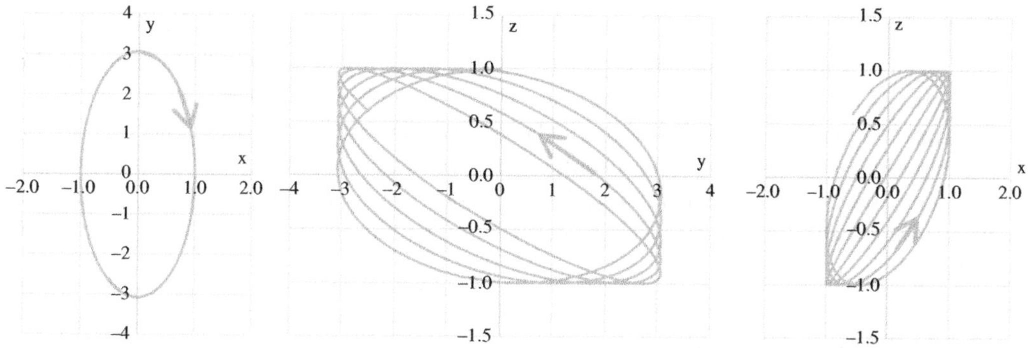


**Figure 3.6** Lissajous orbit in the vicinity of point $L_2$ ($\mu = 0.01$).

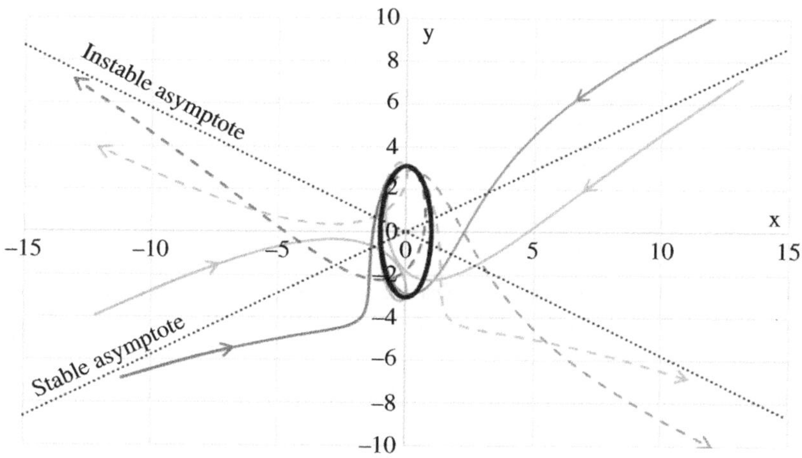


**Figure 3.7** Trajectories in the vicinity of point $L_2$ ($\mu = 0.01$).

define asymptotes of equations $y = \pm k_\sigma x$. For $\mu$ small, the asymptote slope assessed from (3.42) and (3.44) is about $k_\sigma \approx 4.498$.

Figure 3.6 shows the (x,y), (y,z), and (x,z) projections of a Lissajous orbit in the vicinity of the point $L_2$ for $\mu = 0.01$. The amplitudes in x and z are set to 1. The arrows indicate the motion direction in each plane.

Figure 3.7 shows the same Lissajous orbit (black thick line) with stable and unstable components of amplitudes $A_s = A_i = 10^{-3}$. The four trajectories are generated by changing the phase $\varphi_p$ and the sign of $A_s$. The converging (resp. diverging) branches are in solid (resp. dotted) lines. These branches are associated respectively with the stable and unstable asymptotes in the plane Oxy.

The linear model makes it possible to understand the dynamics trends in the vicinity of $L_1$ or $L_2$, but it is not sufficient to construct a precise orbit. Indeed, even if the initial conditions (3.48) are met, the nonlinear effects generate perturbations and divergence from the Lissajous orbit. These effects must be taken into account in constructing general solutions of the motion.

### 3.3.2 Periodic Solution

The linearized model in the vicinity of $L_1$ and $L_2$ gives nonperiodic solutions due to the difference between the in-plane $\omega_p$ and out-of-plane $\omega_z$ pulsations (3.46).

However, nonplanar periodic orbits can be achieved owing to nonlinear effects.

We present successively Richardson's method based on analytical developments and Poincaré's method based on numerical simulations. These methods are complementary, since Richardson's analytical solution provides an initial guess for Poincaré's numerical method.

#### 3.3.2.1 Richardson's Method

The solution proposed by Richardson [R31] is based on the usual series expansion with Legendre polynomials [R1].

$$\frac{1}{\|\vec{r}-\vec{d}\|} = \frac{1}{d}\sum_{n=0}^{\infty}\left(\frac{r}{d}\right)^n P_n\left(\frac{\vec{r}\cdot\vec{d}}{rd}\right) \tag{3.49}$$

where $\vec{r}, \vec{d}$ are any two vectors and $P_n$ is the Legendre polynomial of degree n.

This formula is applied to the gravitational potential $U_g$ (3.27) associated with the two bodies $C_1$ and $C_2$. This potential is expressed as:

$$U_g = \frac{1-\mu}{r_1} + \frac{\mu}{r_2} = \frac{1-\mu}{\|\vec{r}-\vec{d}_1\|} + \frac{\mu}{\|\vec{r}-\vec{d}_2\|} \tag{3.50}$$

The calculations are carried out in **Richardson's frame** noted $RL_e$. The origin of this frame is the equilibrium point studied $L_e$ ($L_1$ or $L_2$). The unit of distance is the distance $\gamma_e$ between the body $C_2$ and the point $L_e$.

The coordinates in $RL_e$ are noted $(\tilde{x}, \tilde{y}, \tilde{z})$. They are deduced from the coordinates in RC by a translation along x and a homothetic ratio $1/\gamma_e$.

$$\tilde{x} = \frac{x-(1-\mu+\varepsilon_e\gamma_e)}{\gamma_e},\ \tilde{y} = \frac{y}{\gamma_e},\ \tilde{z} = \frac{z}{\gamma_e} \quad \text{with} \quad \begin{cases} \varepsilon_e = -1 \ \text{for } RL_1 \\ \varepsilon_e = +1 \ \text{for } RL_2 \end{cases} \tag{3.51}$$

The coordinates $(\tilde{x}, \tilde{y}, \tilde{z})$ follow the equations deduced from (3.26) and (3.27).

$$\begin{cases} \ddot{\tilde{x}} - 2\dot{\tilde{y}} = \tilde{x} + \gamma_e^{-3}\nabla_{\tilde{x}}\tilde{U}_g \\ \ddot{\tilde{y}} + 2\dot{\tilde{x}} = \tilde{y} + \gamma_e^{-3}\nabla_{\tilde{y}}\tilde{U}_g \\ \ddot{\tilde{z}} = \gamma_e^{-3}\nabla_{\tilde{z}}\tilde{U}_g \end{cases} \quad \text{with} \quad \tilde{U}_g = \frac{1-\mu}{\tilde{r}_1} + \frac{\mu}{\tilde{r}_2} = \gamma_e U_g \tag{3.52}$$

The expansion (3.49) is applied to the potential $\tilde{U}_g$ with the coordinates of the attracting bodies $C_1$ and $C_2$ in $RL_e$ given by

$$\begin{cases} C_1(d_1 = -\mu, 0, 0) \ \rightarrow \ (\tilde{d}_1 = -(1+\varepsilon_e\gamma_e)/\gamma_e, 0, 0) \\ C_2(d_2 = 1-\mu, 0, 0) \ \rightarrow \ (\tilde{d}_2 = -\varepsilon_e, 0, 0) \end{cases} \tag{3.53}$$

We obtain after grouping terms of the same power

$$\tilde{U}_g = \sum_{n=0}^{\infty} c_n \tilde{r}^n P_n\left(\frac{\tilde{x}}{\tilde{r}}\right) \quad \text{with} \quad c_n = \frac{1}{\gamma_e^3}\left((-\varepsilon_e)^n\mu + \frac{(-1)^n\gamma_e^{n+1}}{(1+\varepsilon_e\gamma_e)^{n+1}}(1-\mu)\right) \tag{3.54}$$

By separating the linear terms ($n \le 2$), the equations (3.52) are written as:

$$\begin{cases} \ddot{\tilde{x}} - 2\dot{\tilde{y}} - (1 + 2c_2)\tilde{x} = \nabla_{\tilde{x}} S \\ \ddot{\tilde{y}} + 2\dot{\tilde{x}} - (1 - c_2)\tilde{y} = \nabla_{\tilde{y}} S \\ \ddot{\tilde{z}} + c_2 \tilde{z} = \nabla_{\tilde{z}} S \end{cases} \quad \text{with} \quad S(\tilde{x}, \tilde{y}, \tilde{z}) = \sum_{n=3}^{\infty} c_n \tilde{r}^n P_n \left(\frac{\tilde{x}}{\tilde{r}}\right) \tag{3.55}$$

The first coefficient $c_2$ of the series expansion is the one introduced in (3.29). Its value is $c_2 = \frac{1-\mu}{r_1^3} + \frac{\mu}{r_2^3} = \frac{1-\mu}{(1+\varepsilon_e \gamma_e)^3} + \frac{\mu}{\gamma_e^3}$.

The Lindstedt–Poincaré method [R31] consists in looking for a series solution.

$$\begin{cases} \tilde{x} = \sum_{n=1}^{\infty} \tilde{x}_n \ \text{ with } \ \tilde{x}_n = \sum_{k=0}^{n} A_p^k A_z^{n-k} \left( \sum_{i+j \le n} \alpha_{ijn} \cos(n\omega t + \varphi_p) \right) \\ \tilde{y} = \sum_{n=1}^{\infty} \tilde{y}_n \ \text{ with } \ \tilde{y}_n = \sum_{k=0}^{n} A_p^k A_z^{n-k} \left( \sum_{i+j \le n} \beta_{ijn} \sin(n\omega t + \varphi_p) \right) \\ \tilde{z} = \sum_{n=1}^{\infty} \tilde{z}_n \ \text{ with } \ \tilde{z}_n = \sum_{k=0}^{n} A_p^k A_z^{n-k} \left( \sum_{i+j \le n} \gamma_{ijn} \cos(n\omega t + \varphi_z) \right) \\ \omega = \sum_{n=0}^{\infty} \omega_n \ \text{ with } \ \omega_n = \sum_{k=0}^{n} A_p^k A_z^{n-k} \nu_{kn} \end{cases} \tag{3.56}$$

The terms $\tilde{x}_n, \tilde{y}_n, \tilde{z}_n$ are of order n with respect to amplitudes $A_p$, $A_z$. They are periodic of pulsation $n\omega$, where the pulsation $\omega$ is also expressed as a series of the amplitudes. The amplitudes $A_p$, $A_z$ and the phases $\varphi_p$, $\varphi_z$ are given. The unknowns in order n are the coefficients $\alpha_{ijn}, \beta_{ijn}, \gamma_{ijn}, \nu_{kn}$.

The solution method consists in injecting the solution (3.56) of order n into the equations (3.55) also limited to order n, and then identifying the coefficients of the terms $A_p^k A_z^{n-k}$. This resolution is performed iteratively by increasing the order starting from $n = 0$.

The tedious calculations have been developed by Richardson up to order 3 [R31]. They require rewriting equation (3.55) in z under the form

$$\ddot{\tilde{z}} + \omega_{p0}^2 \tilde{z} = \left(\omega_{p0}^2 - \omega_{z0}^2\right) \tilde{z} + \frac{\partial}{\partial \tilde{z}} \sum_{n=3}^{\infty} c_n \tilde{r}^n P_n \left(\frac{\tilde{x}}{\tilde{r}}\right) \tag{3.57}$$

where $\omega_{p0}, \omega_{z0}$ are the pulsations of the linear solution (3.43) recalled here

$$\omega_{p0}^2 = 1 + \frac{\sqrt{c_2(9c_2 - 8)} - c_2}{2}, \quad \omega_{z0}^2 = c_2 \tag{3.58}$$

The formulation (3.57) makes it possible to impose the pulsation of order 0 in z to $\omega_0 = \omega_{p0}$ by considering that the term $\left(\omega_{p0}^2 - \omega_{z0}^2\right)\tilde{z}$ is of order two. The orbit is thus made periodic, by synchronizing the in-plane and out-of-plane motions. On the other hand, the third-order identification imposes constraints on the amplitudes and phases of the form (see [R31] for the detailed calculations)

$$\begin{cases} l_1 A_p^2 + l_2 A_z^2 = \omega_{p0}^2 - \omega_{z0}^2 \\ 2(\varphi_z - \varphi_p) = m\pi, \quad m = 1 \text{ or } 3 \end{cases} \tag{3.59}$$

Due to these third-order constraints, nonplanar periodic orbits can exist only when the amplitude $A_p$ is at least 200 000 km for the Sun–Earth system and 6500 km for the Earth–Moon system. The Richardson third-order solution was used to define the orbit of the ISEE3 solar wind observation mission (1978) in the vicinity of the point $L_1$.

This analytical solution by series expansion gives approximate initial conditions of a periodic orbit. To construct a precise solution, the cross-section method of Poincaré based on the complete equations of motion is used.

#### 3.3.2.2 Poincaré's Method

This method starts from the observation that for any solution $(x(t), y(t), z(t))$ to equations (3.7), there is a "mirror" solution $(x(-t), -y(-t), z(-t))$ symmetrical with respect to the plane Oxz. To obtain a closed orbit, it is sufficient to find initial conditions starting from the plane Oxz and intersecting it with an orthogonal velocity at each end. This arc then represents a half-orbit and connects in position and velocity to its mirror solution.

The initial state vector is of the form $X_0 = (x_0, y_0, z_0, v_{x0}, v_{y0}, v_{z0})$:

- the components $y_0$, $v_{x0}$, $v_{z0}$ are zero since the point is in the plane Oxy with a velocity along y;
- the components $x_0$, $z_0$, $v_{y0}$ are to be determined.

The equations of motion integrated from $X_0$ at time $t_0$ until a given time $t_f$ give the final state $X_f = (x_f, y_f, z_f, v_{xf}, v_{yf}, v_{zf})$. The components $y_f$, $v_{xf}$, $v_{zf}$ must then be zero to achieve a closed orbit of period $T = 2t_f$.

This process is illustrated in Figure 3.8. The trajectory on the right and the one on the left both start orthogonally from the plane Oxz, but only the one on the left again crosses the plane orthogonally and corresponds to a closed orbit.

The plane Oxz is called a **section of Poincaré**. This technique reduces the search space to three state vector components instead of six. It is also used for the construction of transfers presented in Section 3.4.3.

The nonlinear system to solve has four unknowns, which are $x_0$, $z_0$, $v_{y0}$, and $t_f$, but only three equations, which are $y_f = v_{xf} = v_{zf} = 0$. It is thus possible to construct different shapes of orbits by fixing one of the unknowns.

In particular, the usual choices are either to fix the period $T = 2t_f$, or to fix the coordinate $z_0$. This coordinate is the maximal out-of-plane distance from the two-body reference plane, since the component $v_{z0}$ is zero. The value $z_0$ defines thus the orbit amplitude (called excursion), which can be either southward if $z_0 < 0$ or northward if $z_0 > 0$.

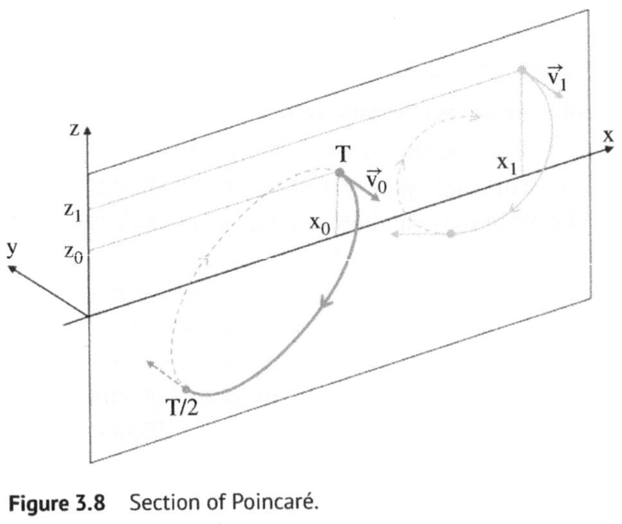


**Figure 3.8** Section of Poincaré.

#### 3.3.2.3 Shooting Method

The nonlinear system is called a shooting problem because it involves the integration of differential equations to pass from the initial state (unknowns of the problem) to the final state (constraints to satisfy). It is generally solved by Newton's method, called shooting method in this context.

Richardson's solution can be used as an initial guess. Convergence issues may occur due to the starting point (if Richardson's solution is not accurate enough) or to the system sensitivity with respect to initial conditions.

- The initial guess issue can be overcome by homotopy methods that solve a series of neighboring problems to pass progressively from a previously known orbit to the desired one.
- The sensitivity issue can be alleviated using variable step integration methods with embedded error control, such as Runge–Kutta fifth- or eighth-order methods.

More details on these numerical methods can be found in [R18] and [R23].

### 3.3.3 Halo Orbits

Poincaré method is used to generate different kinds of orbits in the vicinity of the points $L_1$ and $L_2$, namely Lyapunov, Lissajous, or Halo orbits.

**Lyapunov orbits** are closed planar orbits in the plane Oxy. The addition of an oscillatory component in z leads to **Lissajous orbits**. At low amplitudes, these unclosed orbits are almost symmetrical with respect to the Oxy and Oxz planes. At high amplitudes, they lose symmetry with respect to Oxy, but may become closed **Halo orbits** still symmetrical with respect to the plane Oxz. Halo orbits are the most interesting for space missions because of their periodicity and geometric properties.

The Halo orbits are of the Northern or Southern type depending on which side of the plane Oxy their extreme point is located. Halo orbits cut the plane Oxy into two points symmetrical with respect to the axis Ox. There are also vertical Lyapunov orbits having an eight-shape, symmetrical with respect to the plane Oxy and cutting this plane into a single point. Detailed descriptions of these families of orbits can be found in [R18] and [R23].

#### 3.3.3.1 Earth–Moon System

The Halo orbits of the Earth–Moon system have an amplitude in y that prevents them from being occulted by the Moon when observed from the Earth. They were so named by Farquhar [R17] because of their appearance from the Earth forming a concentric halo around the Moon.

Figure 3.9 shows the family of Southern Halo orbits around the points $L_1$ and $L_2$ of the Earth–Moon system. The Northern Halo orbits (not plotted on this figure) are symmetrical to the Southern orbits with respect to the plane Oxy. The Halo orbits exist from the vicinity of the points $L_1$ and $L_2$ (60 000 km from the Moon) to close to the Moon (placed at the origin of the frame). It can be observed that:

- the orbits around $L_1$ and $L_2$ have not the same shape, since the Earth located on the side of $L_1$ does not have the same influence;
- the orbits close to the Lagrange point have the smallest size (diameter of about 40 000 km). For $L_2$, they are nearly lying in the orbital plane of the Moon (they transform in the limit into Lyapunov orbits);
- the orbits close to the Moon have the largest size (diameter up to 70 000 km) and they become nearly planar in the plane Oyz orthogonal to the Earth–Moon direction.

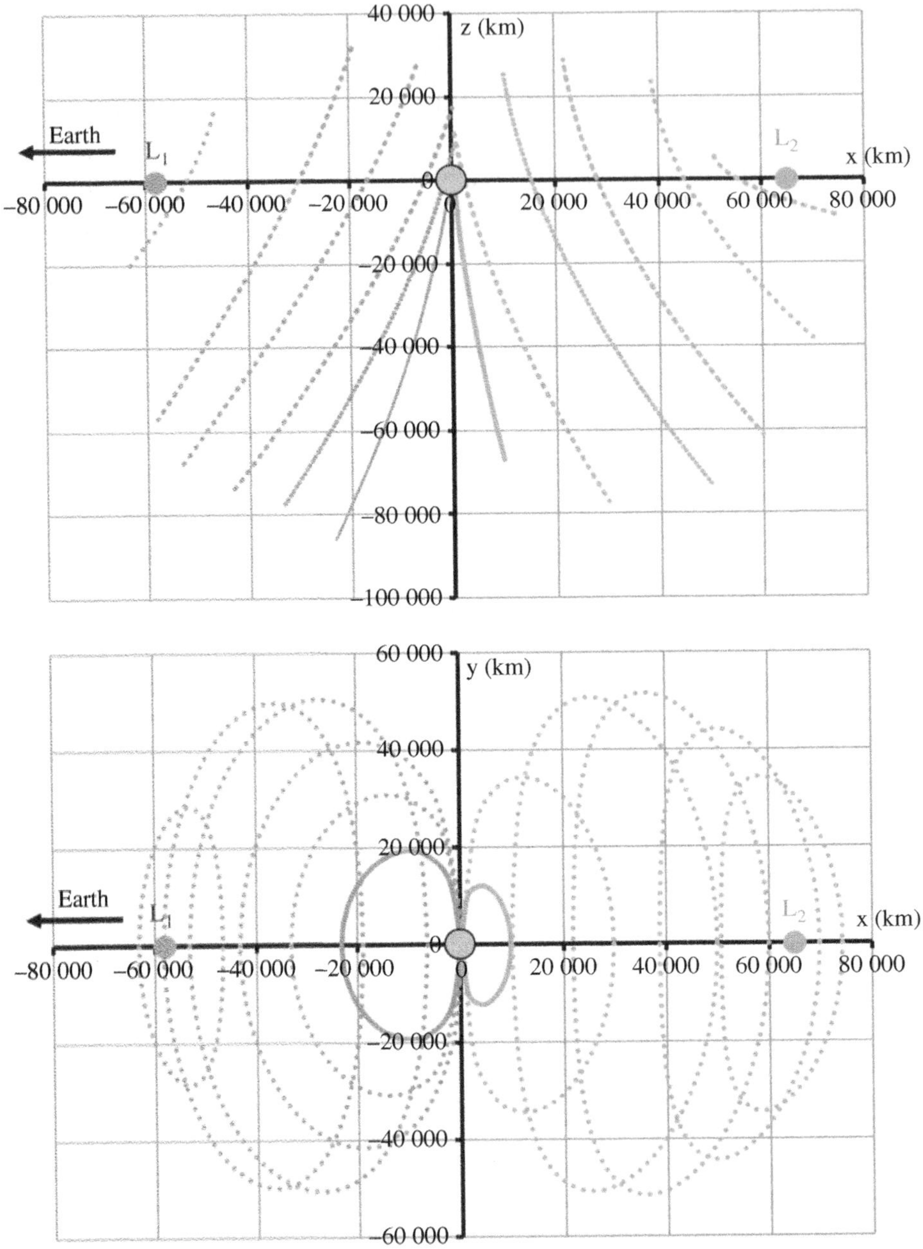


**Figure 3.9** Halo orbits around $L_1$ and $L_2$ for the Earth–Moon system.

Figure 3.10 shows the relation between the orbit period and the minimum distance from the Moon. The period ranges from one to two weeks, which is half the period of the lunar revolution, as estimated by (3.46).

The Moon closest orbits on $L_1$ side and on $L_2$ side are shown in Figure 3.11. These very elongated orbits, with a ratio of four between their width in y and their height in z, are called

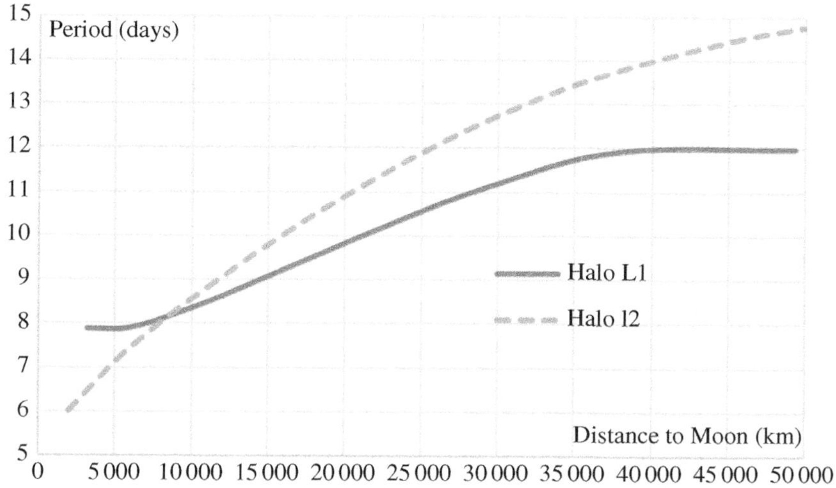


**Figure 3.10** Halo orbit period as a function of the distance to the Moon.

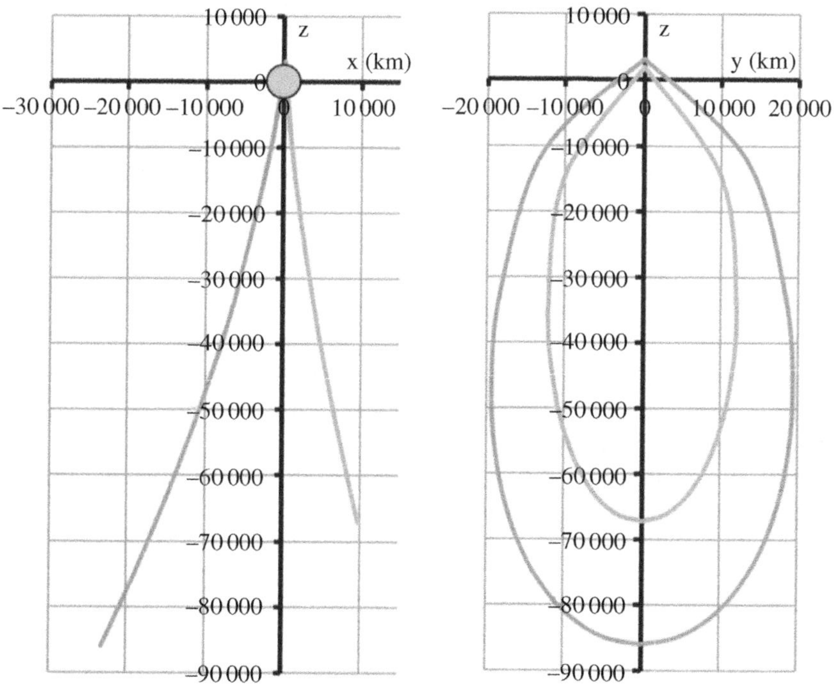


**Figure 3.11** Near-rectilinear Halo orbits.

Near Rectilinear Halo Orbits (NRHO). They are nearly polar with a perilune at about 2000 km above the Moon North pole. The $L_2$ NRHO, with a period of 6.56 days (that is not reachable on $L_1$ side), is moreover synchronized with the Moon synodic period (29.53 days) at a ratio 4:1. It therefore maintains constant lightning conditions, making it particularly suitable for a permanent lunar station like the Lunar Gateway.

## 3.4 Transfers

The linearization in the vicinity of the points $L_1$ and $L_2$ (Section 3.3.1) has shown stable and unstable components superimposed on the periodic components. In a similar way, the linearization in the vicinity of a Halo orbit reveals stable and unstable components called invariant manifolds of the orbit.

### 3.4.1 Linearization in the Orbit Vicinity

The equations (3.26) form an autonomous differential system with the state vector X = (x, y, z, $v_x$, $v_y$, $v_z$) of the position and velocity components in RC.

$$\begin{cases} \ddot{x} - 2\dot{y} = U_x \\ \ddot{y} + 2\dot{x} = U_y \\ \ddot{z} = U_z \end{cases} \Leftrightarrow \begin{cases} \dot{x} = v_x, \quad \dot{v}_x = U_x + 2v_y \\ \dot{y} = v_y, \quad \dot{v}_y = U_y - 2v_x \\ \dot{z} = v_z, \quad \dot{v}_z = U_z \end{cases} \Leftrightarrow \dot{X} = f(X) \tag{3.60}$$

Let us consider the state $X_t$ obtained at any time t from the initial state $X_0$ defined at time $t_0 = 0$, and apply an initial perturbation $\delta X_0$. The resulting perturbation $\delta X_t$ is linked to the initial one $\delta X_0$ by the transition matrix $\Phi$. This matrix of size $6 \times 6$ accounts for first-order effects.

$$X_t = \varphi(X_0, t) \quad \rightarrow \quad \delta X_t = \Phi \delta X_0 \quad \text{with } \Phi = \frac{\partial \varphi}{\partial X}(X_0, t) \tag{3.61}$$

On the other hand, the initial perturbation $\delta X_0$ propagates according to the variational equations developed in (3.28), which have the form

$$\delta \dot{X} = F \delta X \quad \text{with } F \underset{\text{def}}{=} \frac{\partial f}{\partial X} = \begin{pmatrix} 0 & I \\ \nabla^2 U & 2A \end{pmatrix}, \quad A = \begin{pmatrix} 0 & 1 & 0 \\ -1 & 0 & 0 \\ 0 & 0 & 0 \end{pmatrix} \tag{3.62}$$

The matrix F is of size $6 \times 6$. It is composed of the dynamics derivatives assessed along the unperturbed trajectory. It can be split into four submatrices of size $3 \times 3$, among which the matrix $\nabla^2 U$ that has been calculated in (3.29).

By substituting the perturbation $\delta X_t$ given by (3.61) into the variational equations (3.62), we obtain the propagation equation of the transition matrix $\Phi$.

$$\dot{\Phi} = F\Phi, \quad \Phi_0 = I \tag{3.63}$$

The eigenvalues of the transition matrix determine whether a perturbation is amplified or attenuated.

The formal study consists of putting the system (3.60) in Hamiltonian form. The details of these theoretical developments can be found in [R18] and [R23].

In order to alleviate the presentation, we will directly consider a coordinate change (coming in fact from the analysis of the Hamiltonian form).

#### 3.4.1.1 Coordinate Change

Let us define the following coordinate change

$$X = PX' \quad \text{with } P = \begin{pmatrix} I & 0 \\ A & I \end{pmatrix} \quad \rightarrow \quad P^{-1} = \begin{pmatrix} I & 0 \\ -A & I \end{pmatrix} \tag{3.64}$$

where the matrix A is defined in (3.62).

The matrix F′ in the new coordinates is

$$F' = P^{-1}FP = \begin{pmatrix} A & I \\ \nabla^2 U + A^2 & A \end{pmatrix} \tag{3.65}$$

The new transition matrix $\Phi' = P^{-1}\Phi P$ follows the propagation equation

$$\dot{\Phi}' = F'\Phi', \quad \Phi'_0 = I \tag{3.66}$$

The matrix Φ′ has the same characteristic polynomial as Φ. Indeed, we have

$$\det(\Phi' - \lambda I) = \det(P^{-1}\Phi P - \lambda I) = \det[P^{-1}(\Phi - \lambda I)P] = \det(\Phi - \lambda I) \tag{3.67}$$

This change of variable simplifies the eigenvalue study because the matrix Φ′ has several interesting properties.

#### 3.4.1.2 Symplectic Matrix

The matrix Φ′ is **symplectic**, which means that it satisfies the identity

$$\Phi'^t J \Phi' = J \quad \text{with } J = \begin{pmatrix} 0 & I \\ -I & 0 \end{pmatrix} \tag{3.68}$$

This is demonstrated by deriving (3.68), then using (3.66) and (3.65).

$$\frac{d}{dt}(\Phi'^t J \Phi') = \dot{\Phi}'^t J \Phi' + \Phi'^t J \dot{\Phi}' = \Phi'^t (F'^t J + JF')\Phi' = 0 \tag{3.69}$$

The matrix $\Phi'^t J \Phi'$ is thus constant, equal to its initial value J, because $\Phi'_0 = I$.

Like any symplectic matrix, the matrix Φ′ has the following properties [R23].

#### 3.4.1.3 Determinant

From (3.68), we have: $\det(\Phi'^t J \Phi') = \det(J) \Rightarrow \det(\Phi')^2 = 1$.

Since the transition matrices $\Phi'_{12}$ from $t_1$ to $t_2$ and $\Phi'_{23}$ from $t_2$ to $t_3$ are composed according to: $\Phi'_{13} = \Phi'_{12}\Phi'_{23}$, their determinant can only be equal to +1.

#### 3.4.1.4 Characteristic Polynomial

The characteristic polynomial satisfies the following identity:

$$\begin{aligned} \det(\Phi' - \lambda I) &= \det(J^{-1}\Phi'^{-t}J - \lambda I) = \det[J^{-1}(\Phi'^{-t} - \lambda I)J] = \det(\Phi'^{-1} - \lambda I) \\ &= \det\left[-\lambda\Phi'^{-1}\left(\Phi' - \frac{1}{\lambda}I\right)\right] = \lambda^6 \det\left(\Phi' - \frac{1}{\lambda}I\right) \end{aligned} \tag{3.70}$$

where we have used $\det(\Phi'^{-1}) = \frac{1}{\det(\Phi')} = 1$ in the last equality.

#### 3.4.1.5 Eigenvalues

The expression (3.70) of the characteristic polynomial shows that if λ is an eigenvalue of Φ′, then $\lambda^{-1}$ is also an eigenvalue.

Moreover, since the characteristic polynomial has real coefficients, its complex roots are conjugated. If λ is a complex eigenvalue of Φ′, then $\bar{\lambda}$ is also a complex eigenvalue, and by applying (3.70), its inverse $\bar{\lambda}^{-1} = \lambda/|\lambda|^2$ is also a complex eigenvalue. By repeating this reasoning applied to $\lambda/|\lambda|^2$, we deduce that $\lambda/|\lambda|^2, \ldots, \lambda/|\lambda|^{2k}$ are all eigenvalues whatever k.

Since the total number of eigenvalues of Φ′ is 6, we necessarily have $|\lambda| = 1$.

From this analysis, we conclude that the eigenvalues of the transition matrix form either pairs of inverse reals or pairs of conjugate complexes of module 1.

#### 3.4.1.6 Monodromy Matrix

In the case of a periodic orbit of the three-body system, the transition matrix over a period is called the **monodromy matrix**.

This monodromy matrix admits a real double eigenvalue worth 1.

To demonstrate it, let us consider the state $X_{0+\varepsilon}$ obtained at the time $t_0 + \varepsilon$ from the initial state $X_0$ defined at time $t_0 = 0$. For a small time increment $\varepsilon$, the state $X_{0+\varepsilon}$ can be approximated at first order by $X_{0+\varepsilon} = X_0 + \varepsilon \dot{X}_0$.

Since the orbit is periodic of period T, we have $\varphi(X_{0+\varepsilon}, T) = X_{0+\varepsilon}$. By expanding this condition to order 1 we obtain

$$X_0 + \varepsilon \dot{X}_0 = \varphi(X_0 + \varepsilon \dot{X}_0, T) = \varphi(X_0, T) + \frac{\partial \varphi}{\partial X} \varepsilon \dot{X}_0 = X_0 + \varepsilon \Phi \dot{X}_0 \Rightarrow \dot{X}_0 = \Phi \dot{X}_0 \tag{3.71}$$

This relation shows that the vector $\dot{X}_0$ tangent to the orbit is an eigenvector for the eigenvalue 1.

Moreover, by deriving Jacobi prime integral (3.13) with respect to $X_0$, we have

$$K(X_T) = K(X_0) \Rightarrow \left(\frac{\partial X_T}{\partial X_0}\right)^t \frac{\partial K}{\partial X}(X_0) = \frac{\partial K}{\partial X}(X_0) \Rightarrow \Phi^t K_{X0} = K_{X0} \tag{3.72}$$

The gradient vector $K_{X0}$ is also an eigenvector associated with the eigenvalue 1.

It is found in practice that the eigenvalues of the monodromy matrix are generally of the form $(1, 1, \lambda, \lambda^{-1}, \nu, \overline{\nu})$ with $\lambda$ real and $\nu$ complex of module 1. This structure is consistent with that obtained by the linear approximation (3.42).

Although this statement applies to most Halo orbits, the systematic existence of the real eigenvalue $\lambda$ is not proven, and counter-examples have been published.

### 3.4.2 Invariant Manifolds

The eigenvalues of the monodromy matrix determine the behavior in the vicinity of the Halo orbit.

- The two neutral eigenvalues (worth 1) are related to the orbit periodicity and to the Jacobi prime integral.
- The two complex conjugate eigenvalues $(\nu, \overline{\nu})$ create oscillations around the Halo orbit and generate so-called quasi-Halo orbits.
- The two real inverse eigenvalues define the stable ($\lambda < 1$) and the unstable ($\lambda^{-1} > 1$) components. At each point in the orbit, the stable component produces a trajectory converging toward the orbit and the unstable component produces a trajectory diverging from the orbit.

Together, these trajectories form the **invariant manifolds** of the orbit.

These manifolds are constructed numerically at each point of the orbit as follows:

- calculation of the monodromy matrix $\Phi$ (3.61) by finite differences (perturbation of a single component of the state $X_0$, integration over a period, and difference observed in the initial state);
- calculation of the stable $U_s$ and unstable $U_i$ eigenvectors associated with the smallest ($\lambda_s < 1$) and largest ($\lambda_i > 1$) real eigenvalue;
- application of a perturbation $X_0 \pm \varepsilon U_s$ and retrograde integration to generate the stable manifolds in the point $X_0$;

- application of a perturbation $X_0 \pm \varepsilon U_i$ and direct integration to generate the unstable manifolds in the point $X_0$.

Each of these steps requires careful settings regarding the increment magnitude for finite differences, the modulus $\varepsilon$ of the perturbation to generate the manifold and the numerical accuracy of the integration method. The perturbation magnitude should ideally be set so that the trajectory leaves the orbit vicinity after one period, in order to avoid too long propagation times. This depends whether the real eigenvalue is close to one or not. The integration accuracy can be controlled using a variable-step integration method with embedded error control. This control can be enforced by checking the Jacobi constant at each time step.

These settings vary according to the orbit considered and even according to the point of the orbit [R23].

At each point of the orbit, two stable and two unstable manifolds can be generated according to the sign of the perturbation $\varepsilon$. The two manifolds are named according to the resulting perturbation along the x-component of the state vector:

- if the initial x perturbation is directed toward the secondary body $C_2$, we get an **interior** manifold that will continue approaching this body;
- if the x perturbation is directed away from the secondary body $C_2$, we get an **exterior** manifold that will continue moving away from this body.

Figure 3.12 shows some stable interior manifolds of a Halo orbit at point $L_1$ of the Sun–Earth system. The orbit excursion is set to $A_z = 120\,000$ km. The frame origin is in $L_1$ at about 1.5 Mkm from Earth; the Sun is toward the left of the figure. The manifolds shown in this figure pass close to the Earth (less than 5000 km for the one in solid line). The injection of a vehicle on such a manifold makes it possible to converge to Halo orbit in about one year without further maneuvering. In the same way, unstable manifolds can be constructed at each point. These stable and unstable manifolds form gravitational tubes entering and leaving the orbit.

### 3.4.3 Transfer Strategies

The invariant manifolds of a Halo orbit can be exploited in various ways for orbit transfers. The following discussion is restricted to $L_1$ and $L_2$ Halo orbits of the Sun–Earth or the Earth–Moon system, and to transfers starting from the Earth.

In the case of the Sun–Earth system, the Earth is the secondary body and it is located between the $L_1$ and $L_2$ points. It is generally possible to find manifolds from an $L_1$ or $L_2$ orbit passing close to the Earth, as depicted in Figure 3.12. A direct injection on this manifold will bring the vehicle into the Halo orbit for free.

In the case of the Earth–Moon system, the Earth is the primary body. There are generally no manifold from Halo orbits passing close to the Earth. Figure 3.13 depicts some possibilities for transferring a vehicle from a circular orbit (green solid line) around the Earth (body $C_1$) to a Halo orbit around the point $L_1$ or $L_2$ (blue solid lines).

#### 3.4.3.1 Direct Orbit Targeting

The first strategy is a transfer in two maneuvers targeting directly the Halo orbit. The first maneuver injects on a trajectory intercepting the Halo orbit (green dotted starting from $C_1$), and the second maneuver (Halo Orbit Insertion, HOI) corrects the velocity for insertion into the orbit. This scenario allows for a short transfer time.

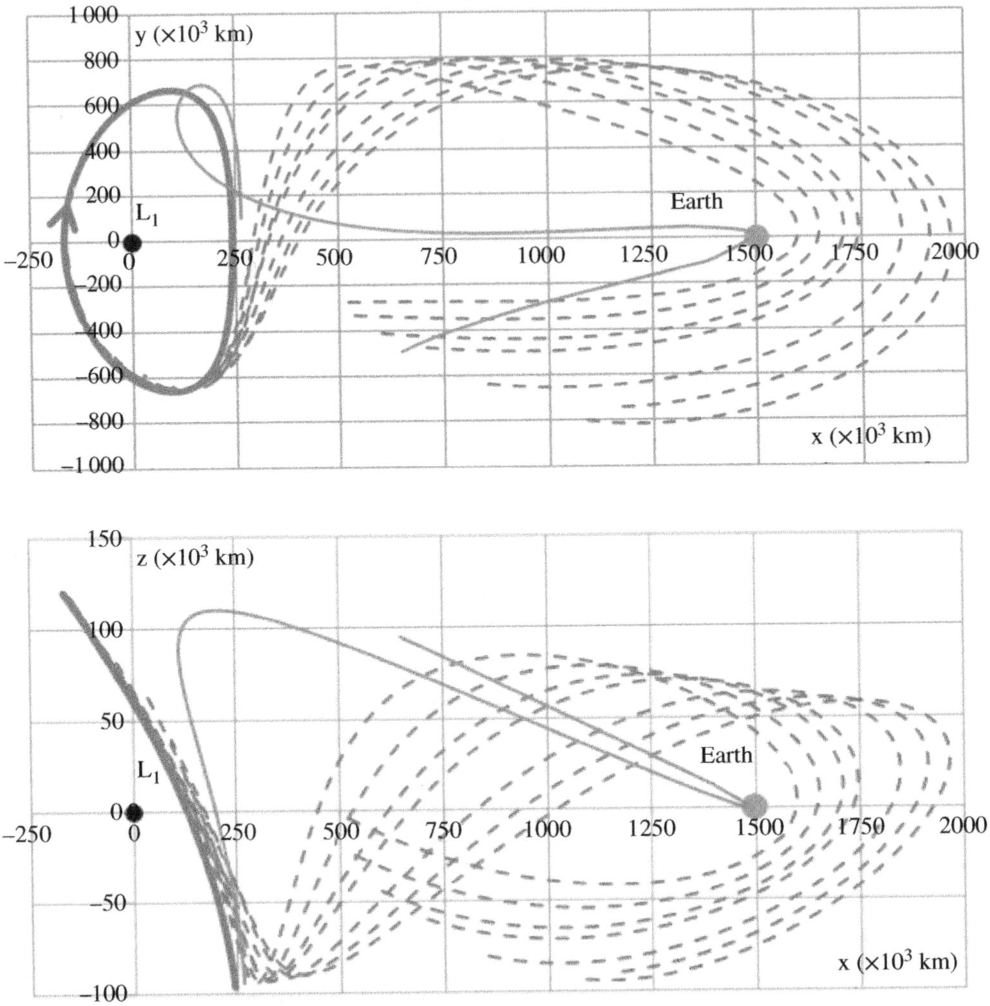


**Figure 3.12** Stable manifolds of a Sun–Earth L1 Halo orbit ($A_z = 120\,000$ km).

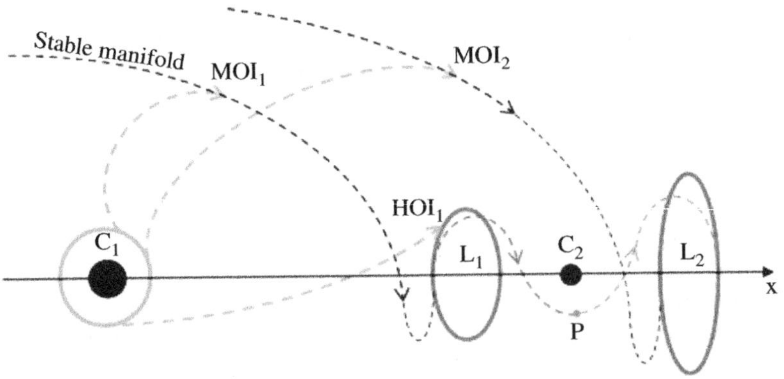


**Figure 3.13** Transfers using manifolds.

#### 3.4.3.2 Manifold Targeting

The second strategy is similar, but targets a point on a stable manifold of the Halo orbit instead of targeting the orbit itself. The second maneuver (Manifold Orbit Insertion, MOI) corrects the velocity for insertion on the manifold. This scenario can reduce the transfer consumption at the expense of a longer duration, provided that the stable manifold is less costly to reach than the orbit itself.

#### 3.4.3.3 Intermediate Halo Orbit

The third strategy is to pass through an intermediate Halo orbit. In Figure 3.13, the $L_1$ orbit is left by an unstable manifold (red dots) and the $L_2$ orbit is reached by a stable manifold (blue dotted line). This type of transfer exploits possible intersections (at some point P) between the manifolds of different orbits. These intersections are searched using Poincaré's section method. If the orbits have the same Jacobi's constant, the transfer is possibly achievable without any maneuver. Otherwise, a velocity impulse will be required at the intersection point P.

Connections between manifolds with equal Jacobi constant are called **heteroclinic** (resp. **homoclinic**) if the departure and arrival orbits are different (resp. the same).

#### 3.4.3.4 Four-Body Problem

The previous idea of manifold connection can be extended to manifolds from two different three-body systems. For example, it is possible to go from the Earth to an Earth–Moon Halo orbit by the following path:

- start on a stable manifold of a Sun–Earth Halo orbit, passing close to the Earth;
- free drift on the stable manifold until the Sun–Earth Halo orbit (or close to it);
- switch to an unstable manifold of the Sun–Earth Halo orbit, that comes back to Moon and crosses a stable manifold of the targeted Earth–Moon Halo orbit;
- free drift on the stable manifold until the Earth–Moon Halo orbit.

This strategy requires finding an adequate Sun–Earth Halo orbit, having favorable stable and unstable manifolds. The stable manifold must pass close to the Earth and the unstable manifold must cross some stable Earth–Moon manifold, with the minimum velocity difference. The solution depends on the relative configuration of the Sun, Earth, and Moon at the transfer beginning. With well-chosen manifolds, the overall transfer from the Earth to the Earth–Moon Halo orbit can in some cases be achieved without propelled maneuver, but at the expense of a much longer transfer duration (several months). The region in the vicinity of the Sun–Earth Halo orbit is called the WSB (Weak Stability Boundary) region. In this region, the Sun's gravitational pull helps increase the energy, so that the return path comes close to the Moon. This WSB strategy was first used by the Hiten spacecraft (1991) to reach a $L_2$ Halo orbit of the Earth–Moon system via a passage near the $L_2$ point of the Sun–Earth system.

Figure 3.14 (retrieved from *Transfers between the Earth–Moon and Sun–Earth systems using manifolds and transit orbits, Kathleen C. Howell, Masaki Kakoi, Acta Astronautica 59*) shows a family of unstable manifolds of a $L_2$ Sun–Earth Halo orbit (in red) crossing a family of stable manifolds of a $L_2$ Earth–Moon Halo orbit (in blue). The existence of a natural junction (no maneuver) depends on the relative configuration of the Earth, Sun, and Moon.

Figure 3.15 depicts ballistic transfers starting from the Earth, going in the vicinity of the $L_1$ point of the Sun–Earth system, then naturally coming back close to the Moon. The perigee of the initial trajectory leaving the Earth is increased up to the Moon orbit owing to the Sun's gravitational pull when passing at the apogee.

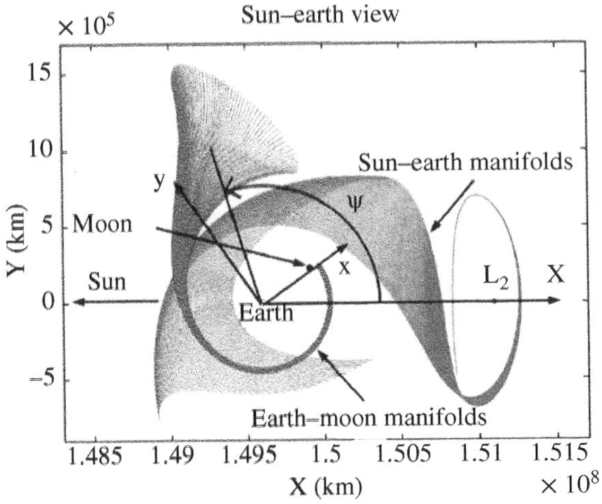


**Figure 3.14** Manifold intersection from Sun–Earth and Earth–Moon orbits.

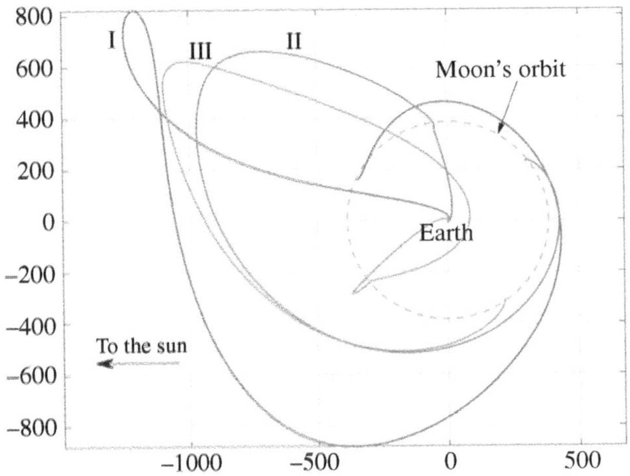


**Figure 3.15** Ballistic transfers from Earth to Moon.

#### 3.4.3.5 Lunar Gravity Assist

The previous strategies can be combined with a lunar gravity assist (LGA), which consists of passing close to the Moon (fly-by) in order to benefit from a swing-by effect, and reduce the need for propelled maneuvers. The dynamics of a gravity assist during a fly-by is described in Chapter 14 on interplanetary missions.

Figure 3.16 illustrates a transfer from the Earth to the Earth–Moon $L_2$ NRHO depicted in Figure 3.11.

These various options (manifolds, LGA, fourth body) offer many possibilities for low-cost transfers, especially for lunar missions. A compromise must be sought between the sparing of propellant and the increase of the transfer duration (see [R18], [R23] for a detailed analysis of the various strategies).

The Halo orbits and their manifolds only exist under the assumptions of the Circular Restricted Three-Body problem, namely that the attracting bodies are in circular orbits, with

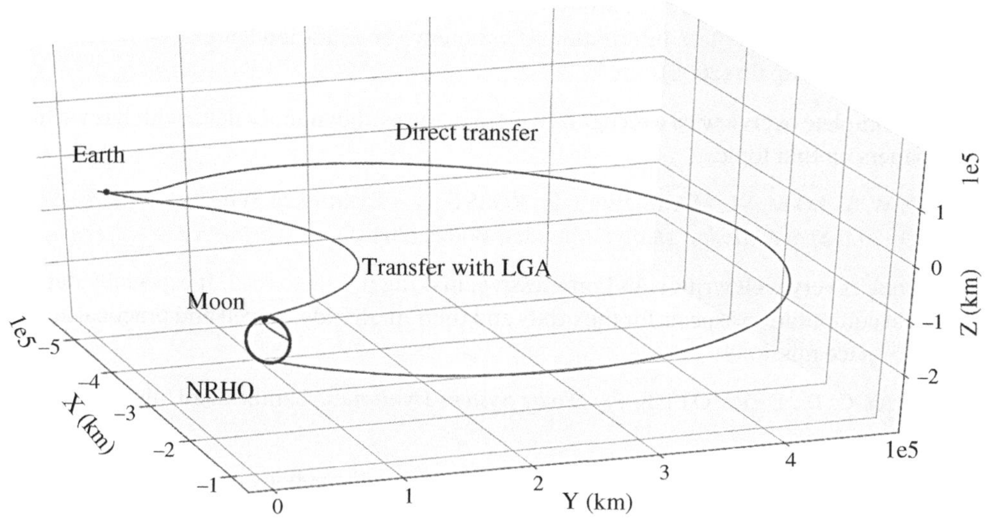


**Figure 3.16** Transfer Earth-NRHO with or without lunar gravity assist.

a purely central gravitational attraction. In the real world, no orbit is perfectly circular and many perturbating forces must be considered. In that context, the Halo orbits and there manifold no longer exist, and the trajectories must be assessed with complete ephemeris and force models. The CR3BP solution nevertheless provides a sufficiently good approximation that can be used to initiate a mission scenario.

## 3.5 Conclusion

### 3.5.1 The Key Points

- The restricted circular three-body problem considers a primary body, a secondary body in circular orbit around the first one, and the motion of a third body of negligible mass.
- The third-body motion admits five equilibrium points called Lagrange points: the three collinear points ($L_1$, $L_2$, and $L_3$) are unstable, whereas the two triangular points ($L_4$ and $L_5$) are stable.
- The Halo orbits are periodic orbits around the points $L_1$ and $L_2$, which are located on either side of the less massive body.
- In each point of a Halo orbit, there exists two stable (resp. unstable) manifolds that are trajectories naturally converging to (resp. diverging from) the orbit.
- The manifolds offer the possibility of free maneuver transfers from the Earth to Halo orbits, either in the Sun–Earth system or in the Earth–Moon system.

### 3.5.2 To Go Further

- FARQUHAR R. W. – *The Control and Use of Libration Point Satellites*. Goddard Space Flight Center, NASA TR R-346 (1970).
  One of the first practical application of Halo orbits.

- GOMEZ G., MASDEMONT J. J., MONDELO J. M. – *Libration Point Orbits: A Survey from the Dynamical Point of View*. International Conference on Libration Point Orbits and Applications, Girona, Spain (2002).

A very complete overview by a recognized team in the astrodynamics field, with has written many papers on that topic.

- KOON W. S., LO M. W., MARSDEN J. E., ROSS S. D. – *Dynamical Systems, the Three-Body Problem and Space Mission Design*. Marsden Books (2008).

This book is very well written and progressive, making it easy to read. It especially details numerical computation aspects for the orbits and their manifolds, as well and practical applications to space missions.

- MURRAY C. D., DERMOTT S. F. – *Solar System Dynamics*. Cambridge University Press (1999).

Chapter 3 presents the analysis of the circular restricted three-body problem with applications in the solar system.

- RICHARDSON D. L. – *Analytic Construction of Periodic Orbits about the Collinear Points*. Celestial Mechanics 22 (1980).

This paper is the original reference for the analytical solution of Halo orbits.

- SZEBEHELY V. G. – *Theory of Orbits: The Restricted Problem of Three Bodies*. Academic Press (1967).

A very complete book detailing the calculations of orbits and manifolds, and the extension to non-circular three-body problems.

- VINTI J. P. – *Orbital and Celestial Mechanics*. AIAA (1998).

Chapters 24 and 25 present detailed calculations regarding the collinear point location and stability analysis.

# 4

# Orbit Determination

## 4.1 Introduction

Space-based positioning or observation applications require a precise knowledge of the position of satellites. The orbit of a satellite is constantly changing under the influence of many perturbations and it needs to be regularly re-estimated.

The first methods of orbitography were developed for astronomy. They used optical observations without distance measurements. Notable successes include Halley's prediction in 1705 of the comet return by 1758 and Gauss' calculation of the orbit of the asteroid Ceres, which allowed it to be reobserved eleven months after its discovery in 1801.

Since the 1950s, the means of observation have developed considerably. Satellite tracking may be carried out by telescopes, radars, or lasers providing direction, distance, and/or velocity measurements. The measurements must be corrected for geometric, optical, and atmospheric perturbations that affect the path of an electromagnetic signal. A preliminary orbit is calculated from a reduced number of observations and then adjusted by differential correction. Least squares and filtering techniques make it possible to estimate orbital parameters and also parameters of the dynamic model. As an example, the observation of Sputnik I orbit in 1957 made it possible to evaluate the gravitational term $J_2$ related to the Earth's flattening.

The proliferation of space debris creates new needs. Surveillance systems such as the SSN (Space Surveillance Network) or Graves (French radar system) and orbitography techniques are constantly improving. The accuracy achieved is a few meters in real time for navigation satellites such as global positioning system (GPS). For scientific satellites such as topography experiment (TOPEX), which is designed to measure the level of the ocean, the accuracy of the radial distance drops to a few centimeters after least squares smoothing.

The first section of this chapter presents the various observation systems and the measurement processing in view of an orbit determination.

The second section presents the main methods for determining a preliminary orbit estimate from a few observations, either with range measurements, or with direction measurements only.

The third section is dedicated to continuous orbit determination, either by smoothing in delayed time or by filtering in real time.

*Space Trajectories: Basic and Advanced Topics*, First Edition. Max Cerf.

## 4.2 Measurements

The determination of an orbit is based on observations of the target from a sensor. This chapter presents the various observation systems, the modeling of measurements with respect to orbital motion and the corrections to be applied.

### 4.2.1 Observation System

This section introduces the categories of sensors, the reference frames used to define the observations, and the orbital motion properties useful for orbit determination.

#### 4.2.1.1 Sensor

The sensor is on the ground in most cases, or it can be in orbit for some specific applications. It receives an electromagnetic signal coming from an orbiting target. Depending on the observation system (optical, radar, or laser), the signal provides information about the direction, distance, and/or velocity of the target.

An optical system receives a light signal from the target and produces an image of the target against a background of stars. From the measurement on the image, the target direction can be reconstituted after applying geometric, optical, and atmospheric corrections. Optical observation from the ground requires favorable weather conditions and it is only possible if the target is illuminated by the Sun, while the sensor is not.

A radar system emits a wave returned by the target and measures the return time of the signal. The returned wave can be reflected (passive target) or retransmitted (active target). Radar measurement provides simultaneously direction, distance, and velocity information. A radar system can be used regardless of lighting conditions, but it is limited in range for passive targets.

A laser system emits a wave reflected by a cube-corner reflector on the target and measures the return time of the signal. The SLR (Satellite Laser Ranging) system uses more than 50 stations for the orbitography of observation satellites such as TOPEX or geodesy satellites such as laser geodynamics satellite (LAGEOS). The accuracy is of the order of centimeter, but this system can only be used on a target equipped with reflectors.

#### 4.2.1.2 Earth Frame

Defining the target orbit, the sensor position and the measurements involves inertial, rotating, and local Earth frames. We first recall briefly the definition of the reference frames used for ephemeris, namely Geocentric Celestial Reference Frame (GCRF) and International Terrestrial Reference Frame (ITRF).

The GCRF is the Earth's equatorial frame on January 1, 2000 at 12:00 (J2000). This frame is considered to be quasi-inertial.

The geocentric ITRF is a rotating Earth frame based on the positions of stations on the Earth's surface. This frame defines the "true" Earth's Equator and pole at a given date [R24], [R32], [R39].

The transition from GCRF (J2000) to ITRF (at a given epoch) is achieved by four successive rotations:

- the precession due to the lunar–solar torque on the equatorial bulge yields the "mean" equatorial frame noted MOD (Mean of Date) at the given epoch;
- the nutation due to the lunar torque, which evolves with the Moon orbit, adds an oscillation around the mean equatorial frame;

- the sidereal rotation takes into account the Earth's rotation around the pole axis;
- the polhodie takes into account the variation of the axis of rotation relative to the geoid.

To simplify the following developments, only the mean equatorial frame and the sidereal rotation are considered in the sequel. We will use three geocentric frames noted RA, RT, and RG, which are defined as follows.

The RA frame with axes $(X_a, Y_a, Z_a)$ is the mean equatorial frame at a given epoch. The $Z_a$ axis is directed toward the North Pole, and the $X_a$ axis is directed toward the vernal point (intersection of the Equator and the ecliptic, which is the plane of the Earth's orbit around the Sun). This quasi-inertial frame is used to define the target orbit.

The RT frame with axes $(X_t, Y_t, Z_t)$ is the rotating frame taking into account the Earth's sidereal rotation around the polar axis $Z_t$, which is identical to $Z_a$. The plane $(X_t, Z_t)$ is the Greenwich meridian. The sidereal period of rotation of the Earth is 86 164 s, corresponding to the angular velocity $\Omega_T = 7.292\,12 \cdot 10^{-5}$ rad/s.

The Earth's shape, also called geoid, is approximated by an ellipsoid of revolution. In the WGS84 model (World Geodetic System), the ellipsoid parameters are:

- $R_e = 6\,378\,137$ m for the equatorial radius;
- $R_p = 6\,356\,752$ m for the polar radius;
- $e_g = 0.081\,819$ for the eccentricity.

The RG frame with axes $(X_g, Y_g, Z_g)$ is the local geodetic frame in point M.

Its axes are $X_g$ to the East, $Y_g$ to the North, and $Z_g$ along the local vertical, which is the normal to the ellipsoid. The geodetic latitude $\varphi_g$ is the angle between $Z_g$ and the Equator $(X_t, Y_t)$, and the longitude $\lambda$ is the angle between $X_t$ and $X_l$, where $(X_l, Z_t)$ is the local meridian of M. The geodetic altitude $h_g$ is the distance to the reference ellipsoid measured along the local vertical $Z_g$

Figure 4.1 depicts the geodetic frame RG and the geodetic coordinates $(h_g, \varphi_g, \lambda)$.

The point M is equivalently defined in RT by its Cartesian coordinates (x, y, z), its spherical coordinates $(r_c, \varphi_c, \lambda)$, or its geodetic coordinates $(h_g, \varphi_g, \lambda)$.

These coordinates are linked by the formulas (1.124) recalled hereafter.

$$\begin{cases} x = r_c \cos\varphi_c \cos\lambda = (N + h_g)\cos\varphi_g \cos\lambda \\ y = r_c \cos\varphi_c \sin\lambda = (N + h_g)\cos\varphi_g \sin\lambda \\ z = r_c \sin\varphi_c \quad = \left[N\left(1 - e_g^2\right) + h_g\right] \sin\varphi_g \end{cases} \quad \text{with } N = \frac{R_e}{\sqrt{1 - e_g^2 \sin^2\varphi_g}} \tag{4.1}$$

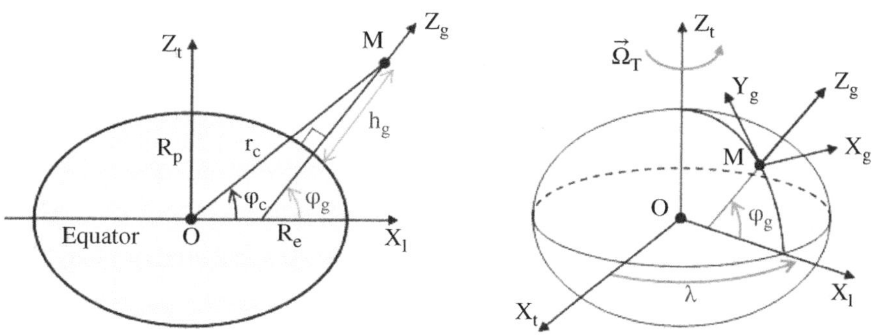


**Figure 4.1** Local geodetic frame.

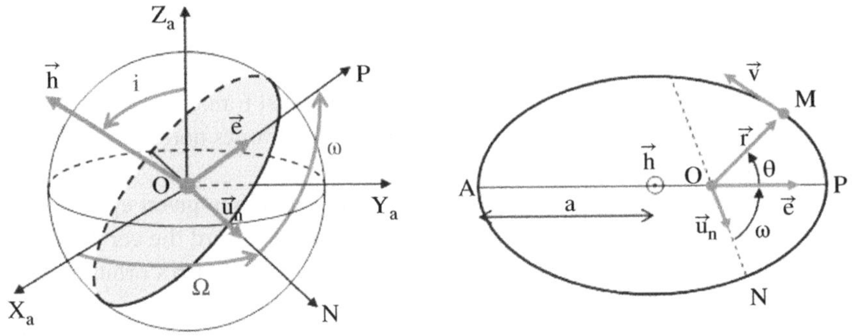


**Figure 4.2** Orbital parameters.

#### 4.2.1.3 Orbital Motion

The orbit of the target is, in first approximation, a Keplerian conic defined by its orbital parameters (a, e, i, Ω, ω, θ) in the inertial geocentric frame RA defined at a given epoch.

The inclination is the angle between the equatorial plane ($X_a$, $Y_a$) and the orbital plane. The right ascension of the ascending node Ω is the angle between $X_a$ and the ascending node N (point of the Equator flown over from south to north). The argument of the perigee is the angle between the node N and the perigee P.

The conic shape is defined by the semi-major axis a and the eccentricity e. The true anomaly θ is the angle between the perigee and the current position on the orbit (Figure 4.2). In the Keplerian model, all orbital parameters are constant except for the anomaly, which is time dependent.

The Keplerian motion has been described in Chapter 1. We recall here the main formulas and properties useful for orbit determination.

The angular momentum $\vec{h}$ and eccentricity $\vec{e}$ vectors are defined from the position $\vec{r}$ and velocity $\vec{v}$ at a given time.

$$\vec{h} = \vec{r} \wedge \vec{v}, \quad \vec{e} = \frac{1}{\mu}\vec{v} \wedge \vec{h} - \frac{\vec{r}}{r} \tag{4.2}$$

where $\mu = 3.986\,005 \cdot 10^{14}\ \text{m}^3/\text{s}^2$ is the Earth gravitational constant.

The angular momentum vector oriented by the angles (i, Ω) is normal to the orbital plane. The eccentricity vector of modulus e points to the perigee defined by the angle ω. These two vectors are constant throughout the Keplerian motion.

The Keplerian orbit that would be followed from a time $t_0$ in a central gravitational field is called the osculating orbit. It may be defined in an equivalent way:

- by the initial position $\vec{r}_0$ and velocity $\vec{v}_0$;
- or by the orbital parameters (a, e, i, Ω, ω) and the initial true anomaly $\theta_0$;
- or by the vectors $(\vec{h}, \vec{e})$ and the initial true anomaly $\theta_0$.

The radius vector depends on the true anomaly and the parameter p of the conic.

$$r = \frac{p}{1 + e\cos\theta} \quad \text{with } p = \frac{h^2}{\mu} = a(1 - e^2) \tag{4.3}$$

The anomaly is time dependent by Kepler's equation. For an elliptical orbit, we have the following relations.

$$E - e\sin E = M \quad \text{with } \cos E = \frac{e + \cos\theta}{1 + e\cos\theta}, \quad M = n(t - t_p), \quad n = \sqrt{\frac{\mu}{a^3}} \tag{4.4}$$

where E is the eccentric anomaly, M is the mean anomaly, $t_p$ is the time at perigee, and n is the pulsation (or mean motion).

The position and velocity on the orbit can be expressed at any time as a function of the initial conditions at time $t_0$, by using Lagrange coefficients F and G defined in Section 1.3.4.

$$\begin{cases} \vec{r} = F\vec{r}_0 + G\vec{v}_0 \\ \vec{v} = F_t\vec{r}_0 + G_t\vec{v}_0 \end{cases} \tag{4.5}$$

Lagrange coefficients (F, G) and their time derivatives $(\dot{F}, \dot{G})$ depend on the anomalies $(\theta_0, \theta)$ by the formulas (1.71) recalled here.

$$\begin{cases} F = 1 - \dfrac{r}{p}(1 - \cos\Delta\theta), & F_t = \dfrac{\sqrt{\mu}}{pr_0}[\sigma_0(1 - \cos\Delta\theta) - \sqrt{p}\sin\Delta\theta] \\ G = \dfrac{rr_0}{h}\sin\Delta\theta, & G_t = 1 + \dfrac{r_0}{p}(1 - \cos\Delta\theta) \end{cases}$$

$$\text{with} \quad \Delta\theta = \theta - \theta_0, \quad \sigma_0 = \frac{\vec{r}_0 \cdot \vec{v}_0}{\sqrt{\mu}} \tag{4.6}$$

These formulas are in closed form because they depend on the current position through r and θ. This is not suitable for orbit determination, when only positions are measured and the orbital parameters are sought (the anomaly θ is then not known). A more suitable form for Lagrange coefficients (F, G) consists of a time-dependent expansion.

For that purpose, we start with the two-body problem dynamic equation

$$\ddot{\vec{r}} = -\frac{\mu}{r^3}\vec{r} \tag{4.7}$$

and differentiate it to assess the third-order derivatives of $\vec{r}$, noted $\vec{r}^{(3)}$.

$$\vec{r}^{(3)} = -\frac{\mu}{r^3}\dot{\vec{r}} + \frac{3\mu\dot{r}}{r^4}\vec{r} \tag{4.8}$$

The derivative $\dot{r}$ is given by: $r^2 = \vec{r} \cdot \vec{r} \;\Rightarrow\; \dot{r} = \frac{\vec{r} \cdot \dot{\vec{r}}}{r}$, so that (4.8) takes the form

$$\vec{r}^{(3)} = -\frac{\mu}{r^3}\dot{\vec{r}} + \frac{3\mu}{r^5}(\vec{r} \cdot \dot{\vec{r}})\vec{r} \tag{4.9}$$

Differentiating (4.9) yields after calculation, and using again $\dot{r} = \frac{\vec{r} \cdot \dot{\vec{r}}}{r}$

$$\vec{r}^{(4)} = \frac{6\mu}{r^5}(\vec{r} \cdot \dot{\vec{r}})\dot{\vec{r}} - \left[\frac{2\mu^2}{r^6} - \frac{12\mu}{r^7}(\vec{r} \cdot \dot{\vec{r}})^2\right]\vec{r} \tag{4.10}$$

The Taylor series of $\vec{r}$ from initial conditions given at time $t_0$ is to order four

$$\vec{r} = \vec{r}_0 + (t - t_0)\dot{\vec{r}}_0 + \frac{1}{2}(t - t_0)^2\ddot{\vec{r}}_0 + \frac{1}{6}(t - t_0)^3\vec{r}_0^{(3)} + \frac{1}{24}(t - t_0)^4\vec{r}_0^{(4)} \tag{4.11}$$

In order to alleviate the formulas, we define

$$\tau = t - t_0, \quad \gamma = \frac{\mu}{r_0^3}, \quad s = \frac{\vec{r}_0 \cdot \vec{v}_0}{r_0^2} \tag{4.12}$$

Substituting into (4.11) the derivatives given by (4.7), (4.9), and (4.10), with τ, γ, s defined by (4.12), we obtain an expression of $\vec{r}$ depending on $\vec{r}_0$ and $\dot{\vec{r}}_0 = \vec{v}_0$.

$$\vec{r} = \vec{r}_0\left[1 - \frac{1}{2}\gamma\tau^2 + \frac{1}{2}\gamma s\tau^3 - \frac{1}{12}\gamma(\gamma + 6s^2)\tau^4\right] + \vec{v}_0\left[\tau - \frac{1}{6}\gamma\tau^3 + \frac{1}{4}\gamma s\tau^4\right] \tag{4.13}$$

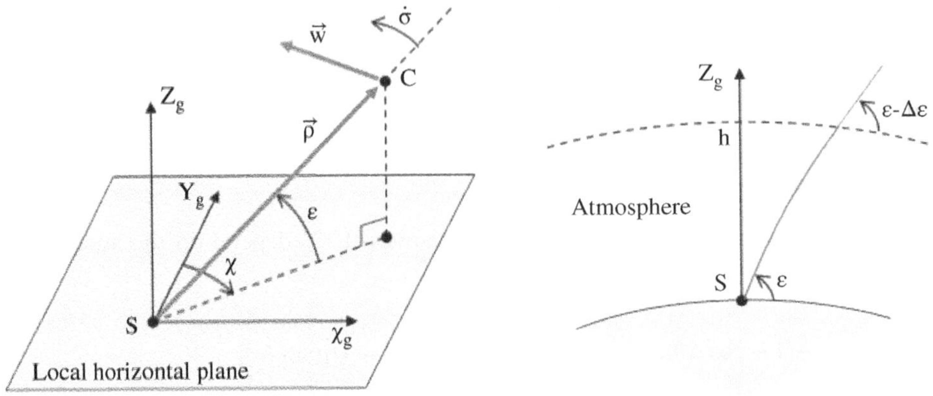


**Figure 4.3** Elevation and azimuth.

By identification with (4.5), we get the expansion of Lagrange coefficients up to order four.

$$\begin{cases} F = 1 - \frac{1}{2}\gamma\tau^2 + \frac{1}{2}\gamma s\tau^3 - \frac{1}{12}\gamma(\gamma + 6s^2)\tau^4 \\ G = \tau - \frac{1}{6}\gamma\tau^3 + \frac{1}{4}\gamma s\tau^4 \end{cases} \quad \text{with } \gamma = \frac{\mu}{r_0^3}, \quad s = \frac{\vec{r}_0 \cdot \vec{v}_0}{r_0^2} \tag{4.14}$$

These formulas will be used by the Gauss method presented in Section 4.3.2.

### 4.2.2 Measurements

Depending on their nature, the sensors can provide measurements of direction, distance, angular velocity, or radial velocity. The reduction process consists of expressing the measurements in a geocentric frame. This process includes frame changes and geometric, optical, atmospheric, and instrumental corrections.

#### 4.2.2.1 Angular Measurements

The direction of observation between the sensor S and the target C is defined by the elevation and azimuth angles in the local geodetic frame RG of the sensor. The elevation ε is the angle between the line of sight SC and the local horizontal plane ($X_g$, $Y_g$), the azimuth χ is the angle between the North direction $Y_g$ and the projection of SC on the horizontal plane (Figure 4.3).

The sensor-target vector noted $\vec{\rho} = \vec{SC}$ has for components in RG

$$\vec{\rho} = \rho\vec{u}_\rho \quad \text{with } \vec{u}_\rho = \begin{pmatrix} \cos\varepsilon\ \sin\chi \\ \cos\varepsilon\ \cos\chi \\ \sin\varepsilon \end{pmatrix} \tag{4.15}$$

An optical sensor measures the angle differences between the target and a star background. The specific corrections applied to star directions are described in Section 4.2.3.

The elevation shall be corrected for atmospheric refraction. Laplace's formula [R24], [R32] is based on a spherical atmosphere and a refractive index varying linearly with density. The correction to be applied to the zenithal distance $\zeta = 90 - \varepsilon$ is estimated by

$$\Delta\zeta = A\tan\zeta - B\tan^3\zeta \tag{4.16}$$

with $A = 57.085''$ and $B = 0.066\,6''$ under standard atmospheric conditions.

#### 4.2.2.2 Distance Measurements

A radar or laser sensor estimates the distance from the travel time of a signal sent back by the target. The speed of the signal in the atmosphere depends on the atmospheric index n by

$$c(n) = \frac{c_{vacuum}}{n} \tag{4.17}$$

where $c_{vacuum} = 299\,792\,458$ m/s is the speed of light in vacuum ($n = 1$).

If the index is a linear function of the density, the distance refraction correction is of the form

$$\Delta\rho = \frac{C}{\cos\zeta} \tag{4.18}$$

The signal speed is also affected by the atmosphere ionization. The correction is proportional to the electronical load encountered on the path and it also depends on the signal frequency f.

$$\Delta\rho = \frac{E}{f^2} \tag{4.19}$$

More accurate empirical models of angular refraction and distance refraction take into account the temperature, the pressure, the atmospheric humidity, and the signal wavelength [R24], [R42]. The parameters of these models can be estimated by comparing two signals of different wavelengths. This technique eliminates the main sources of errors related to atmospheric refraction.

#### 4.2.2.3 Velocity Measurements

The velocity $\vec{w}$ of the target relative to the sensor is the derivative of $\vec{\rho}$ in the rotating frame RG. It is projected along the radial and normal directions defined by the unit vectors $\vec{u}_\rho$ and $\vec{u}_\sigma$, respectively.

$$\vec{w} = \dot{\vec{\rho}} = \dot{\rho}\vec{u}_\rho + \rho\dot{\sigma}\vec{u}_\sigma \tag{4.20}$$

The angular rate $\dot{\sigma}$ of the line of sight $\vec{u}_\rho$ is the norm of the derivative of the unit vector $\vec{u}_\rho$, which is defined in (4.15). It is expressed as a function of the angular rates in elevation $\dot{\varepsilon}$ and in azimuth $\dot{\chi}$.

$$\dot{\sigma} = \sqrt{\dot{\varepsilon}^2 + \dot{\chi}^2\cos^2\varepsilon} \tag{4.21}$$

The radial velocity $w_\rho = \dot{\rho}$ can be measured by the Doppler effect. The variation $\Delta\lambda$ of the emitted wavelength $\lambda_0$ is given by

$$\frac{\Delta\lambda}{\lambda_0} = \frac{w_\rho}{c} \tag{4.22}$$

#### 4.2.2.4 Measurement Reduction

The reduction consists of expressing the relations between $\vec{\rho}$ and $\vec{w}$ (which are related to the measurements) and the position $\vec{r}_c$ and velocity $\vec{v}_c$ of the target in the inertial geocentric frame RA. The position $\vec{r}_c$ of the target is given by

$$\vec{r}_c = \vec{r}_s + \vec{\rho} \tag{4.23}$$

where the position $\vec{r}_s$ of the sensor is defined by its geodetic coordinates through the formulas (4.1).

The target velocity $\vec{v}_c$ in RA is expressed in the RG frame, which is rotating at the Earth's angular rate $\vec{\Omega}_T$. Since the sensor on the ground has a null velocity in the RG frame, we get

$$\vec{v}_c = \left.\frac{d\vec{r}_c}{dt}\right|_{RA} = \left.\frac{d\vec{r}_c}{dt}\right|_{RG} + \vec{\Omega}_T \wedge \vec{r}_c = \left.\frac{d(\vec{r}_s + \vec{\rho})}{dt}\right|_{RG} + \vec{\Omega}_T \wedge \vec{r}_c = \vec{w} + \vec{\Omega}_T \wedge \vec{r}_c \tag{4.24}$$

For preliminary orbit determination (Section 4.3), the relation (4.23) directly gives the position $\vec{r}_c$ of the target from the observation $\vec{\rho}$. For the differential correction (Section 4.4), the relations (4.23) and (4.24) will be reversed to express the expected measurements $(\vec{\rho}, \vec{w})$ depending on the estimated target state $(\vec{r}_c, \vec{v}_c)$.

### 4.2.3 Directions of Stars

In the case of an optical sensor (telescope), the elevation $\varepsilon$ and azimuth $\chi$ of the target are estimated by comparison with a background of stars. This type of measurement requires specific corrections, firstly to pass from the star's absolute coordinates to their apparent coordinates, which depend on the observer location, secondly to project these coordinates onto the two-dimensional image produced by the telescope.

#### 4.2.3.1 Apparent Coordinates

Catalogs of stars (such as Fundamental Katalog 5, Hipparcos, Tycho, US Naval Observatory, …) give the equatorial coordinates (right ascension $\alpha$ along the Equator, declination $\delta$ along the meridian) of celestial bodies in the mean reference frame of a given epoch. The **apparent** equatorial coordinates at the time of observation are obtained by taking into account the proper motion of the stars and corrections for parallax, aberration, and gravitational deviation. These corrections are expressed as deviations in the line of sight.

The **parallax** is due to the observer position, which moves with time:

- the heliocentric or annual parallax is related to the revolution of the Earth around the Sun;
- the geocentric or diurnal parallax is related to the rotation of the Earth around the pole axis.

Due to this observer displacement from the frame origin, the star line of sight varies by $\Delta\sigma = \Delta r_t / r_s$, where $\Delta r_t$ denotes the observer displacement transverse to the line of sight and $r_s$ is the distance to the star.

The **aberration** is due to the observer motion during the travel of the light signal from the star to the observer. The star line of sight varies by $\Delta\sigma = v_t/c$, where $v_t$ is the observer velocity transverse to the line of sight and c is the speed of the signal in the vacuum. For more precision, a relativistic light-time delay must also be applied, especially for laser-ranging measurements, or when the signal passes close to a massive attracting body.

The **gravitational deviation** is a relativistic effect related to an attracting body (Sun, Earth, …) located near the signal path. The star line of sight varies by $\Delta\sigma = \frac{2GM}{r_M c^2} \cot \frac{\theta_M}{2}$, where G is the universal gravitational constant, M is the attracting body mass, $r_M$ is the distance between the observer and the attracting body, and $\theta_M$ is the angle between the direction of the body and that of the star.

For more details on the aforementioned corrections, the interested reader may consult [R24], [R32], and [R42].

Once the apparent equatorial coordinates have been determined, they can be transformed into elevation and azimuth angles relative to the time and place of observation.

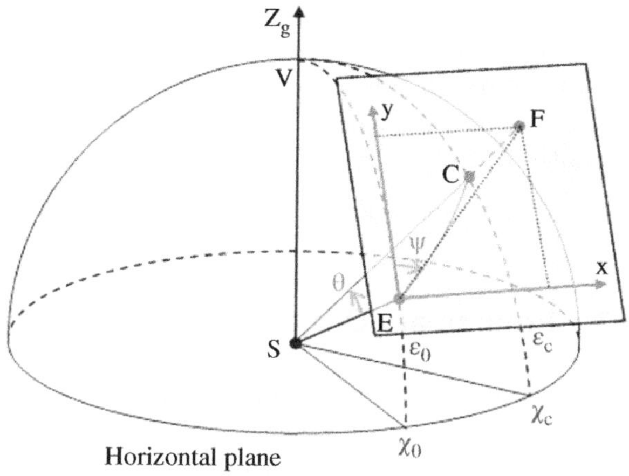


**Figure 4.4** Gnomonic (or central) projection.

#### 4.2.3.2 Gnomonic or Central Projection

The optical sensor S is pointed at a reference star E near the target C and it delivers a bi-dimensional image of the celestial sphere in the plane tangent to the unitary sphere in E. The axes Ex and Ey in this plane are along the local parallel and the local meridian.

The gnomonic (or central) projection is the transformation of the angular coordinates on the celestial sphere into Cartesian coordinates in the tangent plane.

This projection is depicted in Figure 4.4. The elevation and azimuth of the star E and of the target C are noted $(\varepsilon_0, \chi_0)$ and $(\varepsilon_c, \chi_c)$, respectively, $\theta$ is the angle between SE and SC, and $\psi$ is the spherical angle VEC, where SV is the local vertical of the sensor S. The point F is the central projection (from center S) of the target C on the plane Exy.

We have the following relations in the spherical triangle (VEC).

$$\begin{cases} \cos\theta = \sin\varepsilon_0 \sin\varepsilon_c + \cos\varepsilon_0 \cos\varepsilon_c \cos(\chi_c - \chi_0) \\ \sin\varepsilon_c = \sin\varepsilon_0 \cos\theta + \cos\varepsilon_0 \sin\theta \cos\psi \\ \sin\theta \sin\psi = \cos\varepsilon_c \sin(\chi_c - \chi_0) \end{cases} \tag{4.25}$$

The distance EF is $\tan\theta$ and the coordinates of F in the plane Exy are

$$\begin{cases} x = \tan\theta \sin\psi \\ y = \tan\theta \cos\psi \end{cases} \tag{4.26}$$

Using relations (4.25) to eliminate $\theta$ and $\psi$ gives

$$\begin{cases} x = \dfrac{\sin(\chi_c - \chi_0)}{\sin\varepsilon_0 \tan\varepsilon_c + \cos\varepsilon_0 \cos(\chi_c - \chi_0)} \\[2ex] y = \dfrac{\tan\varepsilon_c - \tan\varepsilon_0 \cos(\chi_c - \chi_0)}{\tan\varepsilon_0 \tan\varepsilon_c + \cos(\chi_c - \chi_0)} \end{cases} \tag{4.27}$$

If the angular deviations between E and C are small, these relations can be expanded at first order, resulting in

$$\begin{cases} x \approx (\chi_c - \chi_0)\cos\varepsilon_0 \\ y \approx \varepsilon_c - \varepsilon_0 \end{cases} \tag{4.28}$$

The x and y measurements on the image thus make it possible to retrieve the target elevation and azimuth $(\varepsilon_c, \chi_c)$ by comparison with stars of known coordinates defined by $(\varepsilon_0, \chi_0)$.

Because the field of view of an optical sensor is small, the refraction is almost identical for the direction of the star and the target. In this case, the refraction correction can be applied in differential form by expanding Laplace's formula (4.16). Denoting $\zeta = 90 - \varepsilon$ the zenithal distance of the target and $\zeta_0 = 90 - \varepsilon_0$ that of the star, the refraction correction is at first order

$$\Delta\zeta = A\tan\zeta_0 - B\tan^3\zeta_0 + (1+\tan^2\zeta_0)(A - 3B\tan^2\zeta_0)(\zeta - \zeta_0) \tag{4.29}$$

## 4.3 Preliminary Orbit Estimation

Three observations are sufficient to define a Keplerian orbit. On the basis of this preliminary orbit estimate, it is possible to predict the target trajectory in order to reobserve it later and improve the orbit estimate. This chapter presents the classical methods of determination based either on positions or on directions.

### 4.3.1 Position Measurements

When the sensor-target distance is measured, the position $\vec{\rho}$ of the target relative to the observer is completely known. The position vector is then given by (4.23). Three coplanar positions $\vec{r}_1, \vec{r}_2, \vec{r}_3$ allow to define the Keplerian orbit either by the orbital parameters or by the angular momentum and eccentricity vectors $\vec{h}, \vec{e}$.

#### 4.3.1.1 Determination by Orbital Parameters

The three position vectors $\vec{r}_1, \vec{r}_2, \vec{r}_3$ are in the orbital plane, which is normal to the angular momentum $\vec{h}$. The unit vector $\vec{u}_h$ along $\vec{h}$ is therefore determined by the vector product $\vec{r}_1 \wedge \vec{r}_2$. The components of this unit vector $\vec{u}_h$ in the inertial geocentric frame RA (Figure 4.2) are

$$\vec{u}_h = \begin{pmatrix} \sin i\ \sin\Omega \\ -\sin i\ \cos\Omega \\ \cos i \end{pmatrix} \tag{4.30}$$

Knowing $\vec{u}_h$ components allows us to determine the angles $(i, \Omega)$.

The remaining unknowns are the semi-major axis a, the eccentricity e, and the argument of the perigee $\omega$ of the conic passing through the three points.

The node line is oriented by the unit vector $\vec{u}_n$ directed along $\vec{u}_z \wedge \vec{u}_h$ (or along $\vec{u}_x$ if the orbit is equatorial). The latitude argument $\nu = (\vec{u}_n, \vec{r})$ can then be computed for all three positions. By substituting into (4.3) with $\nu = \omega + \theta$, we get a system of three equations with three unknowns $(p, e, \omega)$.

$$r_k = \frac{p}{1 + e\cos(\nu_k - \omega)} \quad \text{for} \quad k = 1, 2, 3 \tag{4.31}$$

This system is nonlinear, which complicates the resolution. It can be put into a linear form by introducing the eccentricity vector components $(e_x, e_y)$ defined by

$$\begin{cases} e_x = e\cos\omega \\ e_y = e\sin\omega \end{cases} \tag{4.32}$$

By this way, we obtain the following linear system of dimension three with the unknowns $(p, e_x, e_y)$.

$$\begin{cases} p - e_x r_1 \cos\nu_1 - e_y r_1 \sin\nu_1 = r_1 \\ p - e_x r_2 \cos\nu_2 - e_y r_2 \sin\nu_2 = r_2 \\ p - e_x r_3 \cos\nu_3 - e_y r_3 \sin\nu_3 = r_3 \end{cases} \tag{4.33}$$

This system is solved by Cramer's method based on the determinants.

The principal determinant noted D is given by

$$D = \begin{vmatrix} 1 & -r_1 \cos\nu_1 & -r_1 \sin\nu_1 \\ 1 & -r_2 \cos\nu_2 & -r_2 \sin\nu_2 \\ 1 & -r_3 \cos\nu_3 & -r_3 \sin\nu_3 \end{vmatrix}$$
$$= r_2 r_3 \sin(\nu_3 - \nu_2) + r_3 r_1 \sin(\nu_1 - \nu_3) + r_1 r_2 \sin(\nu_2 - \nu_1) \tag{4.34}$$

For three distinct observations, this determinant is non-null and the solution is

$$\begin{cases} p = r_1 r_2 r_3 [\sin(\nu_3 - \nu_2) + \sin(\nu_1 - \nu_3) + \sin(\nu_2 - \nu_1)]/D \\ e_x = [(r_2 - r_3) r_1 \sin\nu_1 + (r_3 - r_1) r_2 \sin\nu_2 + (r_1 - r_2) r_3 \sin\nu_3]/D \\ e_y = [(r_3 - r_2) r_1 \cos\nu_1 + (r_1 - r_3) r_2 \cos\nu_2 + (r_2 - r_1) r_3 \cos\nu_3]/D \end{cases} \tag{4.35}$$

The eccentricity e and the perigee argument $\omega$ are then retrieved from (4.32).

#### 4.3.1.2 Determination by Vectors

The vector method of Gibbs (1839–1903) consists of directly calculating the angular momentum vector $\vec{h}$ and the eccentricity vector $\vec{e}$ without going through the orbital parameters.

Since the three position vectors are by assumption coplanar, we can express $\vec{r}_2$ as a linear combination of $\vec{r}_1$ and $\vec{r}_3$.

$$\vec{r}_2 = c_1 \vec{r}_1 + c_3 \vec{r}_3 \tag{4.36}$$

Let us define the vector $\vec{w} = \vec{r}_1 \wedge \vec{r}_3$, which is normal to the orbital plane.

The coefficients $c_1$ and $c_3$ are obtained by calculating the following products.

$$\begin{cases} (\vec{r}_2 \wedge \vec{r}_3) \cdot (\vec{r}_1 \wedge \vec{r}_3) = c_1 \|\vec{r}_1 \wedge \vec{r}_3\|^2 \Rightarrow c_1 = \dfrac{(\vec{r}_2 \wedge \vec{r}_3) \cdot \vec{w}}{w^2} \\ (\vec{r}_1 \wedge \vec{r}_2) \cdot (\vec{r}_1 \wedge \vec{r}_3) = c_3 \|\vec{r}_1 \wedge \vec{r}_3\|^2 \Rightarrow c_3 = \dfrac{(\vec{r}_1 \wedge \vec{r}_2) \cdot \vec{w}}{w^2} \end{cases} \tag{4.37}$$

The scalar product of the eccentricity vector $\vec{e}$ (not known yet) with (4.36) gives

$$\vec{e} \cdot \vec{r}_2 = c_1 \vec{e} \cdot \vec{r}_1 + c_3 \vec{e} \cdot \vec{r}_3 \tag{4.38}$$

The scalar product $\vec{e} \cdot \vec{r}$ is developed using the relations (4.2).

$$\vec{e} \cdot \vec{r} = \left( \frac{1}{\mu} \vec{v} \wedge \vec{h} - \frac{\vec{r}}{r} \right) \cdot \vec{r} = \frac{1}{\mu} (\vec{r} \wedge \vec{v}) \cdot \vec{h} - r = \frac{h^2}{\mu} - r = p - r \tag{4.39}$$

By substituting the three scalar products into (4.38), we deduce successively the parameter p and the angular momentum $\vec{h}$.

$$p = \frac{c_1 r_1 + c_3 r_3 - r_2}{c_1 + c_3 - 1}, \quad \vec{h} = \frac{\sqrt{\mu p}}{w}\vec{w} \tag{4.40}$$

It remains to find the eccentricity vector $\vec{e}$. By forming the double vector product

$$(\vec{w} \wedge \vec{e}) \wedge \vec{w} = w^2 \vec{e} \quad \text{since } \vec{w} \cdot \vec{e} = 0 \tag{4.41}$$

and using (4.39) to calculate $\vec{w} \wedge \vec{e}$

$$\vec{w} \wedge \vec{e} = (\vec{r}_1 \wedge \vec{r}_3) \wedge \vec{e} = (\vec{e} \cdot \vec{r}_1)\vec{r}_3 - (\vec{e} \cdot \vec{r}_3)\vec{r}_1 = (p - r_1)\vec{r}_3 - (p - r_3)\vec{r}_1 \tag{4.42}$$

we get the eccentricity vector $\vec{e}$

$$\vec{e} = [(p - r_1)\vec{r}_3 - (p - r_3)\vec{r}_1] \wedge \frac{\vec{w}}{w^2} \tag{4.43}$$

#### 4.3.1.3 Use of the Observation Times

The aforementioned methods determine the exact Keplerian orbit passing through three coplanar points. Nevertheless, they may suffer from numerical inaccuracy if the successive observations are close together and the angular deviations are small.

In such cases, the method of Herrick (1911–1974) using the three observation times $t_1, t_2, t_3$ in addition to the three positions $\vec{r}_1, \vec{r}_2, \vec{r}_3$ may give better results.

This method is based on an expansion of the position as a function of the time.

The time difference between two observations made at $t_i$ and $t_j$ is noted

$$t_{ij} \underset{\text{def}}{=} t_j - t_i \tag{4.44}$$

The times are numbered in increasing order ($t_1 < t_2 < t_3$).

The extreme positions $\vec{r}_1, \vec{r}_3$ are expressed as fourth-order expansions with respect to time around the intermediate time $t_2$.

$$\begin{cases} \vec{r}_1 = \vec{r}_2 + \vec{v}_2 t_{21} + \frac{1}{2}\vec{a}_2 t_{21}^2 + \frac{1}{6}\vec{c}_3 t_{21}^3 + \frac{1}{24}\vec{c}_4 t_{21}^4 \\ \vec{r}_3 = \vec{r}_2 + \vec{v}_2 t_{23} + \frac{1}{2}\vec{a}_2 t_{23}^2 + \frac{1}{6}\vec{c}_3 t_{23}^3 + \frac{1}{24}\vec{c}_4 t_{23}^4 \end{cases} \tag{4.45}$$

The unknowns of the problem are the velocity $\vec{v}_2$ at $t_2$, the acceleration $\vec{a}_2$ at $t_2$, and the two vectors $\vec{c}_3, \vec{c}_4$.

The goal is to eliminate $\vec{a}_2, \vec{c}_3, \vec{c}_4$ to express the velocity $\vec{v}_2$ as a function of the positions $\vec{r}_1, \vec{r}_2, \vec{r}_3$ and times $t_1, t_2, t_3$. The orbit will then be defined at $t_2$ by $(\vec{r}_2, \vec{v}_2)$.

The two equations (4.45) are first combined to eliminate $\vec{a}_2$.

$$\begin{aligned} \vec{r}_1 t_{23}^2 - \vec{r}_3 t_{21}^2 &= \vec{r}_2 \left(t_{23}^2 - t_{21}^2\right) + \vec{v}_2 \left(t_{21} t_{23}^2 - t_{23} t_{21}^2\right) \\ &\quad + \frac{1}{6}\vec{c}_3 \left(t_{21}^3 t_{23}^2 - t_{23}^3 t_{21}^2\right) + \frac{1}{24}\vec{c}_4 \left(t_{21}^4 t_{23}^2 - t_{23}^4 t_{21}^2\right) \end{aligned} \tag{4.46}$$

Using the relation $t_{23} - t_{21} = t_{13}$, equation (4.46) takes the form

$$\begin{aligned} \vec{r}_1 t_{23}^2 - \vec{r}_3 t_{21}^2 &= \vec{r}_2 t_{13}(t_{21} + t_{23}) + \vec{v}_2 t_{21} t_{23} t_{13} \\ &\quad - \frac{1}{6}\vec{c}_3 t_{21}^2 t_{23}^2 t_{13} - \frac{1}{24}\vec{c}_4 t_{21}^2 t_{23}^2 t_{13}(t_{21} + t_{23}) \end{aligned} \tag{4.47}$$

The vectors $\vec{c}_3, \vec{c}_4$ are then expressed from the accelerations $\vec{a}_1, \vec{a}_3$ at $t_1, t_3$.

$$\begin{cases} \vec{a}_1 = \vec{a}_2 + \vec{c}_3 t_{21} + \dfrac{1}{2}\vec{c}_4 t_{21}^2 \\ \vec{a}_3 = \vec{a}_2 + \vec{c}_3 t_{23} + \dfrac{1}{2}\vec{c}_4 t_{23}^2 \end{cases} \Rightarrow \begin{cases} \vec{c}_3 = \dfrac{\vec{a}_1 t_{23}^2 - \vec{a}_3 t_{21}^2 - \vec{a}_2 (t_{21} + t_{23}) t_{13}}{t_{21} t_{23} t_{13}} \\ \vec{c}_4 = -2\dfrac{\vec{a}_1 t_{23} - \vec{a}_3 t_{21} - \vec{a}_2 t_{13}}{t_{21} t_{23} t_{13}} \end{cases} \tag{4.48}$$

In the Keplerian model, the central acceleration is given by

$$\vec{a} = -\frac{\mu}{r^3}\vec{r} \underset{\text{noted}}{=} -\gamma\vec{r} \quad \text{with } \gamma = \frac{\mu}{r^3} \tag{4.49}$$

By substituting into (4.47) the vectors $\vec{c}_3, \vec{c}_4$ given by (4.48) with the accelerations expressed by (4.49), we obtain after simplifications the velocity $\vec{v}_2$ as a function of the three positions and times.

$$\vec{v}_2 = -\vec{r}_1 t_{23}\left(\frac{1}{t_{12}t_{13}} + \frac{\gamma_1}{12}\right) + \vec{r}_3 t_{12}\left(\frac{1}{t_{13}t_{23}} + \frac{\gamma_3}{12}\right) + \vec{r}_2(t_{23} - t_{12})\left(\frac{1}{t_{12}t_{23}} + \frac{\gamma_2}{12}\right) \tag{4.50}$$

If the times are equidistant ($t_{12} = t_{23} = \Delta t$), the formula is reduced to

$$\vec{v}_2 = \frac{1}{2\Delta t}(\vec{r}_3 - \vec{r}_1) - \frac{\Delta t}{12}(\vec{a}_3 - \vec{a}_1) \tag{4.51}$$

The position $\vec{r}_2$ and velocity $\vec{v}_2$ at time $t_2$ completely define the Keplerian orbit.

This method assumes close observations and small angular deviations (typically less than 1 deg). For angular deviations greater than 5 deg, the Gibbs' method is more accurate. More elaborate methods combine Gibbs' or Herrick's approaches with the solution of Lambert's problem, which gives the Keplerian orbit passing through two given positions at two given times [R16].

### 4.3.2 Direction Measurements

When the distance is not measured (in the case of an optical sensor), only the target direction is known. The previous methods based on the target position are no longer applicable. The methods of Laplace and Gauss construct an approximate solution based on three directions of observation.

#### 4.3.2.1 Formulation of the Problem

The positions of the sensor S and the target C relative to the attracting center O are noted $\vec{r}_s$ and $\vec{r}_c$ respectively. The position $\vec{r}_s$ of the sensor is known at any time. The sensor-target vector is $\vec{\rho} = \vec{r}_c - \vec{r}_s = \rho\vec{u}$. The observations made at the successive times $t_1, t_2, t_3$ are the directions defined by the unit vectors $\vec{u}_1, \vec{u}_2, \vec{u}_3$. The distances $r_1, r_2, r_3$ from the sensor to the target are not known.

Figure 4.5 shows the geometry of the problem with some relations that will be used for the orbit estimation.

Laplace and Gauss methods consist of estimating the target position $\vec{r}_{c2}$ and velocity $\vec{v}_{c2}$ at the intermediate time $t_2$, from which the orbital parameters will be deduced.

#### 4.3.2.2 Laplace Method

The method of Laplace (1780) starts from the acceleration of the target C.

$$\begin{cases} \vec{r}_c = \vec{r}_s + \rho\vec{u} \\ \vec{v}_c = \vec{v}_s + \dot{\rho}\vec{u} + \rho\dot{\vec{u}} \\ \vec{a}_c = \vec{a}_s + \ddot{\rho}\vec{u} + 2\dot{\rho}\dot{\vec{u}} + \rho\ddot{\vec{u}} \end{cases} \tag{4.52}$$

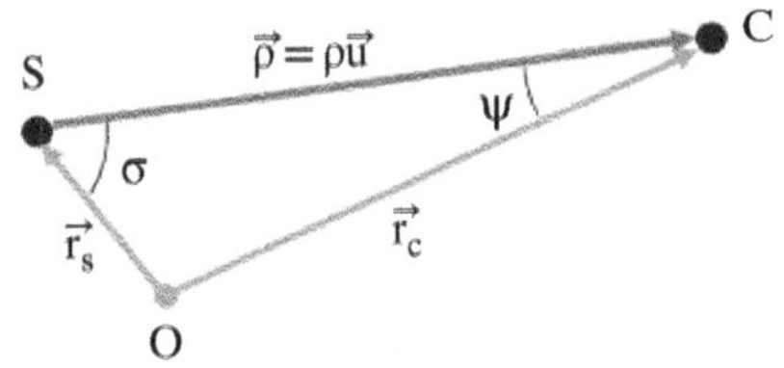


$$\vec{r}_c = \vec{r}_s + \rho\vec{u}$$
$$r_c^2 = r_s^2 + \rho^2 + 2\rho r_s \cos\sigma$$
$$r_c \sin\psi = r_s \sin\sigma$$
$$\rho \sin\psi = r_s \sin(\psi + \sigma)$$

**Figure 4.5** Sensor and target positions.

Since the target follows a Keplerian motion, we have

$$\vec{a}_c = -\frac{\mu}{r_c^3}\vec{r}_c = -\gamma(\vec{r}_s + \rho\vec{u}) \quad \text{with } \gamma = \frac{\mu}{r_c^3} \tag{4.53}$$

Equating the two expressions of $\vec{a}_c$ (4.52) and (4.53) leads to

$$(\ddot{\rho} + \gamma\rho)\vec{u} + 2\dot{\rho}\dot{\vec{u}} + \rho\ddot{\vec{u}} = -\gamma\vec{r}_s - \vec{a}_s \tag{4.54}$$

By taking the scalar product with $(\vec{u} \wedge \dot{\vec{u}})$, then with $(\vec{u} \wedge \ddot{\vec{u}})$, we get the system

$$\begin{cases} \rho\ddot{\vec{u}} \cdot (\vec{u} \wedge \dot{\vec{u}}) = -(\gamma\vec{r}_s + \vec{a}_s) \cdot (\vec{u} \wedge \dot{\vec{u}}) \\ 2\dot{\rho}\dot{\vec{u}} \cdot (\vec{u} \wedge \ddot{\vec{u}}) = -(\gamma\vec{r}_s + \vec{a}_s) \cdot (\vec{u} \wedge \ddot{\vec{u}}) \end{cases} \Rightarrow \begin{cases} \rho = k_1 + k_2\gamma \\ \dot{\rho} = k_3 + k_4\gamma \end{cases} \tag{4.55}$$

where the coefficients $k_1$, $k_2$, $k_3$, $k_4$ depend on the known vectors $(\vec{r}_s, \vec{a}_s, \vec{u})$ and on the unknown vectors $(\dot{\vec{u}}, \ddot{\vec{u}})$ by

$$\begin{cases} k_1 = -\dfrac{\vec{a}_s \cdot (\vec{u} \wedge \dot{\vec{u}})}{\ddot{\vec{u}} \cdot (\vec{u} \wedge \dot{\vec{u}})}, \quad k_2 = -\dfrac{\vec{r}_s \cdot (\vec{u} \wedge \dot{\vec{u}})}{\ddot{\vec{u}} \cdot (\vec{u} \wedge \dot{\vec{u}})} \\ k_3 = -\dfrac{\vec{a}_s \cdot (\vec{u} \wedge \ddot{\vec{u}})}{2\dot{\vec{u}} \cdot (\vec{u} \wedge \ddot{\vec{u}})}, \quad k_4 = -\dfrac{\vec{r}_s \cdot (\vec{u} \wedge \ddot{\vec{u}})}{2\dot{\vec{u}} \cdot (\vec{u} \wedge \ddot{\vec{u}})} \end{cases} \tag{4.56}$$

In order to eliminate the derivatives $(\dot{\vec{u}}, \ddot{\vec{u}})$, the unit vector $\vec{u}$ is expressed using Lagrange interpolation formula.

$$\vec{u}(t) = \sum_{i=1}^{n} \vec{u}(t_i) \prod_{k \neq i} \frac{t - t_k}{t_i - t_k} \tag{4.57}$$

Deriving this formula twice, with $n = 3$ observations, and with the notation already introduced in (4.44): $t_{ij} \underset{\text{def}}{=} t_j - t_i$, we get at time $t_2$

$$\begin{cases} t_{21}t_{13}t_{23}\dot{\vec{u}}_2 = \quad \vec{u}_1 t_{23}^2 - \vec{u}_3 t_{21}^2 - \vec{u}_2 t_{13}(t_{21} + t_{23}) \\ t_{21}t_{13}t_{23}\ddot{\vec{u}}_2 = -2\vec{u}_1 t_{23} + 2\vec{u}_3 t_{21} + 2\vec{u}_2 t_{13} \end{cases} \tag{4.58}$$

The derivatives $(\dot{\vec{u}}_2, \ddot{\vec{u}}_2)$ at time $t_2$ are thus known as a function of the observed unit vectors $(\vec{u}_1, \vec{u}_2, \vec{u}_3)$ and their times. This makes it possible to assess the coefficients $k_1$, $k_2$, $k_3$, $k_4$ by (4.56).

Our goal is to find $(\rho, \dot{\rho})$ in order to assess the target position and velocity at the intermediate time $t_2$ by (4.52). At this step, we are left with three unknowns $(\rho, \dot{\rho}, \gamma)$, or equivalently $(\rho, \dot{\rho}, r_c)$, linked by the two equations (4.55).

A third equation comes from the relation between the sides of the triangle (OSC) shown in Figure 4.5. We can thus form the following nonlinear system

$$\begin{cases} \rho = k_1 + k_2\gamma \\ r_c^2 = r_s^2 + \rho^2 + 2\rho r_s \cos\sigma \end{cases} \quad \text{with } \gamma = \frac{\mu}{r_c^3} \tag{4.59}$$

where the remaining unknowns are $(\rho, r_c)$ whereas the values of $(\sigma, r_s, k_1, k_2)$ are known at time $t_2$. The angle $\sigma$ between $\vec{r}_s$ and $\vec{u}$ is indeed deduced from the elevation angle $\varepsilon$ of the target by $\sigma = 90 - \varepsilon$.

Substituting $\rho$ and $\gamma$ into the first relation (4.59) would yield a polynomial equation of degree eight in $r_c$. In order to ease the resolution, the unknowns $(\rho, r_c)$ are expressed as functions of the angle $\psi$ with the formulas shown in Figure 4.5.

We obtain thus a trigonometric equation in $\psi$ of the form

$$\sin^4\psi - A\,\sin(\psi + \beta) = 0 \tag{4.60}$$

where the coefficient A is assessed by

$$A = \frac{B r_s^3 \sin^3\sigma}{\mu k_2} \quad \text{with} \quad \begin{cases} B\,\sin\beta = r_s \sin\sigma \\ B\,\cos\beta = r_s \cos\sigma - k_1 \end{cases} \tag{4.61}$$

The trigonometric equation (4.60) must be solved numerically, giving the value of the angle $\psi$. There may be multiple roots that should be considered separately in order to retain the appropriate solution (in terms of orbital parameters).

From $\psi$, we can assess $(\rho, r_c, \gamma)$ by

$$\rho = r_s \frac{\sin(\psi + \sigma)}{\sin\psi}, \quad r_c = r_s \frac{\sin\sigma}{\sin\psi}, \quad \gamma = \frac{\mu}{r_c^3} \tag{4.62}$$

whereas $\dot{\rho}$ is assessed by (4.55) using the coefficients $k_1, k_2, k_3, k_4$ given by (4.56).

Formulas (4.52) finally give the target position and velocity at time $t_2$, which makes it possible to estimate the Keplerian orbit.

#### 4.3.2.3 Gauss Method

The method of Gauss is based on Lagrange coefficients F and G defined in (4.5). The positions $\vec{r}_{c1}, \vec{r}_{c3}$ of the target are expressed as a function of the position $\vec{r}_{c2}$ and velocity $\vec{v}_{c2}$ at the intermediate time $t_2$.

$$\begin{cases} \vec{r}_{c1} = F_1\,\vec{r}_{c2} + G_1\,\vec{v}_{c2} \\ \vec{r}_{c3} = F_3\,\vec{r}_{c2} + G_3\,\vec{v}_{c2} \end{cases} \tag{4.63}$$

The exact formulas for F and G recalled in (4.6) are not usable because they are in closed form and depend on the unknown orbit parameters. On the other hand, F and G can be approximated by the time series expansion given in (4.14).

For the Gauss method, the following approximations are used.

$$\begin{cases} F_1 \approx 1 - t_{21}^2\gamma/2, \quad G_1 \approx t_{21} - t_{21}^3\gamma/6 \\ F_3 \approx 1 - t_{23}^2\gamma/2, \quad G_3 \approx t_{23} - t_{23}^3\gamma/6 \end{cases} \quad \text{with} \quad \gamma = \frac{\mu}{r_2^3} \tag{4.64}$$

Since the three positions are coplanar (by assumption), $\vec{r}_{c2}$ can be expressed as a linear combination of $(\vec{r}_{c1}, \vec{r}_{c3})$ with coefficients noted $(c_1, c_3)$.

$$\vec{r}_{c2} = c_1\vec{r}_{c1} + c_3\vec{r}_{c3} \Rightarrow (c_1F_1 + c_3F_3 - 1)\vec{r}_{c2} + (c_1G_1 + c_3G_3)\vec{v}_{c2} = \vec{0} \tag{4.65}$$

If the conic is nondegenerate, the position and velocity are not collinear. Their coefficients in (4.65) must then be null so that $c_1$ and $c_3$ can be expressed as functions of Lagrange coefficients $(F_1, G_1, F_3, G_3)$. Using (4.64), we get after calculation

$$\begin{cases} c_1 \approx \dfrac{t_{23}}{t_{13}}\left[1 + \dfrac{1}{6}\gamma\left(t_{13}^2 - t_{23}^2\right)\right] \\ c_3 \approx -\dfrac{t_{21}}{t_{13}}\left[1 + \dfrac{1}{6}\gamma\left(t_{13}^2 - t_{21}^2\right)\right] \end{cases} \tag{4.66}$$

The relation (4.65) is now written by substituting $\vec{r}_c = \vec{r}_s + \rho\vec{u}$.

$$c_1\rho_1\vec{u}_1 - \rho_2\vec{u}_2 + c_3\rho_3\vec{u}_3 = -c_1\vec{r}_{s1} + \vec{r}_{s2} - c_3\vec{r}_{s3} \tag{4.67}$$

The unknowns in this vector equation are the three distances $(\rho_1, \rho_2, \rho_3)$.

By taking the scalar products with $\vec{u}_1$, $\vec{u}_2$, and $\vec{u}_3$, we form the linear system

$$\begin{cases} \rho_1 c_1 - \rho_2\vec{u}_1\cdot\vec{u}_2 + \rho_3 c_3\vec{u}_1\cdot\vec{u}_3 = -c_1\vec{u}_1\cdot\vec{r}_{s1} + \vec{u}_1\cdot\vec{r}_{s2} - c_3\vec{u}_1\cdot\vec{r}_{s3} \\ \rho_1 c_1\vec{u}_1\cdot\vec{u}_2 - \rho_2 + \rho_3 c_3\vec{u}_2\cdot\vec{u}_3 = -c_1\vec{u}_2\cdot\vec{r}_{s1} + \vec{u}_2\cdot\vec{r}_{s2} - c_3\vec{u}_2\cdot\vec{r}_{s3} \\ \rho_1 c_1\vec{u}_1\cdot\vec{u}_3 - \rho_2\vec{u}_2\cdot\vec{u}_3 + \rho_3 c_3 = -c_1\vec{u}_3\cdot\vec{r}_{s1} + \vec{u}_3\cdot\vec{r}_{s2} - c_3\vec{u}_3\cdot\vec{r}_{s3} \end{cases} \tag{4.68}$$

Solving this system yields $\rho_2$ as a linear function of the coefficients $c_1$ and $c_3$, which are themselves linear functions of $\gamma$ by (4.66).

This results in an expression of the form $\rho_2 = k_1 + k_2\gamma$, similar to that of Laplace's method (4.55), but with different coefficients. The resolution is finalized by the same formulas from (4.59) to (4.62).

---

**Ceres Discovery**

The asteroid Ceres was first observed on January 1, 1801 by Piazzi and could be followed until the beginning of February 1801 when it disappeared behind the Sun. The observations available during this short time interval were used by Gauss to estimate its orbit and predict its reappearance at the end of the year. Ceres could indeed be reobserved at a position close to the one predicted by Gauss. This success established Gauss' reputation in astrodynamics.

---

Gauss method yields the same system structure as the Laplace method, but the coefficients of the system are different and the solutions are therefore different.

The main difference lies in the local approximation:

- either by Lagrange polynomial to interpolate the direction $\vec{u}$ (4.57) in the Laplace method;
- or by Lagrange coefficients F and G to express the position and velocity (4.63) in the Gauss method.

Although the Gauss method seems more physical, since it is based on a local approximation of the Keplerian motion, no method is systematically better than the other. The main sources of errors include sensitivity to time spacing between measurements, imperfect sensor measurements and non-Keplerian effects.

Gauss and Laplace methods are complementary and they can be customized to different situations: several sensors, number of observations greater than three, velocity measurements .... More details can be found in [R16] and [R39].

## 4.4 Continuous Orbit Estimation

The methods presented in the previous section provide an initial estimate of the orbit of a space object. As the actual motion is not Keplerian, the object must be tracked in order to regularly update its orbital parameters and accurately predict its future position (with the goal of reobserving it later). The orbit update is made by comparing the prediction with the observations made. It can be processed either in delayed time (by least squares) or in real time (by differential correction).

### 4.4.1 Least Squares

Global fitting (or smoothing) consists of finding the trajectory that is the closest to a series of past observations. The goal may be to accurately determine the position of a scientific satellite at the time of its observations. It may also involve estimating parameters of the force models, such as gravitational perturbations, drag, and radiation pressure, ... in order to improve the model's reliability.

#### 4.4.1.1 Trajectory Modeling

The target undergoes the central attraction $\vec{a}_c$ and perturbing accelerations, whose sum is noted $\vec{a}_p$. In the absence of perturbations, the trajectory is a Keplerian conic. The perturbations induce variations of the osculating orbital parameters. The motion equations are formulated in position-velocity or in orbital parameters.

- The formulation in position-velocity is directly given by Newton's second law

$$\begin{cases} \dot{\vec{r}} = \vec{v} \\ \dot{\vec{v}} = \vec{a}_c + \vec{a}_p \end{cases} \quad \text{with} \quad \vec{a}_c = -\frac{\mu}{r^3}\vec{r} \tag{4.69}$$

- The formulation in orbital parameters is given by Gauss equations developed in Section 2.3.3, which depend on the perturbing acceleration components $(a_r, a_\theta, a_h)$ along the radial, orthoradial, and normal directions. The position on the osculating orbit is defined by the mean anomaly M, from which the true anomaly $\theta$ and the latitude argument $\nu = \omega + \theta$ can be computed.

$$\begin{cases} \dfrac{da}{dt} = \dfrac{2a^2}{h}[a_r e \sin\theta + a_\theta(1 + e\cos\theta)] \\ \dfrac{de}{dt} = \dfrac{p}{h}\left[a_r \sin\theta + a_\theta\left(\cos\theta + \dfrac{e + \cos\theta}{1 + e\cos\theta}\right)\right] \\ \dfrac{di}{dt} = a_h \dfrac{r\cos\nu}{h}, \quad \dfrac{d\Omega}{dt} = a_h \dfrac{r\sin\nu}{h\sin i} \\ \dfrac{d\omega}{dt} = \dfrac{r}{he}[-a_r\cos\theta(1 + e\cos\theta) + a_\theta \sin\theta(2 + e\cos\theta)] - a_h\dfrac{r\sin\nu}{h\tan i} \\ \dfrac{dM}{dt} = n + \dfrac{b}{aeh}[a_r(p\cos\theta - 2re) - a_\theta(p + r)\sin\theta] \end{cases} \tag{4.70}$$

The integration of the differential equations (4.69) or (4.70) requires a numerical method. The formulation in orbital parameters is preferred when the perturbations are small. Indeed, the second member of the equations (4.70) is then close to zero and the numerical integration can progress with large time steps. In other cases, the formulation (4.69) in position-velocity is more stable, as the actual motion may be quite different from a Keplerian motion.

#### 4.4.1.2 Measurement Modeling

The target is represented by a state vector x either in position-velocity $x = (\vec{r}, \vec{v})$ or in orbital parameters $x = (a, e, i, \Omega, \omega, \theta)$. The trajectory starting from an initial condition $x_0 = x(t_0)$ in the presence of perturbations is obtained by numerical integration of equations (4.69) or (4.70). These state equations take the form

$$\dot{x} = f(x, t, \gamma), \quad x(t_0) = x_0 \tag{4.71}$$

where $\gamma$ represents the parameters of the perturbation models.

The observation of the trajectory by sensors provides a set of measurements at discrete times $(t_k)_{k=1,\cdots,N}$. The measurements noted $(z_k)_{k=1,\cdots,N}$ can be distances, angles, or velocities depending on the sensor (several types of measurements can be used simultaneously in the trajectory estimation process).

Each measurement is a function of the target state (position, velocity) at the time of observation. The measurement equation takes the form

$$z_k = h_k(x_k, t_k) \quad \text{for } k = 1 \text{ to } N \tag{4.72}$$

For a given initial condition $x_0$ and a given set of perturbations $\gamma$, the state equations (4.71) are integrated, and the measurement equations (4.72) are applied at the observation times. This yields the expected measurements (also called **pseudo-measurements**) associated with $x_0$ and $\gamma$. This set of measurements is grouped together in a vector noted $z(x_0, \gamma) = (z_1, \cdots, z_N)$. The deviations between the expected measurements $z(x_0, \gamma)$ and the actual measurements $z_{obs}$ are the residuals noted $r = z(x_0, \gamma) - z_{obs}$. Low values reflect a good compliance of the models with the actual trajectory.

#### 4.4.1.3 Least Squares Problem

The fitting process consists of finding the values $(x_0, \gamma)$ minimizing the norm of the residuals. As not all measurements have the same accuracy or reliability (depending on the sensor and on the conditions of observation), the deviations are weighted by coefficients forming a positive diagonal matrix W.

Noting the unknowns $\xi = (x_0, \gamma)$, the least squares problem is formulated as:

$$\min_{\xi} \|r\|_W^2 = r^t W r \tag{4.73}$$

where $\|r\|_W^2$ denotes the quadratic norm weighted by the matrix W.

The solution $\xi^*$ is obtained by canceling the gradient.

$$\nabla r(\xi^*) W r(\xi^*) = 0 \tag{4.74}$$

This set of equations cannot be solved analytically because the residuals r depend on the unknowns $\xi$ by the function $z(x_0, \gamma)$, which is assessed by a numerical integration of the trajectory equations.

Nevertheless, this problem can be solved efficiently by **Gauss–Newton** method, which is Newton's method customized to the least squares problem (this method was proposed by Gauss in 1809).

It consists of solving the system (4.74) by successive linearizations starting from an initial guess $\xi_0$. The Jacobian matrix $\nabla r(\xi_0)$ is computed by finite differences, perturbing the components of $\xi_0$ one by one, integrating (4.71) and applying (4.72) to obtain the perturbed measurements and assess the partial derivatives.

The system (4.74) is then linearized by expanding $r(\xi)$ to order one in the neighborhood of $\xi_0$.

$$AW(r_0 + A^t \delta\xi) = 0 \quad \text{with } A = \nabla r(\xi_0), \quad r_0 = r(\xi_0) \tag{4.75}$$

The matrix A has dimension $N \times n$, where the number of measurements N must be greater than the number of unknowns n. If the matrix $AWA^t$ is inversible, the variation $\delta\xi$ is given by

$$\delta\xi = -(AWA^t)^{-1} AW r_0 \tag{4.76}$$

The process is repeated in the new point $\xi_1 = \xi_0 + \delta\xi$ and it is continued until stabilization ($\delta\xi = 0$). The convergence depends on the quality of the initial guess, which is generally quite good for orbit determination problems (e.g. by using the methods of Section 4.3).

**Remark 4.1** *When the function r is linear, the cost function in (4.73) is quadratic. In that case, the equations (4.75) are called the* ***normal equations*** *and their solution gives directly the solution* $\xi^*$ *of the least square problem (4.73).*

### 4.4.2 Differential Correction

For real-time processing, the orbit estimate is updated with each new measurement by differential correction. The correction is based on a probabilistic modeling of the deviations from a reference trajectory.

#### 4.4.2.1 Deviation Modeling

Let us assume that we have a reference trajectory noted $x_{ref}(t)$ resulting from previous computations. Our aim is to estimate the current deviation $\delta x(t)$ using new measurements z made at time t. The measurement vector of dimension m is a function of the state and it is subject to a state-independent random noise v of zero mean and of covariance matrix R.

$$z = h(x, t) + v \tag{4.77}$$

The expected measurement at time t is $z_{ref} = h(x_{ref}, t)$, which would be obtained on the reference trajectory and without noise.

The state deviation $\delta x$ produces a measurement deviation $\delta z$. By expanding the measurement equation (4.77) to order one, we have

$$\delta z = H\delta x + v \quad \text{with } H = \frac{\partial h}{\partial x}(x_{ref}, t) \tag{4.78}$$

The deviation $\delta x$ before measurement is considered as a random variable of mean $\delta\overline{x}$ and covariance M. Let us assume that a deviation $\delta z$ is observed, and search for the "most probable" values of $\delta x$ and v that would explain this deviation.

We define these "most probable" values of $\delta x$ and v as the solutions to the following linear quadratic optimization problem

$$\min_{\delta x, v} J(\delta x, v) = \frac{1}{2}(\delta x - \delta\overline{x})^t M^{-1}(\delta x - \delta\overline{x}) + \frac{1}{2} v^t R^{-1} v \quad \text{s.t. } \delta z = H\delta x + v \tag{4.79}$$

The quadratic cost function J to be minimized is called the Mahalanobis distance (1936) of the pair of independent variables $(\delta x, v)$. The linear constraint expresses that the values chosen for $\delta x$ and v must yield the observed deviation $\delta z$.

**Remark 4.2** *In the case where* $\delta x$ *and v are independent Gaussian variables, the probability density of the pair* $(\delta x, v)$ *is proportional to* $e^{-J(\delta x, v)}$. *The solution of the problem (4.79) is then the estimator of the maximum likelihood [R6].*

#### 4.4.2.2 Deviation Estimate

The linear quadratic optimization problem (4.79) is solved by writing the necessary optimality conditions of Karush, Kuhn, and Tucker (KKT), based on the Lagrangian. The Lagrangian of the problem (4.79) is

$$L(\delta x, v, \lambda) = \frac{1}{2}(\delta x - \delta\overline{x})^t M^{-1}(\delta x - \delta\overline{x}) + \frac{1}{2} v^t R^{-1} v + \lambda^t(\delta z - H\delta x - v) \tag{4.80}$$

where $\lambda \in \mathbb{R}^m$ denotes the multiplier of the constraint.

The necessary conditions for optimality at order one (KKT conditions) are

$$\begin{cases} \nabla_{\delta x} L = 0 \\ \nabla_v L = 0 \\ \nabla_\lambda L = 0 \end{cases} \Rightarrow \begin{cases} \delta x = \delta\overline{x} + MH^t\lambda \\ v = R\lambda \\ \delta z = H\delta x + v \end{cases} \tag{4.81}$$

The unknowns of this system are $(\delta x, v, \lambda)$. The KKT system (4.81) is now solved by two different approaches, each giving useful formulas.

- First solution method

The first method consists of substituting into the third equation $\delta x$ and v given by the first two equations. The resulting equation in $\lambda$ is then solved

$$\delta z = H(\delta\overline{x} + MH^t\lambda) + R\lambda \quad \Rightarrow \quad \lambda = (HMH^t + R)^{-1}(\delta z - H\delta\overline{x}) \tag{4.82}$$

and $\lambda$ is substituted in the first equation to obtain the solution $\delta x$.

$$\delta x = \delta\overline{x} + K(\delta z - H\delta\overline{x}) \quad \text{with} \quad K = MH^t(HMH^t + R)^{-1} \tag{4.83}$$

- Second solution method

The second method consists of expressing $\lambda$ as a function of $\delta x$ from the last two equations and substituting it into the first equation.

$$\lambda = R^{-1}v = R^{-1}(\delta z - H\delta x) \quad \Rightarrow \quad \delta x = \delta\overline{x} + MH^tR^{-1}(\delta z - H\delta x) \tag{4.84}$$

This equation is rearranged to make the term $\delta z - H\delta\overline{x}$ appear.

$$(I + MH^tR^{-1}H)\delta x = (I + MH^tR^{-1}H)\delta\overline{x} + MH^tR^{-1}(\delta z - H\delta\overline{x}) \tag{4.85}$$

Premultiplying by $M^{-1}$ gives the solution in a form similar to (4.83).

$$\delta x = \delta\overline{x} + K(\delta z - H\delta\overline{x}) \quad \text{with} \quad K = (M^{-1} + H^tR^{-1}H)^{-1}H^tR^{-1} \tag{4.86}$$

Both approaches give the same solution noted $\delta\hat{x}$, but with different expressions of the matrix K.

$$\begin{aligned} \delta\hat{x} &= \delta\overline{x} + K(\delta z - H\delta\overline{x}) \\ \text{with} \quad K &= MH^t(HMH^t + R)^{-1} = (M^{-1} + H^tR^{-1}H)^{-1}H^tR^{-1} \end{aligned} \tag{4.87}$$

The **innovation** term $\delta z - H\delta\overline{x}$ is the difference between the observed and the expected measurement, which is associated with the mean state deviation $\delta\overline{x}$. The innovation contains the information brought by the measurement.

The **gain** matrix K (n×m) weights the mean $\delta\overline{x}$ of the state before measurement and the information brought by the measurement. This weight depends on the uncertainties (covariances) on the state (M) and on the measurement (R).

#### 4.4.2.3 Error on the Estimate

The estimation error noted e is the difference between the deviation estimate $\delta\hat{x}$ and the true deviation $\delta x$, which is unknown.

$$e = \delta\hat{x} - \delta x = -(I - KH)(\delta x - \delta\overline{x}) + Kv \tag{4.88}$$

The error mean E[e] is null because we have $E[\delta x] = \delta\overline{x}$ (by definition) and $E[v] = 0$ (by assumption on the noise). The estimate $\delta\hat{x}$ is therefore **unbiased**.

The error covariance noted P is given by

$$\begin{aligned} P = E[ee^t] &= (I - KH)M(I - KH)^t + KRK^t \\ &= (I - KH)M - MH^tK^t + KHMH^tK^t + KRK^t \\ &= (I - KH)M - MH^tK^t + K(HMH^t + R)K^t \end{aligned} \tag{4.89}$$

Using (4.83): $K = MH^t(HMH^t + R)^{-1}$ to replace K, the third term in (4.89) is reduced to $MH^tK^t$, which cancels out with the second term. There remains

$$P = (I - KH)M \tag{4.90}$$

Another expression of P is obtained using the matrix inversion lemma, also called Sherman–Morrison–Woodbury formula ([R6], [R33]).

$$(A + UBV)^{-1} = A^{-1} - A^{-1}U(B^{-1} + VA^{-1}U)^{-1}VA^{-1} \tag{4.91}$$

Applying this lemma to the matrix appearing in (4.86), and using (4.73), we have

$$\begin{aligned} (M^{-1} + H^tR^{-1}H)^{-1} &= M - MH^t(R + HMH^t)^{-1}HM \\ &= M - KHM = (I - KH)M \end{aligned} \tag{4.92}$$

Comparing with (4.90), we retrieve the covariance matrix $P = (I - KH)M$.
The formula (4.92) therefore provides an alternate expression for the covariance.

$$P = (M^{-1} + H^tR^{-1}H)^{-1} \tag{4.93}$$

This formula makes it possible to write the gain K (4.87) in the form

$$K = PH^tR^{-1} \tag{4.94}$$

By substituting into (4.90), we obtain the relation

$$M = P + PH^tR^{-1}HM \tag{4.95}$$

Since all covariance matrices M, P, R are positive, so is the matrix $H^tR^{-1}H$.
For any vector $u \in \mathbb{R}^n$, we have therefore the inequality

$$u^tMu = u^tPu + u^tPH^tR^{-1}HMu \geq u^tPu, \ \forall u \quad \Rightarrow \quad M \geq P \tag{4.96}$$

This inequality means that the covariance P after measurement is "lower" than the covariance M before measurement. The effect of the measurement is to reduce the uncertainty associated with the estimation.

#### 4.4.2.4 Properties of the Deviation Estimate

The deviation estimate (4.87) has several interesting properties.

- Minimum covariance

An alternative approach to the formulation (4.79) is to look for a linear estimator that minimizes the mean of the error norm. In that case, the unknown is the gain matrix K ($n \times m$) and the optimization problem is formulated as:

$$\min_K F(K) = E[e^2] \quad \text{with} \quad \begin{cases} e = \delta\hat{x} - \delta x \\ \delta\hat{x} = \delta\overline{x} + K(\delta z - H\delta\overline{x}), \delta z = H\delta x + v \end{cases} \tag{4.97}$$

It can be observed that the cost function F is the trace of the matrix P.

$$e^2 = \text{Tr}(ee^t) \quad \Rightarrow \quad F(K) = E[e^2] = E[\text{Tr}(ee^t)] = \text{Tr}(E[ee^t]) = \text{Tr}(P) \tag{4.98}$$

The expression of P as a function of K is given by the first line of (4.89).

$$P = (I - KH)M(I - KH)^t + KRK^t \tag{4.99}$$

The optimality condition for the minimization problem (4.97) is $\frac{dF}{dK} = 0$, where $F(K) = Tr[P(K)] = Tr[(I - KH)M(I - KH)^t + KRK^t]$.

By applying the matrix formula: $\frac{\partial Tr(ABA^t)}{\partial A} = AB^t + AB$, we get the condition

$$\frac{\partial F}{\partial K} = 2(I - KH)M(-H^t) + 2KR = 0 \quad \Rightarrow \quad K = MH^t(HMH^t + R)^{-1} \tag{4.100}$$

We retrieve the formula already found for the gain K in (4.87), corresponding to the solution of the weighted least squares problem (4.79).

The weighted least squares estimator is therefore also the linear minimum covariance estimator. It is possible to obtain a lower covariance with a nonlinear estimator, but the formulation becomes more complex [R33].

- Decorrelation between error and estimation

The correlation matrix between the error e and the estimate $\delta\hat{x}$ is

$$E[e\delta\hat{x}^t] = E[(-(I - KH)(\delta x - \delta\overline{x}) + Kv)(\delta\overline{x} + KH(\delta x - \delta\overline{x}) + Kv)^t] \tag{4.101}$$

By expanding, replacing the means of $\delta x$ and v given by $E[\delta x] = \delta\overline{x}$, $E[v] = 0$ and with the assumption of an uncorrelated noise $E[\delta x(Kv)^t] = 0$, we obtain

$$\begin{aligned} E[e\delta\hat{x}^t] &= -(I - KH)E[(\delta x - \delta\overline{x})(\delta x - \delta\overline{x})^t]H^tK^t + KE[vv^t]K^t \\ &= -(I - KH)MH^tK^t + KRK^t \\ &= -PH^tK^t + KRK^t = 0 \quad \text{by using (4.94)} \end{aligned} \tag{4.102}$$

The error e is decorrelated from the estimate $\delta\hat{x}$, which means that the estimate contains all the information brought by the measurement. This allows a sequential update of the estimate after each new measurement, without loss of information. This sequential approach is the basis of the Kalman filter presented in the next section.

### 4.4.3 Kalman Filtering

The Kalman filter combines the differential correction (4.87) and the propagation of the difference between two consecutive measurements.

#### 4.4.3.1 Propagation of Deviation

The deviation equations, also called variational equations, are obtained by linearizing the equation (4.69) or (4.70) in the vicinity of a reference trajectory.

The Keplerian model is generally sufficient for this linearization, even if a more accurate propagation model is used for the reference trajectory.

The variational equations in position-velocity are derived from (4.69) by considering only the central term $\vec{a}_c$. In these equations, the gravity gradient matrix G is assessed on the reference trajectory $\vec{r}_{ref}$.

$$\begin{cases} \delta\dot{\vec{r}} = \delta\vec{v} \\ \delta\dot{\vec{v}} = G(\vec{r}_{ref})\delta\vec{r} \end{cases} \quad \text{with } G = \frac{\delta\vec{a}_c}{\delta\vec{r}} = \frac{\mu}{r^5}(3\vec{r}\,\vec{r}^t - r^2 I) \tag{4.103}$$

The variational equations in orbital parameters are derived from (4.70). By noting K = (a, e, i, Ω, ω) the set of orbital parameters other than the anomaly, we obtain in the Keplerian case (null perturbing acceleration $\vec{a}_p = \vec{0}$)

$$\begin{cases} \dot{K} = 0 \\ \dot{M} = n \end{cases} \quad \text{with } n = \sqrt{\frac{\mu}{a^3}} \quad \Rightarrow \quad \begin{cases} \delta\dot{K} = 0 \\ \delta\dot{M} = -\delta a\left(\frac{3n}{2a}\right)_{ref} \end{cases} \tag{4.104}$$

The deviation δK is constant, whereas the deviation δM has a constant rate (because δa is constant) and varies linearly with time. The variational equations (4.104) are integrated into

$$\begin{cases} \delta K(t) = \delta K(t_0) \\ \delta M(t) = -\delta a(t_0)\left(\frac{3n}{2a}\right)_{ref}(t - t_0) \end{cases} \tag{4.105}$$

In both cases (state in position-velocity or in orbital parameters), the variational equations (4.103) or (4.104) are of the form

$$\delta\dot{x} = F\delta x \quad \text{with } F = \frac{\partial f}{\partial x}(x_{ref}, t) \tag{4.106}$$

The solution of this linear system is expressed with its transition matrix Φ (n × n).

$$\delta x(t) = \Phi(t, t_0)\delta x(t_0) \tag{4.107}$$

Substituting δx given by (4.107) into (4.106), we obtain the differential equation followed by the transition matrix. This equation is linear and can be integrated.

$$\begin{cases} \dot{\Phi}(t, t_0) = F(t)\Phi(t, t_0) \\ \Phi(t_0, t_0) = I \end{cases} \quad \Rightarrow \quad \Phi(t, t_0) = e^{(t-t_0)F(t)} \tag{4.108}$$

Equation (4.107) is not a perfect representation of reality, for two reasons:

- it considers only the first-order term of the state equation (4.71);
- it neglects the perturbations γ (accounted in the reference trajectory, but not in the variational equations).

To compensate for these errors, a fictitious noise w, called model noise, is usually added to equation (4.107). This model noise has mean $\overline{w}$ and covariance Q. The covariance matrix Q shall be adjusted according to the assumed representativeness of the model. The deviation propagation between two consecutive times $t_{k-1}$ and $t_k$ takes the discrete form

$$\delta x_k = \Phi_{k-1}\delta x_{k-1} + w_{k-1} \tag{4.109}$$

#### 4.4.3.2 Linear Kalman Filter

The **Kalman filter** is composed of a propagation step and an update step. These two steps are based on the linearized equations (4.78) and (4.109), as follows:

- The estimated deviation at $t_{k-1}$ is $\delta\widehat{x}_{k-1}$ and it has a covariance matrix $P_{k-1}$.
- The deviation propagation by equation (4.109) yields the estimate $\delta\overline{x}_k$ and the covariance matrix $M_k$ before the measurement at $t_k$.

$$\delta x_k = \Phi_{k-1}\delta x_{k-1} + w_{k-1} \quad \Rightarrow \quad \begin{cases} \delta\overline{x}_k = \Phi_{k-1}\delta\widehat{x}_{k-1} + \overline{w}_{k-1} \\ M_k = \Phi_{k-1}P_{k-1}\Phi_{k-1}^t + Q_{k-1} \end{cases} \tag{4.110}$$

- The differential correction from the measurement deviation (4.78) yields the estimate $\delta\hat{x}_k$ and the covariance matrix $P_k$ after the measurement at $t_k$.

$$\delta z_k = H_k \delta x_k + v_k \quad \Rightarrow \quad \begin{cases} \delta\hat{x}_k = \delta\bar{x}_k + K_k(\delta z_k - H_k \delta\bar{x}_k) \\ P_k = \left(M_k^{-1} + H_k^t R_k^{-1} H_k\right)^{-1} \end{cases} \quad \text{with} \quad K_k = P_k H_k^t R_k^{-1} \tag{4.111}$$

Equations (4.110) and (4.111) form the discrete Kalman filter (1960). Since this filter has proved its practical efficiency, many variants have been developed such as continuous filter (continuous state and measurement), hybrid filter (continuous state and discrete measurements), and colored or state-correlated noise.

The matrix gain calculation and the covariance update can be formulated in different ways in order to limit numerical errors or to reduce the calculation load and/or memory storage [R33].

The Kalman filter can be adapted to update past estimates. **Smoothing** consists of both forward and backward filtering stages so that the estimation at each time takes into account the previous and subsequent measurements. The detailed procedures can be found in [R6] and [R33]. For a linear system, this approach is equivalent to the global fitting by least squares.

Model parameters $\gamma$ can also be estimated by the filter by being incorporated into the state vector (augmented state) with a null evolution equation $\dot{\gamma} = 0$.

#### 4.4.3.3 Nonlinear System

The Kalman filter theory assumes that the reference system is linear and perfectly modeled, with white uncorrelated zero-mean noises, and that the calculations are free of numerical errors. These assumptions are never satisfied in practice and the filter may diverge or converge to an erroneous estimate. The main difficulty comes from the nonlinearities. The general form of the state and measurement equations, with respective noises w and v, is as follows:

$$\begin{cases} \dot{x} = f(x, t, w) \\ z = h(x, t, v) \end{cases} \tag{4.112}$$

The Kalman filter can be adapted to this system using the following approaches.

The *linearized Kalman filter* (LKF) consists of linearizing equations (4.112) in the vicinity of a reference trajectory and applying the linear filter to the deviations (4.110) and (4.111), as presented just above. The reference trajectory is fixed, which allows the matrices $F_k$, $H_k$, $\Phi_k$, $P_k$, and $K_k$ to be calculated *in advance*. The $H_k$ and $K_k$ matrices are stored in memory and only the deviation estimate $\delta\hat{x}_k$ is updated with each new measurement. The LFK reduces on-line computations, but it is not very robust to nonlinearities. The potential modeling errors can be compensated by a large model noise covariance $Q_k$, or by a fading memory strategy by applying a coefficient $\varepsilon \geq 1$ in the covariance propagation (4.110): $M_k = \varepsilon \Phi_{k-1} P_{k-1} \Phi_{k-1}^t + Q_{k-1}$. This experimental setting is often delicate.

The *extended Kalman filter* (EKF) is similar, except that the reference path is no longer fixed. The reference state is updated with each new deviation estimate: $x_{ref}(t_k) \to x_{ref}(t_k) + \delta\hat{x}_k$. This requires a *real-time* reassessment of the matrices $F_k$, $H_k$, $\Phi_k$, $P_k$, and $K_k$. The linearization error is theoretically reduced because the new reference trajectory is closer to the real trajectory. Although more robust, the EKF remains subject to the same issues as the LKF.

The *unscented Kalman filter* (UKF, 1995) consists of estimating the mean and the covariance without linearization, using the following technique.

Let us assume that a random vector x of dimension n, mean $\bar{x}$ and covariance $P_x$ undergoes some transformation of the form $y = g(x)$. The empirical mean and covariance of y are assessed from a sample of 2n points noted $(y_i = g(x_i))_{i=1,\cdots,2n}$.

$$\bar{y} = \frac{1}{2n}\sum_{i=1}^{2n} y_i, \quad P_y = \frac{1}{2n}\sum_{i=1}^{2n}(y_i - \bar{y})(y_i - \bar{y})^t \tag{4.113}$$

By choosing the 2n points such that (other choices are possible)

$$x_i = \sqrt{n}L_i, \quad x_{n+i} = -x_i, \quad i = 1,\cdots,n \tag{4.114}$$

where $L_i$ is the column i of the matrix L derived from the Cholesky factorization: $P_x = LL^t$, a third-order approximation of the mean and covariance of y is obtained (instead of first order by an EKF-type linearization). This technique is applied to the propagation of the state and then to the measurement equation. It does not require Jacobian matrices ($F_k$, $H_k$) and can handle nonlinear noise or correlated noises (by increasing the state dimension).

Nonlinear systems can be treated with many other approaches such as adaptive Kalman filtering (to adjust the model noise), $H_\infty$ filtering (to find the worst-case noise by min/max optimization), particular filtering (based on a population of particles) … Greater robustness is achieved at the cost of a higher computational loads, which must remain compatible with real-time processing. The interested reader can consult [R36] for a detailed review of these variants.

## 4.5 Conclusion

### 4.5.1 The Key Points

- Orbit determination relies on measurements of distance by radar or laser ranging, velocity by the Doppler effect or direction by telescopes.
- The reduction consists of expressing the measurements in the geocentric frame after geometric, optical, atmospheric, and instrumental corrections.
- The Laplace and Gauss methods determine an approximate orbit from three-direction measurements.
- The least squares method is used in delayed time to globally adjust the orbit and dynamic model to the measurements.
- Kalman filtering is used in real time to sequentially correct the orbit estimate with each new measurement.

### 4.5.2 To Go Further

- BATTIN R. – *An Introduction to the Mathematics and Methods of Astrodynamics*. AIAA (1999).

The methods for preliminary orbit determination, based on range or direction measurements, are presented in Chapter 3.7.

- ESCOBAL P. R. – *Methods of Orbit Determination*. Wiley (1965).

This book is dedicated to the various algorithms of orbit determination, depending on the measurement type. It analyzes many variants with application cases.

- KOVALEVSKY J. – *Modern Astrometry – Second Edition*. Springer (2002).

This astronomy-oriented book is one of the best references regarding the reduction process. It details the various types of sensors and the corrections that can be applied to get the most reliable measurements.

- ROY A. E. – *Orbital Motion*. Adam Hilger (1988).

Chapter 3 presents the reduction process, with the different corrections that must be applied to the measurement. Chapter 13 presents the methods for preliminary orbit determination based on direction measurements, which are the most used in astronomy.

- VALLADO D. – *Fundamentals of Astrodynamics and Applications*. Microcosm Press, Springer (2007).

This book contains two chapters dedicated to the orbit determination problem. Chapter 7 presents the various methods for preliminary orbit determination, with algorithms and numerical examples. Chapter 10 details filtering methods that can be applied in real time, with many considerations about their practical implementation, in relation with the sensors used and their performance.

# 5

# Collision Risks

## 5.1 Introduction

The proliferation of space debris creates a growing threat to operational satellites as well as to the International Station. The number of debris larger than one centimeter is estimated to be several hundred thousand. A high-speed collision with even small debris can cause serious damage to a satellite. Among the most spectacular collisions were that of the Cerise satellite with a piece of an Ariane 1 upper stage on July 24, 1996, and that of the Cosmos 2251 and Iridium 33 satellites on February 10, 2009.

Continuous monitoring of debris orbits is necessary to anticipate dangerous conjunctions and plan avoidance maneuvers. The NORAD (North American Aerospace Defense Control) maintains a catalog of several thousand objects in orbit under the TLE (Two-Line Elements) format [R44]. Daily analyses, such as those provided by Celestrak [R45], yield a measurement of the risk in the form of a maximum collision probability, depending on the orbital uncertainties.

When a critical threshold is exceeded, an avoidance maneuver is planned. Such maneuvers are to be performed only if necessary. Indeed, they disrupt the satellite operational service or the Space Station microgravity experiments, and they also induce unexpected propellant consumption.

The risk assessment needs to be both fast, since it is carried out frequently for thousands of objects, and reliable to avoid unjustified warnings. Monte Carlo methods based on numerical simulation are the most reliable, but they are too costly in terms of computation time to handle all the debris at once. Specific faster techniques are necessary with the aim of assessing risks quickly but reliably.

The first section of this chapter recalls the main laws of orbital motion and introduces the modeling of uncertainties.

The second section presents the methods for determining a conjunction. These methods are based first on the geometric distance between the Keplerian orbits and second on the motion time law. They can also be applied to conjunctions between a launch trajectory and a satellite.

The third section is dedicated to the probabilistic formulation of the collision risk. The assessment is based on a simplified modeling of the motion in the vicinity of the conjunction. It provides an analytical expression of the probability depending on the miss-distance and the position uncertainty.

*Space Trajectories: Basic and Advanced Topics*, First Edition. Max Cerf.

## 5.2 Orbit Uncertainties

This chapter recalls the laws of orbital motion and the modeling of uncertainties. These models are used to predict conjunctions and assess the associated risks.

### 5.2.1 Orbital Motion

#### 5.2.1.1 Keplerian Orbit

A Keplerian orbit is the trajectory followed by a material point in a central gravitational field. This orbit is a conic characterized by its orbital parameters ($a$, $e$, $i$, $\Omega$, $\omega$, $\theta$) in the inertial geocentric frame (origin O at the center of the Earth, axis Z toward the North pole, axis X in the meridian of Greenwich). The orbital plane is defined by the inclination $i$ (angle with the equator) and the longitude of the ascending node $\Omega$. The ascending node A is the point where the equator is flown over northward (Figure 5.1). The argument of the perigee $\omega$ is the angle between the node A and the perigee P. The conic shape is defined by the semi-major axis $a$ and the eccentricity $e$. The true anomaly $\theta$ is the angle between the perigee and the current position on the orbit.

The **perifocal frame** $(\vec{u}_p, \vec{u}_q, \vec{u}_h)$ is obtained from the inertial geocentric frame $(\vec{u}_X, \vec{u}_Y, \vec{u}_Z)$ by the three successive rotations $\Omega\,[Z]$, $i\,[X']$, $\omega\,[Z']$. The first letter denotes the rotation angle, whereas the rotation axis is in brackets. By noting $R(\omega, i, \Omega)$ the transition matrix from $(\vec{u}_X, \vec{u}_Y, \vec{u}_Z)$ to $(\vec{u}_p, \vec{u}_q, \vec{u}_h)$, we have

$$(\vec{u}_p, \vec{u}_q, \vec{u}_h) = R(\omega, i, \Omega)(\vec{u}_X, \vec{u}_Y, \vec{u}_Z) \tag{5.1}$$

The rotation matrix $R(\omega, i, \Omega)$ is the product of the matrices of the three rotations.

$$R(\omega, i, \Omega) = \begin{pmatrix} \cos\omega & \sin\omega & 0 \\ -\sin\omega & \cos\omega & 0 \\ 0 & 0 & 1 \end{pmatrix}\begin{pmatrix} 1 & 0 & 0 \\ 0 & \cos i & \sin i \\ 0 & -\sin i & \cos i \end{pmatrix}\begin{pmatrix} \cos\Omega & \sin\Omega & 0 \\ -\sin\Omega & \cos\Omega & 0 \\ 0 & 0 & 1 \end{pmatrix} \tag{5.2}$$

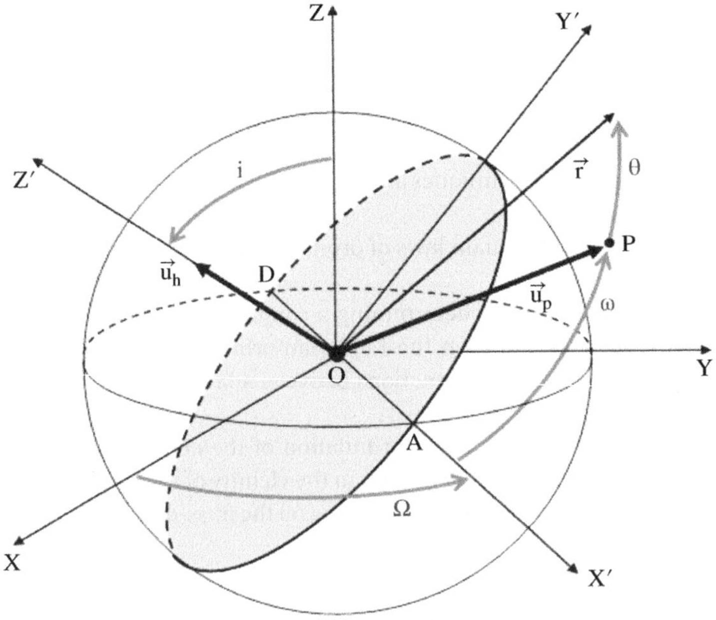


**Figure 5.1** Orbital parameters.

The unit vector $\vec{u}_p$ is directed toward the perigee (or toward the ascending node for a circular orbit). The unit vector $\vec{u}_h$ orthogonal to the orbital plane is oriented along the angular momentum $\vec{h} = \vec{r} \wedge \vec{v}$, where $\vec{r}$ and $\vec{v}$ denote the position and velocity, respectively.

The position is expressed in the perifocal frame by

$$\vec{r} = r(\vec{u}_p \cos\theta + \vec{u}_q \sin\theta) \quad \text{with } r = \frac{p}{1 + e\cos\theta}, \quad p = a(1 - e^2) = \frac{h^2}{\mu} \tag{5.3}$$

where p is the conic parameter and μ is the Earth's gravitational constant.

The true anomaly evolves as a function of time according to Kepler's equation. For an elliptical orbit, Kepler's equation takes the form (1.48)

$$E - e\sin E = n(t - t_p) \quad \text{with } \cos E = \frac{e + \cos\theta}{1 + e\cos\theta}, \quad n = \sqrt{\frac{\mu}{a^3}} = \frac{2\pi}{T} \tag{5.4}$$

where E is the eccentric anomaly, $t_p$ is the time of passage at the perigee, n is the pulsation (or mean motion), and T is the period of revolution given by Kepler's third law (1.41).

#### 5.2.1.2 Actual Trajectory

A Keplerian orbit is fully defined either by the position $\vec{r}_0$ and velocity $\vec{v}_0$ at a time $t_0$ or equivalently by the orbital parameters $(a, e, i, \Omega, \omega, \theta_0)$, where $\theta_0$ is the true anomaly at time $t_0$. The Keplerian orbit that would be followed from an initial time $t_0$ in the case of a central gravitational field $\vec{a}_c$ is called the **osculating orbit** at time $t_0$. Due to perturbing accelerations $\vec{a}_p$ (gravitational terms, other forces …), the osculating orbit changes with time. This evolution is assessed by differential equations in position–velocity $(\vec{r}, \vec{v})$ or in orbital parameters (a, e, i, Ω, ω, θ).

- The evolution in position–velocity follows Newton's second law.

$$\begin{cases} \dot{\vec{r}} = \vec{v} \\ \dot{\vec{v}} = \vec{a}_c + \vec{a}_p \end{cases} \quad \text{with } \vec{a}_c = -\frac{\mu}{r^3}\vec{r} \tag{5.5}$$

- The evolution in orbital parameters follows Gauss equations given in (2.45) and recalled hereafter.

$$\begin{cases} \dfrac{da}{dt} = \dfrac{2a^2}{h}[a_r e\sin\theta + a_\theta(1 + e\cos\theta)] \\ \dfrac{de}{dt} = \dfrac{p}{h}\left[a_r\sin\theta + a_\theta\left(\cos\theta + \dfrac{e + \cos\theta}{1 + e\cos\theta}\right)\right] \\ \dfrac{di}{dt} = a_h\dfrac{r\cos\nu}{h}, \quad \dfrac{d\Omega}{dt} = a_h\dfrac{r\sin\nu}{h\sin i} \\ \dfrac{d\omega}{dt} = \dfrac{r}{he}[-a_r\cos\theta(1 + e\cos\theta) + a_\theta\sin\theta(2 + e\cos\theta)] - a_h\dfrac{r\sin\nu}{h\tan i} \\ \dfrac{dM}{dt} = n + \dfrac{b}{aeh}[a_r(p\cos\theta - 2re) - a_\theta(p + r)\sin\theta] \end{cases} \tag{5.6}$$

In Gauss equations, $(a_r, a_\theta, a_h)$ are the perturbating acceleration components in the local orbital frame noted $(\vec{u}_r, \vec{u}_\theta, \vec{u}_h)$. The unit vector $\vec{u}_r$ is radial; the unit vector $\vec{u}_\theta$ is orthoradial in the orbital plane. The latitude argument $\nu = \omega + \theta$ is the angle between the ascending node and the current position.

More generally, the orbit is represented by a state vector x, either in Cartesian coordinates $x = (\vec{r}, \vec{v})$ or in orbital parameters $x = (a, e, i, \Omega, \omega, \theta)$.

This vector varies with time according to a state equation (also called evolution or dynamics equation) of the form

$$\dot{x} = f(x, t, \gamma)\,, \quad x(t_0) = x_0 \tag{5.7}$$

where $\gamma$ represents parameters of the perturbation models.

Integrating this differential system requires a numerical method. The formulation in orbital parameters (5.6) is more efficient when the perturbations are small and the propagation time is long. In this case, the second member of the equations (5.7) is close to zero and the numerical integration can be achieved with large time steps. In other cases, the formulation in position–velocity (5.5) is more stable.

## 5.2.2 Ellipsoid of Uncertainty

### 5.2.2.1 Deviation Propagation

Integrating the state equation (5.7) determines the trajectory associated with initial conditions $x_0$ and perturbations $\gamma$. Since the values of $x_0$ (initial state estimate) and $\gamma$ (force models) cannot be perfectly known, the propagation generates a growing state uncertainty over time.

The deviation from the reference (or nominal) trajectory is noted $\delta x$. Its evolution follows the variational equations obtained by expanding the state equations (5.7) to the first order.

$$\delta\dot{x} = \frac{\partial f}{\partial x}\delta x \underset{\text{noted}}{=} F\delta x\,, \quad \delta x(t_0) = \delta x_0 \tag{5.8}$$

In this equation, the gradient matrix F is evaluated along the nominal trajectory.

In the case of a state in position–velocity (5.5), the variational system (5.8) takes the simple form

$$\begin{cases} \delta\dot{\vec{r}} = \delta\vec{v} \\ \delta\dot{\vec{v}} = G\delta\vec{r} \end{cases} \quad \text{with } G = \frac{\partial \vec{a}_c}{\partial \vec{r}} \tag{5.9}$$

where G is the gravity gradient matrix accounting only for the central term $\vec{a}_c$.

Indeed, the perturbation $\vec{a}_p$ is generally small compared to the central term $\vec{a}_c$. The gradient of $\vec{a}_p$ is then of second order relative to the gradient of $\vec{a}_c$ and it can be neglected in the variational system.

The expression of the gravity gradient matrix G already established in (2.86) is

$$G = \frac{\mu}{r^5}\begin{pmatrix} 3x^2 - r^2 & 3xy & 3xz \\ 3xy & 3y^2 - r^2 & 3yz \\ 3xz & 3yz & 3z^2 - r^2 \end{pmatrix} = \frac{\mu}{r^5}(3\vec{r}\,\vec{r}^{\,t} - r^2 I_3) \tag{5.10}$$

where $I_3$ is the $3 \times 3$ identity matrix.

In the case of a state in orbital parameters (5.6), the variational equations are more complex. Their analytical expression can be found in [R16] or [R39].

The evolution of the deviation $\delta x$ depends on the initial condition $\delta x_0$, which is not perfectly known. This initial deviation is considered as a random variable. The resulting deviation $\delta x$ along the trajectory is therefore also a random variable with a covariance matrix noted C. The evolution equation of this covariance matrix is directly deduced from (5.8).

$$C \underset{\text{def}}{=} E[\delta x \delta x^t] \;\Rightarrow\; \dot{C} = FC + CF^t \tag{5.11}$$

#### 5.2.2.2 Circular Orbit

In the frequent case of a circular (or quasi-circular) orbit of semi-major axis a, the expression of the gravity gradient matrix G can be simplified to

$$\vec{a}_c = -\frac{\mu}{r^3}\vec{r} \approx \frac{\mu}{a^3}\vec{r} \quad \Rightarrow \quad G = -\frac{\partial \vec{a}_c}{\partial \vec{r}} \approx -\frac{\mu}{a^3}\frac{\partial \vec{r}}{\partial \vec{r}} = -n^2 I_3 \tag{5.12}$$

where n is the orbital pulsation.

With the matrix G being constant, equations (5.9) can be analytically integrated giving an explicit formula for the covariance $C_r$ related to the position.

$$\delta\vec{r} = e^{nt}\delta\vec{r}_0 \quad \Rightarrow \quad C_r = E[\delta\vec{r}\,\delta\vec{r}^t] = e^{2nt}E\left[\delta\vec{r}_0\,\delta\vec{r}_0^t\right] = e^{2nt}C_{r0} \tag{5.13}$$

This simplified formula shows that the position uncertainty grows exponentially over time at a rate equal to the orbital pulsation.

#### 5.2.2.3 Characteristic Dimensions

The position uncertainty is represented by an ellipsoid whose main axes are along the tangential, radial, and normal directions. The characteristic dimensions of this ellipsoid can be estimated for a circular orbit by considering that the main uncertainties come from the initial velocity. The velocity v and the period T depend on the orbit radius a by

$$v = \sqrt{\frac{\mu}{a}}\,, \quad T = 2\pi\sqrt{\frac{a^3}{\mu}} \tag{5.14}$$

Let us consider successively errors on the initial velocity modulus, then on the initial velocity direction.

- Error on the initial velocity modulus

A tangential error $\delta v$ induces an error $\delta a$ on the radius and an error $\delta T$ on the period. These errors are expressed by expanding (5.14) to the first order.

$$\delta a = -\frac{2a}{v}\delta v\,, \quad \delta T = \frac{3\pi}{v}\delta a \tag{5.15}$$

This results after one orbital period in tangential $\delta r_t$ and radial $\delta r_r$ position errors.

$$\begin{cases} \delta r_t = v\delta T = 3\pi\delta a \\ \delta r_r = \delta a \end{cases} \quad \Rightarrow \quad \frac{\delta r_t}{\delta r_r} \approx 10 \tag{5.16}$$

The ratio between the tangential and radial error is 10 after one period.

- Error on the initial velocity direction

A normal error $\delta v$ in the orbital plane induces an error $\delta i$ on the inclination. This results in a normal position error $\delta r_n$ (normal to the orbital plane), which reaches a maximum after one-quarter orbit and comes back to zero after a half orbit.

For $\delta i$ small, we have at the first order

$$\delta i = \frac{\delta v}{v}\,, \quad \delta r_n = a\delta i \tag{5.17}$$

Let us compare this normal error $\delta r_n$ to the tangential error $\delta r_t$ after one-quarter orbit (due to an error on the initial velocity modulus).

$$\begin{cases} \delta r_t = \frac{1}{4}v\delta T = \frac{3\pi}{2}\frac{a}{v}\delta v \\ \delta r_n = a\delta i = \frac{a}{v}\delta v \end{cases} \quad \Rightarrow \quad \frac{\delta r_t}{\delta r_n} \approx 5 \tag{5.18}$$

The ratio between the tangential and normal error is 5 after one-quarter period.

These rough assessments show that the uncertainties are in a ratio 10:2:1 along the tangential, normal, and radial directions, respectively. The uncertainty ellipsoid is mostly elongated in a direction close to the velocity. This tangential uncertainty ranges from 10 m to 10 km depending on the available measurement accuracy.

For the calculation of collision risks, the position will be considered as a Gaussian random vector whose covariance matrix C is defined by the ellipsoid at 1σ. In practical applications, this matrix is assumed to be diagonal in the local orbital frame with variances $\sigma_t^2, \sigma_n^2, \sigma_r^2$ in the tangential, normal, and radial directions.

### 5.2.3 Gaussian Model

This section recalls some properties of the Gaussian model that will be used for the assessment of collision risks.

#### 5.2.3.1 Gaussian Vector

A Gaussian vector x of dimension n follows the normal law N(μ, C) of density

$$p(x) = \frac{1}{\sqrt{(2\pi)^n \det C}} e^{-\frac{1}{2}(x-\mu)^t C^{-1}(x-\mu)} \tag{5.19}$$

The mean is $\mu = E(x)$ and the covariance is $C = E[(x-\mu)(x-\mu)^t]$.

The covariance matrix C is positive definite and it admits a Cholesky factorization into the form $C = LL^t$, where the matrix L is lower triangular [R5]. The variable change defined by $x = \mu + Ly$ transforms the Gaussian vector x of law N(μ, C) into a Gaussian vector y of law N(0, I) with independent components.

#### 5.2.3.2 Equiprobability Ellipsoid

The equiprobability surface of equation $(x-\mu)^t C^{-1}(x-\mu) = s^2$ defines an ellipsoid of volume V in dimension n. The probability of being inside this ellipsoid is

$$P(V) = \frac{1}{\sqrt{(2\pi)^n \det C}} \iiint_V e^{-\frac{1}{2}(x-\mu)^t C^{-1}(x-\mu)} dV_x \quad \text{with} \quad dV_x = dx_1 \ldots dx_n \tag{5.20}$$

This multiple integral is assessed in two stages.

The first stage is to transform this integral by the variable change $x = \mu + Ly$, where L is the triangular matrix coming from the Cholesky factorization of C.

- The ellipsoid becomes a sphere of radius s: $(x-\mu)^t C^{-1}(x-\mu) = y^t y = s^2$.
- The volume element is transformed by: $dV_x = (\det L) dV_y = \sqrt{\det C}\, dV_y$.

The integral (5.20) passed in variable y takes the form

$$P(V) = \frac{1}{\sqrt{(2\pi)^n}} \iiint_{y_1^2+\cdots+y_n^2 \le s^2} e^{-\frac{1}{2}(y_1^2+\cdots+y_n^2)} dy_1 \cdots dy_n \tag{5.21}$$

The second stage is to pass in spherical coordinates with $r^2 = y_1^2 + \cdots + y_n^2$.

The multiple integral (5.21) becomes then a simple integral. The Cartesian volume element $dV_y = dy_1 \cdots dy_n$ is replaced by the spherical volume element $dV_r$, whose expression depends on the dimension n. The results are [R5]

$$\begin{aligned}
n = 1:\ dV_r &= dr & \rightarrow P(V) &= \frac{1}{\sqrt{2\pi}}\int_0^s e^{-\frac{1}{2}r^2}dr = \mathrm{Erf}\left(\frac{s}{\sqrt{2}}\right)\\
n = 2:\ dV_r &= 2\pi r dr & \rightarrow P(V) &= \int_0^s e^{-\frac{1}{2}r^2} r\,dr = 1 - e^{-\frac{1}{2}s^2}\\
n = 3:\ dV_r &= 4\pi r^2 dr & \rightarrow P(V) &= \sqrt{\frac{2}{\pi}}\int_0^s e^{-\frac{1}{2}r^2} r^2 dr = \mathrm{Erf}\left(\frac{s}{\sqrt{2}}\right) - \sqrt{\frac{2}{\pi}} s e^{-\frac{1}{2}s^2}
\end{aligned} \tag{5.22}$$

#### 5.2.3.3 Error Function

The distribution function of the normal law given by (5.22) is expressed with the error function defined as

$$\mathrm{Erf}\,(s) = \frac{2}{\sqrt{\pi}}\int_0^s e^{-u^2}du \tag{5.23}$$

This function is evaluated numerically by a series expansion, as in [R5] and [R11].

$$\mathrm{Erf}\,(s) = \frac{2}{\sqrt{\pi}}\sum_{k=0}^{\infty} u_k \quad \text{with } u_0 = s\,,\quad u_{k+1} = -u_k s^2 \frac{(2k+1)}{(k+1)(2k+3)} \tag{5.24}$$

For values of $s > 4$, an approximation to within $10^{-10}$ is given by

$$\mathrm{Erf}\,(s) \approx 1 - \frac{e^{-s^2}}{s\sqrt{\pi}} \quad \text{for } s > 4 \tag{5.25}$$

For $s = 6$, Erf(s) is worth 1 to within $10^{-17}$, which is beyond the numerical accuracy in double precision calculation. The probabilities (5.22) for the ellipsoids associated with the usual standard deviation values ($s = 1, 2, 3$) are [R5]

| | s = 1 | s = 2 | s = 3 |
|---|---|---|---|
| n = 1 | 0.683 | 0.955 | 0.997 |
| n = 2 | 0.394 | 0.865 | 0.989 |
| n = 3 | 0.200 | 0.739 | 0.971 |

#### 5.2.3.4 Marginal Law

Let us assume that the Gaussian vector x of dimension n is partitioned into two sub-vectors u and v of dimensions $n_u$ and $n_v$, respectively. The mean $\mu$, the covariance C, and its inverse $C^{-1}$ are partitioned accordingly.

$$x = \begin{pmatrix} u \\ v \end{pmatrix},\quad \mu = \begin{pmatrix} \mu_u \\ \mu_v \end{pmatrix},\quad C = \begin{pmatrix} C_{uu} & C_{uv} \\ C_{uv}^t & C_{vv} \end{pmatrix},\quad C^{-1} \underset{\text{noted}}{=} A = \begin{pmatrix} A_{uu} & A_{uv} \\ A_{uv}^t & A_{vv} \end{pmatrix} \tag{5.26}$$

The submatrices of C and A are linked by the relations

$$CA = I \Rightarrow \begin{cases} C_{uu}A_{uu} + C_{uv}A_{uv}^t = I \\ C_{uu}A_{uv} + C_{uv}A_{vv} = 0 \end{cases} \Rightarrow C_{uu}\left(A_{uu} - A_{uv}A_{vv}^{-1}A_{uv}^t\right) = I \tag{5.27}$$

The marginal law of u is obtained by integrating the law of x with respect to v.

$$p_u(u) = \int_{v=-\infty}^{v=+\infty} p(u, v)\,dv \tag{5.28}$$

To assess this integral, the quadratic form $Q = (x - \mu)^t C^{-1}(x - \mu)$ appearing in the exponential (5.19) is expanded in u and v. The corresponding centered variables are noted $u_c = u - \mu_u$ and $v_c = v - \mu_v$.

Applying the variable change $w = v_c + A_{vv}^{-1} A_{uv}^t u_c$ and using the formula (5.27) to make $C_{uu}^{-1}$ appear, the quadratic form Q is written as:

$$\begin{aligned} Q = (x-\mu)^t C^{-1}(x-\mu) &= \begin{pmatrix} u_c \\ v_c \end{pmatrix}^t \begin{pmatrix} A_{uu} & A_{uv} \\ A_{uv}^t & A_{vv} \end{pmatrix} \begin{pmatrix} u_c \\ v_c \end{pmatrix} \\ &= u_c^t A_{uu} u_c + 2u_c^t A_{uv} v_c + v_c^t A_{vv} v_c \\ &= u_c^t \left(A_{uu} - A_{uv} A_{vv}^{-1} A_{uv}^t\right) u_c + w^t A_{vv} w \\ &= u_c^t C_{uu}^{-1} u_c + w^t A_{vv} w \end{aligned} \tag{5.29}$$

This allows to express the integral (5.28) with the variable w instead of v.

$$p_u(u) = \frac{1}{\sqrt{(2\pi)^n \det C}} e^{-\frac{1}{2} u_c^t C_{uu}^{-1} u_c} \int_{w=-\infty}^{w=+\infty} e^{-\frac{1}{2} w^t A_{vv} w}\, dw \tag{5.30}$$

The variable w, which is a linear combination of $u_c$ and $v_c$, is itself Gaussian and centered. The integral in (5.30) is that of a normal law of dimension $n_v$ and covariance $A_{vv}^{-1}$. Its value is therefore

$$\int_{w=-\infty}^{w=+\infty} e^{-\frac{1}{2} w^t A_{vv} w}\, dw = \sqrt{(2\pi)^{n_v} \det A_{vv}^{-1}} \tag{5.31}$$

Substituting the integral (5.31) into (5.30), we find that the marginal law of u is the normal law of dimension $n_u$ $(= n - n_v)$, deduced from the full law of x by simply removing the $n_v$ components of v.

$$p_u(u) = \frac{1}{\sqrt{(2\pi)^{n_u} \det C_{uu}}} e^{-\frac{1}{2}(u-\mu_u)^t C_{uu}^{-1}(u-\mu_u)} \tag{5.32}$$

In particular, the covariance matrix of u is directly the submatrix $C_{uu}$.

This property will be used in Section 5.4.1 to calculate the collision probability.

#### 5.2.3.5 Maximum Value

When the covariance matrix C is not known (as often happens), one can look for the matrix C that will maximize the density (5.19) in a given point $x_0$.

$$\max_C p(x_0) = \frac{1}{\sqrt{(2\pi)^n \det C}} e^{-\frac{1}{2}(x_0-\mu)^t C^{-1}(x_0-\mu)} \tag{5.33}$$

By Cholesky factorization, the positive definite matrix C can be put under the form $C = LDL^t$, where L is a lower triangular matrix and D is the diagonal matrix of the eigenvalues $\sigma_1^2, \cdots, \sigma_n^2$ of C.

Applying the variable change $x = \mu + Ly$, the density (5.33) becomes

$$p(y_0) = \frac{1}{\sqrt{(2\pi)^n \sigma_1 L \sigma_n}} e^{-\frac{1}{2}\left[\left(\frac{y_{01}}{\sigma_1}\right)^2 + \cdots + \left(\frac{y_{0n}}{\sigma_n}\right)^2\right]} \tag{5.34}$$

The values $\sigma_1, \cdots, \sigma_n$ maximizing the density (5.34) in the point $y_0$ (associated with the given point $x_0$ by $x_0 = \mu + Ly_0$) are then

$$\frac{\partial p}{\partial \sigma_k}(y_0) = 0 \quad \Rightarrow \quad \sigma_k = y_{0k} \quad \text{for } k = 1, \cdots, n \tag{5.35}$$

This result extends to the limit case $y_0 = 0$ by considering that the normal law of zero standard deviation is a Dirac function.

The formula (5.35) will be used in Section 5.4.4 to estimate the maximum probability of collision when the covariance matrices are not known.

## 5.3 Conjunction

To ensure the safety of an operational space vehicle, it is necessary to anticipate possible conjunctions with other orbiting objects, which can be active satellites or space debris. The collision risk also exists for launchers when the ascent trajectory reaches high altitudes above the Low Earth Orbits (LEO) region. The alert shall be given as early as possible, so that an avoidance procedure (maneuver or launch delay) can be planned if the collision risk is too high. This section presents the conjunction assessment methods for satellites and for launchers.

### 5.3.1 Numerical Simulation

In order to detect a conjunction, it is necessary to integrate numerically the motion equations (5.5) or (5.6) taking into account all forces. The distance between two objects in orbit varies rapidly due to their respective velocities, which are of the order of 8 km/s. The integration step must be sufficiently small (typically one second at most) to detect a miss-distance within 10 km.

Consequently, numerical simulation cannot be applied systematically to predict conjunctions with numerous objects over several days. It would lead to prohibitive computation times. To reduce the calculations, a preliminary analysis is done from the osculating orbits, which can be considered as sufficiently good approximations for short-term predictions.

In the first stage, the analysis is based only on the geometric distance (called **MOID** for Minimum Orbit Intersection Distance):

- between two Keplerian ellipses in the case of an operational satellite, with the method presented in Section 5.3.2;
- between the launch trajectory and a Keplerian ellipse in the case of a launcher, with the method presented in Section 5.3.3.

If the geometric distance is below a given threshold, the second stage consists of comparing the dates of passage of the two objects in the conjunction point. This yields the **miss-distance** accounting for the temporal motion of each object. This miss-distance, still based on the assumption of Keplerian motions, is necessarily higher than the MOID, which is based solely on geometrical considerations.

If the miss-distance is lower than a given threshold (large enough to account for the imprecision of the Keplerian model), the conjunction is critical and it is further analyzed by numerical simulation with an appropriate time step, as explained above. Nevertheless, knowing the time of passage reduces drastically the computation load, since the integration step only needs to be refined in the vicinity of the conjunction point predicted by the Keplerian assessment.

### 5.3.2 Orbit-to-Orbit Distance

In the case of two satellites, the distance is assessed from the Keplerian orbits of both objects. Figure 5.2 shows two elliptical orbits with their respective perifocal frames $(\vec{u}_{p1}, \vec{u}_{q1}, \vec{u}_{h1})$ and

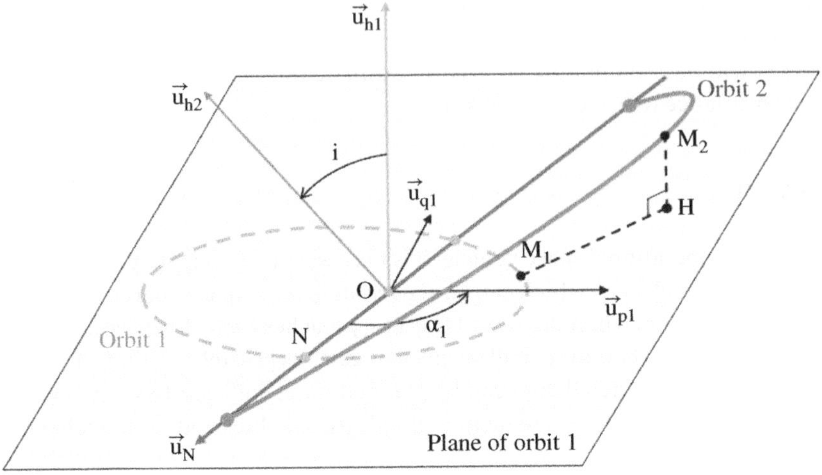


**Figure 5.2** Geometric distance between orbits.

$(\vec{u}_{p2}, \vec{u}_{q2}, \vec{u}_{h12})$. The two orbital planes make an angle i and they intersect along the node axis (solid line) oriented by the unit vector $\vec{u}_N$.

#### 5.3.2.1 Distance Calculation

A point $M_2$ of orbit 2 (blue solid line) is projected in H on the plane of orbit 1. The distance of $M_2$ to any point $M_1$ in orbit 1 is given by

$$(M_1M_2)^2 = (HM_1)^2 + (HM_2)^2 \tag{5.36}$$

because $HM_2$ is orthogonal to the plane of orbit 1.

For a given point $M_2$, the point H is fixed. The point of orbit 1 closest to $M_2$ is then the projection $M_1$ of H on orbit 1 (green dotted line).

The distance $HM_1$ is the distance between a point and an ellipse. It is calculated by **Bowring's formulas**, which are also used to calculate the geodetic altitude (altitude relative to the Earth's shape called geoid).

Bowring's formulas are given hereafter in Figure 5.3. The demonstration can be found in [4].

Bowring's formulas give the distance between the point $H(x_0, y_0)$ and the ellipse of equation $\frac{x^2}{a^2} + \frac{y^2}{b^2} = 1$, eccentricity $e = \frac{\sqrt{a^2-b^2}}{a}$ and flattening $f = 1 - \frac{b}{a}$.

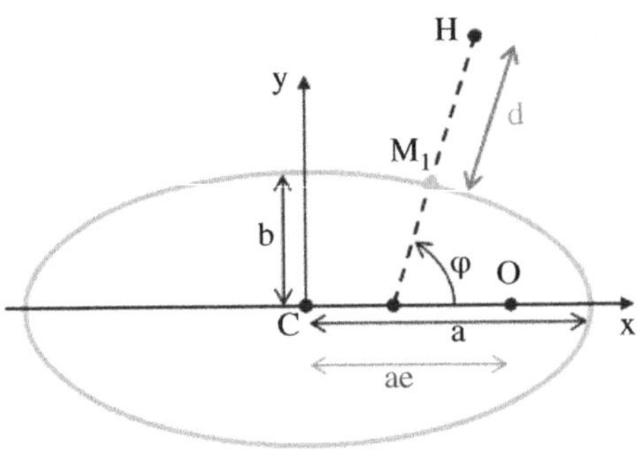


$$\tan\delta = \frac{y_0}{x_0}\left[1 - f + \frac{ae^2}{\sqrt{x_0^2+y_0^2}}\right]$$

$$\tan\varphi = \frac{y_0(1-f) + ae^2\sin^3\delta}{(1-f)(x_0 - ae^2\cos^3\delta)}$$

$$d = x_0\cos\varphi + y_0\sin\varphi - a\sqrt{1-e^2\sin^2\varphi}$$

**Figure 5.3** Distance between a point and an ellipse (Bowring's formulas).

These formulas are given in the frame (Cxy) having for origin the ellipse center and with axes along the ellipse major and minor axes.

For the application to the MOID calculation, the x and y axes are directed along the vectors $\vec{u}_{p1}$ and $\vec{u}_{q1}$ associated with the orbit 1 perifocal frame. The orbit center is on the x-axis at the distance CO = ae from the focus O (which is the Earth's center).

The following procedure yields the distance between any point $M_2$ of anomaly $\theta_2$ on orbit 2 and orbit 1 (Figure 5.2).

- Position of $M_2$: $\overrightarrow{OM_2} \underset{\text{noted}}{=} \vec{r}_2 = \frac{p_2}{1+e_2\cos\theta_2}(\vec{u}_{p2}\cos\theta_2 + \vec{u}_{q2}\sin\theta_2)$, from equation (5.3)
- Projection H on the plane of orbit 1: $\overrightarrow{OH} \underset{\text{noted}}{=} \vec{r}_H = \vec{r}_2 - (\vec{r}_2\cdot\vec{u}_{h1})\vec{u}_{h1}$
- Distance $HM_2$: $d_2 = HM_2 = |\vec{r}_2\cdot\vec{u}_{h1}|$
- Coordinates of H in (Cxy): $x_H = \vec{r}_H\cdot\vec{u}_{p1} + a_1e_1\ ,\ \ y_H = \vec{r}_H\cdot\vec{u}_{q1}$
- Distance from H to orbit 1: $d_1 = HM_1$ (by Bowring's formulas)
- Distance from $M_2$ to orbit 1: $d = \sqrt{d_1^2 + d_2^2}$

With this calculation procedure, the distance depends only on the anomaly $\theta_2$ of the point $M_2$ on orbit 2. The minimum geometric distance MOID between the two orbits is then found by scanning over the anomaly and/or using a one-dimensional minimization method (such as Fibonacci or golden ratio).

#### 5.3.2.2 Optimality Check

The optimality of the result is checked by calculating the gradient of the distance with respect to the variables $(\theta_1, \theta_2)$. To simplify the calculations, the positions are expressed in frames based on the node axis $\vec{u}_N$ of the two orbital planes.

The unit vector $\vec{u}_N$ is along $\vec{u}_{h1}\wedge\vec{u}_{h2}$ and it makes the angle $\alpha_1$ with $\vec{u}_{p1}$ as shown in Figure 5.2. The frame $(\vec{u}_{N1}, \vec{u}_{Q1}, \vec{u}_{h1})$ is derived from the perifocal frame $(\vec{u}_{p1}, \vec{u}_{q1}, \vec{u}_{h1})$ by a rotation of angle $-\alpha_1$ around $\vec{u}_{h1}$. The frame $(\vec{u}_{N1}, \vec{u}_{Q2}, \vec{u}_{h2})$ is defined in the same way for orbit 2. In these frames based on the node vector $\vec{u}_N$, the formula (5.3) applied in $M_1$ and $M_2$ (k = 1,2) becomes

$$\vec{r}_k = r_k(\vec{u}_N\cos\nu_k + \vec{u}_{Qk}\sin\nu_k)\quad\text{with}\quad \nu_k = \theta_k + \alpha_k,\ \ r_k = \frac{p_k}{1+e_k\cos\theta_k} \tag{5.37}$$

The distance $d = \|\vec{r}_2 - \vec{r}_1\|$ is minimum if

$$\frac{\partial}{\partial\theta_k}(\vec{r}_2 - \vec{r}_1)^2 = 0\quad\text{for } k = 1,2 \tag{5.38}$$

Substituting with (5.37) and deriving leads to the following conditions on $(\theta_1, \theta_2)$

$$\begin{cases} e_1 r_1 \sin\theta_1 (r_2\cos\alpha_0 - r_1) - p_1 r_2\cos\beta_0 = 0 \\ e_2 r_2 \sin\theta_2 (r_1\cos\alpha_0 - r_2) - p_2 r_1\cos\beta_0 = 0 \end{cases} \tag{5.39}$$

where the angles $(\alpha_0, \beta_0)$ depend on $(\theta_1, \theta_2)$ by

$$\begin{cases} \cos\alpha_0 = \cos\nu_1\cos\nu_2 + \sin\nu_1\ \sin\nu_2\cos i \\ \cos\beta_0 = \sin\nu_1\cos\nu_2 - \cos\nu_1\sin\nu_2\cos i \end{cases}\quad\text{with}\quad \begin{cases} \nu_1 = \theta_1 + \alpha_1 \\ \nu_2 = \theta_2 + \alpha_2 \end{cases} \tag{5.40}$$

The conditions (5.39) are used to check that the solution $(\theta_1, \theta_2)$ is optimal.

The algebraic resolution of the system (5.39) is also possible, but more complex. It leads after long calculations to look for the roots of a polynomial of degree 16. The detailed procedure developed by G.F. Gronchi can be found in [R19].

Once the anomalies $(\theta_1, \theta_2)$ of the conjunction point are known for both orbits, the time of passage of each object at this point is given by Kepler's equation (5.4) and the miss-distance can be calculated.

#### 5.3.2.3 Collision Avoidance

When the miss-distance with another orbiting object (generally a debris) is below a prescribed threshold of a few kilometers, a collision avoidance maneuver must be planned. The goal is to increase the miss-distance up to a sufficient value ensuring the satellite safety. The calculation of the maneuver is based on the variational equations as follows.

The position and velocity of the satellite on its nominal orbit are noted $(\vec{r}_{nom}, \vec{v}_{nom})$ and the deviations $(\delta\vec{r}, \delta\vec{v})$ follow the variational equations (5.9).

$$\begin{cases} \delta\dot{\vec{r}} = \delta\vec{v} \\ \delta\dot{\vec{v}} = G\delta\vec{r} \end{cases} \quad \text{with } G = \frac{\partial\vec{a}_c}{\partial\vec{r}} = \frac{\mu}{r^5}(3\vec{r}\vec{r}^t - r^2 I_3) \tag{5.41}$$

As mentioned in Section 5.2.2, even if the nominal trajectory is computed with the full force model, it is sufficient to consider only the central term for the gravity gradient matrix G, which is assessed along the nominal trajectory.

These equations form a linear differential system of matrix F(t).

$$\begin{pmatrix} \delta\dot{\vec{r}} \\ \delta\dot{\vec{v}} \end{pmatrix} = F(t)\begin{pmatrix} \delta\vec{r} \\ \delta\vec{v} \end{pmatrix} \quad \text{with } F(t) = \begin{pmatrix} 0 & I \\ G[\vec{r}_{nom}(t)] & 0 \end{pmatrix} \tag{5.42}$$

The solution is expressed with the transition matrix $\Phi(t, t_0)$

$$\begin{pmatrix} \delta\vec{r} \\ \delta\vec{v} \end{pmatrix} = \Phi(t, t_0)\begin{pmatrix} \delta\vec{r}_0 \\ \delta\vec{v}_0 \end{pmatrix} \tag{5.43}$$

which follows the differential equation

$$\dot{\Phi}(t, t_0) = F(t)\Phi(t, t_0) \quad \text{with } \Phi(t_0, t_0) = I \tag{5.44}$$

In the case of a two-body problem (central term only), the transition matrix can be calculated analytically from Lagrange coefficients introduced in Section 1.3.4.

$$\begin{cases} \vec{r} = f_r\vec{r}_0 + g_r\vec{v}_0 \\ \vec{v} = f_v\vec{r}_0 + g_v\vec{v}_0 \end{cases} \tag{5.45}$$

The Lagrange coefficients are noted here $(f_r, g_r, f_v, g_v)$ in order not to be confused with the F and G matrices. Their analytic expressions depending on the position, the velocity, and the orbital parameters are given either by (1.71) using the true anomaly or by (1.77) using the eccentric anomaly.

By deriving the Lagrange coefficients with respect to the initial conditions $(\vec{r}_0, \vec{v}_0)$, an explicit expression of the transition matrix can be obtained.

The tedious calculations are detailed in [R1] and they yield the transition matrix as decomposition into four sub-matrices, related to position and velocity.

$$\Phi(t, t_0) = \begin{pmatrix} \dfrac{\partial\vec{r}}{\partial\vec{r}_0} & \dfrac{\partial\vec{r}}{\partial\vec{v}_0} \\ \dfrac{\partial\vec{v}}{\partial\vec{r}_0} & \dfrac{\partial\vec{v}}{\partial\vec{v}_0} \end{pmatrix} \underset{\text{noted}}{=} \begin{pmatrix} \Phi_{rr} & \Phi_{rv} \\ \Phi_{vr} & \Phi_{vv} \end{pmatrix} \tag{5.46}$$

This matrix allows to assess the effect of an impulsive maneuver $\delta\vec{v}(t)$ at any time t on the final position deviation $\delta\vec{r}(t_f)$, where the final time $t_f$ corresponds to the predicted conjunction. The formula (5.43) gives

$$\delta\vec{r}(t_f) = \Phi_{rv}(t_f, t)\delta\vec{v}(t) \tag{5.47}$$

By inverting this relation and using the composition property to use only transition matrices referring to the initial time $t_0$, we get

$$\delta\vec{v}(t) = [\Phi(t_f, t_0)\Phi^{-1}(t, t_0)]_{rv}^{-1}\delta\vec{r}(t_f) \tag{5.48}$$

This formula yields the required impulse to apply at any time t in order to achieve the desired position deviation at the conjunction time $t_f$.

The required impulse can be minimized by choosing the direction of $\delta\vec{r}(t_f)$ along the smallest eigenvector of the matrix $[\Phi(t_f, t_0)\Phi^{-1}(t, t_0)]_{rv}^{-1}$.

When approaching the final time, the transition matrix $\Phi(t_f, t)$ tends to identity. The sub-matrix $\Phi_{rv}(t_f, t)$ in (5.46) tends therefore to zero, meaning that the required impulse increases as the conjunction approaches. Indeed, the sooner the avoidance maneuver is performed, the more effective it is.

### 5.3.3 Trajectory-to-Orbit Distance

In the case of a launcher, the possible conjunctions are assessed from the launch trajectory and from the Keplerian orbit of the space object (satellite or debris).

#### 5.3.3.1 Launcher Trajectory

The launch trajectory, starting at a nominal take-off time $t_0$, has been computed beforehand and it provides the launcher spherical coordinates $(t_k, r_k, \phi_k, \lambda_k)_{k=1,\cdots,n}$ at discrete times in the inertial geocentric frame (OXYZ) depicted in Figure 5.1.

The nominal launch trajectory is the result of a specific optimization process, taking into account several path constraints and final constraints, as will be described in Chapter 11. It cannot be changed except for the take-off time, which can be shifted within a prescribed launch window. The launch trajectory will then be exactly the same as the nominal one, with a rotation in longitude from an angle $\Delta\lambda = \Omega_T\Delta t$, where $\Omega_T$ is the Earth's rotation rate and $\Delta t$ is the take-off time shift.

#### 5.3.3.2 Satellite Trajectory

The satellite follows, in first approximation, a Keplerian orbit defined by its orbital parameters $(a, e, i, \Omega, \omega, \theta_0)$ in the inertial geocentric frame (OXYZ). This orbit is defined at some time $\tau_0$ and $\theta_0$ is the true anomaly at that time.

The true anomaly $\theta$ varies with time according to Kepler's equation (5.4). The spherical coordinates of the satellite are determined by the conic equation for the radius and by the formulas (1.121) for the latitude $\varphi$ and the longitude $\lambda$.

$$r = \frac{p}{1 + e\cos\theta}, \quad \begin{cases} \sin\varphi & = \sin(\omega + \theta)\sin i \\ \cos\varphi\ \sin(\lambda - \Omega) & = \sin(\omega + \theta)\cos i \\ \cos\varphi\ \cos(\lambda - \Omega) & = \cos(\omega + \theta) \end{cases} \tag{5.49}$$

#### 5.3.3.3 Distance Calculation

The goal is to find if a geometric conjunction is possible or not, depending on the take-off time. This analysis is based on the spherical coordinates of the nominal launch trajectory, which are followed in the same way whatever the take-off time.

For each discrete launcher position $(t_k, r_k, \varphi_k, \lambda_k)_{k=1,\cdots,n}$, we examine if the satellite can pass at the same radius or at the same latitude.

- The satellite will pass at the radius $r_k$ for $\cos\theta_k = \frac{1}{e}\left(\frac{p}{r_k} - 1\right)$.
  This determines 0 or 2 possible anomalies $\theta_k$ of conjunctions.
- The satellite will pass at the latitude $\varphi_k$ for $\sin(\omega + \theta_k) = \dfrac{\sin\varphi_k}{\sin i}$.
  This determines 0 or 2 additional possible anomalies $\theta_k$ of conjunctions.

For each anomaly $\theta_k$ of potential conjunction, we assess the satellite time $\tau_k$ by Kepler's equation and its spherical coordinates noted $(\tilde{r}_k, \tilde{\varphi}_k, \tilde{\lambda}_k)$ by (5.49). By construction, one of the coordinates $(\tilde{r}_k, \tilde{\varphi}_k)$ matches that of the launcher $(r_k, \varphi_k)$.

On the nominal trajectory taking off at $t_0$, the launcher passes in the conjunction point at time $t_k$, whereas the satellite passes at time $\tau_k$. The time of passage could be the same if the launch were shifted by $\Delta t_k = \tau_k - t_k$. The longitude would be shifted by $\Delta\lambda_k = \Omega_T \Delta t_k$ and the distance $d_k$ at the conjunction point would be

$$d_k^2 = [\tilde{r}_k - r_k]^2 + [r_k(\tilde{\varphi}_k - \varphi_k)]^2 + [r_k \cos\varphi_k(\tilde{\lambda}_k - \lambda_k - \Omega_T \Delta t_k)]^2 \tag{5.50}$$

This distance is calculated from the vertical component $\tilde{r}_k - r_k$, the northward component along the local meridian of radius $r_k$ and the eastward component along the local parallel of radius $r_k \cos\phi_k$.

#### 5.3.3.4 Collision Avoidance

If the distance $d_k$ is lower than a given threshold $d_{min}$ of a few kilometers, the conjunction is critical and the launch cannot occur at the take-off time $t_0 + \Delta t_k$. The launch window must be temporarily closed as long as the miss-distance remains below $d_{min}$. The corresponding closure times $[t_0 + \Delta t_{min}; t_0 + \Delta t_{max}]$ can be estimated from the velocity difference between the launcher and the satellite at the conjunction point.

This preliminary analysis considers a Keplerian motion for the satellite and a nominal trajectory for the launcher. It identifies the forbidden launch slots and the dangerous space objects that can be met along the ascent trajectory. It must be refined by numerical simulations accounting for the non-Keplerian motion of the satellite and for the launcher trajectory envelope in the presence of perturbations. This refined assessment can be processed either by covariance analysis or by Monte Carlo simulations, with the goal of determining the statistical risk associated with the launch time.

### 5.3.4 Combined Covariance

When an estimated conjunction time (noted $t_0$) has been found by simulation with reduced time steps, the accurate time $t_m$ of minimum distance is calculated by considering that the object motion is locally close to a straight line. This assumption is justified in most cases by the very short duration of the conjunction and the corresponding risk assessment method is presented in Section 5.4.1. Long conjunctions, which are possible for objects in close orbits, are dealt with by a specific approach developed in Section 5.4.5.

The two objects have respective positions $\vec{r}_{10}, \vec{r}_{20}$ and velocities $\vec{v}_1, \vec{v}_2$ at the estimated conjunction time $t_0$. By assuming constant velocities in the vicinity of the conjunction, the time $t_m$ of minimum distance is defined by

$$\min_t \|\vec{r}_1(t) - \vec{r}_2(t)\| \quad \text{with} \quad \begin{cases} \vec{r}_1(t) = \vec{r}_{10} + (t - t_0)\vec{v}_1 \\ \vec{r}_2(t) = \vec{r}_{20} + (t - t_0)\vec{v}_2 \end{cases} \tag{5.51}$$

Noting $\delta\vec{r} = \vec{r}_1 - \vec{r}_2$ and $\delta\vec{v} = \vec{v}_1 - \vec{v}_2$ the deviations in position and velocity, we get

$$t_m = t_0 + \frac{\delta\vec{v} \cdot \delta\vec{r}_0}{\delta v^2} \quad \text{and} \quad \delta\vec{r}(t_m) \cdot \delta\vec{v} = 0 \tag{5.52}$$

The positions of the two objects at the conjunction time $t_m$ are considered as Gaussian independent variables with respective means $\vec{r}_{1m}, \vec{r}_{2m}$ and covariance matrices $C_{1/L}, C_{2/L}$. These matrices are assumed to be diagonal in the local orbital frame associated with each object, noted $RL_1$ and $RL_2$, respectively.

$$C_{1/L} = \begin{pmatrix} \sigma_{t1}^2 & 0 & 0 \\ 0 & \sigma_{n1}^2 & 0 \\ 0 & 0 & \sigma_{r1}^2 \end{pmatrix}, \quad C_{2/L} = \begin{pmatrix} \sigma_{t2}^2 & 0 & 0 \\ 0 & \sigma_{n2}^2 & 0 \\ 0 & 0 & \sigma_{r2}^2 \end{pmatrix} \tag{5.53}$$

Figure 5.4 shows the straight-line trajectories of the two objects, with their respective positions at the initial time $t_0$ and at the conjunction time $t_m$, and their uncertainty ellipsoids at $1\sigma$.

The position deviation $\delta\vec{r} = \vec{r}_1 - \vec{r}_2$ at the conjunction time $t_m$ is a Gaussian variable of mean $\delta\vec{r}_m = \vec{r}_{1m} - \vec{r}_{2m}$ and of covariance matrix C defined by

$$\begin{aligned} C &= E\left[(\delta\vec{r} - \delta\vec{r}_m)(\delta\vec{r} - \delta\vec{r}_m)^t\right] \\ &= E\left[((\vec{r}_1 - \vec{r}_{1m}) - (\vec{r}_2 - \vec{r}_{2m}))((\vec{r}_1 - \vec{r}_{1m}) - (\vec{r}_2 - \vec{r}_{2m}))^t\right] \\ &= E\left[(\vec{r}_1 - \vec{r}_{1m})(\vec{r}_1 - \vec{r}_{1m})^t\right] + E\left[(\vec{r}_2 - \vec{r}_{2m})(\vec{r}_2 - \vec{r}_{2m})^t\right] = C_1 + C_2 \end{aligned} \tag{5.54}$$

The cross terms vanish because the random variables $\vec{r}_1, \vec{r}_2$ are assumed to be independent. This hypothesis is questionable, since the sources of uncertainty are usually the same for both objects, namely the modeling of perturbing forces and the measurement errors of the observation system. The independence assumption, like subsequent assumptions, is intended to simplify the probability calculations.

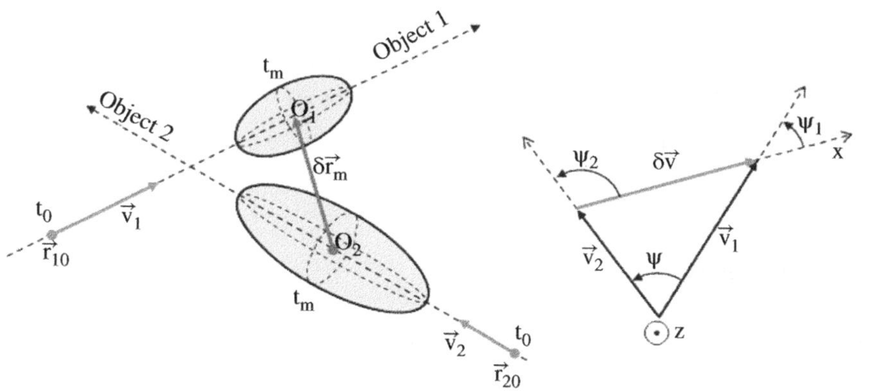


**Figure 5.4** Uncertainty ellipsoids and conjunction frame.

To calculate the sum (5.54) of the covariance matrices $C_1$ and $C_2$, we need to express them in the same frame. The conjunction frame noted RC has its x-axis along the velocity deviation $\delta\vec{v} = \vec{v}_1 - \vec{v}_2$, its z-axis along $\vec{v}_1 \wedge \vec{v}_2$ (axis normal to the velocity plane), and its y-axis completing the direct trihedron. This frame is depicted in Figure 5.4.

A change of coordinates $X = PX'$ with the transition matrix P transforms the covariance matrix according to $C = E[XX^t] = E[PX'X'^tP^t] = PC'P^t$.

In the case of arbitrary orbits, this yields a full matrix C in the RC frame

$$C = \begin{pmatrix} \sigma_x^2 & \rho_{xy}\sigma_x\sigma_y & \rho_{xz}\sigma_x\sigma_z \\ \rho_{xy}\sigma_x\sigma_y & \sigma_y^2 & \rho_{yz}\sigma_y\sigma_z \\ \rho_{xz}\sigma_x\sigma_z & \rho_{yz}\sigma_y\sigma_z & \sigma_z^2 \end{pmatrix} \tag{5.55}$$

where $\sigma_x$, $\sigma_y$, $\sigma_z$ are the standard deviations along the axes of RC and $\rho_{xy}$, $\rho_{xz}$, $\rho_{yz}$ are the correlation coefficients between the components.

In the frequent case of quasi-circular orbits, both velocities $\vec{v}_1, \vec{v}_2$ are almost horizontal and the z-axis is almost parallel to both position vectors $\vec{r}_1, \vec{r}_2$. The frame RC is deduced from $RL_1$ (resp. $RL_2$) by a rotation of angle $-\psi_1$ (resp. $-\psi_2$) around the z-axis (Figure 5.4). The transition matrix $P_1$ from $RL_1$ to RC is

$$P_1 = \begin{pmatrix} \cos\psi_1 & \sin\psi_1 & 0 \\ -\sin\psi_1 & \cos\psi_1 & 0 \\ 0 & 0 & 1 \end{pmatrix} \tag{5.56}$$

The matrix $C_{1/L}$ is assumed to be diagonal in $RL_1$ (5.53). Passing to the RC frame, we get the matrix $C_{1/C} = P_1^t C_{1/L} P_1$, whose components are

$$C_{1/C} = \begin{pmatrix} \sigma_{t1}^2\cos^2\psi_1 + \sigma_{n1}^2\sin^2\psi_1 & (\sigma_{t1}^2 - \sigma_{n1}^2)\cos\psi_1\sin\psi_1 & 0 \\ (\sigma_{t1}^2 - \sigma_{n1}^2)\cos\psi_1\sin\psi_1 & \sigma_{t1}^2\sin^2\psi_1 + \sigma_{n1}^2\cos^2\psi_1 & 0 \\ 0 & 0 & \sigma_{r1}^2 \end{pmatrix} \tag{5.57}$$

The formula for $C_{2/C}$ is similar. The matrices $C_{1/C}$ and $C_{2/C}$ in the frame RC can then be summed to yield the combined covariance matrix C (5.54).

For objects in circular orbits, the approximations $\|\vec{v}_1\| = \|\vec{v}_2\|$ and $C_{1/L} = C_{2/L}$ can reasonably be made. In this case, the angles $\psi_1$, $\psi_2$ have a sum of 180 deg and the combined covariance C is diagonal in the RC frame. Its components are expressed as a function of the angle $\psi$ of the velocities $\vec{v}_1, \vec{v}_2$ (Figure 5.4).

$$C = \begin{pmatrix} \sigma_x^2 & 0 & 0 \\ 0 & \sigma_y^2 & 0 \\ 0 & 0 & \sigma_z^2 \end{pmatrix} \quad \text{with} \quad \begin{cases} \sigma_x^2 = \sigma_t^2(1-\cos\psi) + \sigma_n^2(1+\cos\psi) \\ \sigma_y^2 = \sigma_t^2(1+\cos\psi) + \sigma_n^2(1-\cos\psi) \\ \sigma_z^2 = 2\sigma_r^2 \end{cases} \tag{5.58}$$

## 5.4 Risk of Collision

A conjunction presents a level of risk depending on the size of the objects, the distance of passage, and the uncertainties on the positions. This section presents the methods used to estimate the collision probability for short conjunctions. The methods suited to long conjunctions are presented in Section 5.4.5.

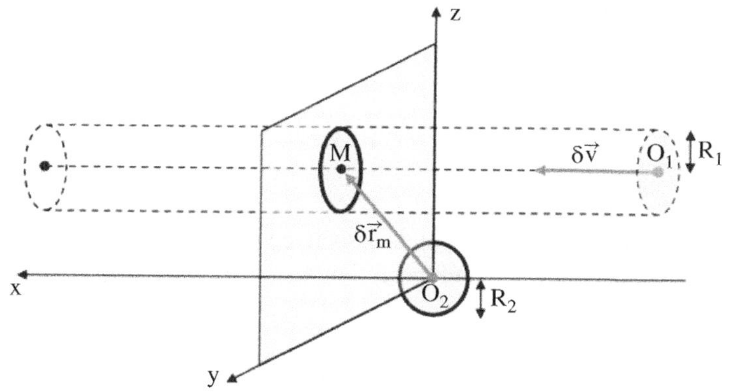


Figure 5.5 Relative motion in the frame RC.

### 5.4.1 Short Conjunction

In the case of objects moving in different orbits, the conjunction lasts at most a few seconds due to the high relative velocity. The relative motion can be considered as a straight line during the conjunction. The two objects are assumed to be spherical with centers $O_1$, $O_2$ and radii $R_1$, $R_2$, respectively. This assumption makes it possible to ignore their attitudes.

The object $O_2$ is taken as the origin of the conjunction frame RC, which is defined at the time $t_m$ of the closest approach (5.52). The motion of $O_1$ relative to $O_2$ is a uniform translation with the velocity $\delta\vec{v}$ parallel to the x-axis of the RC frame. The closest approach occurs at the point M of coordinates $(0, y_m, z_m)$.

The conjunction geometry is depicted in Figure 5.5.

Let us now define the assumptions of the probabilistic model for the assessment of the collision risk.

#### 5.4.1.1 Probabilistic Model

The position $\delta\vec{r}$ of the object $O_1$ relatively to the object $O_2$ is a Gaussian random variable of mean $\delta\vec{r}_m$ and of combined covariance C expressed in the conjunction frame RC. The calculation of C has been detailed in Section 5.3.4. This covariance determined at the time $t_m$ of the closest approach is assumed to be constant over the duration of the conjunction.

A collision occurs if $\delta r < R_1 + R_2$. The goal is to assess this event probability under the above assumptions regarding the object's motion and uncertainties.

For that purpose, we consider that:

- the first object $O_1$ is of combined radius $R_s = R_1 + R_2$ and it moves on a *deterministic* straight-line trajectory in RC;
- the second object $O_2$ is point-like and its position is a Gaussian *random* variable centered at the origin of RC and of combined covariance C.

The collision probability during the conjunction is the probability that the point object $O_2$ comes within the cylindrical volume of radius $R_s$ swept by the spherical object $O_1$ (Figure 5.5).

This probability is assessed as the integral of the normal law N(0, C) on the volume $V_s$ of the cylinder.

$$P = \frac{1}{\sqrt{(2\pi)^3 \det C}} \iiint_{V_s} e^{-\frac{1}{2}\delta\vec{r}^t C^{-1} \delta\vec{r}} dxdydz \tag{5.59}$$

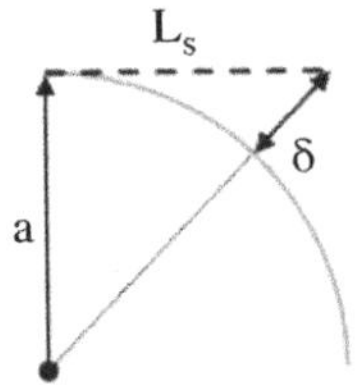


**Figure 5.6** Error due to the straight-line motion approximation.

At this stage, we need to specify the integration bounds associated with the cylinder. The cylindrical volume $V_s$ is defined by its length $2L_s$ and its section $A_s$.

$$\begin{cases} -L_s < x < +L_s & \rightarrow \text{ length } 2L_s \\ (y - y_m)^2 + (z - z_m)^2 \le R_s^2 & \rightarrow \text{ section } A_s = \pi R_s \end{cases} \tag{5.60}$$

The bound $L_s$ represents the limit of validity of the straight trajectory model.

This limit can be estimated over a circular orbit of radius a, by calculating the altitude error δ associated with the straight-line motion approximation. This calculation is depicted in Figure 5.6.

$$\delta = \sqrt{a^2 + L_s^2} - a \quad \rightarrow \quad L_s \approx \sqrt{2a\delta} \tag{5.61}$$

For a circular orbit at the altitude of 400 km and an acceptable altitude error of 100 m, a length of the order of 35 km is obtained. The uncertainty $\sigma_x$ is usually less than 5 km for this type of orbit, so that we have $L_s \approx 7\sigma_x$.

The formulas (5.22) and (5.25) show that increasing the integration bound from $7\sigma_x$ to $+\infty$ gives an approximation within $10^{-12}$ for a normal law in dimension 3. The passage to the limit $L_s \rightarrow +\infty$, which is equivalent to considering a cylinder of infinite length, is therefore correct in most practical cases [R11].

In this way, the calculation of the integral (5.59) is greatly simplified. Indeed, it was shown in Section 5.2.3 that integrating the normal law of dimension three with respect to one variable x leads to the marginal law of dimension two reduced to the remaining variables y and z. The covariance $C_0$ of the marginal law is simply the submatrix of C associated with the components y and z (5.32).

Using this property, the probability (5.59) is reduced to a double integral on the cylinder section $A_s$.

$$P = \frac{1}{2\pi\sqrt{\det C_0}} \iint_{A_s} e^{-\frac{1}{2}\binom{y}{z}^t C_0^{-1} \binom{y}{z}} dydz \tag{5.62}$$

### 5.4.2 Probability of Collision

The integral (5.62) is that of a two-dimensional normal law on a disk whose center is displaced from the origin. In order to calculate it, a three-step geometric transformation is applied as depicted in Figure 5.7. In this figure, the covariance matrix C and the collision section $A_s$ are colored in gray (light) and orange (dark), respectively.

#### 5.4.2.1 Step 1: Passage in Principal Axes

The first step consists in rotating the covariance matrix $C_0$ toward the principal axes. In the frequent case of quasi-circular orbits, the matrix $C_0$ obtained by integrating the x component is already diagonal (5.58).

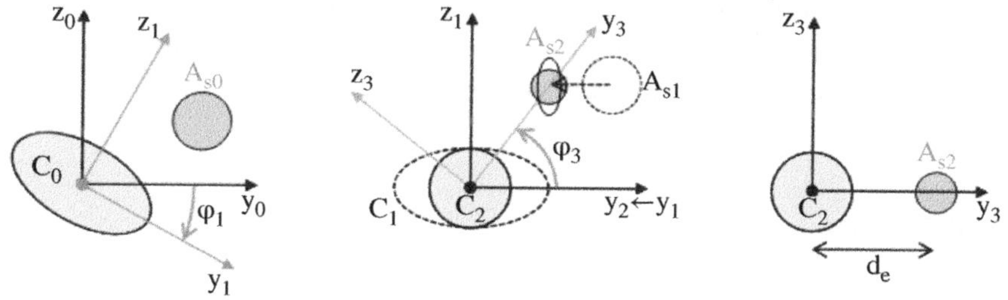


**Figure 5.7** Frame changes.

In the case of a full matrix $C_0$, the following rotation of angle $\varphi_1$ (chosen such that $\sigma_{y1} \geq \sigma_{z1}$) yields a diagonal matrix $C_1$.

$$C_0 = \begin{pmatrix} \sigma_y^2 & \rho_{yz}\sigma_y\sigma_z \\ \rho_{yz}\sigma_y\sigma_z & \sigma_z^2 \end{pmatrix} \xrightarrow[\tan 2\varphi_1 = \frac{2\rho_{yz}\sigma_y\sigma_z}{(\sigma_y^2 - \sigma_z^2)}]{\text{rotation } \varphi_1} C_1 = \begin{pmatrix} \sigma_{y1}^2 & 0 \\ 0 & \sigma_{z1}^2 \end{pmatrix} \tag{5.63}$$

The section $A_{s1}$ in axes $(y, z)_1$ is a circle of center $(y_{m1}, z_{m1})$ and of radius $R_s$.

$$A_s = (y_m, z_m, R_s) \xrightarrow[\begin{cases} y_{m1} = y_m \cos\varphi_1 + z_m \sin\varphi_1 \\ z_{m1} = -y_m \sin\varphi_1 + z_m \cos\varphi_1 \end{cases}]{\text{rotation } \varphi_1} A_{s1} = (y_{m1}, z_{m1}, R_s) \tag{5.64}$$

The matrix $C_1$ is dotted on the second diagram of Figure 5.7, as is the section $A_{s1}$.

#### 5.4.2.2 Step 2: Isotropic Covariance

The second step consists of applying a homothety of ratio $k_y$ along the y-axis to obtain an isotropic matrix $C_2$.

$$C_1 = \begin{pmatrix} \sigma_{y1}^2 & 0 \\ 0 & \sigma_{z1}^2 \end{pmatrix} \xrightarrow[k_y = \frac{\sigma_{z1}}{\sigma_{y1}}]{\text{homothety } k_y} C_2 = \sigma_{z1}^2 \begin{pmatrix} 1 & 0 \\ 0 & 1 \end{pmatrix} \tag{5.65}$$

The section $A_{s2}$ in axes $(y, z)_2$ is an ellipse of center $(k_y y_{m1}, z_{m1})$, of semi-major axis $R_s$ along z and of semi-minor axis $k_y R_s$ along y. This ellipse is then replaced arbitrarily by a circle of the same center and same surface. The radius of the equivalent circle is $R_e = \sqrt{k_y} R_s$. The goal of this approximation is to allow the analytical integration in Section 5.4.3.

#### 5.4.2.3 Step 3: Collision Section on the y-Axis

The third step consists of a rotation of an angle $\varphi_3$ to place the collision section on the y-axis. The isotropic matrix $C_2$ and the circular section $A_{s2}$ are unchanged by this rotation. The center of the section is at the distance $d_e = \sqrt{k_y^2 y_{m1}^2 + z_{m1}^2}$ from the origin.

At the end of these three transformations, the probability of collision (5.62) is expressed by the integral

$$P = \frac{1}{2\pi\sigma_e^2} \iint\limits_{(y-d_e)^2+z^2 \leq R_e^2} e^{-\frac{1}{2\sigma_e^2}(y^2+z^2)} dydz \tag{5.66}$$

where $\sigma_e^2 = \sigma_{z1}^2$ is the smallest eigenvalue of the initial matrix $C_0$ (step 1), $R_e$ is the radius of the equivalent circle (step 2), and $d_e$ is the distance from the center to the origin (step 3).

This integral of a Gaussian distribution on a disk eccentric from the origin is called a **Rice integral** [R5], [R11]. It also appears in problems of signal detection in noise or ballistic impact accuracy. It is calculated analytically by a series expansion as explained in the following text.

### 5.4.3 Analytical Formula

#### 5.4.3.1 Rice Integral

The Rice integral obtained in (5.66) is passed into polar coordinates.

$$P = \frac{1}{2\pi\sigma_e^2} \iint_{\substack{0 \le r \le R_e \\ -\pi \le \theta \le \pi}} e^{-\frac{1}{2\sigma_e^2}(r^2 + d_e^2 + 2d_e r\cos\theta)} r\, dr d\theta \quad \text{with} \quad \begin{cases} y = d_e + r\cos\theta \\ z = r\sin\theta \end{cases} \tag{5.67}$$

The calculation uses the Bessel function of order 0. This real function noted $J_0$ is defined equivalently by an integral or by a series [R5].

$$J_0(u) = \frac{1}{\pi}\int_0^{\pi} e^{iu\cos\theta} d\theta = \sum_{k=0}^{\infty} \frac{(-1)^k}{(k!)^2}\left(\frac{u}{2}\right)^{2k} \tag{5.68}$$

The modified Bessel function of order 0 noted $B_0$ is defined by $B_0(u) = J_0(iu)$.

$$B_0(u) = J_0(iu) = \frac{1}{\pi}\int_0^{\pi} e^{-u\cos\theta} d\theta = \sum_{k=0}^{\infty} \frac{1}{(k!)^2}\left(\frac{u}{2}\right)^{2k} \tag{5.69}$$

With the function $B_0$, the probability of collision (5.67) can be expressed as:

$$P = \int_0^{R_e} \frac{r}{\sigma_e^2} e^{-\frac{r^2 + d_e^2}{2\sigma_e^2}} B_0\left(\frac{r d_e}{\sigma_e^2}\right) dr \tag{5.70}$$

#### 5.4.3.2 Series Expansion

The integral (5.70) is computed by replacing $B_0$ with its series expansion (5.69), then by inverting integration and summation (this is possible because the series is convergent). These calculations give a series expression of P [R11].

$$P = e^{-\delta_e} \sum_{m=0}^{\infty} \frac{\delta_e^m}{m!}\left(1 - e^{-\rho_e}\sum_{k=0}^{m}\frac{\rho_e^k}{k!}\right)$$

$$\text{with } 2\delta_e = \left(\frac{d_e}{\sigma_e}\right)^2 = \left(\frac{y_{m1}}{\sigma_{y1}}\right)^2 + \left(\frac{z_{m1}}{\sigma_{z1}}\right)^2, \quad 2\rho_e = \left(\frac{R_e}{\sigma_e}\right)^2 = \frac{R_S^2}{\sigma_{y1}\sigma_{z1}} \tag{5.71}$$

This series is convergent for all values of $\delta_e$ and $\rho_e$. It can be written in the following form, which allows a recurrent calculation of the successive terms.

$$P = e^{-\delta_e}\sum_{m=0}^{\infty} t_m - e^{-(\delta_e+\rho_e)}\sum_{m=0}^{\infty} t_m S_m \quad \text{with} \quad \begin{cases} t_m = \delta_e t_{m-1}/m, & t_0 = 1 \\ s_m = \rho_e s_{m-1}/m, & s_0 = 1 \\ S_m = S_{m-1} + s_m, & S_0 = 1 \end{cases} \tag{5.72}$$

Since the terms of the series are positive, a truncation gives a rounded down value for the probability of collision. Truncating the series to the first (or second) term gives an estimate of the probability with an error bounded by $T_1$ (or $T_2$).

$$T_1 < \delta_e \rho_e^2 e^{-\delta_e(1-\rho_e)}, \quad T_2 < \frac{1}{4}\delta_e^2 \rho_e^3 e^{-\delta_e(1-\rho_e)} \tag{5.73}$$

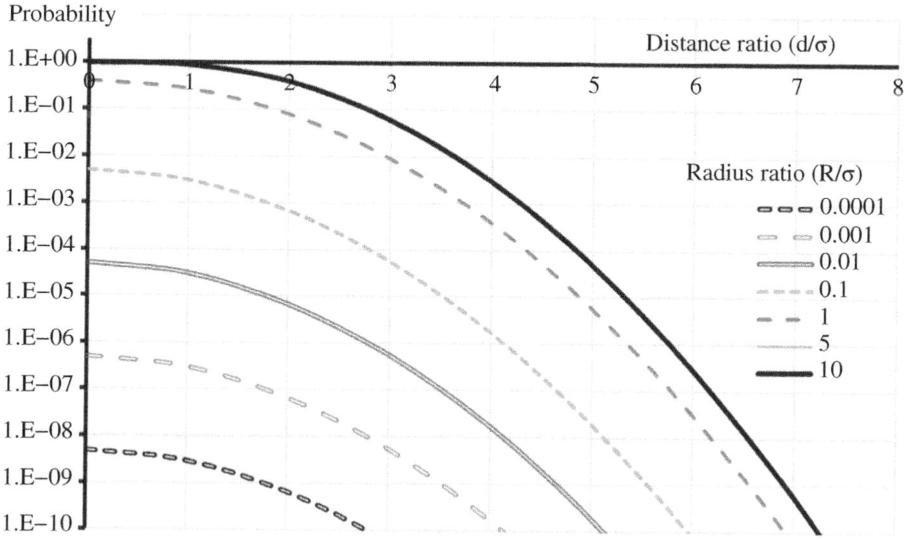


**Figure 5.8** Probability as a function of distance and radius.

Figure 5.8 plots the collision probability calculated by (5.72) (logarithmic scale) as a function of the distance ratio $d_e/\sigma_e$ for different radius ratios $R_e/\sigma_e$ ranging from $10^{-4}$ to 10.

#### 5.4.3.3 Two-Term Approximation

If we consider only the first two terms of the expansion (5.71), we obtain

$$P \approx e^{-\delta_e}\left[1 - e^{-\rho_e} + \delta_e\left(1 - (1+\rho_e)e^{-\rho_e}\right)\right]$$

$$\text{with } 2\delta_e = \left(\frac{d_e}{\sigma_e}\right)^2 = \left(\frac{y_{m1}}{\sigma_{y1}}\right)^2 + \left(\frac{z_{m1}}{\sigma_{z1}}\right)^2, \quad 2\rho_e = \left(\frac{R_e}{\sigma_e}\right)^2 = \frac{R_s^2}{\sigma_{y1}\sigma_{z1}} \tag{5.74}$$

This formula gives the collision probability as a function of the total radius $R_s$, the conjunction coordinates $(y_{m1}, z_{m1})$ and the standard deviations $(\sigma_{y1}, \sigma_{z1})$.

These parameters are expressed in the principal axes of the combined covariance.

The parameters $\delta_e$, $\rho_e$ are in practice included in the following ranges.

$$0 \le \delta_e \le 100\,, \quad 0 \le \rho_e \le 1 \tag{5.75}$$

In these intervals, the formula (5.74) is correct to 1% compared to the precise numerical calculation of the integral (5.62) [R11]. This approximation is sufficient to estimate a collision probability. For a better approximation, the next terms of the expansion (5.71) can be computed to achieve the desired precision.

### 5.4.4 Maximum Probability

The formula (5.74) necessitates the standard deviations $(\sigma_{y1}, \sigma_{z1})$ of the combined covariance. In most practical applications, these values are not (or poorly) known.

In such cases, the alternative is to look for the values $(\sigma_{y1}, \sigma_{z1})$ that maximize the probability of collision. This "worst case" approach results in a majorant of the collision probability.

#### 5.4.4.1 General Case

The maximum probability can be calculated from the formula (5.71) reduced to the first term: $P = e^{-\delta_e}(1 - e^{-\rho_e})$. We look for the values $(\sigma_{y1}, \sigma_{z1})$ that cancel out the partial derivatives.

$$\frac{\partial P}{\partial \sigma_{y1}} = \frac{\partial P}{\partial \sigma_{z1}} = 0 \quad \Rightarrow \quad \begin{cases} e^{-\rho_e}\left(2y_{m1}^2\sigma_{z1} + R_s^2\sigma_{y1}\right) = 2y_{m1}^2\sigma_{z1} \\ e^{-\rho_e}\left(2z_{m1}^2\sigma_{y1} + R_s^2\sigma_{z1}\right) = 2z_{m1}^2\sigma_{y1} \end{cases} \tag{5.76}$$

By eliminating $e^{-\rho_e}$ between these equations, we get the relation: $\frac{\sigma_{y1}}{y_{m1}} = \frac{\sigma_{z1}}{z_{m1}}$.

The values of $(\sigma_{y1}, \sigma_{z1})$ and the maximum probability are then given by

$$\frac{\sigma_{y1}}{y_{m1}} = \frac{\sigma_{z1}}{z_{m1}} = \sqrt{\frac{\alpha}{\ln(1+\alpha)}} \quad \text{and } P_{max} = \frac{\alpha}{1+\alpha}\left(\frac{1}{1+\alpha}\right)^{\frac{1}{\alpha}} \quad \text{with } \alpha = \frac{R_s^2}{2y_{m1}z_{m1}} \tag{5.77}$$

#### 5.4.4.2 Small Collision Section

If the collision section is small compared to the distance of passage, the density of the normal law can be considered as constant over this section, equal to its value at the point of passage $(y_{m1}, z_{m1})$.

In this case, the probability of collision is reduced to

$$P = A_s p(y_{m1}, z_{m1}) \tag{5.78}$$

The values $(\sigma_{y1}, \sigma_{z1})$ maximizing the density $p(y_{m1}, z_{m1})$ are given by (5.35).

$$\begin{cases} \sigma_{y1} = y_{m1} \\ \sigma_{z1} = z_{m1} \end{cases} \Rightarrow P_{max} = \frac{R_s^2}{2ey_{m1}z_{m1}} \tag{5.79}$$

This formula is valid only if $y_{m1}$ and $z_{m1}$ are not zero.

It can be observed that the specific formula (5.79) is consistent with the general one (5.77) when the ratio $\alpha$ is small. Indeed, $\alpha \approx 0$ is equivalent to considering that the collision section is negligible with respect to the distance of passage. Passing to the limit $\alpha \to 0$ in formula (5.77) yields

$$\begin{cases} \sigma_{y1} = y_{m1} \\ \sigma_{z1} = z_{m1} \end{cases} \text{and } P_{max} = \frac{\alpha}{e} = \frac{R_s^2}{2ey_{m1}z_{m1}} \tag{5.80}$$

If $y_{m1}$ or $z_{m1}$ is zero, the simplification (5.78) remains applicable, considering that the density along the null component is a Dirac function (as said at the end of Section 5.2.3) and that the circular collision section $A_s$ "flattens" into a segment of length $2R_s$. The formula (5.79) is then replaced by (e.g. in the case $y_{m1} = 0$)

$$\sigma_{z1} = z_{m1} \Rightarrow P_{max} = \sqrt{\frac{2}{\pi e}}\frac{R_s}{z_{m1}} \tag{5.81}$$

#### 5.4.4.3 Alert Threshold

For a passage at a point of coordinates $y_{m1} = z_{m1}$, the miss-distance is $d_s = \sqrt{2}y_{m1}$, and the maximum probability (5.79) is reduced to

$$P_{max} = \frac{1}{e}\left(\frac{R_s}{d_s}\right)^2 \tag{5.82}$$

For a miss-distance of $100\,R_s$, the maximum probability is about $10^{-5}$. This threshold $P_{max} > 10^{-5}$ is generally considered as critical for raising a collision alert. A more detailed study of the conjunction is then necessary by taking into account the covariance matrices of the two objects if they can be estimated. An analysis by Monte Carlo method (simulations with random sampling of the uncertain parameters and statistical analysis of the results) should be carried out in case of doubt in order to avoid an unnecessary avoidance maneuver by the satellite.

### 5.4.5 Long Conjunction

For objects in very close orbits (satellites in formation flying, debris close to the geostationary orbit), the conjunction may last for a period or longer. The straight-line motion model is no longer suitable. These long conjunctions occur for objects in quasi-circular orbits, particularly in the vicinity of the geostationary orbit. The following method, similar to that developed for short conjunctions, estimates the probability of collision over one orbital period.

The point object $O_2$ has a Gaussian random position centered on a circular orbit. By transformations strictly similar to those made in Section 5.4.2, we obtain an equivalent combined covariance of isotropic form $C = \sigma_e^2 I$. The object $O_1$ is a sphere of equivalent radius $R_e$ in an orbit close to $O_2$, coplanar, and of the same period (this is required for a long conjunction). The motion of $O_1$ relative to $O_2$ is described by the Hill–Clohessy–Wiltshire equations (presented in Section 7.3) in the local orbital frame (origin $O_2$, x opposite to velocity, y radial, and z normal).

The relative free motion is an ellipse centered in $(d_e, 0)$ on the x-axis, having a semi-major axis $a_e$ along x, a semi-minor axis $a_e/2$ along y, and traversed at the orbital period in the retrograde direction (Figure 5.9).

The probability of collision during one revolution is the integral of the centered isotropic normal law $N\left(0, \sigma_e^2 I\right)$ on the volume swept by the sphere of radius $R_e$. The center of the sphere moves on an ellipse, making the integration difficult.

In order to obtain an analytical formula, the ellipse is replaced by an equivalent circle of the same center, same surface (radius $R_c = a_e/\sqrt{2}$) and traversed at constant velocity. The integration volume becomes a circular torus comprised between the radii $R_c - R_e$ and $R_c + R_e$. The section of this torus is bounded by the two vertical dotted lines shown in Figure 5.9b.

The integral of the normal law $N\left(0, \sigma_e^2 I\right)$ on the torus takes the form

$$P = \frac{1}{\sqrt{(2\pi)^3 \sigma_e^3}} \int_{z=-R_e}^{z=+R_e} \left( \iint\limits_{R_{int}(z) \le \sqrt{(x-d_e)^2+y^2} \le R_{ext}(z)} e^{-\frac{1}{2\sigma_e^2}(x^2+y^2+z^2)} dxdy \right) dz \tag{5.83}$$

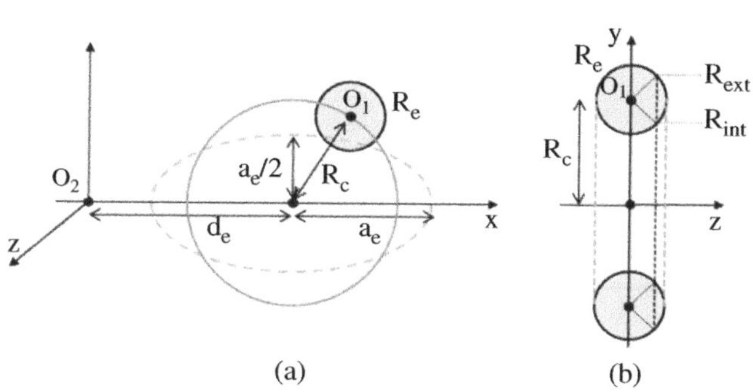


Figure 5.9 Relative motion in circular orbit.

The integral in brackets is the integral of the normal law on a ring of inner radius $R_{int} = R_c - \sqrt{R_e^2 - z^2}$ and outer radius $R_{ext} = R_c + \sqrt{R_e^2 - z^2}$. Expressing it as the difference of the integrals on the outer and inner circle, two Rice integrals appear as in Section 5.4.3. The formula (5.71) limited to the first term gives the analytical expression of these two integrals for $R = R_{int}$ and $R = R_{ext}$.

$$\frac{1}{2\pi\sigma_e^2} \iint\limits_{\sqrt{(x-d_e)^2+y^2} \le R} e^{-\frac{x^2+y^2}{2\sigma_e^2}} dxdy = e^{-\delta_e}(1 - e^{-\rho_e}) \text{ with } \delta_e = \frac{d_e^2}{2\sigma_e^2}, \quad \rho_e = \frac{R^2}{2\sigma_e^2} \tag{5.84}$$

After substituting these expressions in (5.83), there remains an integral in z, which becomes by the change of variable $z = -R_e \cos\varphi$

$$P = \sqrt{\frac{2}{\pi}} \frac{R_e}{\sigma_e} e^{-\frac{R_c^2+R_e^2+d_e^2}{2\sigma_e^2}} \int_{\varphi=0}^{\pi} \sinh\left(\frac{R_c R_e}{\sigma_e^2} \sin\varphi\right) \sin\varphi d\varphi \tag{5.85}$$

This integral is calculated by series expansion using the following formulas [R5].

$$\sinh x = \sum_{m=0}^{\infty} \frac{x^{2m+1}}{(2m+1)!} \quad \text{and} \quad \int_0^{\pi} (\sin x)^{2m+2} dx = \pi \frac{(2m+1)!}{2^{2m+1} m!(m+1)!} \tag{5.86}$$

The collision probability is thus expressed as a convergent series [R11].

$$P = \sqrt{2\pi} \frac{R_c R_e^2}{\sigma_e^3} e^{-\frac{R_c^2+R_e^2+d_e^2}{2\sigma_e^2}} \sum_{m=0}^{\infty} \frac{1}{2^{2m+1} m!(m+1)!} \left(\frac{R_c R_e}{\sigma_e^2}\right)^{2m} \tag{5.87}$$

In most cases, the first term ($m = 0$) is sufficient, because $R_c, R_e \ll \sigma_e$.

$$P \approx \sqrt{\frac{\pi}{2}} \frac{R_c R_e^2}{\sigma_e^3} e^{-\frac{R_c^2+R_e^2+d_e^2}{2\sigma_e^2}} \tag{5.88}$$

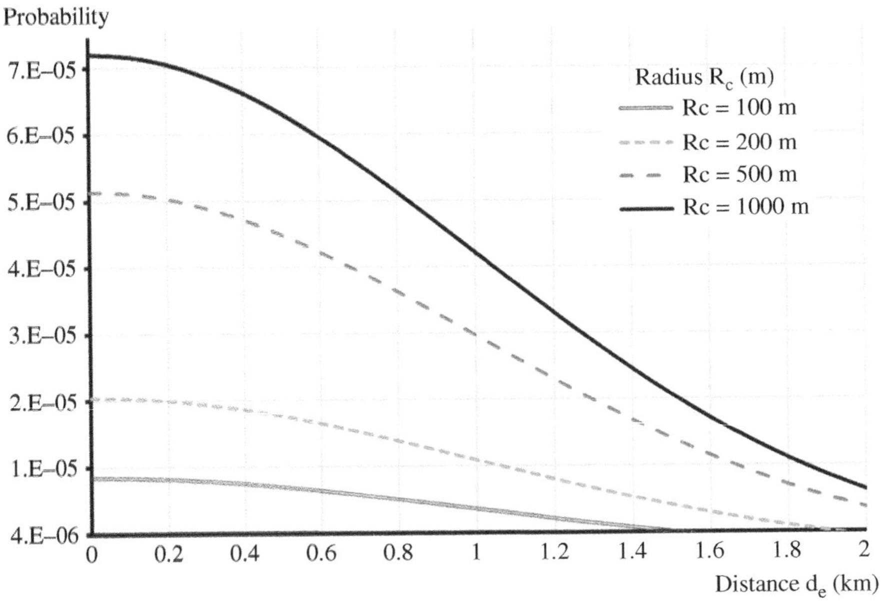


**Figure 5.10** Probability as a function of distance and radius.

This formula gives the probability of collision over one orbital period as a function of the collision radius $R_e$, the distance $d_e$, the radius $R_c$ of the circle equivalent to the ellipse, and the standard deviation $\sigma_e$ of the relative position.

Figure 5.10 plots the probability of collision as a function of the distance $d_e$ for different radii $R_c$. The plots correspond to typical values $R_e = 10$ m, $\sigma_e = 1$ km.

## 5.5 Conclusion

### 5.5.1 The Key Points

- The position uncertainty in orbit is modeled by an ellipsoid whose tangential, normal, and radial dimensions are generally in the ratios 10:2:1.
- The conjunction detection is based first on the geometric distance between the Keplerian orbits, then on the time of passage at the conjunction point.
- For short or long conjunctions, the collision probability is assessed by a Rice integral expanded to the first two terms, giving an approximation of 1%.
- Collision risks at launch are assessed by a geometric approach, which determines the forbidden time slots within a launch window.
- A collision probability greater than $10^{-5}$ is considered as critical and an avoidance maneuver may be necessary.

### 5.5.2 To Go Further

- CHAN F. K. – *Spacecraft Collision Probability*. The Aerospace Press (2008).

This book is the most valuable reference on the topics of collision in space. It describes in detail the different methods, with their modeling assumptions depending on the type of conjunction. All methods are illustrated by numerical applications examples, with a thorough discussion of the result's validity.

- GRONCHI G.F. – *An Algebraic Method to Compute the Critical Points of the Distance Function Between Two Keplerian Orbits*. Celestial Mech Dyn Astr 93, 295–329 (2005).

This article presents a fully analytical method for determining the minimum geometric distance between two Keplerian orbits.

- NORAD Two-Line Element Sets: https://celestrak.com/NORAD/elements/

The website of the NORAD, updated with the ephemerides of space objects in TLE format.

- SOCRATES Conjunction Reports: https://celestrak.com/SOCRATES/

The website of CELESTRAK, updated with the most risky conjunctions.

# Part II

## Controlled Orbital Motion

# 6

# Impulsive Transfer

## 6.1 Introduction

Artificial satellites are placed in specific orbits, chosen according to the mission objectives (telecommunication, navigation, observation, etc.). The satellite must perform propelled maneuvers, first to reach the operational orbit after launch, then to stay precisely on it for several years, and finally to evacuate it at its end of life.

In the first approximation, the maneuvers can be modeled as velocity impulses. This assumption is well suited to high thrust levels produced by chemical propulsion. Transfers with low-thrust engines are addressed in Chapter 8.

With the assumption of instantaneous velocity changes, the transfer problem consists in finding the impulse times, directions, and magnitudes in order to reach the target while minimizing the sum of the impulse magnitudes. This total magnitude defines indeed directly the required consumption of propellants.

A transfer problem can be formulated in different ways, depending on the number of impulses allowed and on the target definition: by orbital parameters (partially or totally specified), by position, or by position and time. Analytical solutions exist only for the simplest cases.

The first section considers targets defined by orbital parameters. The classical analytical solutions are presented. In the case of a single impulse, the goal is to change one orbit parameter related either to the orbital plane (inclination or right ascension of ascending node), to the orbit orientation (argument of perigee), or to the orbit shape (apogee or perigee). In the case of two or three impulses, analytical solutions exist for circular orbits, namely the Hohmann transfer in two impulses and the bi-elliptic transfer in three impulses.

The second section considers targets defined by position. The analytical solutions for the minimum-energy and minimum-eccentricity transfers are presented. For the minimum-impulse transfer, an approximate solution is derived by minimizing the sum of the squared impulses.

The third section considers targets defined by position and time. This defines Lambert's problem, which is of fundamental importance in space applications. Lambert's theorem is presented and the transfer time equation is derived in several forms. The various solution methods are then described.

*Space Trajectories: Basic and Advanced Topics*, First Edition. Max Cerf.

## 6.2 Orbit Target

This section deals with a target defined by its orbital parameters. The classical solutions in one, two, or three impulses are presented.

### 6.2.1 Problem Formulation

Impulsive transfers are based on Keplerian orbits. We first recall briefly the definition of the orbital parameters and the Keplerian formulas that will be used in this chapter, then we formulate the general transfer problem.

#### 6.2.1.1 Keplerian Orbit

A Keplerian orbit is a conic positioned in the inertial geocentric frame (O Earth center, Z pole axis, and X meridian of Greenwich) by its orbital parameters.

The orbital plane (Figure 6.1) is obtained from the equatorial plane OXY by a rotation of axis Z and angle $\Omega$ (right ascension of ascending node), followed by a rotation of axis X′ and angle i (inclination). The ascending node AN, opposite the descending node DN, is the point at which the equator is crossed from south to north. The axis X′ is the line of nodes. The axis Z′ is directed along the angular momentum s. The eccentricity vector $\vec{e}$ directed toward the perigee is obtained from the axis X′ by a rotation of axis Z′ and angle $\omega$ (argument of perigee).

The orbit shape (Figure 6.2) is determined by the semi-major axis a and the eccentricity e, or equivalently by the radii at perigee $r_P$ and apogee $r_A$. The position on the orbit is determined by the true anomaly $\theta$, which varies with time according to Kepler's equation. The argument of latitude $\nu = \omega + \theta$ is the angular distance from the ascending node.

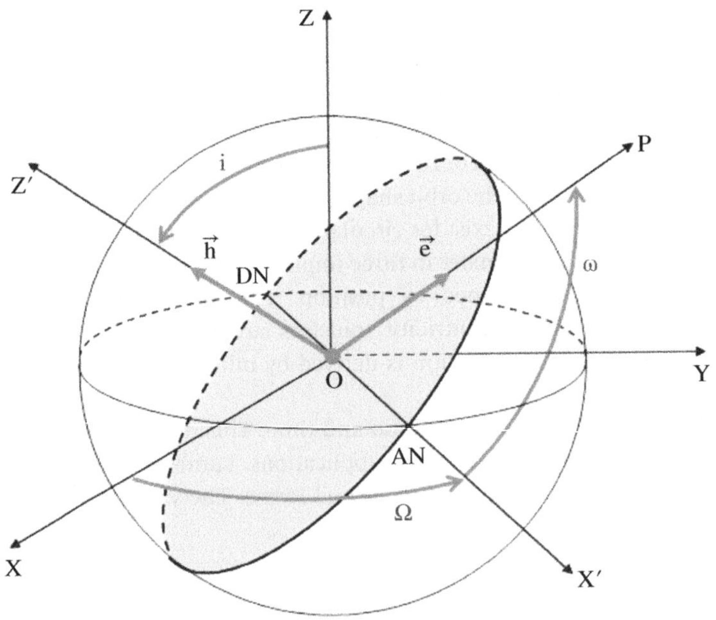


**Figure 6.1** Orbital plane.

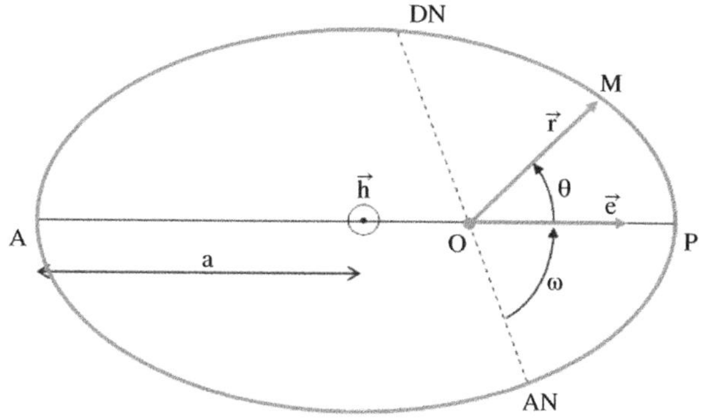


**Figure 6.2** Orbit shape (elliptical case).

A Keplerian conic can be defined by the five orbital parameters $(a, e, i, \Omega, \omega)$, or equivalently by the two vectors $(\vec{h}, \vec{e})$. The following relations demonstrated in Chapter 1 will be used for orbital transfers.

$$r = \frac{p}{1 + e\cos\theta}, \quad v^2 = \mu\left(\frac{2}{r} - \frac{1}{a}\right), \quad p = \frac{h^2}{\mu} = a(1 - e^2), \quad n = \sqrt{\frac{\mu}{a^3}}$$

$$\vec{h} = \vec{r} \wedge \vec{v}, \quad \vec{e} = \frac{\vec{v} \wedge \vec{h}}{\mu} - \vec{u}_r, \quad \vec{v} = \frac{1}{p}\vec{h} \wedge (\vec{e} + \vec{u}_r) \quad \text{with } \vec{u}_r = \frac{\vec{r}}{r} \tag{6.1}$$

p is the conic parameter, μ is the gravitational constant of the attracting body, n is the pulsation (or mean motion), and $\vec{u}_r$ is the unit radial vector.

#### 6.2.1.2 Transfer Problem

An impulse is an instantaneous change in velocity $\Delta\vec{v}$ applied at a position $\vec{r}$.

It modifies the vectors $(\vec{h}, \vec{e})$ and the orbital parameters. The propellant consumption depends directly on the impulse modulus.

The transfer consists of moving from initial orbital parameters $(a_1, e_1, i_1, \Omega_1, \omega_1)$ to final orbital parameters $(a_2, e_2, i_2, \Omega_2, \omega_2)$. The maximum number n of impulses is prescribed. The goal is to find the impulse vectors $(\Delta\vec{v}_k)_{k=1,\ldots,n}$ and the anomalies $(\theta_k)_{k=1,\ldots,n}$ or times $(t_k)_{k=1,\ldots,n}$ to which they are applied, by minimizing the sum of the impulse modules.

In the general case (any orbits, any number of impulses), this optimization problem is complex and it must be solved by numerical methods.

In special cases, analytical solutions are available. The following sections present the classical one-, two-, and three-impulse transfers that can be solved explicitly.

### 6.2.2 Transfer in One Impulse

Elementary one-impulse maneuvers involve modifying either the orbital plane, the orbit orientation in its plane, or an apsis (apogee or perigee).

#### 6.2.2.1 Modifying the Orbital Plane

The orbital plane is defined by the right ascension of the ascending node Ω and the inclination i. Figure 6.3 shows the initial orbit $(\Omega_1, i_1)$ of node $N_1$, the final orbit $(\Omega_2, i_2)$ of node $N_2$, and the angle α between the orbital planes. The point M is located at the intersection of the two orbits. The maneuver consists of a rotation of the velocity by an angle α around the position vector $\vec{r}$.

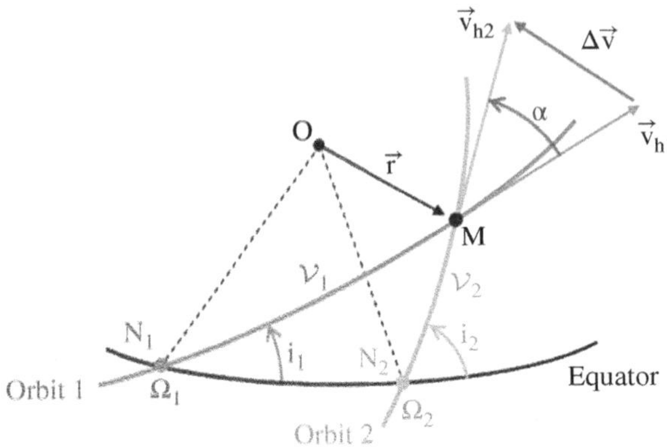


**Figure 6.3** Plane change.

We first establish the general formula valid for any plane change; then we apply it to the specific cases where only one angle (inclination or right ascension of the ascending node) is to be changed.

Let us consider the spherical triangle ($N_1N_2M$). By noting $\Delta\Omega = \Omega_2 - \Omega_1$, we have from spherical trigonometry the following relations:

$$\begin{cases} \cos\nu_1 = \cos\nu_2\cos\Delta\Omega - \sin\nu_2\sin\Delta\Omega\cos i_2 \\ \cos\nu_2 = \cos\nu_1\cos\Delta\Omega + \sin\nu_1\sin\Delta\Omega\cos i_1 \\ \sin i_1\sin\nu_1 = \sin i_2\sin\nu_2 \end{cases} \tag{6.2}$$

Using the third relation to substitute $\nu_2$ in the first equation and $\nu_1$ in the second equation, we obtain the latitude arguments $\nu_1$ and $\nu_2$ of the point M on orbit 1 and on orbit 2 as functions of the orbital parameters $i_1$, $i_2$ and $\Delta\Omega = \Omega_2 - \Omega_1$.

$$\begin{cases} \tan\nu_1 = \dfrac{\tan i_2\sin\Delta\Omega}{\tan i_2\cos i_1\cos\Delta\Omega - \sin i_1} \\ \tan\nu_2 = \dfrac{-\tan i_1\sin\Delta\Omega}{\tan i_1\cos i_2\cos\Delta\Omega - \sin i_2} \end{cases} \tag{6.3}$$

For the two orbits to intersect in M, the radius vectors at $\nu_1$ (on orbit 1) and at $\nu_2$ (on orbit 2) must be equal, which imposes an additional condition.

$$r_1 = r_2 \quad \Leftrightarrow \quad \frac{p_1}{1 + e_1\cos(\nu_1 - \omega_1)} = \frac{p_2}{1 + e_2\cos(\nu_2 - \omega_2)} \tag{6.4}$$

There may be zero, one, or two points of intersection, depending on the initial and final orbital parameters.

Let us consider the case of orbits having the same shape ($p_1 = p_2$, $e_1 = e_2$).

If a pair ($\nu_1$, $\nu_2$) satisfies the relations (6.3) and (6.4), then the pair ($\nu_1 + \pi$, $\nu_2 + \pi$) also satisfies them. There are two diametrically opposed points of intersection in which the maneuver can be performed. The best solution (minimum impulse) is to perform the maneuver at the highest point. Indeed, the velocity modulus will be lower from (6.1) and the velocity vector can be changed in direction with a smaller impulse.

The angle α between the initial orbit ($\Omega_1$, $i_1$) and the final orbit ($\Omega_2$, $i_2$) is given by the following trigonometric relation in the spherical triangle ($N_1N_2M$).

$$\cos\alpha = \cos i_1\cos i_2 + \sin i_1\sin i_2\cos\Delta\Omega \tag{6.5}$$

To achieve the plane change in M, the velocity must be rotated by the angle $\alpha$ around the axis of rotation defined by the position vector $\vec{r}$. This changes only the horizontal component of the velocity and the required impulse modulus is

$$\Delta v = 2v_h \sin\frac{\alpha}{2} \tag{6.6}$$

The point of maneuver M has an anomaly $\theta_1 = \nu_1 - \omega_1$. In that point, the velocity components (radial $\vec{v}_r$ and horizontal $\vec{v}_h$) are deduced from (6.1).

$$\vec{v} = \vec{v}_r + \vec{v}_h \quad \text{with} \quad v_r = \vec{v} \cdot \vec{u}_r = \frac{he \sin\theta_1}{p} \tag{6.7}$$

Formula (6.6) is valid for orbits having the same shape. If the transfer involves simultaneous plane and shape changes, the velocity impulse must be computed as the difference between the velocities in M on orbit 1 and orbit 2.

Let us now look at two special cases of practical interest.

- Pure inclination change: $\Omega_1 = \Omega_2$

Putting $\Delta\Omega = 0$ in (6.3) gives $\nu_1 = 0$ or $\pi$ and $\nu_2 = 0$ or $\pi$. The maneuver takes place at a node and from (6.5), the angle of the planes is $\alpha = i_2 - i_1$.

For a circular orbit of radius r, the maneuver impulse is from (6.6)

$$\Delta v = 2\sqrt{\frac{\mu}{r}} \sin\frac{i_2 - i_1}{2} \tag{6.8}$$

- Pure ascending node change: $i_1 = i_2 \underset{\text{noted}}{=} i$

Putting $i_1 = i_2$ in (6.3) gives $\tan\nu_1 = \frac{\sin\Delta\Omega}{\cos i(1-\cos\Delta\Omega)} = -\tan\nu_2$.

From (6.2) the only valid solution is $\nu_2 = -\nu_1$, and from (6.5), the angle of the planes is given by $\cos\alpha = \cos^2 i + \sin^2 i \cos\Delta\Omega$.

In the case of polar orbits, this angle is directly $\alpha = \Omega_2 - \Omega_1$ and the maneuver takes place at a pole ($\nu_1 = \pm\frac{\pi}{2}$).

#### 6.2.2.2 Drift Strategy

For changing the right ascension of ascending node, a more economical strategy consists of using the natural precession due to Earth flattening (term in $J_2$ of the gravitational potential). Formula (2.103) gives the nodal precession rate depending on the parameters (a, e, i).

$$\frac{d\Omega}{dt} = -\frac{3J_2R_e^2 n}{2p^2}\cos i \tag{6.9}$$

The transfer strategy consists of moving to a drift orbit by modifying a, e, and/or i, waiting until the desired shift $\Delta\Omega$ is achieved, then returning to the initial orbital parameters (a, e, i). This drift strategy is more time consuming, but much more economical than the direct change of $\Omega$ by a propelled impulse as in (6.6). It will be developed in Chapter 10, which deals with the problem of space debris cleaning.

#### 6.2.2.3 Modifying the Major Axis Orientation

The orbit orientation in its plane is defined by the argument of the perigee $\omega$. Figure 6.4 shows two coplanar orbits of the same shape, whose major axes make an angle $\Delta\omega$. Condition (6.4) gives the anomaly $\theta_1$ of the intersection point M.

$$\cos\theta_1 = \cos\theta_2 \quad \text{with} \quad \theta_2 = \theta_1 - \Delta\omega \;\Rightarrow\; \theta_1 = \frac{\Delta\omega}{2} \tag{6.10}$$

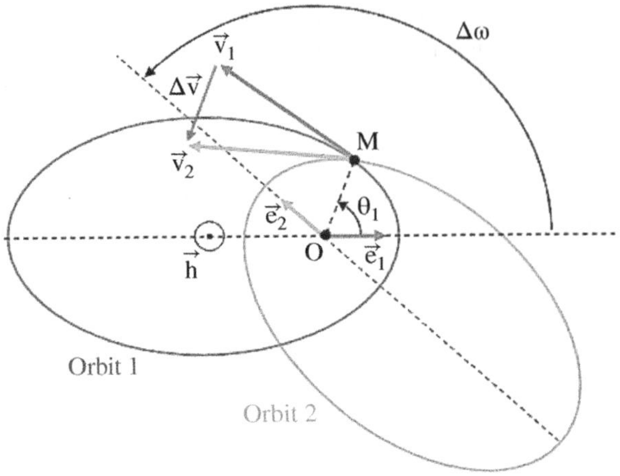


**Figure 6.4** Major axis orientation change.

The initial and final orbits have the same angular momentum $\vec{h}$, whereas their eccentricity vectors $\vec{e}_1$ and $\vec{e}_2$ have the same modulus and make an angle $\Delta\omega$.

The impulse $\Delta\vec{v}$ is the difference of the velocities in the intersection point M. Using formula (6.1), it is expressed as $\Delta\vec{v} = \vec{v}_2 - \vec{v}_1 = \frac{1}{p}\vec{h} \wedge (\vec{e}_2 - \vec{e}_1)$.

On the other hand, the common angular momentum, which is orthogonal to both vectors $\vec{e}_1$ and $\vec{e}_2$, can be written as $\vec{h} = \frac{h}{e^2 \sin \Delta\omega}(\vec{e}_1 \wedge \vec{e}_2)$.

We obtain therefore a purely radial impulse assessed by

$$\Delta\vec{v} = -\frac{2\mu e}{h} \sin\left(\frac{\Delta\omega}{2}\right) \vec{u}_r \tag{6.11}$$

As with the node change, it is possible to use the natural apsidal precession due to the Earth flattening. Formula (2.103) gives the apsidal precession rate depending on the parameters (a, e, i).

$$\frac{d\omega}{dt} = \frac{3J_2R_e^2n}{4p^2}(4 - 5\sin^2 i) \tag{6.12}$$

Here also, a transfer strategy via a drift orbit is longer, but more economical than the direct modification of $\omega$ by a propelled impulse.

#### 6.2.2.4 Modifying an Apsis

The modification of one apsis (apogee or perigee) takes place in the opposite apsis.

Figure 6.5 shows two coplanar orbits having the same major axis and one common apsis (apogee A in Figure 6.5a and perigee P in Figure 6.5b). The semi-major axes are noted $a_1$ for the first orbit and $a_2$ for the second orbit.

In Figure 6.5a, the perigee is increased from $P_1$ to $P_2$ by a tangential impulse $\Delta\vec{v}_A$ applied at the apogee A. The impulse modulus is the difference of the velocities.

$$\Delta v_A = v_{A2} - v_{A1} = \sqrt{\mu\left(\frac{2}{r_A} - \frac{1}{a_2}\right)} - \sqrt{\mu\left(\frac{2}{r_A} - \frac{1}{a_1}\right)} \tag{6.13}$$

In Figure 6.5b, the apogee is increased from $A_1$ to $A_2$ by a tangential impulse $\Delta\vec{v}_P$ applied at the perigee P. The impulse modulus is the difference of the velocities.

$$\Delta v_P = v_{P2} - v_{P1} = \sqrt{\mu\left(\frac{2}{r_P} - \frac{1}{a_2}\right)} - \sqrt{\mu\left(\frac{2}{r_P} - \frac{1}{a_1}\right)} \tag{6.14}$$

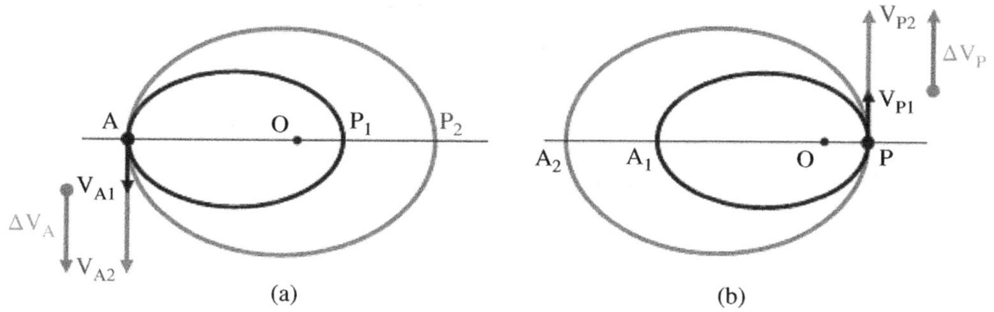


**Figure 6.5** (a) Perigee or (b) apogee change.

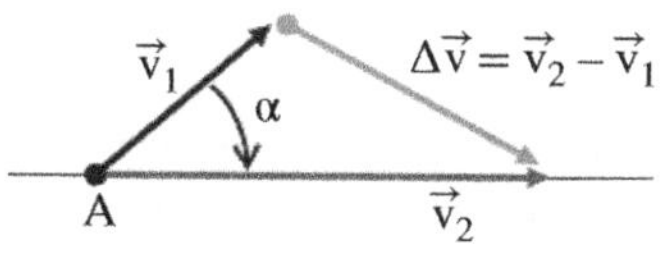


$$\Delta v = \sqrt{v_1^2 + v_1^2 - 2v_1v_2\cos\alpha}$$

**Figure 6.6** Perigee and inclination change.

In case of an apsis decrease, the same formulas (6.13) and (6.14) apply, but the impulse is opposite to the velocity.

#### 6.2.2.5 Coupled Apsis and Plane Change

The modification of an apsis can be coupled with a change of plane. In this case, the impulse modifies both the velocity modulus and its orientation. Figure 6.6 shows a coupled maneuver increasing the velocity modulus from $v_1$ to $v_2$ and rotating the velocity by an angle $\alpha$. The velocity impulse is calculated from the sides of lengths $v_1$ and $v_2$ in the velocity triangle.

This type of maneuver is used to circularize a satellite in the geostationary Earth orbit (GEO, circular equatorial orbit at 35 786 km altitude, described in Section 1.4.4) from a geostationary transfer orbit (GTO) obtained after launch. The GTO orbit is generally inclined due to the launch pad latitude. The circularization in a single maneuver is possible only if the GTO apogee A is at the equator, since this point must belong to the final equatorial orbit. This necessitates an argument of the perigee $\omega$ worth 0 or $\pi$.

Starting from a GTO orbit (35 786 km/200 km/10 deg), the velocity at apogee is $v_1 = 1597$ m/s and the targeted GEO velocity is $v_2 = 3075$ m/s. The impulse for the circularization maneuver with inclination change is then worth 1527 m/s.

### 6.2.3 Transfer in Two or Three Impulses

When the initial and final orbits do not intersect, a single impulse is not sufficient and the transfer requires at least two impulses. In the case of circular orbits, the classical solutions are the Hohmann transfer and the bi-elliptic transfer.

#### 6.2.3.1 Hohmann Transfer

The Hohmann transfer (1925) consists in moving from a circular orbit of radius $r_1$ to a coplanar circular orbit of radius $r_2$ in two impulses (Figure 6.7).

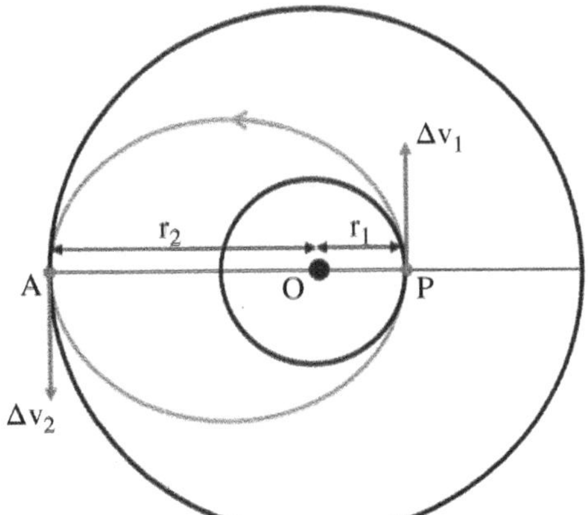


**Figure 6.7** Hohmann transfer.

The following description assumes that $r_2 > r_1$, which is the most frequent case in practice. The calculations in the case $r_2 < r_1$ are similar by inverting the apsis roles and applying impulses opposite to the velocity.

The transfer is composed of three phases.

- The first impulse $\Delta v_1$ is applied tangentially at some point P on the initial circular orbit. It puts the satellite in an elliptical transfer orbit with a perigee in P (radius $r_1$) and an apogee in A (radius $r_2$).
- The satellite coasts during a half revolution along the elliptical transfer orbit until the apogee A.
- The second impulse $\Delta v_2$ is applied tangentially at the apogee A of the transfer orbit. It puts the satellite in the targeted circular orbit of radius $r_2$.

The impulses are assessed by formulas (6.13) and (6.14), with the semi-major of the elliptical transfer orbit given by $a = \frac{r_1+r_2}{2}$.

$$\begin{cases} \Delta v_1 = \sqrt{\mu\left(\frac{2}{r_1} - \frac{2}{r_1+r_2}\right)} - \sqrt{\frac{\mu}{r_1}} = \sqrt{\frac{\mu}{r_1}}\left(\sqrt{\frac{2r_2}{r_1+r_2}} - 1\right) \\ \Delta v_2 = \sqrt{\frac{\mu}{r_2}} - \sqrt{\mu\left(\frac{2}{r_2} - \frac{2}{r_1+r_2}\right)} = \sqrt{\frac{\mu}{r_2}}\left(1 - \sqrt{\frac{2r_1}{r_1+r_2}}\right) \end{cases} \tag{6.15}$$

They can be expressed more concisely as functions of the radius ratio $\rho = \frac{r_2}{r_1}$ and the initial velocity $v_1 = \sqrt{\frac{\mu}{r_1}}$. The total impulse is noted $\Delta v_t = \Delta v_1 + \Delta v_2$.

$$\begin{aligned} &\Delta v_1 = v_1\left(\sqrt{\frac{2\rho}{1+\rho}} - 1\right), \quad \Delta v_2 = \frac{v_1}{\sqrt{\rho}}\left(1 - \sqrt{\frac{2}{1+\rho}}\right) \\ &\Rightarrow \frac{\Delta v_t}{v_1} = \sqrt{\frac{2\rho}{1+\rho}} - \sqrt{\frac{2}{\rho(1+\rho)}} + \frac{1}{\sqrt{\rho}} - 1 \end{aligned} \tag{6.16}$$

Figure 6.8 plots the variation of the ratio $\Delta v_t/v_1$ depending on the radius ratio $\rho$ in logarithmic scale.

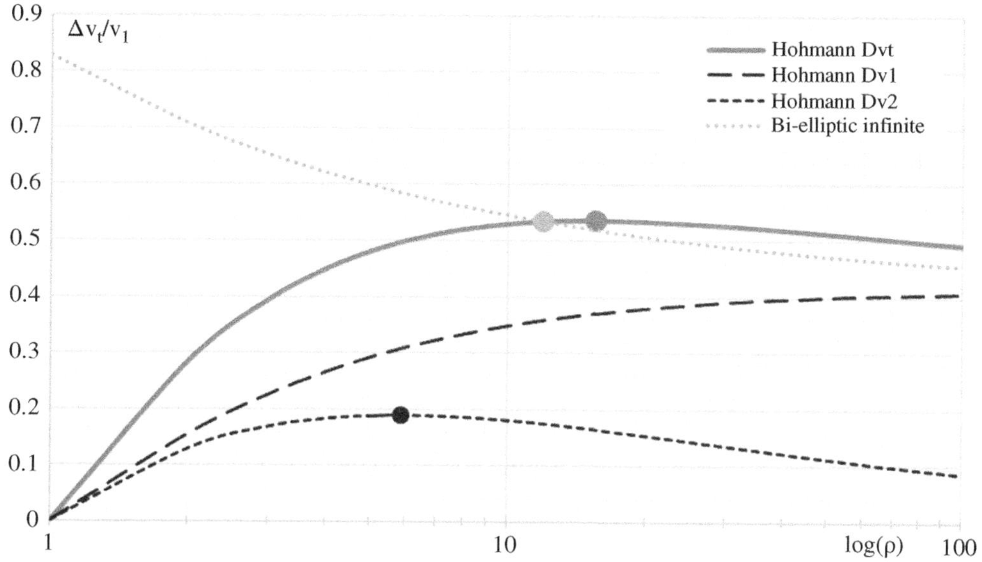


**Figure 6.8** Hohmann transfer cost depending on the radius ratio.

The ratio $\Delta v_t/v_1$ is maximum for $\rho = 15.58$. It decreases until a minimum limit worth $\sqrt{2}-1$ when $\rho \to \infty$. This comes from the decrease of $\Delta v_2$ for $\rho \geq 5.88$.

The Hohmann transfer is optimal (for minimizing the total impulse) if the number of impulses is limited to two. The transfer time, which is half the period of the transfer orbit, can be reduced by targeting an apogee greater than $r_2$. The final orbit is thus reached more quickly, at the expense of a larger total impulse $\Delta v_t$.

#### 6.2.3.2 Hohmann Transfer with Plane Change

Hohmann's strategy can be adapted to the case of noncoplanar circular orbits by splitting the plane change of angle $\alpha$ between the two impulses.

The transfer orbit makes an angle $s\alpha$ with the initial orbit and an angle $(1-s)\alpha$ with the final orbit, with $0 \leq s \leq 1$. The value of s must be chosen to minimize the total impulse $\Delta v_t$, which is assessed by

$$\Delta v_t = \Delta v_1 + \Delta v_2 \quad \text{with} \quad \begin{cases} \Delta v_1 = \sqrt{v_1^2 + v_P^2 - 2v_1 v_P \cos[s\alpha]} \\ \Delta v_2 = \sqrt{v_2^2 + v_A^2 - 2v_2 v_A \cos[(1-s)\alpha]} \end{cases} \tag{6.17}$$

In this formulation, $v_1$ and $v_2$ denote the velocities on the initial and final orbits, whereas $v_P$ and $v_A$ denote the velocities at perigee and apogee of the transfer orbit, assuming that the targeted orbit is higher than the initial one.

The calculation of $\Delta v_1$ and $\Delta v_2$ is similar to the one depicted in Figure 6.6.

The optimal value of s must be determined by a numerical method.

An approximate analytical solution can be obtained using Lisowski method (1995). It consists in minimizing the sum of squared impulses $\Delta v_1^2 + \Delta v_2^2$ instead of minimizing the sum of their modulus $\Delta v_1 + \Delta v_2$. This eliminates the square roots that hinder the analytical solution. By canceling the derivative with respect to s

$$\frac{d}{ds}\left(\Delta v_1^2 + \Delta v_2^2\right) = 0 \quad \Rightarrow \quad v_1 v_P \sin[s\alpha] - v_2 v_A \sin[(1-s)\alpha] = 0 \tag{6.18}$$

and expressing all velocities as functions of $\rho = \frac{r_2}{r_1}$ and $v_1 = \sqrt{\frac{\mu}{r_1}}$ as in (6.16)

$$v_P = v_1\sqrt{\frac{2\rho}{1+\rho}}, \quad v_A = v_1\sqrt{\frac{2}{\rho(1+\rho)}}, \quad v_2 = \frac{v_1}{\sqrt{\rho}} \tag{6.19}$$

we obtain the factor s as a function of the ratio ρ and the angle α.

$$\tan(s\alpha) = \frac{\sin\alpha}{\cos\alpha + \rho^{3/2}} \quad \text{with} \quad \rho = \frac{r_2}{r_1} \tag{6.20}$$

Numerical comparisons of this approximate solution with the optimal one (when minimizing the sum $\Delta v_t = \Delta v_1 + \Delta v_2$) show a quite good agreement, so that formula (6.20) can be used with confidence for preliminary evaluations.

If the initial and final radius are the same ($\rho = 1$), formula (6.20) yields $s = 0.5$. The plane change is evenly distributed between the two maneuvers.

#### 6.2.3.3 Bi-elliptic Transfer

The bi-elliptic transfer consists of moving from a circular orbit of radius $r_1$ to a coplanar circular orbit of radius $r_2$ in three impulses.

- The first impulse $\Delta v_1$ injects into an ellipse with perigee $r_1$ and apogee $r_A$.
- After a half revolution, the second impulse $\Delta v_A$ at apogee $r_A$ injects into an ellipse with perigee $r_2$ and apogee $r_A$.
- After another half revolution, the third impulse $\Delta v_2$ at perigee $r_2$ circularizes at the radius $r_2$.

The limit cases of the bi-elliptic transfer are the Hohmann transfer for $r_A = r_2$, and the infinite bi-elliptic transfer for $r_A \to \infty$.

Figure 6.9 illustrates the general bi-elliptic transfer, including a plane change (to be discussed after the planar case).

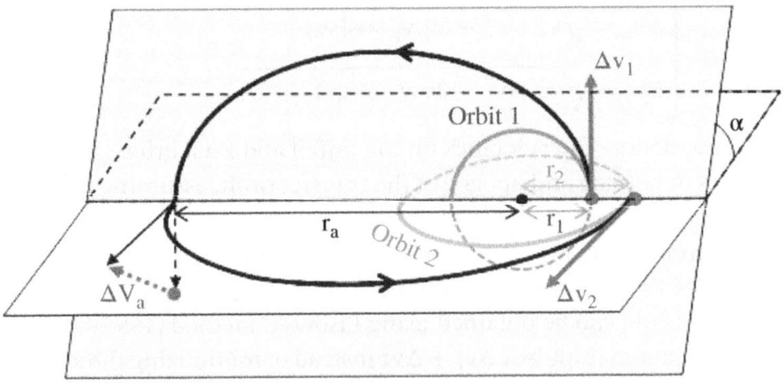


**Figure 6.9** Bi-elliptic transfer with plane change.

As for the Hohmann transfer, it is convenient to express the three impulses as functions of the radius ratios $\rho = \frac{r_2}{r_1}$ and $\rho_A = \frac{r_A}{r_1}$.

$$\begin{cases} \Delta v_1 = v_1 \left( \sqrt{\dfrac{2\rho_A}{1+\rho_A}} - 1 \right) \\ \Delta v_A = \dfrac{v_1}{\sqrt{\rho_A}} \left( \sqrt{\dfrac{2\rho}{\rho+\rho_A}} - \sqrt{\dfrac{2}{1+\rho_A}} \right) \\ \Delta v_2 = \dfrac{v_1}{\sqrt{\rho}} \left( \sqrt{\dfrac{2\rho_A}{\rho+\rho_A}} - 1 \right) \end{cases} \tag{6.21}$$

The intermediate apogee $r_A$ can be freely chosen. A high apogee reduces the total impulse $\Delta v_t = \Delta v_1 + \Delta v_A + \Delta v_2$ at the expense of a longer transfer.

The minimum is obtained for the (fictitious) infinite bi-elliptic transfer $r_A \to \infty$, for which the intermediate impulse $\Delta v_A$ vanishes.

$$\lim_{r_A \to \infty} \Delta v_t = v_1(\sqrt{2} - 1) \left( 1 + \sqrt{\frac{1}{\rho}} \right) \tag{6.22}$$

The bi-elliptic transfer is less expensive than the Hohmann transfer for $\rho \geq 11.94$ as shown in Figure 6.8. Nevertheless, the gain is too small (< 8%) to be interesting since it results in a huge increase in the transfer time.

On the other hand, the bi-elliptic transfer can prove interesting for noncoplanar orbits. The plane change is indeed carried out at a higher altitude during the second impulse, which reduces its cost. For different initial and final altitudes ($r_2 \neq r_1$), a numerical method is needed to find the optimum apogee $r_A$. In the case of a pure plane change ($r_2 = r_1$), an analytical solution can be found as follows:

- The first impulse $\Delta v_1$ injects into an ellipse with perigee $r_1$ and apogee $r_A$.
- After a half revolution, the second impulse $\Delta v_A$ at the apogee $r_A$ performs the plane change of angle $\alpha$.
- After another half revolution, the third impulse $\Delta v_2$ at perigee $r_1$ circularizes at the radius $r_1$.

Let us express the three impulses as functions of the ratio $\rho_A = \frac{r_A}{r_1}$.

$$\begin{cases} \Delta v_1 = v_1 \left( \sqrt{\dfrac{2\rho_A}{1+\rho_A}} - 1 \right) = \Delta v_2 \\ \Delta v_A = 2 v_A \sin \dfrac{\alpha}{2} \quad \text{with} \quad v_A = v_1 \sqrt{\dfrac{2}{\rho_A(1+\rho_A)}} \end{cases} \tag{6.23}$$

The optimum apogee minimizes the total impulse.

$$\min_{\rho_A} \Delta v_1 + \Delta v_A + \Delta v_2 \quad \Rightarrow \quad \frac{d}{d\rho_A} \left( \sqrt{\frac{\rho_A}{1+\rho_A}} + \frac{1}{\sqrt{\rho_A(1+\rho_A)}} \sin \frac{\alpha}{2} \right) = 0 \tag{6.24}$$

After derivation and simplification, we obtain

$$\rho_A = \frac{\sin \frac{\alpha}{2}}{1 - 2 \sin \frac{\alpha}{2}} \tag{6.25}$$

The total impulse is then assessed by substituting the solution (6.25) into (6.23).

This solution makes sense only if $\rho_A \geq 1$, which corresponds to $\alpha = 38.94$ deg.

On the other hand, the optimum apogee goes to infinity for an angle $\alpha = 60$ deg.

Above that value, the solution is always the infinite bi-elliptic transfer and the cost remains identical whatever the angle $\alpha$.

There remains to check whether the three-impulse strategy is better than the single-impulse strategy, depending on the angle $\alpha$. For that purpose, we calculate the difference between their respective impulses and study the sign.

To alleviate the formulas, we note $x = \sin\frac{\alpha}{2}$.

- The cost of the three-impulse strategy is

$$\Delta v_{3-imp} = 2v_1\left(\sqrt{\frac{2\rho_A}{1+\rho_A}} - 1 + x\sqrt{\frac{2}{\rho_A(1+\rho_A)}}\right) \quad \text{with} \quad \rho_A = \frac{x}{1-2x} \tag{6.26}$$

- The cost of the single impulse strategy is

$$\Delta v_{1-imp} = 2v_1 x \tag{6.27}$$

The sign of the difference is that of

$$D = \sqrt{\frac{2\rho_A}{1+\rho_A}}\left(1 + \frac{x}{\rho_A}\right) - 1 - x \tag{6.28}$$

which gives after substituting $\rho_A = \frac{x}{1-2x}$

$$D = 2\sqrt{\frac{2x}{1-x}}(1-x) - (1+x) = 2\sqrt{2x(1-x)} - (1+x) \tag{6.29}$$

This function of x is null for $x = \frac{1}{3}$ and negative for $x > \frac{1}{3}$.

We retrieve the lower bound $\alpha = 38.94$ deg found above for the validity of the three-impulse solution. This strategy is therefore optimal if $\alpha$ exceeds 38.94 deg. In such cases, the loss due to the first and three impulses (to reach a sufficiently high apogee) is more than compensated by the gain on the plane change maneuver.

The reader can consult [R12] for further analyses of the bi-elliptic strategy.

## 6.3 Point Target

This section deals with a target defined by its position. The minimum-energy and minimum-eccentricity transfers are presented, before studying the minimum-impulse problem in the general case.

### 6.3.1 Problem Formulation

The problem consists of finding a Keplerian conic joining two points $P_1$ and $P_2$. The triangle formed by the attracting center O and the points $P_1$ and $P_2$ has sides of lengths $OP_1 = r_1$, $OP_2 = r_2$, $P_1P_2 = c$ (chord), respectively. The transfer takes place in the plane $OP_1P_2$ (if the points are not aligned), and the transfer angle is noted $\theta$. The half-perimeter of the triangle $(OP_1P_2)$ is noted $s = \frac{1}{2}(r_1 + r_2 + c)$.

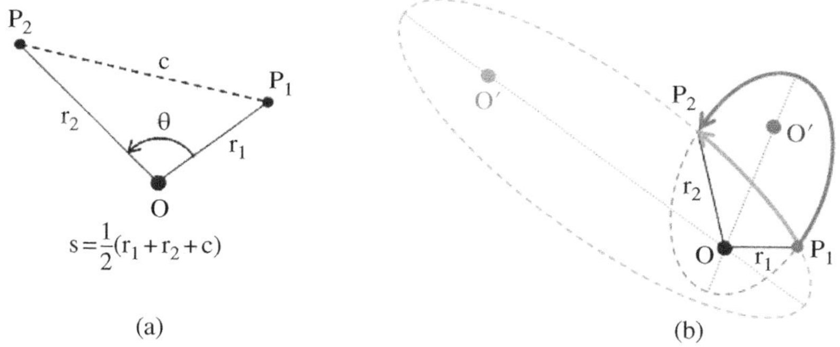


**Figure 6.10** (a) Transfer geometry. (b) Transfer solutions.

The transfer geometry is depicted in Figure 6.10a.

Figure 6.10b shows two different ellipses joining $P_1$ and $P_2$. Each ellipse has its own semi-major axis a, the same real focus in O (attracting center) and a different virtual focus in O′. For an ellipse of semi-major axis a and foci O and O′, we have

$$\begin{cases} OP_1 + O'P_1 = 2a \\ OP_2 + O'P_2 = 2a \end{cases} \tag{6.30}$$

### 6.3.2 Minimum-Energy Transfer

The energy of a Keplerian conic depends only on the semi-major axis a, as defined in (1.14): $w = -\frac{\mu}{2a}$.

For two given points $P_1$ and $P_2$, a first interesting conic passing through $P_1$ and $P_2$ is the conic of minimum energy (equivalent to minimum semi-major axis).

Adding the equalities (6.30), with the triangular inequality: $O'P_1 + O'P_2 \geq P_1P_2$, we obtain a minimum bound $a_m$ on the semi-major axis.

$$a \geq a_m = \frac{r_1 + r_2 + c}{4} = \frac{s}{2} \tag{6.31}$$

The ellipse of semi-major axis $a_m$ is the **minimum-energy** ellipse joining $P_1$ to $P_2$.

Let us calculate the eccentricity and the parameter of this minimum-energy ellipse.

From (6.30) and (6.31), we have by noting $OP_1' = r_1'$, $OP_2' = r_2'$

$$\begin{cases} r_1 + r_1' = 2a_m = s \\ r_2 + r_2' = 2a_m = s \end{cases} \Rightarrow r_1' + r_2' = 2s - (r_1 + r_2) = c \tag{6.32}$$

This equality implies that the points $P_1$, $P_2$, and O′ are aligned. The virtual focus O′ of the minimum-energy ellipse is located between $P_1$ and $P_2$ at distances given by

$$\begin{cases} r_1' = s - r_1 \\ r_2' = s - r_2 \end{cases} \tag{6.33}$$

These points and distances are shown in Figure 6.11.

Let us calculate the cosine of the angle α at vertex $P_1$ in the triangle $(OP_1P_2)$.

Starting from $r_2^2 = r_1^2 + c^2 - 2r_1c\cos\alpha$, and substituting $r_2 = 2s - r_1 - c$, we get after expansion $2s(r_1 + c - s) = r_1c(1 + \cos\alpha)$.

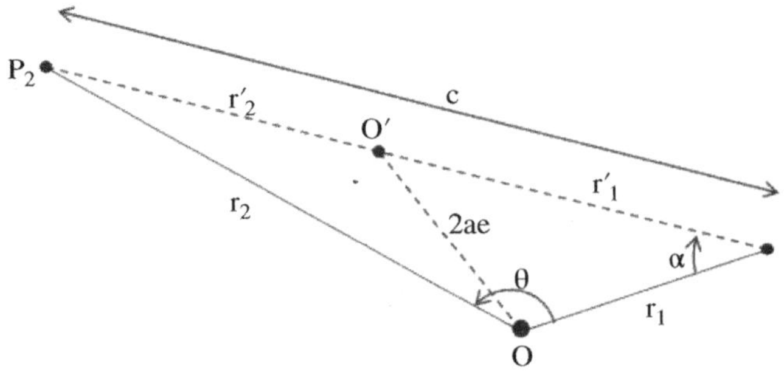


**Figure 6.11** Minimum-energy geometry.

Substituting then $r_1 + c = 2s - r_2$ into the first parenthesis, we obtain

$$\cos\alpha = \frac{2s(s - r_2)}{r_1 c} - 1 \tag{6.34}$$

The minimum-energy ellipse has an eccentricity noted $e_m$. The distance between the two foci O and O′ is $OO' = 2a_m e_m$. This distance is calculated in the triangle $(OO'P_1)$ by $(2a_m e_m)^2 = r_1^2 + r_1'^2 - 2r_1 r_1' \cos\alpha$. Substituting $r_1' = s - r_1$ and $\cos\alpha$ given by (6.34), we obtain after simplifications

$$(2a_m e_m)^2 = s^2 - \frac{4s}{c}(s - r_1)(s - r_2) \tag{6.35}$$

Let us first assess the ellipse parameter noted $p_m$. From $p_m = a_m\left(1 - e_m^2\right)$, we have $a_m^2 e_m^2 = a_m^2 - a_m p_m$ and with $a_m = \frac{s}{2}$, we have $4a_m^2 e_m^2 = s^2 - 2sp_m$.

Substituting into (6.35) yields the parameter of the minimum-energy ellipse.

$$p_m = \frac{2}{c}(s - r_1)(s - r_2) \tag{6.36}$$

An alternative expression for $p_m$ is

$$p_m = \frac{r_1 r_2}{c}(1 - \cos\theta) \tag{6.37}$$

Indeed, starting from $c^2 = r_1^2 + r_2^2 - 2r_1 r_2 \cos\theta$ in the triangle $(OP_1P_2)$, we have

$$2r_1 r_2 (1 - \cos\theta) = c^2 - r_1^2 - r_2^2 + 2r_1 r_2 = c^2 - (r_1 - r_2)^2 = (c - r_1 + r_2)(c + r_1 - r_2)$$
$$= (2s - 2r_1)(2s - 2r_2) = 4(s - r_1)(s - r_2)$$

so that we retrieve the formula (6.36).

We finally get the eccentricity of the minimum-energy ellipse from $e_m^2 = 1 - \frac{p_m}{a_m}$.

$$e_m = \sqrt{1 - \frac{4}{sc}(s - r_1)(s - r_2)} \tag{6.38}$$

### 6.3.3 Minimum-Eccentricity Transfer

Another interesting conic for the transfer problem between two points $P_1$ and $P_2$ is the conic of minimum eccentricity passing through $P_1$ and $P_2$.

To find this conic, we start by taking the scalar product of the eccentricity vector defined in (1.10) with $\vec{r}$.

$$\vec{e} = \frac{\vec{v} \wedge \vec{h}}{\mu} - \frac{\vec{r}}{r} \Rightarrow \vec{e} \cdot \vec{r} = \frac{(\vec{v} \wedge \vec{h}) \cdot \vec{r}}{\mu} - r = \frac{(\vec{r} \wedge \vec{v}) \cdot \vec{h}}{\mu} - r = \frac{h^2}{\mu} - r = p - r \tag{6.39}$$

Applying (6.39) in $P_1$ and $P_2$, and subtracting yields

$$\begin{cases} \vec{e} \cdot \vec{r}_1 = p - r_1 \\ \vec{e} \cdot \vec{r}_2 = p - r_2 \end{cases} \Rightarrow \vec{e} \cdot (\vec{r}_2 - \vec{r}_1) = \vec{e} \cdot \vec{c} = r_2 - r_1 \tag{6.40}$$

The minimum eccentricity occurs when $\vec{e}$ is collinear to $\vec{c}$.

This eccentricity noted $e_f$ (for "fundamental" ellipse as in [R1]) shall not be confused with $e_m$ (6.38), which is the eccentricity of the minimum-energy ellipse. Its value, which is always lower than 1, is calculated from

$$\vec{e}_f = \frac{r_1 - r_2}{c^2}\vec{c} = \frac{r_1 - r_2}{c^2}(\vec{r}_2 - \vec{r}_1) \tag{6.41}$$

The vector $\vec{e}_f$ is directed along $+\vec{c}$ if $r_1 > r_2$, or along $-\vec{c}$ if $r_1 < r_2$, so that

$$e_f = \frac{|r_2 - r_1|}{c} \tag{6.42}$$

Let us calculate the semi-major axis $a_f$ of the minimum-eccentricity ellipse.

Applying (6.39) in $P_1$ and $P_2$ with $\vec{e}_f$ given by (6.41), and summing yields

$$\begin{cases} \vec{e}_f \cdot \vec{r}_1 = p_f - r_1 \\ \vec{e}_f \cdot \vec{r}_2 = p_f - r_2 \end{cases} \Rightarrow \frac{r_1 - r_2}{c^2}(\vec{r}_2 - \vec{r}_1) \cdot (\vec{r}_1 + \vec{r}_2) = 2p_f - (r_1 + r_2) \tag{6.43}$$

After expansion and simplification, we obtain

$$2p_f = (r_1 + r_2)\left[1 - \left(\frac{r_2 - r_1}{c}\right)^2\right] = (r_1 + r_2)\left(1 - e_f^2\right) \tag{6.44}$$

The semi-major axis of the minimum-eccentricity ellipse joining $P_1$ and $P_2$ is thus

$$a_f = \frac{1}{2}(r_1 + r_2) \tag{6.45}$$

The geometry of the minimum-eccentricity ellipse is depicted in Figure 6.12.

### 6.3.4 Noncollinear Transfer

This section deals with the transfer between two points $P_1$ and $P_2$ when the transfer angle $\theta$ is different from 180 deg. We first establish a relation between the initial and final velocities, then consider the minimum-impulse transfer from $P_1$ to $P_2$.

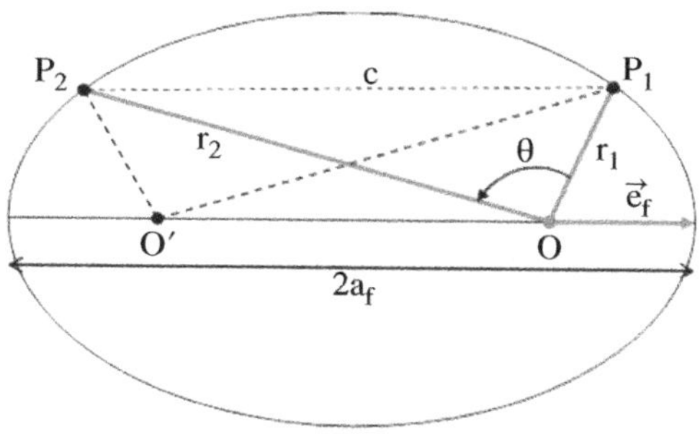


**Figure 6.12** Minimum-eccentricity ellipse.

#### 6.3.4.1 Initial and Final Velocities

For any Keplerian conic joining $P_1$ to $P_2$, the final position is expressed from the initial position and velocity with Lagrange coefficients defined in Section 1.3.4. By noting p, the conic parameter, and with the index 12 for the transfer going from $P_1$ to $P_2$, we have from (1.71)

$$\vec{r}_2 = F_{12}\vec{r}_1 + G_{12}\vec{v}_1 \quad \text{with} \quad \begin{cases} F_{12} = 1 - \dfrac{r_2}{p}(1 - \cos\theta_{12}) \\ G_{12} = \dfrac{r_1 r_2}{\sqrt{\mu p}}\sin\theta_{12} \end{cases} \tag{6.46}$$

The initial velocity can thus be expressed as:

$$\vec{v}_1 = \frac{F_{12}\vec{r}_1 - \vec{r}_2}{G_{12}} = \frac{\sqrt{\mu p}}{r_1 r_2 \sin\theta_{12}}\left[(\vec{r}_2 - \vec{r}_1) + \frac{r_2(1 - \cos\theta_{12})}{p}\vec{r}_1\right] \tag{6.47}$$

Conversely, the same conic of parameter p joins $P_2$ to $P_1$, with the opposite transfer angle $\theta_{21} = -\theta_{12}$. Applying the formulas (6.46) and (6.47) by swapping the indexes 1 and 2 and changing the angle sign yields the final velocity under the form

$$\vec{v}_2 = \frac{F_{21}\vec{r}_2 - \vec{r}_1}{G_{21}} = \frac{\sqrt{\mu p}}{r_2 r_1 \sin\theta_{21}}\left[(\vec{r}_1 - \vec{r}_2) + \frac{r_1(1 - \cos\theta_{21})}{p}\vec{r}_2\right] \tag{6.48}$$

Let us rewrite the formulas (6.47) and (6.48) by introducing the unit vectors along the initial and final positions and along the chord $P_1P_2$, defined by

$$\vec{u}_1 = \frac{\vec{r}_1}{r_1}, \quad \vec{u}_2 = \frac{\vec{r}_2}{r_2}, \quad \vec{u}_c = \frac{\vec{r}_2 - \vec{r}_1}{\|\vec{r}_2 - \vec{r}_1\|} = \frac{\vec{r}_2 - \vec{r}}{c} \tag{6.49}$$

Dropping the index for the transfer angle ($\theta = \theta_{12}$), we obtain

$$\begin{cases} \vec{v}_1 = v_c\vec{u}_c + v_\rho\vec{u}_1 \\ \vec{v}_2 = v_c\vec{u}_c - v_\rho\vec{u}_2 \end{cases} \quad \text{with} \quad v_c = \frac{c\sqrt{\mu p}}{r_1 r_2 \sin\theta} \quad \text{and} \quad v_\rho = \sqrt{\frac{\mu}{p}}\,\frac{1 - \cos\theta}{\sin\theta} \tag{6.50}$$

The initial and final velocities have thus identical components when projected along skewed axes oriented along the initial and final radii and the chord. The notation $v_\rho$ is chosen in order to stress that this component is not the projection of the velocity on the radius vector (usually noted $v_r$), since the axes of projection are not orthogonal. These components $v_c$ and $v_\rho$ are depicted in Figure 6.13.

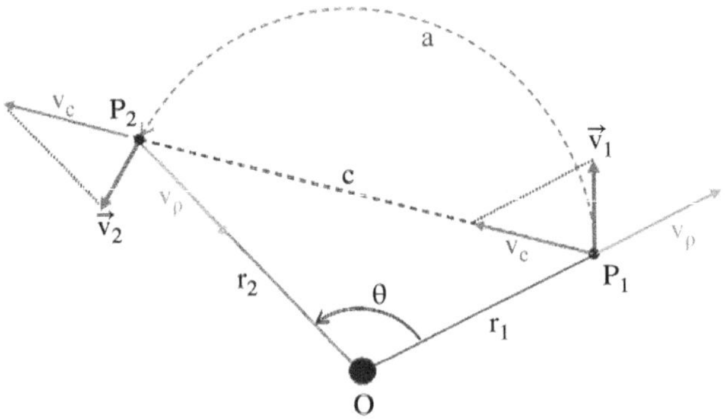


**Figure 6.13** Velocity projection on skewed axes.

The components $v_c$ and $v_\rho$ depend only on the parameter p of the conic and their product depends only on the problem geometry (and not on the transfer conic).

$$v_c v_\rho = \frac{\mu c}{r_1 r_2 (1+\cos\theta)} \tag{6.51}$$

The components $v_c$ and $v_\rho$ can also be expressed as functions of the semi-major axis instead of the parameter.

For that purpose, let us assess the velocity modulus in $P_1$ and $P_2$ from (6.50).

$$\begin{cases} \vec{v}_1 = v_c\vec{u}_c + v_\rho\vec{u}_1 \\ \vec{v}_2 = v_c\vec{u}_c - v_\rho\vec{u}_2 \end{cases} \Rightarrow \begin{cases} v_1^2 = v_c^2 + v_\rho^2 + 2v_c v_\rho \vec{u}_c\cdot\vec{u}_1 = \mu\left(\dfrac{2}{r_1} - \dfrac{1}{a}\right) \\ v_2^2 = v_c^2 + v_\rho^2 - 2v_c v_\rho \vec{u}_c\cdot\vec{u}_2 = \mu\left(\dfrac{2}{r_2} - \dfrac{1}{a}\right) \end{cases} \tag{6.52}$$

Summing member by member gives

$$v_c^2 + v_\rho^2 = \mu\left(\frac{1}{r_1} + \frac{1}{r_2} - \frac{1}{a}\right) + v_c v_\rho(\vec{u}_c\cdot\vec{u}_2 - \vec{u}_c\cdot\vec{u}_1) \tag{6.53}$$

With the scalar products assessed from (6.49)

$$\vec{u}_c\cdot\vec{u}_1 = \frac{\vec{r}_2-\vec{r}}{c}\cdot\frac{\vec{r}_1}{r_1} = \frac{r_2\cos\theta - r_1}{c} \quad \text{and} \quad \vec{u}_c\cdot\vec{u}_2 = \frac{\vec{r}_2-\vec{r}}{c}\cdot\frac{\vec{r}_2}{r_2} = \frac{r_2 - r_1\cos\theta}{c} \tag{6.54}$$

equation (6.53) becomes

$$v_c^2 + v_\rho^2 = \mu\left(\frac{1}{r_1} + \frac{1}{r_2} - \frac{1}{a}\right) + v_c v_\rho \frac{r_1+r_2}{c}(1-\cos\theta) \tag{6.55}$$

Let us now assess the sum and the difference between the components $v_c$ and $v_\rho$.

Using the product $v_c v_\rho$ given by (6.51), we have

$$\begin{aligned} (v_c \pm v_\rho)^2 &= v_c^2 + v_\rho^2 \pm 2v_c v_\rho \\ &= \mu\left(\frac{1}{r_1} + \frac{1}{r_2} - \frac{1}{a}\right) + v_c v_\rho\left[\frac{r_1+r_2}{c}(1-\cos\theta) \pm 2\right] \\ &= \mu\left(\frac{1}{r_1} + \frac{1}{r_2} - \frac{1}{a}\right) + \frac{\mu}{r_1 r_2(1+\cos\theta)}[(r_1+r_2)(1-\cos\theta) \pm 2c] \end{aligned} \tag{6.56}$$

and after rearranging terms

$$(v_c \pm v_\rho)^2 = \frac{2\mu}{1+\cos\theta}\left(\frac{1}{r_1} + \frac{1}{r_2} \pm \frac{c}{r_1 r_2}\right) - \frac{\mu}{a} \tag{6.57}$$

To eliminate the term in $\cos\theta$, we use the two expressions (6.36) and (6.37) obtained for the parameter of the minimum-energy ellipse.

$$p_m = \frac{r_1 r_2}{c}(1-\cos\theta) = \frac{2}{c}(s-r_1)(s-r_2) \quad \text{with} \quad 2s = r_1 + r_2 + c \tag{6.58}$$

which yields

$$1+\cos\theta = \frac{(r_1+r_2+c)(r_1+r_2-c)}{2r_1 r_2} \tag{6.59}$$

Substituting into (6.57)

$$(v_c \pm v_\rho)^2 = 4\mu\frac{r_1+r_2 \pm c}{(r_1+r_2+c)(r_1+r_2-c)} - \frac{\mu}{a} \tag{6.60}$$

then taking the square root and separating it in two equations with the + or − sign

$$\begin{cases} v_c + v_\rho = \pm\sqrt{\dfrac{4\mu}{r_1 + r_2 - c} - \dfrac{\mu}{a}} = \pm\sqrt{\dfrac{2\mu}{s - c} - \dfrac{\mu}{a}} \\ v_c - v_\rho = \pm\sqrt{\dfrac{4\mu}{r_1 + r_2 + c} - \dfrac{\mu}{a}} = \pm\sqrt{\dfrac{2\mu}{s} - \dfrac{\mu}{a}} \end{cases} \tag{6.61}$$

we obtain $v_c$ and $v_\rho$ as functions of the semi-major axis.

$$\begin{cases} v_c = P + Q \\ v_r = P - Q \end{cases} \quad \text{with} \quad P \underset{\text{def}}{=} \pm\sqrt{\frac{\mu}{2}\left(\frac{1}{s-c} - \frac{1}{2a}\right)},\ Q \underset{\text{def}}{=} \pm\sqrt{\frac{\mu}{2}\left(\frac{1}{s} - \frac{1}{2a}\right)} \tag{6.62}$$

The signs of P and Q give four possible combinations, corresponding to the two virtual foci (Figure 6.10) satisfying (6.30) and the two senses of the transfer (direct with angle $\theta$, or retrograde with angle $2\pi - \theta$).

Let us now formulate the minimum-impulse transfer problem using these results.

#### 6.3.4.2 Minimum-Impulse Transfer

Let us assume that the velocity before maneuver in $P_1$ is a given vector $\vec{w}_1$ and that the targeted velocity after maneuver in $P_2$ is a given vector $\vec{w}_2$.

The minimum-impulse transfer from $P_1$ $(\vec{r}_1, \vec{w}_1)$ to $P_2$ $(\vec{r}_2, \vec{w}_2)$ is the solution of the following optimization problem.

$$\min_p \|\Delta\vec{v}_1\| + \|\Delta\vec{v}_2\| \quad \text{with} \quad \begin{cases} \Delta\vec{v}_1 = \vec{v}_1 - \vec{w}_1 = (v_c\vec{u}_c + v_\rho\vec{u}_1) - \vec{w}_1 \\ \Delta\vec{v}_2 = \vec{w}_2 - \vec{v}_2 = \vec{w}_2 - (v_c\vec{u}_c - v_\rho\vec{u}_2) \end{cases} \tag{6.63}$$

The transfer plane is orthogonal to the unit vector $\vec{u}_n \underset{\text{def}}{=} \pm\frac{\vec{u}_1 \wedge \vec{u}_2}{\sin\theta}$.

This vector is well defined for a noncollinear transfer and the sign depends on the angle $\theta$ (sign + if $0 < \theta < \pi$, sign − if $\pi < \theta < 2\pi$).

The solution of (6.63) is obtained by canceling the gradient of the cost function.

Unfortunately, this one-variable equation is complicated by square roots in the norms and in $v_c$ and $v_\rho$ expressions (6.50). The optimal value of p cannot be expressed analytically and it must be sought by a numerical method such as dichotomy or Newton.

Nevertheless, a useful approximation of the solution can be found by minimizing the sum of the squared impulses, instead of their norms.

This modified cost function is noted K and the problem becomes

$$\min_p K = \|\Delta\vec{v}_1\|^2 + \|\Delta\vec{v}_2\|^2 = (\vec{v}_1 - \vec{w}_1)^2 + (\vec{v}_2 - \vec{w}_2)^2 \tag{6.64}$$

The optimal value p cancels the gradient of K. Deriving with respect to p, we have

$$\frac{dK}{dp} = 2(\vec{v}_1 - \vec{w}_1)\cdot\frac{d\vec{v}_1}{dp} + 2(\vec{v}_2 - \vec{w}_2)\cdot\frac{d\vec{v}_2}{dp} = 0 \tag{6.65}$$

The derivatives of $\vec{v}_1$ and $\vec{v}_2$ are assessed from (6.50). By observing that

$$\frac{dv_c}{dp} = \frac{v_c}{2p} \quad \text{and} \quad \frac{dv_\rho}{dp} = -\frac{v_\rho}{2p} \tag{6.66}$$

we obtain

$$\frac{d\vec{v}_1}{dp} = \frac{1}{2p}(v_c\vec{u}_c - v_\rho\vec{u}_1) \quad \text{and} \quad \frac{d\vec{v}_2}{dp} = \frac{1}{2p}(v_c\vec{u}_c + v_\rho\vec{u}_2) \tag{6.67}$$

Substituting into (6.65) $\vec{v}_1$ and $\vec{v}_2$ from (6.50) and the derivatives from (6.66) gives

$$(v_c\vec{u}_c + v_\rho\vec{u}_1 - \vec{w}_1)\cdot(v_c\vec{u}_c - v_\rho\vec{u}_1) + (v_c\vec{u}_c - v_\rho\vec{u}_2 - \vec{w}_2)\cdot(v_c\vec{u}_c + v_\rho\vec{u}_2) = 0 \tag{6.68}$$

and after expansion and grouping of terms

$$2\left(v_c^2 - v_\rho^2\right) + B_c v_c + B_\rho v_\rho = 0 \quad \text{with} \quad \begin{cases} B_c \underset{\text{def}}{=} (\vec{w}_2 - \vec{w}_1)\cdot\vec{u}_c \\ B_\rho \underset{\text{def}}{=} \vec{w}_1\cdot\vec{u}_1 - \vec{w}_2\cdot\vec{u}_2 \end{cases} \tag{6.69}$$

where the constants $B_c$ and $B_\rho$ depend only on the problem geometry.

On the other hand, the components $v_c$ and $v_\rho$ given in (6.50) take the form

$$\begin{cases} v_c = A_c\sqrt{p} \quad \text{with} \quad A_c \underset{\text{def}}{=} \dfrac{c\sqrt{\mu}}{r_1 r_2 \sin\theta} \\ v_\rho = \dfrac{A_\rho}{\sqrt{p}} \quad \text{with} \quad A_\rho \underset{\text{def}}{=} \sqrt{\mu}\dfrac{1-\cos\theta}{\sin\theta} \end{cases} \tag{6.70}$$

where the constants $A_c$ and $A_\rho$ depend only on the problem geometry.

By substituting into (6.69), and making the variable change $x = \sqrt{p}$, the equation to solve for x is finally

$$2A_c^2x^4 + A_cB_cx^3 + A_\rho B_\rho x - 2A_\rho^2 = 0 \tag{6.71}$$

This quartic equation with a missing quadratic term is solved by Ferrari's method.

---

**Ferrari's Solution**

Let us consider the quartic equation $x^4 + a_3x^3 + a_1x + a_0 = 0$.

By posing: $\alpha = -\frac{3}{8}a_3^2$, $\beta = \frac{1}{8}a_3^2 + a_1$, $\gamma = -\frac{3}{256}a_3^2 + \frac{1}{4}a_1a_3 + a_0$ the four solutions noted $x_{1,2,3,4}$ depend on the value of β.

- If $\beta = 0$, then $x_{1,2,3,4} = -\dfrac{a_3}{4} \pm \sqrt{\dfrac{-\alpha \pm \sqrt{\alpha^2 - 4\gamma}}{2}}$
- If $\beta \neq 0$, then $x_{1,2,3,4} = -\dfrac{a_3}{4} + \dfrac{\varepsilon}{2}\left[\sqrt{\alpha + 2y} \pm \sqrt{-\left(3\alpha + 2y + \varepsilon\dfrac{2\beta}{\sqrt{\alpha+2y}}\right)}\right]$

with the notations

$$P = -\frac{\alpha^2}{12} - \gamma,\quad Q = -\frac{\alpha^3}{108} + \frac{\alpha\gamma}{3} - \frac{\beta^2}{8},\quad R = \frac{Q}{2} \pm \sqrt{\frac{Q^2}{4} + \frac{P^3}{27}},\quad U = \sqrt[3]{R}$$

$$\text{and } y = \begin{cases} \dfrac{5\alpha}{6} - \sqrt[3]{Q} & \text{if } U = 0, \quad \varepsilon = +1 \text{ for } x_{1,2} \\ \dfrac{5\alpha}{6} + U - \dfrac{P}{3U} & \text{if } U \neq 0, \quad \varepsilon = -1 \text{ for } x_{3,4} \end{cases}$$

---

Equation (6.71) admits at most four solutions corresponding to the two virtual foci (Figure 6.10) satisfying (6.30) and the two senses of transfer (direct with angle θ or retrograde with angle 2π – θ). There is at least one positive solution because the function (6.71) is negative in x = 0 and has limit +∞ for x→ +∞.

Once $x = \sqrt{p}$ is known, the velocities are assessed from (6.70). This solution can be used as initial guess for solving the "exact" minimization problem (6.63).

In the case of null initial and final velocities ($w_1 = w_2 = 0$), the coefficients $B_c$ and $B_p$ in (6.69) cancel, and the solution is $p = x^2 = \frac{A_p}{A_c} = \frac{r_1 r_2}{c}(1 - \cos\theta)$, which corresponds to the minimum-energy ellipse passing through $P_1$ and $P_2$ (6.37).

### 6.3.5 Collinear Transfer

This section deals with the transfer between two points $P_1$ and $P_2$ when the transfer angle $\theta$ is 180 deg. The analysis is similar to the noncollinear case, but different axes are used since the collinear positions no longer define the transfer plane.

#### 6.3.5.1 Initial and Final Velocities

From the conic equation written in $P_1$ and $P_2$, and with $c = r_1 + r_2$ for collinear opposite positions, we have

$$\begin{cases} r_1 = \dfrac{p}{1 + e\cos\theta_1} \\ r_2 = \dfrac{p}{1 + e\cos(\theta_1 + \pi)} \end{cases} \Rightarrow \begin{cases} p = \dfrac{2r_1 r_2}{r_1 + r_2} = \dfrac{2r_1 r_2}{c} \\ e\cos\theta_1 = \dfrac{r_2 - r_1}{r_1 + r_2} = \dfrac{r_2 - r_1}{c} \end{cases} \tag{6.72}$$

This shows that all conics joining $P_1$ to $P_2$ have the same parameter, which depends only on the radius vectors.

The velocity is expressed in the local orbital frame defined in Section 2.2.3 (see Figure 2.2). Formula (2.10) gives, with $h^2 = \mu p$,

$$\vec{v} = \sqrt{\frac{\mu}{p}}[e\sin\theta\,\vec{u}_r + (1 + e\cos\theta)\vec{u}_\theta] \tag{6.73}$$

so that the radial and orthoradial components are

$$v_r = \frac{he\sin\theta}{p}, \quad v_\theta = \frac{\sqrt{\mu p}}{r} \tag{6.74}$$

These formulas are applied to the collinear positions $P_1$ and $P_2$, with the parameter of the conic given by (6.72).

The orthoradial components are

$$v_{1\theta} = \sqrt{\frac{2\mu r_2}{cr_1}}, \quad v_{2\theta} = \sqrt{\frac{2\mu r_1}{cr_2}} \tag{6.75}$$

Using $c = r_1 + r_2$, we observe that these components can also be expressed as:

$$v_{1\theta} = \sqrt{\mu\left(\frac{2}{r_1} - \frac{2}{c}\right)}, \quad v_{2\theta} = \sqrt{\mu\left(\frac{2}{r_2} - \frac{2}{c}\right)} \tag{6.76}$$

The orthoradial components are equal to the velocities on the minimum-energy ellipse (Section 6.3.2). Indeed, the semi-major axis $a_m$ of the minimum-energy ellipse for the collinear case is $a_m = \frac{s}{2} = \frac{r_1+r_2+c}{4} = \frac{c}{2}$.

The radial components are calculated from the velocity modulus

$$v^2 = v_r^2 + v_\theta^2 = \mu\left(\frac{2}{r} - \frac{1}{a}\right) \Rightarrow v_r^2 = \mu\left(\frac{2}{r} - \frac{1}{a}\right) - \frac{\mu p}{r^2} = \mu\left(\frac{2}{c} - \frac{1}{a}\right) \tag{6.77}$$

We observe that the orthoradial components $v_{1\theta}$, $v_{2\theta}$ (6.76) are fixed, whereas the radial components $v_{1r}$, $v_{2r}$ (6.77) depend only on the conic semi-major axis.

We are now in position to formulate the minimum-impulse transfer problem.

#### 6.3.5.2 Minimum-Impulse Transfer

As in Section 6.3.4, the minimum-impulse transfer from $P_1$ $(\vec{r}_1, \vec{w}_1)$ to $P_2$ $(\vec{r}_2, \vec{w}_2)$ is formulated as an optimization problem.

The radial and orthoradial unit vectors in $P_1$ are noted $\vec{u}_r$ and $\vec{u}_\theta$, respectively. Since the transfer is collinear, the unit vectors in $P_2$ are opposite. The velocities are expressed with their orthoradial and radial components (6.76) and (6.77).

By noting $v_r$ the radial component in $P_1$, we have

$$\begin{cases} \vec{v}_1 = v_{1\theta}\vec{u}_{1\theta} + v_{1r}\vec{u}_{1r} = \quad v_{1\theta}\vec{u}_\theta + v_r\vec{u}_r \\ \vec{v}_2 = v_{2\theta}\vec{u}_{2\theta} + v_{2r}\vec{u}_{2r} = -v_{2\theta}\vec{u}_\theta - v_r\vec{u}_r \end{cases} \tag{6.78}$$

The optimization variables for the minimum-impulse transfer are the semi-major axis a, which defines the radial component, and the orthoradial unit vector $\vec{u}_\theta$, which defines the transfer plane. This vector is constrained to be orthogonal to the initial radius vector $\vec{u}_r$. The optimization problem is thus formulated as:

$$\min_{a,\vec{u}_\theta} \|\Delta\vec{v}_1\| + \|\Delta\vec{v}_2\| \ \text{ s.t. } \begin{cases} \vec{u}_r \cdot \vec{u}_\theta = 0 \\ \vec{u}_\theta \cdot \vec{u}_\theta = 1 \end{cases} \text{ with } \begin{cases} \Delta\vec{v}_1 = \vec{v}_1 - \vec{w}_1 \\ \Delta\vec{v}_2 = \vec{w}_2 - \vec{v}_2 \end{cases} \tag{6.79}$$

Since the problem cannot be solved analytically, we search for an approximate solution by minimizing the sum of the squared norms.

$$\min_{a,\vec{u}_\theta} K = \|\Delta\vec{v}_1\|^2 + \|\Delta\vec{v}_2\|^2 \ \text{ s.t. } \begin{cases} \vec{u}_r \cdot \vec{u}_\theta = 0 \\ \vec{u}_\theta \cdot \vec{u}_\theta = 1 \end{cases} \tag{6.80}$$

The Lagrangian of this constrained optimization problem is

$$L = \frac{1}{2}\left(\|\Delta\vec{v}_1\|^2 + \|\Delta\vec{v}_2\|^2\right) + \lambda_1\vec{u}_r \cdot \vec{u}_\theta + \frac{1}{2}\lambda_2(\vec{u}_\theta \cdot \vec{u}_\theta - 1) \tag{6.81}$$

with multipliers $\lambda_1$ and $\lambda_2$ associated with the constraints on $\vec{u}_\theta$.

The cost function and the norm constraint have been halved to simplify the subsequent calculations. The Karush, Kuhn, and Tucker (KKT) necessary optimality conditions are

$$\frac{\partial L}{\partial a} = 0, \quad \frac{\partial L}{\partial \vec{u}_\theta} = \vec{0} \ \text{ with } \begin{cases} \vec{u}_r \cdot \vec{u}_\theta = 0 \\ \vec{u}_\theta \cdot \vec{u}_\theta = 1 \end{cases} \tag{6.82}$$

The first condition in (6.82) gives, using (6.78)

$$\frac{\partial \Delta\vec{v}_1}{\partial a} \cdot \Delta\vec{v}_1 + \frac{\partial \Delta\vec{v}_2}{\partial a} \cdot \Delta\vec{v}_2 = 0 \ \text{ with } \begin{cases} \Delta\vec{v}_1 = v_{1\theta}\vec{u}_\theta + v_r\vec{u}_r - \vec{w}_1 \\ \Delta\vec{v}_2 = \vec{w}_2 + v_{2\theta}\vec{u}_\theta + v_r\vec{u}_r \end{cases} \tag{6.83}$$

Since $v_{1\theta}$, $v_{2\theta}$ (6.76) do not depend on a, and observing from (6.77) that

$$\frac{dv_r}{da} = \frac{\mu}{2a^2 v_r} \quad \Rightarrow \quad \frac{\partial \Delta\vec{v}_1}{\partial a} = \frac{\partial \Delta\vec{v}_2}{\partial a} = \frac{\mu}{2a^2 v_r}\vec{u}_r \tag{6.84}$$

we obtain by substituting into (6.83)

$$\vec{u}_r \cdot (\Delta\vec{v}_1 + \Delta\vec{v}_2) = 0 \quad \Rightarrow \quad v_r = \frac{1}{2}\vec{u}_r \cdot (\vec{w}_1 - \vec{w}_2) \tag{6.85}$$

The relation determines the semi-major axis. It can be written by introducing the radial components of the velocities $\vec{w}_1$ before the first impulse and $\vec{w}_2$ after the second impulse, taking care that $\vec{u}_{2r} = -\vec{u}_{1r} = -\vec{u}_r$.

$$v_r = \sqrt{\mu\left(\frac{2}{c} - \frac{1}{a}\right)} = \frac{w_{1r} + w_{2r}}{2} \quad \text{with} \quad \begin{cases} w_{1r} = \vec{w}_1 \cdot \vec{u}_{1r} = \ \ \vec{w}_1 \cdot \vec{u}_r \\ w_{2r} = \vec{w}_2 \cdot \vec{u}_{2r} = -\vec{w}_2 \cdot \vec{u}_r \end{cases} \tag{6.86}$$

The conditions (6.82) on $\vec{u}_\theta$ give, with $\frac{\partial \Delta \vec{v}_{1/2}}{\partial \vec{u}_\theta} = v_{1/2\theta} I$ (I being the identity matrix)

$$v_{1\theta}\Delta\vec{v}_1 + v_{2\theta}\Delta\vec{v}_2 + \lambda_1\vec{u}_r + \lambda_2\vec{u}_\theta = \vec{0} \quad \text{with} \quad \begin{cases} \vec{u}_r \cdot \vec{u}_\theta = 0 \\ \vec{u}_\theta \cdot \vec{u}_\theta = 1 \end{cases} \tag{6.87}$$

The unknowns are $\vec{u}_\theta$, $\lambda_1$ and $\lambda_2$. Taking the scalar products of the first equation with $\vec{u}_\theta$, then with $\vec{u}_r$ allows to assess the two multipliers.

$$\begin{cases} \lambda_1 = -v_{1\theta}(v_{1r} - \vec{w}_1 \cdot \vec{u}_r) - v_{2\theta}(\vec{w}_2 \cdot \vec{u}_r + v_{2r}) \\ \lambda_2 = -v_{1\theta}(v_{1\theta} - \vec{w}_1 \cdot \vec{u}_\theta) - v_{2\theta}(\vec{w}_2 \cdot \vec{u}_\theta + v_{2\theta}) \end{cases} \tag{6.88}$$

The multipliers are then substituted in (6.87).

By defining the unit vector normal to the transfer plane $\vec{u}_n = \vec{u}_r \wedge \vec{u}_\theta$, and expressing the velocities $\vec{w}_1$ and $\vec{w}_2$ with their components in the frame $(\vec{u}_r, \vec{u}_\theta, \vec{u}_n)$, most terms cancel and we obtain

$$v_{1\theta} w_{1n} = v_{2\theta} w_{2n} \quad \text{with} \quad \begin{cases} w_{1n} = \vec{w}_1 \cdot \vec{u}_n \\ w_{2n} = \vec{w}_2 \cdot \vec{u}_n \end{cases} \tag{6.89}$$

The vector $\vec{u}_n$ defined from $\vec{u}_\theta$ is unknown. To assess it, let us place ourselves in the local orbital frame associated with the initial conditions in $P_1$ $(\vec{r}_1, \vec{w}_1)$. This frame is noted $(\vec{u}_{1r}, \vec{u}_{1t}, \vec{u}_{1h})$, where $\vec{u}_{1h}$ is the unit vector along the initial angular momentum $\vec{h}_1 = \vec{r}_1 \wedge \vec{w}_1$. The components of $\vec{w}_1$, $\vec{w}_2$, and $\vec{u}_n$ in this frame are

$$\vec{w}_1 = \begin{pmatrix} w_{1r} \\ w_{1t} \\ 0 \end{pmatrix}, \quad \vec{w}_2 = \begin{pmatrix} w_{2r} \\ w_{2t} \\ w_{2h} \end{pmatrix}, \quad \vec{u}_r = \begin{pmatrix} 1 \\ 0 \\ 0 \end{pmatrix}, \quad \vec{u}_\theta = \begin{pmatrix} 0 \\ \cos\delta \\ \sin\delta \end{pmatrix}, \quad \vec{u}_n = \begin{pmatrix} 0 \\ -\sin\delta \\ \cos\delta \end{pmatrix} \tag{6.90}$$

where $\delta$ is the angle between the initial orbit plane and the transfer plane.

Substituting these components into (6.89), with $v_{1\theta}$, $v_{2\theta}$ given by (6.75), we get

$$\tan\delta = \frac{r_1 w_{2h}}{r_1 w_{2t} - r_2 w_{1t}} \tag{6.91}$$

The formulas (6.86) for a and (6.91) for $\delta$ define the transfer orbit from the conditions $(\vec{r}_1, \vec{w}_1)$ in $P_1$ and $(\vec{r}_2, \vec{w}_2)$ in $P_2$. They can be used as initial guess for solving the "exact" minimization problem (6.79).

In the particular case of a Hohmann transfer, with circular initial and final orbits making an angle $\alpha$, the velocities components (6.90) in the initial local orbital frame $(\vec{u}_{1r}, \vec{u}_{1t}, \vec{u}_{1h})$ are (taking care that $P_2$ is at the opposite descending node)

$$\vec{w}_1 = \sqrt{\frac{\mu}{r_1}} \begin{pmatrix} 0 \\ 1 \\ 0 \end{pmatrix}, \quad \vec{w}_2 = -\sqrt{\frac{\mu}{r_2}} \begin{pmatrix} 0 \\ \cos\alpha \\ \sin\alpha \end{pmatrix} \tag{6.92}$$

The formula (6.86) gives

$$v_r = \sqrt{\mu\left(\frac{2}{c} - \frac{1}{a}\right)} = \frac{w_{1r} + w_{2r}}{2} = 0 \quad \Rightarrow \quad a = \frac{c}{2} = \frac{s}{2} = a_m \tag{6.93}$$

The transfer orbit is the minimum-energy ellipse passing through $P_1$ and $P_2$.
The formula (6.91) gives

$$\tan\delta = \frac{\sin\alpha}{\cos\alpha + \rho^{3/2}} \quad \text{with} \quad \rho = \frac{r_2}{r_1} \tag{6.94}$$

We retrieve the formula of Lisowski (6.20) established for a Hohmann transfer with split inclination change, by the same approach of minimizing the sum of the squared impulses.

## 6.4 Point and Time Target

### 6.4.1 Lambert's Problem

Lambert's problem consists of finding a Keplerian conic joining two given points $P_1$ and $P_2$ in a given transfer time $\Delta t$. The transfer geometry has been depicted in Figure 6.10a and the same notations as in Section 6.3.1 will be used.

### 6.4.2 Lambert's Theorem

The theorem conjectured by Lambert (1728–1779) states that the time of transfer from $P_1$ to $P_2$ depends only on the radius sum $r_1 + r_2$, the chord c, and the semi-major axis a of the conic. This theorem was demonstrated by Lagrange in 1778.

We successively consider the case of an ellipse, a hyperbola, and a parabola. In each case, the demonstration starts from Kepler's equation and ends with a transfer time equation depending on $r_1 + r_2$, c, and a.

#### 6.4.2.1 Elliptical Case

For an ellipse, Kepler's equation takes the form (1.48).

$$E - e\sin E = n(t - t_P) \quad \text{with} \quad n = \sqrt{\frac{\mu}{a^3}} \tag{6.95}$$

where E is the eccentric anomaly and $t_P$ is the time of passage at the pericenter.

The transfer time from $P_1$ to $P_2$ is obtained by the difference of Kepler's equations in $P_1$ and $P_2$. In the elliptical case, multiple revolutions are possible. The number N of revolutions is therefore introduced in the equation.

$$\sqrt{\mu}(t_2 - t_1) = \sqrt{a^3}\,[(E_2 - E_1) - e(\sin E_2 - \sin E_1) + 2N\pi] \tag{6.96}$$

Defining two angles $\psi$ and $\varphi$ by

$$\psi = \frac{E_2 - E_1}{2}, \quad \cos\varphi = e\cos\frac{E_1 + E_2}{2} \tag{6.97}$$

equation (6.96) takes the form

$$\sqrt{\mu}(t_2 - t_1) = 2\sqrt{a^3}\,(\psi - \sin\psi\cos\varphi + N\pi) \tag{6.98}$$

To prove Lambert's theorem, we have to show that the two angles $\psi$ and $\varphi$ only depend on $r_1 + r_2$, c and a. For $r_1 + r_2$, we have

$$r_1 + r_2 = a(1 - e\cos E_1) + a(1 - e\cos E_2) = 2a(1 - \cos\psi\ \cos\varphi) \tag{6.99}$$

For c, we have in the triangle $(OP_1P_2)$

$$\begin{aligned} c^2 &= r_1^2 + r_2^2 - 2r_1r_2\ \cos\theta \quad \text{with} \quad \theta = \theta_2 - \theta_1 \\ &= (r_1\cos\theta_1 - r_2\cos\theta_2)^2 + (r_1\sin\theta_1 - r_2\sin\theta_2)^2 \end{aligned} \tag{6.100}$$

The formulas established in (1.24) give in points $P_1$ and $P_2$

$$\begin{cases} r_{1,2}\ \cos\theta_{1,2} = a(\cos E_{1,2} - e) \\ r_{1,2}\ \sin\theta_{1,2} = a\sqrt{1 - e^2}\sin E_{1,2} \end{cases} \tag{6.101}$$

By substituting into (6.100), transforming the sums of sine or cosine in products, and using (6.97) to have an expression in terms of $\psi$ and $\varphi$, we obtain

$$c = 2a\ \sin\psi\sin\varphi \tag{6.102}$$

Formulas (6.99) and (6.102) show that the angles $(\psi, \varphi)$ only depend on $r_1+r_2$, c, and a, so the same holds for the transfer time given by (6.98).

#### 6.4.2.2 Hyperbolic Case

The demonstration in the hyperbolic case is very similar to the elliptical case.

For a hyperbola, Kepler's equation takes the form (1.57).

$$e\sinh\ H - H = n(t - t_P) \quad \text{with} \quad n = \sqrt{\frac{\mu}{(-a)^3}} \tag{6.103}$$

where H is the hyperbolic anomaly and $t_P$ is the time of passage at the pericenter.

The transfer time from $P_1$ to $P_2$ is obtained by the difference of Kepler's equations in $P_1$ and $P_2$. In the hyperbolic case, multiple revolutions are not possible.

$$\sqrt{\mu}(t_2 - t_1) = \sqrt{-a^3}\ [e(\sinh H_2 - \sinh H_1) - (H_2 - H_1)] \tag{6.104}$$

Defining two angles $\psi$ and $\varphi$ by

$$\psi = \frac{H_2 - H_1}{2}, \quad \cosh\ \varphi = e\cosh\ \frac{H_1 + H_2}{2} \tag{6.105}$$

equation (6.104) takes the form

$$\sqrt{\mu}(t_2 - t_1) = 2\sqrt{-a^3}\ (\sinh\ \psi\cosh\varphi - \psi) \tag{6.106}$$

We have now to express $r_1 + r_2$ and c as functions of $\psi$, $\varphi$, and a.

For $r_1 + r_2$, we have

$$r_1 + r_2 = a(1 - e\ \cosh\ H_1) + a(1 - e\ \cosh\ H_2) = 2a(1 - \cosh\ \psi\cosh\varphi) \tag{6.107}$$

For c, formula (6.100) still holds, and the formulas established in (1.31) give in points $P_1$ and $P_2$

$$\begin{cases} r_{1,2}\cos\theta_{1,2} = a(\cosh H_{1,2} - e) \\ r_{1,2}\sin\theta_{1,2} = a\sqrt{e^2 - 1}\sinh H_{1,2} \end{cases} \tag{6.108}$$

By substituting into (6.100), transforming the sums of sine or cosine in products, and using (6.105) to have an expression in terms of $\psi$ and $\varphi$, we obtain

$$c = -2a \sinh \psi \sinh \varphi \tag{6.109}$$

As for the elliptical case, formulas (6.107) and (6.109) show that the angles ($\psi$, $\varphi$) only depend on $r_1 + r_2$, c, and a, so that the same holds for the transfer time given by (6.106).

#### 6.4.2.3 Parabolic Case

For a parabola, Barker's equation takes the form (1.62).

$$B^3 + 3B = 6n(t - t_P) \quad \text{with} \quad n = \sqrt{\frac{\mu}{p^3}} \tag{6.110}$$

where B is the parabolic anomaly, $t_P$ is the time of passage at the pericenter, and p is the parabola parameter. The relations among the parabolic anomaly B, the true anomaly $\theta$, and the radius vector r are retrieved from (1.34).

$$B = \tan\frac{\theta}{2}, \quad r = \frac{1}{2}p(1 + B^2) \tag{6.111}$$

The transfer time from $P_1$ to $P_2$ is obtained by the difference of Barker's equations in $P_1$ and $P_2$. In the parabolic case, multiple revolutions are not possible.

$$\begin{aligned} 6\sqrt{\mu}(t_2 - t_1) &= \sqrt{p^3}\left[B_2^3 - B_1^3 + 3(B_2 - B_1)\right] \\ &= \sqrt{p^3}(B_2 - B_1)\left[B_2^2 + B_1B_2 + B_1^2 + 3\right] \\ &= \sqrt{p^3}(B_2 - B_1)\left[(B_2 - B_1)^2 + 3(B_1B_2 + 1)\right] \end{aligned} \tag{6.112}$$

Let us calculate successively the terms $B_1B_2 + 1$ and $B_2 - B_1$.

For the term $B_1B_2 + 1$, we start from $s(s - c)$, with $2s = r_1 + r_2 + c$.

$$4s(s - c) = (r_1 + r_2 + c)(r_1 + r_2 - c) = (r_1 + r_2)^2 - c^2 = 2r_1r_2(1 + \cos\theta) \tag{6.113}$$

by using $c^2 = r_1^2 + r_2^2 - 2r_1r_2 \cos\theta$ in the triangle ($OP_1P_2$).

We express $\cos\theta = \cos(\theta_2 - \theta_1)$ in terms of $B_1 = \tan\frac{\theta_1}{2}$ and $B_2 = \tan\frac{\theta_2}{2}$ by using the trigonometric identities with the tangent of the half angle.

$$\cos\theta = \cos(\theta_2 - \theta_1) = \cos\theta_1\cos\theta_2 + \sin\theta_1\sin\theta_2 = \frac{\left(1 - B_1^2\right)\left(1 - B_2^2\right) + 4B_1B_2}{\left(1 + B_1^2\right)\left(1 + B_2^2\right)} \tag{6.114}$$

Substituting $\cos\theta$ into (6.113) with $r_1 = \frac{1}{2}p\left(1 + B_1^2\right)$ and $r_2 = \frac{1}{2}p\left(1 + B_2^2\right)$ yields

$$4s(s - c) = p^2(B_1B_2 + 1)^2 \tag{6.115}$$

Taking the square root, we obtain the first useful relation to assess (6.112).

$$B_1B_2 + 1 = \pm\frac{2}{p}\sqrt{s(s - c)} \tag{6.116}$$

To determine the sign, we assess $B_1B_2 + 1$ in terms of the transfer angle $\theta$.

$$B_1B_2 + 1 = \tan\frac{\theta_1}{2}\tan\frac{\theta_2}{2} + 1 = \frac{\sin\frac{\theta_1}{2}\sin\frac{\theta_2}{2} + \cos\frac{\theta_1}{2}\cos\frac{\theta_2}{2}}{\cos\frac{\theta_1}{2}\cos\frac{\theta_2}{2}} = \frac{\cos\frac{\theta_2 - \theta_1}{2}}{\cos\frac{\theta_1}{2}\cos\frac{\theta_2}{2}} \tag{6.117}$$

From the parabolic orbit equation $r = \frac{p}{1+\cos\theta}$ (with eccentricity $e = 1$), we have

$$\cos\frac{\theta_1}{2}\cos\frac{\theta_2}{2} = \frac{1}{2}\sqrt{1+\cos\theta_1}\sqrt{1+\cos\theta_2} = \frac{1}{2}\sqrt{\frac{p}{r_1}}\sqrt{\frac{p}{r_2}} = \frac{p}{2\sqrt{r_1 r_2}} \tag{6.118}$$

so that (6.117) becomes

$$B_1B_2 + 1 = \frac{2}{p}\sqrt{r_1 r_2}\cos\frac{\theta}{2} \tag{6.119}$$

This equation determines the sign of $B_1B_2 + 1$ depending on $\theta = \theta_2 - \theta_1$:

- $B_1B_2 + 1 > 0$ if the transfer angle $\theta$ is lower than 180 deg;
- $B_1B_2 + 1 < 0$ if the transfer angle $\theta$ is greater than 180 deg.

For the term $B_2 - B_1$, we have

$$\begin{aligned}(B_2 - B_1)^2 &= B_1^2 + B_2^2 - 2B_1B_2 = \frac{2}{p}(r_1 + r_2) - 2(B_1B_2 + 1)\\ &= \frac{2}{p}(2s - c) \mp \frac{4}{p}\sqrt{s(s-c)} = \frac{2}{p}[s \mp 2\sqrt{s(s-c)} + (s-c)]\\ &= \frac{2}{p}(\sqrt{s} \mp \sqrt{s-c})^2\end{aligned} \tag{6.120}$$

The parabolic anomaly is strictly increasing along the parabolic motion. The term $B_2 - B_1$ is then always positive and it is given by

$$B_2 - B_1 = \sqrt{\frac{2}{p}}(\sqrt{s} \mp \sqrt{s-c}) \tag{6.121}$$

We are now able to evaluate the transfer time in (6.112) using (6.119) and (6.121).

$$6\sqrt{\mu}(t_2 - t_1) = \left(\sqrt{2s} \mp \sqrt{2(s-c)}\right)\left[\left(\sqrt{2s} \mp \sqrt{2(s-c)}\right)^2 \pm 3\sqrt{2s \cdot 2(s-c)}\right] \tag{6.122}$$

By expanding the term in brackets, then the product in the second member

$$\begin{aligned}6\sqrt{\mu}(t_2 - t_1) &= \left(\sqrt{2s} \mp \sqrt{2(s-c)}\right)\left[2s + 2(s-c) \pm \sqrt{2s \cdot 2(s-c)}\right]\\ &= (2s)^{\frac{3}{2}} + 2\sqrt{2s}(s-c) \pm 2s\sqrt{2(s-c)}\\ &\quad \mp 2s\sqrt{2(s-c)} \mp [2(s-c)]^{\frac{3}{2}} - 2(s-c)\sqrt{2s}\\ &= [2s]^{\frac{3}{2}} \mp [2(s-c)]^{\frac{3}{2}}\end{aligned} \tag{6.123}$$

we obtain Euler's equation (1743) for the transfer time in the parabolic case.

$$6\sqrt{\mu}(t_2 - t_1) = \sqrt{(r_1 + r_2 + c)^3} \mp \sqrt{(r_1 + r_2 - c)^3} \tag{6.124}$$

The theorem of Lambert has been established for all kinds of conics and we have obtained the associated transfer time equations, which are summarized in the following section.

#### 6.4.2.4 Transfer Time Equations with Variables $\psi$ and $\varphi$

- For an ellipse, we have

$$\sqrt{\mu}\Delta t = 2\sqrt{a^3}\,(\psi - \sin\psi\cos\varphi + N\pi) \tag{6.125}$$

with $\begin{cases} r_1 + r_2 = 2a(1 - \cos\psi\cos\varphi) \\ c = 2a\sin\psi\sin\varphi \end{cases}$, $\quad \psi = \frac{E_2 - E_1}{2}$, $\quad \cos\varphi = e\cos\frac{E_1 + E_2}{2}$

- For a hyperbola, we have

$$\sqrt{\mu}\Delta t = 2\sqrt{-a^3}\,(\sinh\psi\cosh\varphi - \psi) \tag{6.126}$$

with $\begin{cases} r_1 + r_2 = 2a(1 - \cosh\psi\cosh\varphi) \\ c = -2a\sinh\psi\sinh\varphi \end{cases}$, $\psi = \frac{H_2 - H_1}{2}$, $\cosh\varphi = e\cosh\frac{H_1 + H_2}{2}$

- For a parabola, we have

$$6\sqrt{\mu}\Delta t = \sqrt{(r_1 + r_2 + c)^3} \mp \sqrt{(r_1 + r_2 - c)^3} \tag{6.127}$$

with the sign + (resp. −) for a transfer angle lower (resp. greater) than 180 deg.

The angles $\psi$ and $\varphi$ are bounded as follows:

- $\psi \in [0; \pi]$ because the anomaly difference ($E_2 - E_1$ or $H_2 - H_1$) related to the transfer angle $\theta$ is comprised between 0 and $2\pi$;
- $\varphi \in [0; \pi]$ for an ellipse, because $c = 2a\sin\psi\sin\varphi > 0$ with $a > 0$;
- $\varphi \geq 0$ for a hyperbola, because $c = -2a\sinh\psi\sinh\varphi > 0$ with $a < 0$.

### 6.4.3 Transfer Time Equation

To solve Lambert's problem, we must find the semi-major axis yielding a given transfer time. The transfer time equations (6.125)–(6.127) cannot be solved analytically, and they are not easy to handle numerically. A major issue is that for a given semi-major axis, there exist two conics passing through $P_1$ and $P_2$, whose virtual foci are symmetrical with respect to the chord. This leads to converting the transfer time equations into forms more suitable for numerical solution methods.

For that purpose, two new variables $\alpha$ and $\beta$ are defined by

$$\begin{cases} \alpha = \varphi + \psi \\ \beta = \varphi - \psi \end{cases} \Leftrightarrow \varphi = \frac{\alpha + \beta}{2} \quad \text{and} \quad \psi = \frac{\alpha - \beta}{2} \tag{6.128}$$

with the bounds:

- $\alpha \in [0; 2\pi]$ and $\beta \in [-\pi; \pi]$ for an ellipse;
- $\alpha \geq 0$ and $\beta \geq -\pi$ for a hyperbola.

A parameter $\lambda$ depending only on the problem geometry is also defined by

$$\lambda = \frac{\sqrt{r_1 r_2}}{s}\cos\frac{\theta}{2} \tag{6.129}$$

We have in the triangle ($OP_1P_2$)

$$c^2 = r_1^2 + r_2^2 - 2r_1r_2\cos\theta = (r_1 + r_2)^2 - 4r_1r_2\cos^2\frac{\theta}{2} \tag{6.130}$$

so that

$$4s^2\lambda^2 = 4r_1r_2\cos^2\frac{\theta}{2} = (r_1 + r_2)^2 - c^2 = (2s - c)^2 - c^2 = 4s(s - c) \tag{6.131}$$

We obtain

$$\lambda^2 = \frac{s - c}{s} \tag{6.132}$$

The parameter $\lambda$ is comprised between −1 and +1 and its sign is determined from the definition (6.129), depending on the transfer angle $\theta$.

$$\begin{cases} 0 \le \lambda \le 1 & \text{if } \theta \le \pi \\ -1 \le \lambda \le 0 & \text{if } \theta \ge \pi \end{cases} \tag{6.133}$$

Let us now reformulate the transfer time equations using $\lambda$, $\alpha$ and $\beta$.
We consider successively the elliptical, hyperbolic, and parabolic cases.

#### 6.4.3.1 Elliptical Case

With the variables $\alpha$ and $\beta$, the transfer time equation (6.125) takes the form derived by Lagrange in 1778.

$$\sqrt{\mu}\Delta t = \sqrt{a^3}\,[(\alpha - \sin\alpha) - (\beta - \sin\beta) + 2N\pi] \tag{6.134}$$

The variables $\alpha$ and $\beta$ can be expressed as functions of the lengths $r_1 + r_2$, c, and s and of the unknown semi-major axis a.

For an ellipse, we have from (6.125): $\begin{cases} r_1 + r_2 = 2a(1 - \cos\psi\ \cos\varphi) \\ c = 2a\ \sin\psi\ \sin\varphi \end{cases}$

$$\Rightarrow \begin{cases} 2s = r_1 + r_2 + c = 2a[1 - \cos(\varphi + \psi)] = 4a\sin^2\dfrac{\varphi + \psi}{2} = 4a\sin^2\dfrac{\alpha}{2} \\ 2(s - c) = r_1 + r_2 - c = 2a[1 - \cos(\varphi - \psi)] = 4a\sin^2\dfrac{\varphi - \psi}{2} = 4a\sin^2\dfrac{\beta}{2} \end{cases}$$

which gives

$$\sin^2\frac{\alpha}{2} = \frac{s}{2a} \quad \text{and} \quad \sin^2\frac{\beta}{2} = \frac{s - c}{2a} = \lambda^2\sin^2\frac{\alpha}{2} \tag{6.135}$$

with the bounds deduced from (6.128): $\alpha \in [0; 2\pi]$ and $\beta \in [-\pi; \pi]$.

The formulas (6.135) do not define the quadrants for $\alpha$ and $\beta$, but the ambiguity can be partially resolved owing to the parameter $\lambda$ defined in (6.129).

For that purpose, we establish a relation between $\lambda$, $\alpha$, and $\beta$.

Starting from (1.24): $\cos\theta = \frac{\cos E - e}{1 - e\cos E}$ and $r = a(1 - e\cos E)$, we have

$$\begin{cases} \cos^2\dfrac{\theta}{2} = \dfrac{1}{2}(1 + \cos\theta) = \dfrac{1}{2}\dfrac{(1 - e)(1 + \cos E)}{1 - e\cos E} = \dfrac{a(1 - e)}{r}\cos^2\dfrac{E}{2} \\ \sin^2\dfrac{\theta}{2} = \dfrac{1}{2}(1 - \cos\theta) = \dfrac{1}{2}\dfrac{(1 + e)(1 - \cos E)}{1 - e\cos E} = \dfrac{a(1 + e)}{r}\sin^2\dfrac{E}{2} \end{cases} \tag{6.136}$$

so that

$$\begin{cases} r\cos^2\dfrac{\theta}{2} = a(1 - e)\cos^2\dfrac{E}{2} \\ r\sin^2\dfrac{\theta}{2} = a(1 + e)\sin^2\dfrac{E}{2} \end{cases} \tag{6.137}$$

These relations are used to assess $\lambda$ defined in (6.129).

$$s\lambda = \sqrt{r_1 r_2}\cos\frac{\theta}{2} = \sqrt{r_1 r_2}\left(\cos\frac{\theta_1}{2}\cos\frac{\theta_2}{2} + \sin\frac{\theta_1}{2}\sin\frac{\theta_2}{2}\right)$$

$$= a(1 - e)\cos\frac{E_1}{2}\cos\frac{E_2}{2} + a(1 + e)\sin\frac{E_1}{2}\sin\frac{E_2}{2}$$

$$= a\cos\frac{E_2 - E_1}{2} - ae\cos\frac{E_2 + E_1}{2}$$

$$= a\cos\psi - a\cos\varphi = a\left(\cos\frac{\alpha-\beta}{2} - \cos\frac{\alpha+\beta}{2}\right) = 2a\sin\frac{\alpha}{2}\sin\frac{\beta}{2} \quad (6.138)$$

giving finally

$$\lambda = \frac{2a}{s}\sin\frac{\alpha}{2}\sin\frac{\beta}{2} \quad (6.139)$$

The sign of $\lambda$ is known from (6.133) and $\sin\frac{\alpha}{2} \geq 0$ because $\alpha \in [0; 2\pi]$.

This allows to determines the sign of $\sin\frac{\beta}{2}$, and since $\beta \in [-\pi; \pi]$, we have

$$\begin{cases} \beta \in \left[0; \frac{\pi}{2}\right] & \text{if } \theta \leq \pi \\ \beta \in \left[-\frac{\pi}{2}; 0\right] & \text{if } \theta \geq \pi \end{cases} \quad (6.140)$$

The square root ambiguity in (6.135) is thus resolved giving

$$\sin\frac{\alpha}{2} = \sqrt{\frac{s}{2a}} \quad \text{and} \quad \sin\frac{\beta}{2} = \lambda\sin\frac{\alpha}{2} = \lambda\sqrt{\frac{s}{2a}} \quad (6.141)$$

The relations (6.141) are now substituted in (6.135) to assess the semi-major axis.

$$\sqrt{a^3} = \frac{\sqrt{a_m^3}}{\sin^3\frac{\alpha}{2}} = \frac{\lambda^3\sqrt{a_m^3}}{\sin^3\frac{\beta}{2}} \quad \text{with } a_m = \frac{s}{2} \quad (6.142)$$

By defining a dimensionless transfer time

$$\Delta\tau \underset{\text{def}}{=} \frac{1}{2}\sqrt{\frac{\mu}{a_m^3}}\Delta t \quad (6.143)$$

we get the transfer time equation (6.134) under the form derived by Battin [R1].

$$\Delta\tau = \frac{\alpha - \sin\alpha + 2N\pi}{2\sin^3\frac{\alpha}{2}} - \lambda^3\frac{\beta - \sin\beta}{2\sin^3\frac{\beta}{2}} \quad (6.144)$$

The formulas (6.140) and (6.141) completely determine $\beta$. Nevertheless, there remains an ambiguity on $\alpha$ when assessing the Arcsin in (6.141). To remove it, the final transformation step consists in defining a new variable from which the transfer time can be calculated unambiguously. This variable x is defined by

$$x = \cos\frac{\alpha}{2} \quad (6.145)$$

For any $x \in [-1; 1]$, $\alpha$ is well determined since $\alpha \in [0; 2\pi] \Rightarrow \frac{\alpha}{2} \in [0; \pi]$.

We can then assess $\beta$ from (6.141). An intermediate variable y is introduced to alleviate the formulas.

$$\begin{cases} \cos\frac{\alpha}{2} = x & \Rightarrow \quad \sin\frac{\alpha}{2} = \sqrt{1-x^2} \geq 0 \\ \sin\frac{\beta}{2} = \lambda\sin\frac{\alpha}{2} = \lambda\sqrt{1-x^2} & \Rightarrow \quad \cos\frac{\beta}{2} = \sqrt{1-\lambda^2(1-x^2)} \underset{\text{noted}}{=} y \end{cases} \quad (6.146)$$

We can also assess unambiguously the angles $\psi$ and $\varphi$ from (6.128).

$$\begin{cases} \cos\varphi = \cos\frac{\alpha+\beta}{2} = xy - \lambda(1-x^2), \quad \sin\varphi = \sin\frac{\alpha+\beta}{2} = (y+\lambda x)\sqrt{1-x^2} \\ \cos\psi = \cos\frac{\alpha-\beta}{2} = xy + \lambda(1-x^2), \quad \sin\psi = \sin\frac{\alpha-\beta}{2} = (y-\lambda x)\sqrt{1-x^2} \end{cases} \quad (6.147)$$

Substituting (6.146) into (6.144), we obtain the transfer time equation under the form derived by Lancaster–Blanchard.

$$\Delta\tau = \frac{1}{1-x^2}\left(\frac{\psi + N\pi}{\sqrt{1-x^2}} + \lambda y - x\right) \underset{\text{noted}}{=} F(x) \tag{6.148}$$

The function F depends unambiguously on the variable $x \in [-1; 1]$. This equation with one unknown can thus be solved easily by classical solution methods.

In the case of the minimum-energy ellipse, we have from (6.141) and (6.146)

$$a = a_m = \frac{s}{2} \quad \Rightarrow \quad \sin\frac{\alpha}{2} = 1 \quad \Rightarrow \quad x = 0 \text{ and } y = \sqrt{1-\lambda^2} \tag{6.149}$$

and from (6.147)

$$\cos\psi = \lambda, \quad \sin\psi = y = \sqrt{1-\lambda^2}^{\,3}\; 0 \quad \Rightarrow \quad \psi = \text{Arccos}\,\lambda \tag{6.150}$$

The transfer time for the minimum-energy ellipse is then

$$\Delta\tau_0 = F(x=0) = \text{Arccos}\,\lambda + \lambda\sqrt{1-\lambda^2} + N\pi \tag{6.151}$$

#### 6.4.3.2 Hyperbolic Case

The hyperbolic case is treated in a very similar manner.

With the variables α and β, the transfer time equation (6.126) takes the form

$$\sqrt{\mu}\Delta t = \sqrt{-a^3}\,[(\sinh\alpha - \alpha) - (\sinh\beta - \beta)] \tag{6.152}$$

The variables α and β can be expressed as functions of the lengths $r_1 + r_2$, c, and s and of the unknown semi-major axis a.

For a hyperbola, we have from (6.126): $\begin{cases} r_1 + r_2 = 2a(1 - \cosh\psi\cosh\varphi) \\ c = -2a\sinh\psi\sinh\varphi \end{cases}$

$$\Rightarrow \begin{cases} 2s = r_1 + r_2 + c = 2a[1 - \cosh(\varphi+\psi)] = -4a\sinh^2\dfrac{\varphi+\psi}{2} = -4a\sinh^2\dfrac{\alpha}{2} \\ 2(s-c) = r_1 + r_2 - c = 2a[1 - \cosh(\varphi-\psi)] = -4a\sinh^2\dfrac{\varphi-\psi}{2} = -4a\sinh^2\dfrac{\beta}{2} \end{cases}$$

which gives

$$\sinh^2\frac{\alpha}{2} = -\frac{s}{2a} \quad \text{and} \quad \sinh^2\frac{\beta}{2} = -\frac{s-c}{2a} \tag{6.153}$$

with the bounds deduced from (6.128): $\alpha \geq 0$ and $\beta \geq -\pi$.

As in the elliptical case, we establish a relation between the parameter λ defined in (6.129), α and β.

Starting from (1.31): $\cos\theta = \frac{e - \cosh H}{e\cosh H - 1}$ and $r = a(1 - e\cosh H)$, we have

$$\begin{cases} \cos^2\dfrac{\theta}{2} = \dfrac{1}{2}(1+\cos\theta) = \dfrac{1}{2}\dfrac{(e-1)(1+\cosh H)}{e\cosh H - 1} = \dfrac{-a(e-1)}{r}\cosh^2\dfrac{H}{2} \\ \sin^2\dfrac{\theta}{2} = \dfrac{1}{2}(1-\cos\theta) = \dfrac{1}{2}\dfrac{(e+1)(\cosh H - 1)}{e\cosh H - 1} = \dfrac{-a(e+1)}{r}\sinh^2\dfrac{H}{2} \end{cases} \tag{6.154}$$

so that

$$\begin{cases} r\cos^2\dfrac{\theta}{2} = -a(e-1)\cosh^2\dfrac{H}{2} \\ r\sin^2\dfrac{\theta}{2} = -a(e+1)\sinh^2\dfrac{H}{2} \end{cases} \tag{6.155}$$

These relations are used to assess $\lambda$ defined in (6.129).

$$
\begin{aligned}
s\lambda &= \sqrt{r_1 r_2}\cos\frac{\theta}{2} = \sqrt{r_1 r_2}\left(\cos\frac{\theta_1}{2}\cos\frac{\theta_2}{2} + \sin\frac{\theta_1}{2}\sin\frac{\theta_2}{2}\right) \\
&= -a(e-1)\cosh\frac{H_1}{2}\cosh\frac{H_2}{2} - a(e+1)\sinh\frac{H_1}{2}\sinh\frac{H_2}{2} \\
&= a\cosh\frac{H_2 - H_1}{2} - ae\cosh\frac{H_2 + H_1}{2} \\
&= a\cosh\psi - a\cosh\varphi = a\left(\cosh\frac{\alpha-\beta}{2} - \cosh\frac{\alpha+\beta}{2}\right) = -2a\sinh\frac{\alpha}{2}\sinh\frac{\beta}{2}
\end{aligned} \tag{6.156}
$$

giving finally

$$
\lambda = -\frac{2a}{s}\sinh\frac{\alpha}{2}\sinh\frac{\beta}{2} \tag{6.157}
$$

The sign of $\lambda$ is known from the angle $\theta$ and $\sinh\frac{\alpha}{2} \geq 0$ because $\alpha \in [0; 2\pi]$.
This defines the sign of $\sinh\frac{\beta}{2}$, and since $\beta \geq -\pi$, we have (with $a < 0$)

$$
\begin{cases}
\beta \geq 0 & \text{if } \theta \leq \pi \\
\beta \in [-\pi; 0] & \text{if } \theta \geq \pi
\end{cases} \tag{6.158}
$$

The square root ambiguity in (6.153) is thus resolved giving

$$
\sinh\frac{\alpha}{2} = \sqrt{\frac{s}{-2a}} \quad \text{and} \quad \sinh\frac{\beta}{2} = \lambda\sinh\frac{\alpha}{2} = \lambda\sqrt{\frac{s}{-2a}} \tag{6.159}
$$

The relations (6.159) are now substituted in (6.153) to assess the semi-major axis.

$$
\sqrt{-a^3} = \frac{\sqrt{a_m^3}}{\sinh^3\frac{\alpha}{2}} = \frac{\lambda^3\sqrt{a_m^3}}{\sinh^3\frac{\beta}{2}} \quad \text{with } a_m = \frac{s}{2} \tag{6.160}
$$

With the dimensionless time (6.143), we get the transfer time equation (6.152) under the form derived by Battin [R1].

$$
\Delta\tau = \frac{\sinh\alpha - \alpha}{2\sinh^3\frac{\alpha}{2}} - \lambda^3\frac{\sinh\beta - \beta}{2\sinh^3\frac{\beta}{2}} \tag{6.161}
$$

The formulas (6.158) and (6.159) completely determine $\beta$. To remove the ambiguity on $\alpha$, the variable x is defined by

$$
x = \cosh\frac{\alpha}{2} \tag{6.162}
$$

For any $x \in [1; +\infty]$, $\alpha$ is well determined since $\alpha \in [0; 2\pi] \Rightarrow \frac{\alpha}{2} \in [0; \pi]$.
We can then assess $\beta$ from (6.153), with an intermediate variable y.

$$
\begin{cases}
\cosh\frac{\alpha}{2} = x & \Rightarrow \sinh\frac{\alpha}{2} = \sqrt{x^2-1} \geq 0 \\
\sinh\frac{\beta}{2} = \lambda\sinh\frac{\alpha}{2} = \lambda\sqrt{x^2-1} & \Rightarrow \cosh\frac{\beta}{2} = \sqrt{1-\lambda^2(x^2-1)} \underset{\text{noted}}{=} y
\end{cases} \tag{6.163}
$$

We can also assess unambiguously the angles $\psi$ and $\varphi$ from (6.128).

$$
\begin{cases}
\cosh\varphi = \cosh\frac{\alpha+\beta}{2} = xy + \lambda(x^2-1), \quad \sinh\varphi = \sinh\frac{\alpha+\beta}{2} = (y+\lambda x)\sqrt{x^2-1} \\
\cosh\psi = \cosh\frac{\alpha-\beta}{2} = xy - \lambda(x^2-1), \quad \sinh\psi = \sinh\frac{\alpha-\beta}{2} = (y-\lambda x)\sqrt{x^2-1}
\end{cases} \tag{6.164}
$$

Substituting (6.163) into (6.161), we obtain the transfer time equation under the form derived by Lancaster–Blanchard.

$$\Delta\tau = \frac{1}{1-x^2}\left(\frac{\psi}{\sqrt{x^2-1}} + \lambda y - x\right) \underset{\text{noted}}{=} F(x) \tag{6.165}$$

#### 6.4.3.3 Parabolic Case

The parabolic case is the limit of the elliptical and hyperbolic cases for $a \to \infty$.

From equation (6.141) or (6.153), we have $\lim_{a\to\infty}\alpha = \lim_{a\to\infty}\beta = 0$, and from the definition of x in (6.145) or (6.162), the parabolic case corresponds to $\lim_{a\to\infty} x = 1$.

Let us assess the function F(x) in the elliptical case (6.148) when $x \to 1^-$.

By noting $x = 1 - \varepsilon$ with $\varepsilon \to 0^+$, we have at the first order

$$\begin{cases} 1 - x^2 \approx 2\varepsilon \\ y = \sqrt{1-\lambda^2(1-x^2)} \approx 1 - \lambda^2\varepsilon \\ \sin\psi = (y - \lambda x)\sqrt{1-x^2} \approx \sqrt{2\varepsilon}\,(1-\lambda)(1+\lambda\varepsilon) \end{cases} \tag{6.166}$$

Since $\sin\psi \underset{\text{noted}}{=} z$ is small, we can approximate $\psi$ by $\psi = \text{Arcsin}(z) \approx z + \frac{1}{6}z^3$.

Substituting these expansions into (6.148) with N = 0 in the parabolic case, we obtain

$$F(1-\varepsilon) \approx \frac{1}{2\varepsilon}\left[\frac{\sqrt{2\varepsilon}\,(1-\lambda)(1+\lambda\varepsilon) + \frac{1}{6}\sqrt{(2\varepsilon)^3}\,(1-\lambda)^3(1+\lambda\varepsilon)^3}{\sqrt{2\varepsilon}} + \lambda(1-\lambda^2\varepsilon) - (1-\varepsilon)\right]$$

$$= \frac{2}{3}(1-\lambda)^3 \tag{6.167}$$

The dimensionless transfer time equation in the parabolic case is then

$$\Delta\tau_1 = F(1) = \frac{2}{3}(1-\lambda^3) \tag{6.168}$$

with the sign of $\lambda$ depending on the transfer angle in (6.133).

This result is consistent with (6.127), which gives

$$6\sqrt{\mu}\Delta t = \sqrt{(2s)^3} \mp \sqrt{(2s-2c)^3} \tag{6.169}$$

with $\Delta\tau = \sqrt{\frac{2\mu}{s^3}}\Delta t$ from (6.143) and $\lambda = \pm\sqrt{\frac{s-c}{s}}$ from (6.132) and (6.133)

In the hyperbolic case (6.165), the calculation of the limit is similar with $x \to 1^+$ and it gives the same result (6.168).

#### 6.4.3.4 Recap

We have obtained the transfer time equation under a general form valid for any conic (ellipse, hyperbola, and parabola), with a single-value function F depending on a variable x.

$$\Delta\tau = F(x) = \frac{1}{1-x^2}\left(\frac{\psi + N\pi}{\sqrt{|1-x^2|}} + \lambda y - x\right) \tag{6.170}$$

- For an ellipse: $-1 < x < 1$ and N is the number of revolutions.
- For a hyperbola: $x > 1$ and $N = 0$.
- For a parabola: $x = 1$ and $N = 0$ giving $\Delta\tau_1 = F(1) = \frac{2}{3}(1-\lambda^3)$.
- Minimum energy: $x = 0$ giving $\Delta\tau_0 = F(0) = \text{Arccos}\,\lambda + \lambda\sqrt{1-\lambda^2} + N\pi$

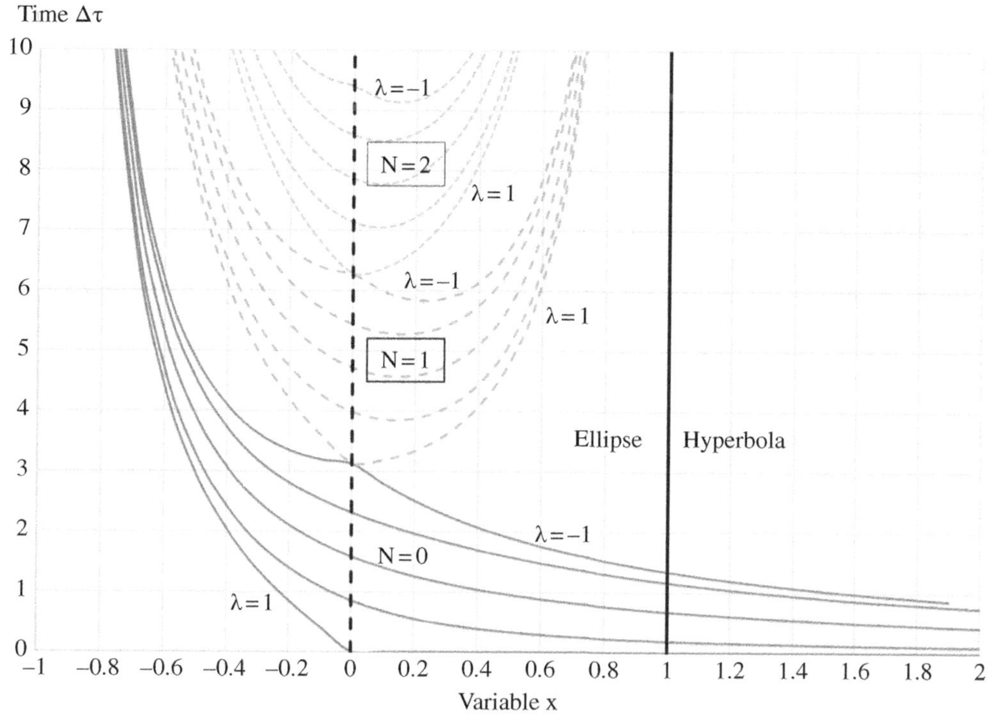


**Figure 6.14** Transfer time function depending on λ and N.

Figure 6.14 shows a plot of the function F(x) for five values of $\lambda = \pm\sqrt{\frac{s-c}{s}}$ set to $\lambda \in \{-0.9999; -0.9; 0; 0.9; 0.9999\}$ and for $N \in \{0; 1; 2\}$ revolutions.

The domain $-1 < x < 1$ corresponds to ellipses, the domain $x > 1$ corresponds to hyperbolas, and the value $x = 1$ corresponds to parabolas.

For $\lambda = \pm 1$, the points $P_1$ and $P_2$ merge: the transfer angle is either $\theta = 0$ if $\lambda = 1$ or $\theta = 2\pi$ if $\lambda = -1$. This explains the smooth junctions observed between the curves for $(\lambda = -1; N)$ and for $(\lambda = 1; N + 1)$. These junctions occur for $x = 0$ corresponding to the minimum-energy ellipse, as can be checked with (6.151).

### 6.4.4 Universal Variable

The transfer time equation can be written using the universal variable (also called generalized anomaly) in a similar way as Kepler's equation in Section 1.3.5.

For that purpose, we retrieve Lagrange coefficients defined in Section 1.3.4.

These coefficients were expressed in (1.71) with the transfer angle θ

$$\begin{cases} F = 1 + \dfrac{r_2}{p}(\cos\theta - 1), & F_t = \sqrt{\dfrac{\mu}{p}}\dfrac{1-\cos\theta}{\sin\theta}\left(\dfrac{1-\cos\theta}{p} - \dfrac{1}{r_1} - \dfrac{1}{r_2}\right) \\ G = \dfrac{r_1 r_2}{h}\sin\theta, & G_t = 1 + \dfrac{r_1}{p}(\cos\theta - 1) \end{cases} \tag{6.171}$$

and in (1.112) with the universal variable $\chi$

$$\begin{cases} F = U_0(\chi,\alpha) + \left(\alpha - \frac{1}{r_0}\right) U_2(\chi,\alpha), & F_t = -\frac{\sqrt{\mu}}{r_0 r} U_1(\chi,\alpha) \\ G = \frac{r_0}{\sqrt{\mu}} U_1(\chi,\alpha) + \frac{\sigma_0}{\sqrt{\mu}} U_2(\chi,\alpha), & G_t = \frac{r_0}{r} U_0(\chi,\alpha) + \frac{\sigma_0}{r} U_1(\chi,\alpha) \end{cases} \tag{6.172}$$

The functions $U_n(\chi, \alpha)$ are defined in (1.93) and their expressions for the three kinds of conics are given in (1.100)–(1.102), where $\alpha$ denotes the inverse of the semi-major axis (1.88).

The formulas (6.172) are first rewritten using Stumpff functions [R14], [R20] defined by

$$S(z) = \begin{cases} \dfrac{\sqrt{z} - \sin\sqrt{z}}{\sqrt{z}^3} \\ \dfrac{\sinh\sqrt{-z} - \sqrt{-z}}{\sqrt{-z}^3} \\ \dfrac{1}{6} \end{cases}, \quad C(z) = \begin{cases} \dfrac{1 - \cos\sqrt{z}}{z} & \text{if } z > 0 \\ \dfrac{\cosh\sqrt{-z} - 1}{-z} & \text{if } z < 0 \\ \dfrac{1}{2} & \text{if } z = 0 \end{cases} \tag{6.173}$$

By noting $z = \alpha\chi^2$, and with (1.100)–(1.102), we obtain more compact formulas for Lagrange coefficients in variable $\chi$.

$$\begin{cases} F = 1 - \frac{\chi^2}{r_1} C(z), & F_t = \frac{\sqrt{\mu}}{r_1 r_2} \chi[zS(z) - 1] \\ G = \Delta t - \frac{\chi^3}{\sqrt{\mu}} S(z), & G_t = 1 - \frac{\chi^3}{r_2} C(z) \end{cases} \tag{6.174}$$

By equating the formulas (6.171) and (6.174) for G, we obtain an expression of the transfer time.

$$G = \Delta t - \frac{\chi^3}{\sqrt{\mu}} S(z) = \frac{r_1 r_2}{h} \sin\theta \quad \Rightarrow \quad \sqrt{\mu}\Delta t = \chi^3 S(z) + \frac{\sqrt{\mu} r_1 r_2}{h} \sin\theta \tag{6.175}$$

The angular momentum is eliminated by equating the formulas for F.

$$F = 1 - \frac{\chi^2}{r_1} C(z) = 1 + \frac{r_2}{p}(\cos\theta - 1) \quad \Rightarrow \quad p = \frac{h^2}{\mu} = \frac{r_1 r_2 (1 - \cos\theta)}{\chi^2 C(z)} \tag{6.176}$$

The transfer time equation (6.175) becomes

$$\sqrt{\mu}\Delta t = \chi^3 S(z) + A\chi\sqrt{C(z)} \quad \text{with } A \underset{\text{def}}{=} \sqrt{\frac{r_1 r_2}{1 - \cos\theta}} \sin\theta \tag{6.177}$$

In this equation, z and $\chi$ are linked by $z = \alpha\chi^2$, and $\alpha$ is the inverse of the unknown semi-major axis. From the formulas for $F_t$, we can find another relation between z and $\chi$ that will not depend on $\alpha$.

By equating the formulas (6.171) and (6.174) for $F_t$, we have

$$F_t = \sqrt{\frac{\mu}{p}} \frac{1 - \cos\theta}{\sin\theta} \left( \frac{1 - \cos\theta}{p} - \frac{1}{r_1} - \frac{1}{r_2} \right) = \frac{\sqrt{\mu}}{r_1 r_2} \chi[zS(z) - 1] \tag{6.178}$$

After substituting h from (6.176) and simplifying, this equation becomes

$$\chi^2 C(z) = r_1 + r_2 + A \frac{zS(z) - 1}{\sqrt{C(z)}} \tag{6.179}$$

The right-hand side is a function of z noted y(z).

$$\chi = \sqrt{\frac{y(z)}{C(z)}} \quad \text{with} \quad y(z) \underset{\text{def}}{=} r_1 + r_2 + A\frac{zS(z) - 1}{\sqrt{C(z)}} \tag{6.180}$$

The transfer time equation in universal variable is finally

$$\sqrt{\mu}\Delta t = \sqrt{\frac{y(z)}{C(z)}}^{3} S(z) + A\sqrt{y(z)} \quad \text{with} \quad A = \sqrt{\frac{r_1 r_2}{1 - \cos\theta}} \sin\theta \tag{6.181}$$

The constant A depends only on the problem geometry. This equation in the variable z in valid for all kinds of conics.

### 6.4.5 Solution Methods

Because of its practical importance for space missions (interception, rendezvous, interplanetary travel, and orbit determination from observations), Lambert's problem has been the subject of much research since Gauss (1809).

Dozens of solution methods have been proposed and tested, most of them iterative. They differ in the formulation of the transfer time equation and the iteration formula. We shortly describe the most widely used, namely Battin's, Gooding's, Sun's, and universal variable methods.

Battin's method is an improvement of Gauss' original method. It introduces the "mean point," which is the intermediate point on the conic where the tangent to the conic is parallel to the chord. This point has some remarkable geometric properties demonstrated in [R1]. The transfer time equation is reformulated as a cubic equation, whose coefficients are evaluated by continued fractions. This equation is solved by a fixed-point method, starting from an initial guess chosen according to the targeted transfer time. The convergence is guaranteed for any kind of conic.

Gooding's method is based on Lancaster–Blanchard formulation of the transfer time equation (6.170). A zero of the function F(x) defined for $x \in [-1; +\infty]$ must be sought. This function is well-behaved, as can be seen in (Figure 6.14). The search interval for x is determined by comparing the targeted transfer time to the parabolic transfer time. Since the first and second derivatives of F(x) are analytically available, it is interesting to apply iterative methods of higher order than Newton's method.

Halley's iteration is derived by considering the equation

$$G(x) \underset{\text{def}}{=} \frac{F(x)}{\sqrt{F'(x)}} = 0 \tag{6.182}$$

Applying Newton's method to this equation gives the iteration formula

$$x_{k+1} = x_k - \frac{2F(x_k)F'(x_k)}{2F'(x_k)^2 - F(x_k)F''(x_k)} \tag{6.183}$$

Laguerre's iteration already derived for Kepler's equation in (1.52) is defined by

$$x_{k+1} = x_k - \frac{nF(x_k)}{F'(x_k) \pm \sqrt{|(n-1)^2 F'(x_k)^2 - n(n-1)F(x_k)F''(x_k)|}} \tag{6.184}$$

where n is an integer freely chosen, usually set to 4.

Halley's or Laguerre's methods involve more calculations per iteration than Newton's method, but they converge faster to the solution. In the case of Lancaster–Blanchard equation, the convergence is guaranteed by a proper initial guess depending on the targeted transfer time.

Sun's method is similar to Gooding's method, but with a reformulation of Lancaster–Blanchard equation that improves the convergence.

The universal variable method is based on the formulation (6.181) of the transfer time equation. This equation can be solved by Halley's or Laguerre's iteration with an adequate initial guess depending on the targeted transfer time. Some care must be taken to ensure the convergence in the vicinity of parabolic solutions.

These methods have been extensively tested and they ensure excellent efficiency and robustness for practical applications. Among many other iterative approaches, we can mention Zarchan's method based on a transfer time equation depending on the initial flight path angle, or iteration method using the eccentricity as unknown. As an alternative to iterative methods, Thorne's series solution detailed in [R39] gives an explicit solution for the semi-major axis. It is based on series expansion of the transfer time equation, depending on the conic kind (hyperbolic, elliptical short, or long way). Three series expansions are necessary to cover all cases. These series are inverted to yield the semi-major axis.

## 6.5 Conclusion

### 6.5.1 The Key Points

- Impulsive transfer in one single maneuver is possible only if the initial and final orbit intersect.
- The Hohmann transfer is the optimal two-impulse transfer between coplanar circular orbits.
- A transfer between circular mutually inclined orbits can be done either in two impulses by a Hohmann transfer with split plane change or in three impulses by a bielliptic transfer.
- The minimum energy orbit between two points has its semi-major axis worth half the semi-perimeter of the triangle joining the two points and the attracting center.
- Lambert's problem (find the conic joining two points in a given time) results in a transfer time equation that must be solved numerically.

### 6.5.2 To Go Further

- BATTIN R. – *An Introduction to the Mathematics and Methods of Astrodynamics*. AIAA (1999).

Chapter 7 presents the transfer time equation and the solution method developed by the author.

- CURTIS H. – *Orbital Mechanics for Engineering Students*. Elsevier (2005).

Chapter 5 provides a very clear presentation of Lambert's problem, with the use of the universal variable and of Lagrange coefficients. Some numerical examples illustrate the solution method.

- IZZO D. – *Revisiting Lambert's Problem*. Celestial Mechanics and Dynamical Astronomy (2014). DOI: 10.1007/s10569-014-9587-y

This very clear paper presents the derivation of Lancaster–Blanchard equation and proposes significant enhancements to the iterative solution methods.

- VALLADO D. – *Fundamentals of Astrodynamics and Applications*. Microcosm Press, Springer (2007).

Section 7.6 details different solution methods for Lambert's problem, including Thorne solution by series.

- WAGNER S., WIE B. D. – *Computational Solutions to Lambert's Problem on Modern Graphics Processing Units*. Journal of Guidance, Control and Dynamics. Vol.38, No.7, July 15.

This paper compares in an exhaustive way the performances of the four most used algorithms for Lambert's problem: Battin's, Gooding's, Sun's, and universal variable.

# 7

# Orbital Rendezvous

## 7.1 Introduction

A space rendezvous consists of reaching position and/or velocity conditions associated with a target at a given time. This type of problem arises for a wide variety of missions: crew transport to an orbital station, supply of propellants or maintenance of operational satellites, geostationary positioning, injection of sun-synchronous satellite or of constellations, capture of space debris, interception of ballistic threats, etc.

The mission scenario involves the launch into a stable orbit, a phasing with the motion of the target, a transfer to the final orbit, and then the actual rendezvous. Controlling these operations requires spacecraft lighting and visibility conditions from ground stations.

When the target is in a circular orbit, which is the most frequent case, the relative dynamics are described by the Hill–Clohessy–Wiltshire equations. The associated analytical solutions make it possible to prepare in a simple way a preliminary scenario and to define the sequence of maneuvers to be carried out.

The main objective of a rendezvous is to reach the target with sufficient precision despite the perturbations encountered in flight. This requires simplified control laws compatible with onboard guidance, such as linear quadratic control, proportional navigation, or velocity-to-gain. These control laws are used for rendezvous, interception, or landing missions. They offer the robustness needed to take account of uncertainties, particularly those linked to dynamic models and orbit determination.

The first section of this chapter describes the main phases of a rendezvous scenario with their constraints.

The second section establishes the Hill–Clohessy–Wiltshire equations for the rendezvous with a target in a circular orbit and presents the way to use them for preparing a preliminary rendezvous scenario.

The third section introduces the usual control laws that are used operationally for on-board control.

*Space Trajectories: Basic and Advanced Topics*, First Edition. Max Cerf.

## 7.2 Phasing and Transfer

This section describes the phasing and transfer phases preceding the rendezvous.

### 7.2.1 Orbital Model

In this chapter, we consider elliptical orbits defined by their osculating orbital parameters noted (a, e, i, Ω, ω, θ). The orbital plane is defined in the Earth frame by the inclination i (angle with the equator) and the right ascension of the ascending node Ω (RAAN).

The ascending node A is the point where the equator is flown over northward (Figure 7.1a). The argument of the perigee is the angle between the node A and the perigee P. The shape of the ellipse is defined by the semi-major axis a and the eccentricity e. The true anomaly θ is the angle between the perigee and the current position on the orbit. In the Keplerian model, the orbital parameters are constant except for the anomaly that evolves as a function of time.

The radius vector and the derivative of the anomaly are expressed as:

$$r = \frac{p}{1 + e \cos \theta}, \quad \dot{\theta} = \frac{h}{r^2} \quad \text{with} \quad p = \frac{h^2}{\mu} = a(1 - e^2) \tag{7.1}$$

where $\mu = 3.986\,005 \cdot 10^{14}\ m^3/s^2$ is the Earth's gravitational constant, h the angular momentum, and p the ellipse parameter.

In reality, the orbital parameters evolve slowly under the effect of perturbations (accelerations other than the central gravitational attraction). In particular, the Earth's flattening causes the precession of the orbital plane around the pole axis. The ascending node A drifts at a constant rate as shown by the arrow in Figure 7.1a. The nodal precession rate depends on the orbital parameters (a, e, i) according to the formula (2.103).

$$\dot{\Omega} = -\frac{3}{2} J_2 R_E^2 \frac{n \cos i}{a^2(1 - e^2)^2} \quad \text{with} \quad n = \sqrt{\frac{\mu}{a^3}} \tag{7.2}$$

where $R_E = 6\,378\,137$ m is the equatorial radius, $J_2 = 1.082\,63 \cdot 10^{-3}$ is the first zonal term related to the Earth's flattening and n the orbital pulsation (or mean motion).

A prograde orbit (rotating in the same direction as the Earth) has an inclination below 90 deg and it undergoes a retrograde precession ($\dot{\Omega} < 0$) as in Figure 7.1a.

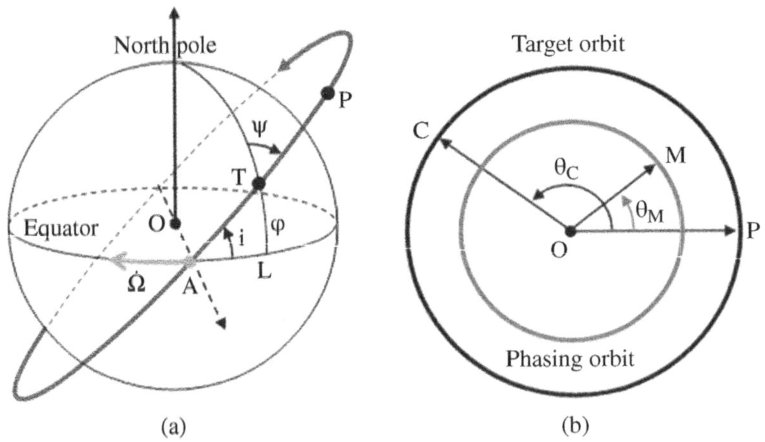


**Figure 7.1** (a) Targeted orbit plane. (b) Anomaly phasing.

### 7.2.2 Phasing

The launcher brings the rendezvous spacecraft into an intermediate orbit, close to the targeted orbit. This orbit, called phasing or drift orbit, must be stable with a perigee altitude greater than 200 km to limit atmospheric braking. Its inclination is usually close to that of the targeted orbit in order to reduce the cost of the subsequent rendezvous maneuvers. To reach the inclination i by taking off from a launch pad T of latitude $\phi$ as depicted in Figure 7.1a, the launch azimuth $\psi$ (angle between the North direction and the launch direction) is calculated in the spherical triangle (triangle ALT in Figure 7.1a) by

$$\cos \mathrm{i} = \cos \varphi \sin \psi \tag{7.3}$$

The phasing orbit has a nodal precession rate $\dot{\Omega}$ and a pulsation n that differ from those of the targeted orbit noted $\dot{\Omega}_c$ and $n_c$, respectively. These different rates allow the phasing in RAAN and anomaly before the transfer to the targeted orbit.

In the frequent cases where the orbits are circular, as in Figure 7.1b, the differential drift rates in RAAN and anomaly are given by

$$\begin{cases} \Delta\dot{\Omega} = \dot{\Omega} - \dot{\Omega}_c = \dfrac{3}{2} J_2 R_E^2 \sqrt{\mu} \left( a_c^{-7/2} \cos i_c - a^{-7/2} \cos i \right) \\ \Delta\dot{\theta} = \dot{\theta} - \dot{\theta}_c = \sqrt{\mu} \left( a^{-3/2} - a_c^{-3/2} \right) \end{cases} \tag{7.4}$$

The phasing in anomaly is fast and it usually does not exceed a few days. For example, a 180 deg catch-up in anomaly with the International Space Station at an altitude of 400 km would require 3 days on a phasing orbit at 350 km.

The phasing in RAAN is much slower (factor $J_2$ of the order of 1/1000) and it can last several weeks. The nodal drift rate (7.4) can be increased by changing either the semi-major axis or the inclination as analyzed hereafter.

On a circular orbit, the nodal precession is expressed as a function of velocity by

$$\dot{\Omega} = -\frac{3 J_2 R_E^2}{2\mu^3} v^7 \cos i \quad \text{with} \quad v = \sqrt{\frac{\mu}{a}} \tag{7.5}$$

A change of inclination $\delta i$ made at the node requires an impulse given by (6.8).

$$\delta v = 2 v \sin \frac{\delta i}{2} \approx v \sin \delta i \quad \text{for } \delta i \text{ small} \tag{7.6}$$

By deriving formula (7.5) with respect to v or i, and substituting $\delta i$ given by (7.6), we can assess the efficiency of changing the precession rate either by semi-major axis or by inclination.

- For a change of semi-major axis: $\frac{\partial \dot{\Omega}}{\partial v} = 7 \frac{\dot{\Omega}}{v}$
- For an inclination change: $\frac{\partial \dot{\Omega}}{\partial v} = \tan i \frac{\dot{\Omega}}{v}$

For inclinations between 82 and 98 deg, it is more economical to increase the nodal drift rate by an inclination change than by a semi-major axis change. For inclination below 82 deg or above 98 deg, the opposite conclusion applies.

The drift strategy based on the formulas (7.4) is much more economical than the transfers by direct maneuvers. It is used in particular to inject constellations of satellites. Constellations used for navigation or telecommunications consist of dozens of satellites covering the entire Earth's surface. Such satellites are to be positioned in planes of different longitudes. In each plane, several satellites are spaced regularly in anomaly. The drift in RAAN being more time and propellant consuming, the injection is generally carried out plane by plane. Each plane is populated from a radius orbit close to the targeted orbit, so as to achieve the anomaly phasing freely. On the other hand, passing from one plane to another requires a drift orbit farther from the targeted orbit in order to limit the duration of the RAAN phasing. The radius and inclination of the phasing orbit result from a compromise between the drift duration and the consumption required by the orbit change maneuvers.

### 7.2.3 Transfer

The transfer consists in moving from the phasing orbit to position and velocity close to that of the target. It is usually achieved by two impulsive maneuvers.

The first impulse $\Delta\vec{v}_1$ is made in a position $P_1(\vec{r}_1)$ on the phasing orbit. The velocity changes from $\vec{w}_1$ on the phasing orbit to $\vec{v}_1 = \vec{w}_1 + \Delta\vec{v}_1$. The vehicle follows a coast arc until reaching a position $P_2(\vec{r}_2)$ close to the target with a velocity $\vec{v}_2$. The second impulse $\Delta\vec{v}_2$ change the velocity from $\vec{v}_2$ on the coast arc to $\vec{w}_2 = \vec{v}_2 + \Delta\vec{v}_2$, which corresponds to the target velocity.

The two-impulse transfer has been presented in Section 6.3.4. It was shown that the initial and final velocities have identical components when projected along skewed axes oriented along the initial and final radii and the chord. These components noted $v_\rho$ and $v_c$, respectively are depicted in Figure 7.2 and we have

$$\begin{cases} \vec{v}_1 = v_c\vec{u}_c + v_\rho\vec{u}_1 \\ \vec{v}_2 = v_c\vec{u}_c - v_\rho\vec{u}_2 \end{cases} \quad \text{with} \quad \vec{u}_1 = \frac{\vec{r}_1}{r_1},\ \vec{u}_2 = \frac{\vec{r}_2}{r_2},\ \vec{u}_c = \frac{\vec{r}_2 - \vec{r}_1}{c} \tag{7.7}$$

where $\vec{u}_1$ and $\vec{u}_2$ are the unit radial vectors in $P_1$ and $P_2$, and $\vec{u}_c$ is the unit vector along the chord $P_1P_2$.

The initial velocity $\vec{w}_1$ before maneuver in $P_1$ depends on the initial orbit and the targeted velocity $\vec{w}_2$ after maneuver in $P_2$ depends on the final orbit.

The minimum-impulse transfer from $P_1$ $(\vec{r}_1, \vec{w}_1)$ to $P_2$ $(\vec{r}_2, \vec{w}_2)$ is the solution of the following optimization problem.

$$\min_{p} \|\Delta\vec{v}_1\| + \|\Delta\vec{v}_2\| \quad \text{with} \quad \begin{cases} \Delta\vec{v}_1 = \vec{v}_1 - \vec{w}_1 = (v_c\vec{u}_c + v_\rho\vec{u}_1) - \vec{w}_1 \\ \Delta\vec{v}_2 = \vec{w}_2 - \vec{v}_2 = \vec{w}_2 - (v_c\vec{u}_c - v_\rho\vec{u}_2) \end{cases} \tag{7.8}$$

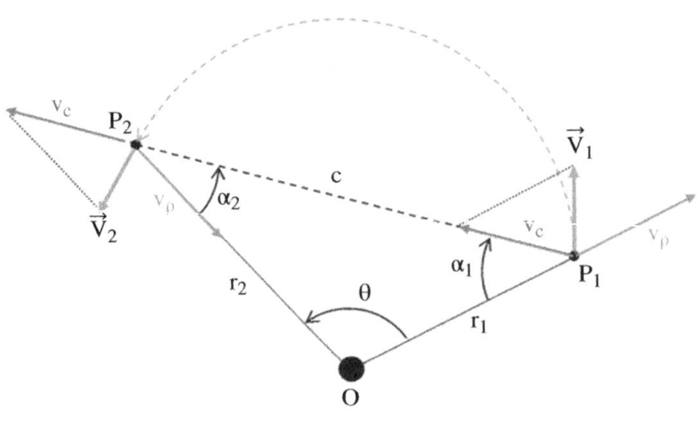


**Figure 7.2** Two-impulse transfer.

In the general case of arbitrary initial and final orbit, this problem cannot be solved analytically. We consider here the frequent case of coplanar circular orbits. The initial and final velocities are then normal to the radius vectors.

By noting $\vec{u}_{1t}$ and $\vec{u}_{2t}$ the unit orthoradial vectors in $P_1$ and $P_2$, we have

$$\vec{w}_1 = \sqrt{\frac{\mu}{r_1}}\vec{u}_{1t}, \quad \vec{w}_2 = \sqrt{\frac{\mu}{r_2}}\vec{u}_{2t} \tag{7.9}$$

In spite of these simplifying assumptions, the problem (7.8) still cannot be solved analytically. A further simplification, which results in a near-optimal solution, is to assume that one of the two boosts is tangential.

Semi-tangential Transfer

This "semi-tangential" strategy is assessed as follows. Let us first assume that the transfer targets a higher orbit ($r_1 < r_2$) and that the initial impulse is tangential. The velocity $\vec{v}_1$ in $P_1$ is then orthoradial as depicted in Figure 7.3.

The initial point $P_1$ becomes the perigee of the transfer orbit. The transfer orbit eccentricity and semi-major axis are assessed by

$$\begin{cases} r_1 = \dfrac{p}{1+e} \\ r_2 = \dfrac{p}{1+e\cos\theta} \end{cases} \Rightarrow \quad e = \frac{r_2 - r_1}{r_1 - r_2\cos\theta}, \quad a = \frac{p}{1-e^2} = \frac{r_1}{1-e} \tag{7.10}$$

In order for the transfer orbit to exist, we must have (with the assumption $r_1 < r_2$)

$$e > 0 \quad \Rightarrow \quad \cos\theta < \frac{r_1}{r_2} \tag{7.11}$$

This lower bound on $\theta$ means that $P_2$ must be to the left of $P_1$ in Figure 7.3.

If we furthermore restrict us to elliptical transfer orbits, we have

$$e < 1 \quad \Rightarrow \quad \cos\theta < 2\frac{r_1}{r_2} - 1 \tag{7.12}$$

This lower bound on $\theta$ is more stringent than (7.11) since $r_1 < r_2$.

The velocity in $P_1$ on the transfer orbit is along $\vec{u}_{1t}$ and its modulus is

$$v_1 = \sqrt{\mu\left(\frac{2}{r_1} - \frac{1}{a}\right)} = \sqrt{\frac{\mu}{r_1}(1+e)} = w_1\sqrt{1+e} \tag{7.13}$$

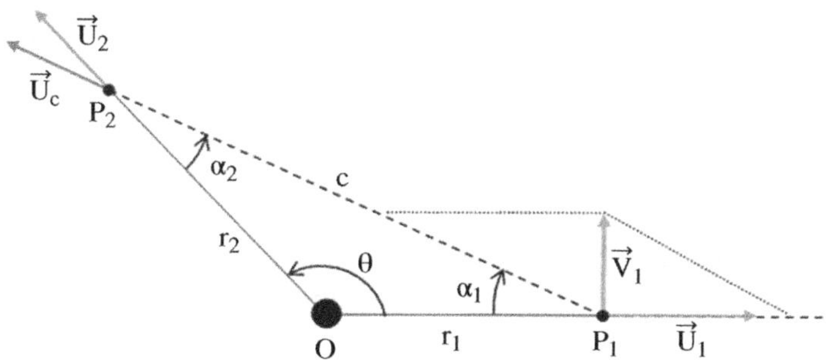


**Figure 7.3** Semi-tangential transfer.

so that the first impulse is

$$\Delta\vec{v}_1 = \vec{v}_1 - \vec{w}_1 = w_1(\sqrt{1+e} - 1)\vec{u}_{1t} \tag{7.14}$$

It remains to assess the second impulse in $P_2$, which is given by (7.8).

$$\Delta\vec{v}_2 = \vec{w}_2 - \vec{v}_2 = \vec{w}_2 - (v_c\vec{u}_c - v_\rho\vec{u}_2) \tag{7.15}$$

The angles at the vertices $P_1$ and $P_2$ in the triangle $(OP_1P_2)$ are noted $\alpha_1$ and $\alpha_2$, as shown in Figure 7.3. The unit vector $\vec{u}_c$ is expressed by

$$\vec{u}_c = -\cos\alpha_1\vec{u}_1 + \sin\alpha_1\vec{u}_{1t} = \cos\alpha_2\vec{u}_2 + \sin\alpha_2\vec{u}_{2t} \tag{7.16}$$

From the parallelogram shown in Figure 7.3 for the initial velocity, we have the following relations between the initial velocity $v_1$ and the components $(v_\rho, v_c)$.

$$v_c = \frac{v_1}{\sin\alpha_1}, \quad v_\rho = \frac{v_1}{\tan\alpha_1} \tag{7.17}$$

Substituting these components into (7.15), with $v_1$ (7.13) and $\vec{u}_c$ (7.16), we obtain

$$\Delta\vec{v}_2 = \frac{w_1\sqrt{1+e}}{\sin\alpha_1}(\cos\alpha_2 - \cos\alpha_1)\vec{u}_2 + w_1\left(\sqrt{1+e}\frac{\sin\alpha_2}{\sin\alpha_1} - \frac{1}{\sqrt{\rho}}\right)\vec{u}_{2t} \tag{7.18}$$

where $\rho \underset{\text{def}}{=} \frac{r_2}{r_1}$ is the radius ratio, so that $w_2 = \sqrt{\frac{\mu}{r_2}} = \sqrt{\frac{\mu}{r_1}}\sqrt{\frac{r_1}{r_2}} = \frac{w_1}{\sqrt{\rho}}$.

The total impulse modulus $\Delta v = \Delta v_1 + \Delta v_2$ is finally given by the scaled formula

$$\frac{\Delta v}{w_1} = (\sqrt{1+e} - 1) + \sqrt{(1+e)\left(\frac{\cos\alpha_2 - \cos\alpha_1}{\sin\alpha_1}\right)^2 + \left(\sqrt{1+e}\frac{\sin\alpha_2}{\sin\alpha_1} - \frac{1}{\sqrt{\rho}}\right)^2} \tag{7.19}$$

The angle $\alpha_1$ is calculated from $(r_1, r_2, \theta)$ by (see Figure 7.3)

$$\begin{cases} \sin\alpha_1 = \dfrac{r_2\sin\theta}{c} \\ \cos\alpha_1 = \dfrac{r_1 - r_2\cos\theta}{c} \end{cases} \quad \text{with} \quad c^2 = r_1^2 + r_2^2 - 2r_1r_2\cos\theta \tag{7.20}$$

and the angle $\alpha_2$ is given by $\alpha_2 = \pi - \alpha_1 - \theta$ in the triangle $(OP_1P_2)$.

The formulas for $\alpha_1$ (7.20) and e (7.10) are scaled by dividing all lengths by $r_1$.

$$\begin{cases} \sin\alpha_1 = \dfrac{\rho}{\sqrt{1+\rho^2 - 2\rho\cos\theta}}\sin\theta \\ \cos\alpha_1 = \dfrac{1 - \rho\cos\theta}{\sqrt{1+\rho^2 - 2\rho\cos\theta}} \end{cases} \quad \text{and} \quad e = \frac{\rho - 1}{1 - \rho\cos\theta} \tag{7.21}$$

The dimensionless formulas (7.21) and (7.19) only depend on the radius ratio $\rho$ and the transfer angle $\theta$.

For a transfer angle of 180 deg, the formula (7.19) is singular since $\alpha_1 = \alpha_2 = 0$. This transfer geometry with tangential impulses at both ends yields a Hohmann transfer. By taking the limit $\theta \to \pi$, we retrieve the formulas established in (6.16).

So far, we have assumed that the targeted orbit is higher $(r_1 < r_2)$ and that the tangential impulse is the first one. The aforementioned results extend to other assumptions, that is, targeting a lower orbit ($r_1 > r_2$) and/or giving the tangential impulse at arrival instead of departure.

Let us assume that the tangential impulse is still the first one, but that the targeted orbit is lower. In this case, the initial point $P_1$ becomes the apogee (instead of perigee) of the transfer orbit.

Formula (7.10) giving the eccentricity and semi-major axis is transformed into

$$\begin{cases} r_1 = \dfrac{p}{1-e} \\ r_2 = \dfrac{p}{1-e\cos\theta} \end{cases} \Rightarrow \quad e = \frac{r_1 - r_2}{r_1 - r_2\cos\theta}, \quad a = \frac{p}{1-e^2} = \frac{r_1}{1+e} \tag{7.22}$$

In this case, there is no bound on the transfer angle θ, since all points of the lower orbit can be reached by an elliptical transfer.

The formula (7.22) for the eccentricity is the same as (7.10) with the opposite sign. Furthermore, we observe that all subsequent calculations from (7.13) to (7.21) apply identically by changing e in −e. In particular, the term $1+e$ (if $\rho > 1$) or $1-e$ (if $\rho < 1$) in formula (7.19) have the same following expression:

$$1 \pm e = 1 \pm \frac{\pm(\rho - 1)}{1 - \rho\cos\theta} = \frac{1-\cos\theta}{\frac{1}{\rho} - \cos\theta} \tag{7.23}$$

The formula giving the total impulse is thus the same whether $\rho > 1$ or $\rho < 1$.

$$\frac{\Delta v}{w_1} = (\sqrt{k} - 1) + \sqrt{k\left(\frac{\cos\alpha_2 - \cos\alpha_1}{\sin\alpha_1}\right)^2 + \left(\sqrt{k}\frac{\sin\alpha_2}{\sin\alpha_1} - \frac{1}{\sqrt{\rho}}\right)^2}$$

$$\text{with} \quad k \underset{\text{def}}{=} \frac{1-\cos\theta}{\frac{1}{\rho} - \cos\theta} \tag{7.24}$$

Let us assume now that the tangential impulse is at arrival instead of departure.

Figure 7.4 depicts a transfer with the tangential impulse on the lowest orbit.

The dotted transfer arc from $Q_2$ ($r_2$) to $Q_1$ ($r_1$) is symmetrical to the solid arc from $P_1$ ($r_1$) to $P_2$ ($r_2$) with respect to the bisector D of the transfer angle. Similarly, Figure 7.5 depicts a transfer with the tangential impulse on the highest orbit.

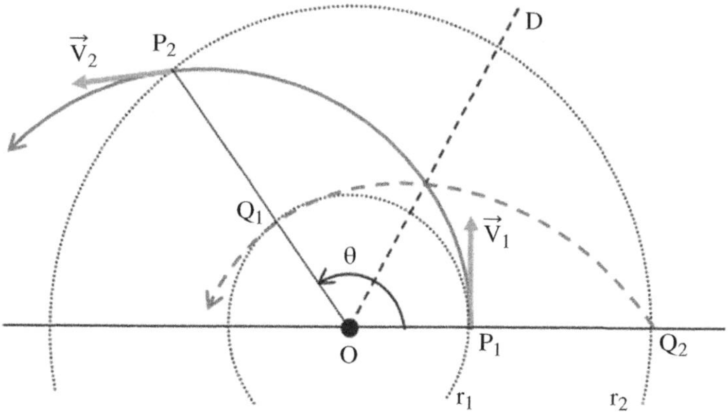


**Figure 7.4** Tangential impulse on the lowest orbit.

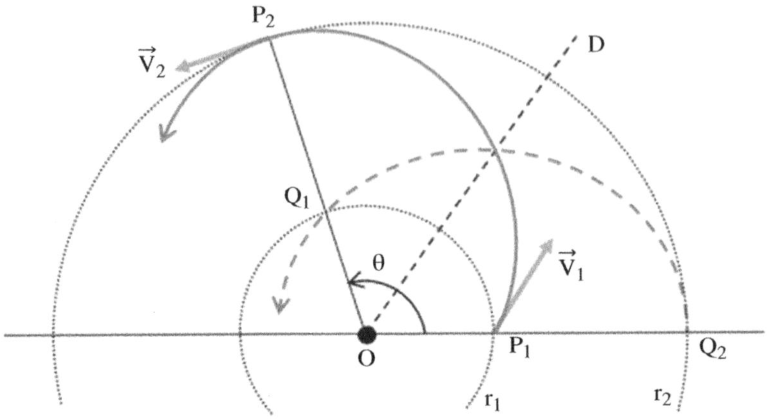


**Figure 7.5** Tangential impulse on the highest orbit.

It is apparent from these two figures that an ascending transfer with tangential impulse at departure is equivalent to a descending transfer with tangential impulse at arrival, and conversely. The formulas established earlier for a departure tangential impulse apply therefore also for an arrival tangential impulse, by simply swapping the variable indexes ($r_1 \leftrightarrow r_2$ and $w_1 \leftrightarrow w_2$).

This is translated in formula (7.24) by changing $\rho$ in $\frac{1}{\rho}$, and dividing the impulse by $\sqrt{\rho}$ (to still scale the impulse with respect to the velocity on the initial orbit).

Let us compare the departure and arrival tangential strategies in the case of an ascending transfer ($\rho > 1$).

- For a departure tangential impulse, the impulse is given by (7.24). The term k is greater than 1 and the term $\frac{1}{\sqrt{\rho}}$ in the last parenthesis is lower than 1.
- For an arrival tangential impulse, by changing $\rho$ into $\frac{1}{\rho}$, the term k becomes lower than 1, and the term $\frac{1}{\sqrt{\rho}}$ is replaced by $\sqrt{\rho}$, which is greater than 1. Furthermore, the impulse is divided by $\sqrt{\rho}$ to be scaled by the initial velocity.

We deduce that the strategy with arrival tangential impulse needs a lower impulse for an ascending transfer, and conversely for a descending transfer.

The conclusion is that the tangential impulse must take place in the highest orbit. This may seem counterintuitive, since changing the velocity direction is less costly at high altitudes, where the velocity is lower. In fact, the gain due to lower velocity is offset by the higher slope on arrival at the highest orbit.

The function $\frac{\Delta v}{w_1}(\theta)$ is plotted in Figure 7.6 for ratios $\rho$ worth 1.1, 1.5, and 2.

The minimum corresponds to a Hohmann transfer for a transfer angle of 180 deg.

These curves are valid regardless of the direction of transfer, either upward or downward. The legend indicates whether the tangential impulse is on the highest or the lowest orbit. With tangential impulse on the highest orbit, all transfer angles are feasible, whereas tangential impulse on the lowest orbit is bounded by (7.12). We observe as established earlier that the tangential impulse on the highest orbit is better, especially as the transfer angle differs from 180 deg.

The corresponding apsis radius on the transfer orbit is plotted in Figure 7.7 for the same values of $\rho$, and the same tangential impulse strategies.

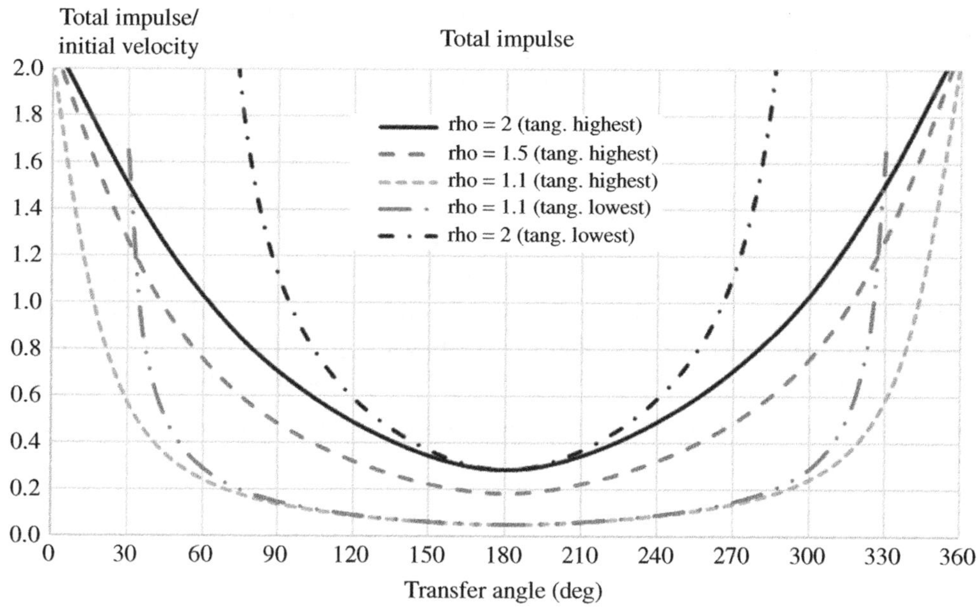


**Figure 7.6** Total impulse (scaled) as a function of the transfer angle.

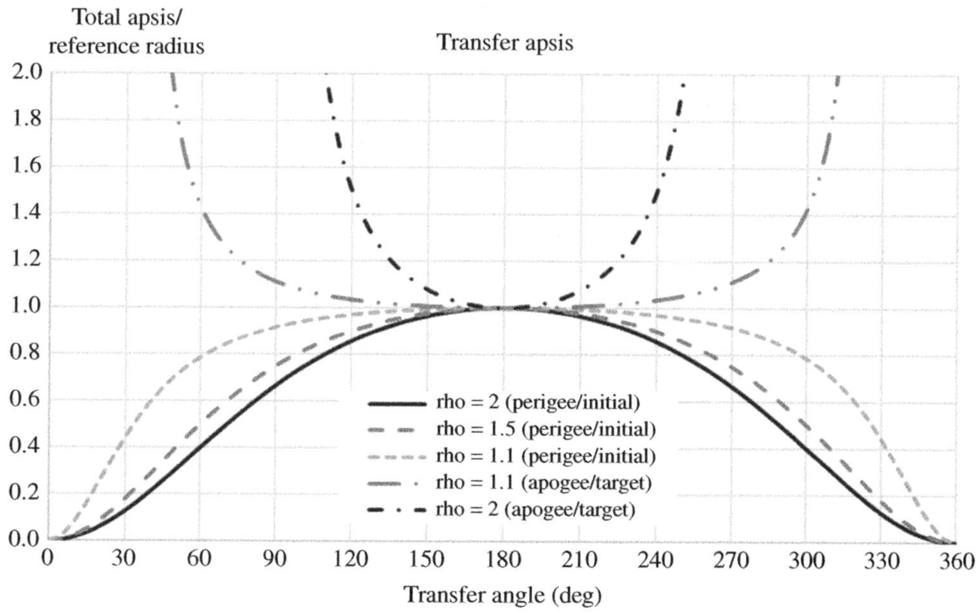


**Figure 7.7** Transfer apogee (scaled) as a function of the transfer angle.

- For a tangential impulse on the highest orbit, the curve plots the perigee scaled by the lowest radius.
- For a tangential impulse on the lowest orbit, the curve plots the apogee scaled by the highest radius.

It can be seen that the apogee ratio increases much faster than the perigee ratio, which explains why the "highest orbit tangential strategy" is better.

The transfer time is calculated from Kepler's equation

$$E - e\sin E = \sqrt{\frac{\mu}{a^3}}(t - t_p) \quad \Leftrightarrow \quad \pi - (E - e\sin E) = \sqrt{\frac{\mu}{a^3}}(t_a - t) \tag{7.25}$$

where $t_p$ and $t_a$ are the times of passage at perigee and apogee respectively and E is the eccentric anomaly assessed from the true anomaly by formulas (1.24).

Depending on the impulse strategy, the transfer time is either $\Delta t = t_a - t$ (for the highest tangential impulse) or $\Delta t = t - t_p$ (for the lowest tangential impulse). Since the transfer angle $\theta$ is the true anomaly difference in $P_1$ and $P_2$, we can use it directly in (1.24) to calculate E and then assess the transfer time by the first formula (7.25) whatever the impulse strategy. To obtain a dimensionless formula, the transfer time is scaled by the target period $T_2 = 2\pi\sqrt{\frac{r_2^3}{\mu}} = 2\pi\sqrt{\frac{a^3}{\mu(1-e)^3}}$, which yields

$$\frac{\Delta t}{T_2} = \frac{E - e\sin E}{2\pi}(1 - e)^{-\frac{3}{2}} \tag{7.26}$$

where e is assessed by (7.21).

The scaled transfer time is plotted in Figure 7.8 as a function of the transfer angle.

With a tangential impulse on the highest orbit, the transfer semi-major axis is less than the target radius and the transfer duration is less than one target period.

With a tangential impulse on the lowest orbit, the transfer is much faster for angles less than 180 deg, but it increases drastically when approaching the upper bound (7.12). This is due to the passage by a high apogee (seen in Figure 7.7).

We can also determine the target anomaly $\theta_0$ at the transfer beginning.

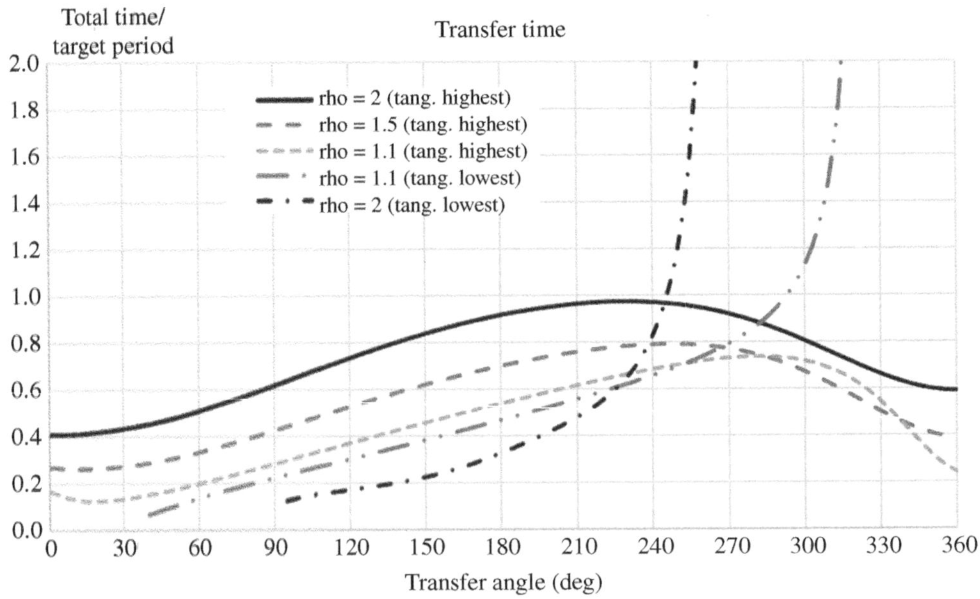


**Figure 7.8** Transfer time (scaled) as a function of the transfer angle.

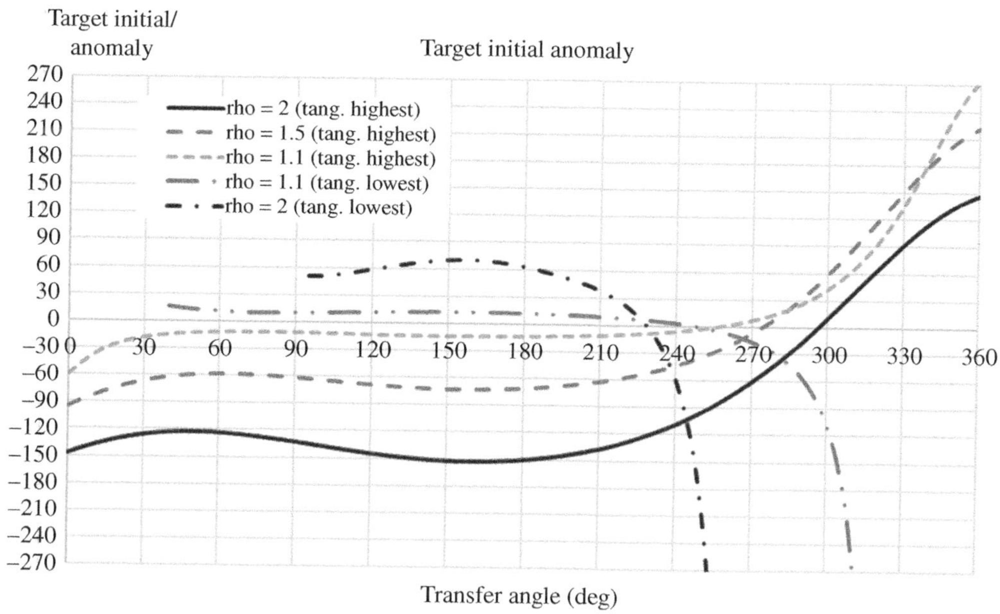


**Figure 7.9** Target initial anomaly as a function of the transfer angle.

From Kepler's equation for the target in circular orbit of radius $r_2$, and using the transfer time given by formula (7.26), we have

$$\theta = \theta_0 + n_2 \Delta t \quad \Rightarrow \quad \theta_0 = \theta - (E - e \sin E)(1 - e)^{-\frac{3}{2}} \tag{7.27}$$

where e is assessed by (7.21) and E is assessed by (1.24).

The target anomaly at the transfer beginning is plotted in Figure 7.9 as a function of the transfer angle. As an example, for $\rho = 2$ and $\theta = 250$ deg, the transfer time is nearly equal to one target period (see Figure 7.8), so the target anomaly at the transfer beginning must be $\theta_0 \approx \theta - 360 = -110$ deg.

For transfer angles approaching the upper bound (7.12), the target must be far behind the vehicle due to large transfer time.

### 7.2.4 Visibility

A space rendezvous requires tracking from ground stations or satellites, as well as Sun illumination if optical sensors are used.

The problem consists of determining if a target C in position $\vec{r}_c$ and a point S (which can be the Sun, a ground station, or a satellite) in position $\vec{r}_s$ are in view from each other (Figure 7.10). The angle between $-\vec{r}_c$ and $\vec{r}_s$ is noted $\varphi$. The Earth cone seen from C has a half-angle $\alpha_e$. The angle between CO and CS is $\alpha_s$. These angles are calculated from the right-angled triangles (CHO) and (CDS).

$$\sin \alpha_e = \frac{R_E}{r_c}, \quad \tan \alpha_s = \frac{r_s \sin \varphi}{r_c + r_s \cos \varphi} \tag{7.28}$$

The condition of visibility is

$$\alpha_s > \alpha_e \quad \text{or} \quad \varphi > \frac{\pi}{2} + \alpha_e \tag{7.29}$$

For the Sun ($r_s = \infty$), the formula (7.28) is simplified and gives $\alpha_s = \varphi$.

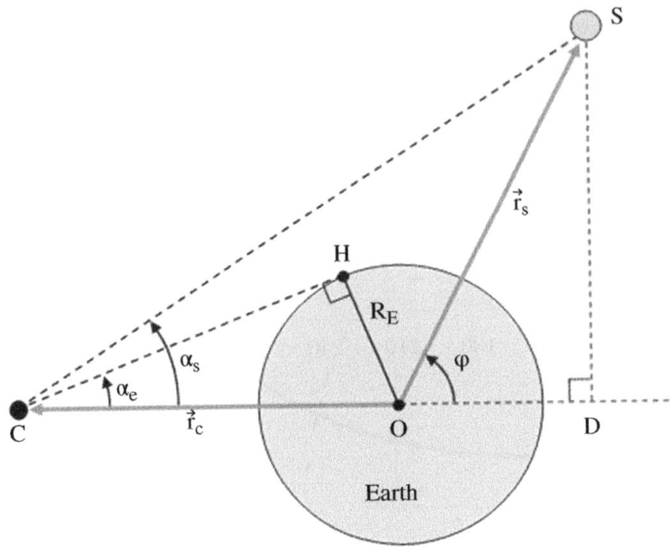


**Figure 7.10** Visibility conditions.

For a ground station ($r_s = R_E$), we have always $\alpha_s \leq \alpha_e$, so that the first condition in (7.29) can never be met.

The positions $\vec{r}_c$ and $\vec{r}_s$ are functions of time, depending on the trajectories of the objects considered. The start and end times $t_v$ of their mutual visibility are obtained by solving the equations

$$\alpha_s(t_v) = \alpha_e(t_v) \quad \text{or} \quad \varphi(t_v) = \frac{\pi}{2} + \alpha_e(t_v) \tag{7.30}$$

## 7.3 Target in Circular Orbit

In most applications, the target moves on a circular orbit, such as the International Space Station. The relative motion is then described by Hill–Clohessy–Wiltshire equations. The general equations of Tschauner–Hempel relative to an elliptical orbit are more complex and will not be discussed here. This section presents Hill–Clohessy–Wiltshire equations with their analytical solution and practical use.

### 7.3.1 Hill–Clohessy–Wiltshire Equations

The vehicle M and the target C are modeled as material points moving around a main attracting body having its center of mass in O. Their respective positions are noted $\vec{r} = \overrightarrow{OM}$ and $\vec{r}_c = \overrightarrow{OC}$. Both are subject to the gravitational attraction $\vec{g}(\vec{r})$ of the main body O. The target C is passive, whereas the vehicle M can control its trajectory owing to a propulsion system. The thrust acceleration is noted $\vec{a}$.

In a Galilean (or inertial) frame RG associated with the attracting body, the motions of M and C follow Newton's second law.

$$\begin{cases} \ddot{\vec{r}} = \vec{g}(\vec{r}) + \vec{a} \\ \ddot{\vec{r}}_c = \vec{g}(\vec{r}_c) \end{cases} \quad \text{with} \quad \vec{g}(\vec{r}) = -\frac{\mu}{r^3}\vec{r} \tag{7.31}$$

where $\mu$ is the gravitational constant of the attracting body.

For relative motion studies, a central gravity model is sufficient. All forces other than central attraction and thrust are neglected. Under these assumptions, the motion of the target C is a Keplerian orbit around O. Its angular momentum $\vec{h}_c$ is constant and it is related to the angular velocity $\vec{\omega}_c$ by

$$\vec{h}_c = r_c^2 \vec{\omega}_c \tag{7.32}$$

The position of M relative to C is noted $\vec{\rho} = \overrightarrow{CM} = \vec{r} - \vec{r}_c$. The differential equation of the relative motion is obtained by subtracting the equations of $\vec{r}$ and $\vec{r}_c$ (7.31).

$$\ddot{\vec{\rho}} = \ddot{\vec{r}} - \ddot{\vec{r}}_c = -\frac{\mu}{r^3}(\vec{r}_c + \vec{\rho}) + \vec{a} + \frac{\mu}{r_c^3}\vec{r}_c = \mu\vec{r}_c\left(\frac{1}{r_c^3} - \frac{1}{r^3}\right) - \frac{\mu}{r^3}\vec{\rho} + \vec{a} \tag{7.33}$$

The distance $\rho = CM$ is assumed to be small compared to $\vec{r}$. The distance r is expanded to order 1 with respect to $\frac{\rho}{r_c}$, by noting $\vec{u}_c$ the unit vector directed toward the target ($\vec{r}_c = r_c\vec{u}_c$).

$$r = \|\vec{r}_c + \vec{\rho}\| = \sqrt{r_c^2 + 2\vec{r}_c \cdot \vec{\rho} + \rho^2} = r_c\sqrt{1 + 2\frac{\vec{u}_c \cdot \vec{\rho}}{r_c} + \frac{\rho^2}{r_c^2}} \approx r_c\left(1 + \frac{\vec{u}_c \cdot \vec{\rho}}{r_c}\right) \tag{7.34}$$

By substituting this distance r into (7.33), we obtain

$$\ddot{\vec{\rho}} \underset{\text{noted}}{=} \ddot{\vec{\rho}}_{|G} = \frac{\mu}{r_c^3}[3(\vec{u}_c \cdot \vec{\rho})\vec{u}_c - \vec{\rho}] + \vec{a} \tag{7.35}$$

Equation (7.35) expresses the acceleration $\ddot{\vec{\rho}}_{|G}$ of the vehicle M with respect to the target C in the inertial frame RG. To analyze the relative motion in a simple way, it is better to pass in the rotating frame RC linked to the target. This local orbital frame has for origin C. The axes Cx and Cy are oriented respectively along the orthoradial and radial directions and the axis Cz is oriented along the angular momentum, normal to the orbital plane of the target (see Figure 7.11a).

**Remark 7.1** *The axis Cx is oriented opposite to the target velocity in the inertial frame. A positive value of x means therefore that the vehicle is behind the target.*

The frame RC rotates around the axis Oz with the angular rate $\vec{\omega}_c$ relatively to the inertial frame RG. The composition of accelerations between RG and RC yields

$$\ddot{\vec{\rho}}_{|G} = \ddot{\vec{\rho}}_{|C} + \dot{\vec{\omega}}_c \wedge \vec{\rho} + \vec{\omega}_c \wedge (\vec{\omega}_c \wedge \vec{\rho}) + 2\vec{\omega}_c \wedge \dot{\vec{\rho}}_{|C} \tag{7.36}$$

where the notation $\dot{\vec{u}}_{|R}$ represents the derivative of $\vec{u}$ in the frame R.

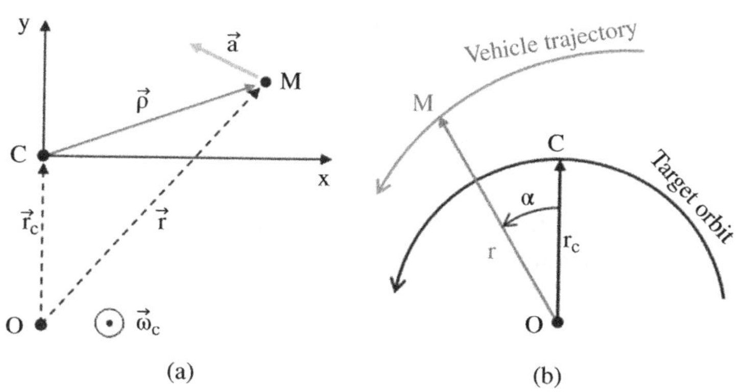


**Figure 7.11** Local orbital frame (a) – Vehicle trajectory and target orbit (b).

Moving on to the components in RC, equation (7.35) gives

$$\vec{\rho} = \begin{pmatrix} x \\ y \\ z \end{pmatrix}, \quad \vec{a} = \begin{pmatrix} a_x \\ a_y \\ a_z \end{pmatrix} \quad \Rightarrow \quad \ddot{\vec{\rho}}_{|G} = \frac{\mu}{r_c^3}\begin{pmatrix} -x \\ 2y \\ -z \end{pmatrix} + \begin{pmatrix} a_x \\ a_y \\ a_z \end{pmatrix} \tag{7.37}$$

Then by substituting into (7.36), we obtain the relative motion equations in RC.

$$\begin{pmatrix} \ddot{x} \\ \ddot{y} \\ \ddot{z} \end{pmatrix} = \begin{pmatrix} \dot{\omega}_c y + \omega_c^2 x + 2\omega_c \dot{y} \\ -\dot{\omega}_c x + \omega_c^2 y - 2\omega_c \dot{x} \\ 0 \end{pmatrix} + \frac{\mu}{r_c^3}\begin{pmatrix} -x \\ 2y \\ -z \end{pmatrix} + \begin{pmatrix} a_x \\ a_y \\ a_z \end{pmatrix} \tag{7.38}$$

These equations apply whatever the target reference orbit, elliptical or circular.
In the case of a target in circular orbit, these equations can be simplified.
The constant angular momentum for a circular orbit is $h_c = r_c v_c = \sqrt{\mu r_c}$ and from equation (7.32), the constant angular velocity noted $\omega$ is

$$\omega = \sqrt{\frac{\mu}{r_c^3}} \tag{7.39}$$

By substituting into (7.38), we obtain the equations of Hill, Clohessy, and Wiltshire.

$$\begin{cases} \ddot{x} = a_x + 2\omega\dot{y} \\ \ddot{y} = a_y - 2\omega\dot{x} + 3\omega^2 y \quad \text{(HCW)} \\ \ddot{z} = a_z - \omega^2 z \end{cases} \tag{7.40}$$

These equations are only an approximation of the actual motion, obtained by linearization in the vicinity of the target. They are useful to assess preliminary rendezvous scenarios. The scenario validation will require an accurate numerical integration of the motion equations (7.31) with more representative models, taking into account the actual target orbit and the most important perturbations.

### 7.3.2 Free Motion

The HCW equations can be solved analytically if the thrust acceleration is null. This analytical solution is derived hereafter and it is used to analyze the influence of the initial conditions on the relative motion.

#### 7.3.2.1 Analytical Solution

The HCW equations integrate analytically in the case of a free motion ($\vec{a} = \vec{0}$).

By deriving the second equation (7.40) for $\ddot{y}$ and substituting $\ddot{x}$ given by the first equation, we have

$$\begin{cases} \ddot{x} = 2\omega\dot{y} \\ \dddot{y} + \omega^2\dot{y} = 0 \\ \ddot{z} + \omega^2 z = 0 \end{cases} \tag{7.41}$$

The out-of-plane (z) and in-plane (xy) motions are decoupled. Both are sinusoidal with the orbital pulsation $\omega$ of the target. The general solution of (7.41) is

$$\begin{cases} x(t) = x_e + \dot{x}_e t - 2A\cos(\omega t + \varphi) \\ y(t) = y_e + A\sin(\omega t + \varphi) \\ z(t) = B\cos(\omega t + \psi) \end{cases} \tag{7.42}$$

The motion in the plane (xy) is an elliptical epicycloid (ellipse in translation) of semi-major axis 2A along x, semi-minor axis A along y. The ellipse is traveled in the opposite direction to the target orbital motion and with the same orbital pulsation $\omega$ as the target. The center of the ellipse, initially in $(x_e, y_e)$, is in translation along x with the drift velocity $\dot{x}_e$.

The integration constants $(A, B, \varphi, \psi)$ in (7.42) depend on the initial conditions of position $(x_0, y_0, z_0)$ and velocity $(\dot{x}_0, \dot{y}_0, \dot{z}_0)$. Applying (7.42) at $t = 0$, we obtain the following relations.

$$x_e = x_0 + \frac{2\dot{y}_0}{\omega}, \quad y_e = 4y_0 - \frac{2\dot{x}_0}{\omega}, \qquad \dot{x}_e = 6\omega y_0 - 3\dot{x}_0$$

$$A = \sqrt{\left(\frac{2\dot{x}_0}{\omega} - 3y_0\right)^2 + \left(\frac{\dot{y}_0}{\omega}\right)^2}, \quad \begin{cases} A\omega\cos\varphi = \dot{y}_0 \\ A\omega\sin\varphi = 2\dot{x}_0 - 3\omega y_0 \end{cases} \tag{7.43}$$

$$B = \sqrt{\left(\frac{\dot{z}_0}{\omega}\right)^2 + (z_0)^2}, \quad \begin{cases} B\omega\cos\psi = \omega z_0 \\ B\omega\sin\psi = -\dot{z}_0 \end{cases}$$

#### 7.3.2.2 Alternative Assessment of the Drift Velocity $\dot{x}_e$

The drift is due to the difference in mean motion between the vehicle and the target. Since the mean motion depends only on the semi-major axis, this result can be retrieved directly from the energy equation as follows.

The position and velocity in the inertial frame are

$$\vec{r}_{IG} = \begin{pmatrix} x_0 \\ r_c + y_0 \\ z_0 \end{pmatrix}, \quad \vec{v}_{IG} = \vec{v}_{IC} + \vec{\omega} \wedge \vec{r} = \begin{pmatrix} \dot{x}_0 \\ \dot{y}_0 \\ \dot{z}_0 \end{pmatrix} + \begin{pmatrix} 0 \\ 0 \\ \omega \end{pmatrix} \wedge \begin{pmatrix} x_0 \\ r_c + y_0 \\ z_0 \end{pmatrix} = \begin{pmatrix} \dot{x}_0 - \omega(r_c + y_0) \\ \dot{y}_0 + \omega x_0 \\ \dot{z}_0 \end{pmatrix}$$

and the energy is $w = \frac{v_{IG}^2}{2} - \frac{\mu}{r_{IG}}$.

By expanding the energy in the vicinity of the target orbit whose energy is $w_c$, we have to order one: $w = \frac{v^2}{2} - \frac{\mu}{r} \approx w_c + \omega r_c(2\omega y_0 - \dot{x}_0) = w_c + 3\omega r_c \dot{x}_e$.

If the drift $\dot{x}_e$ is zero, the vehicle and target have the same energy, therefore the same semi-major axis and the same period.

#### 7.3.2.3 Validity of HCW Equations

Equations (7.42) and (7.43) come from a linearization of the relative motion near the target. They are representative of the actual motion only near the target and over durations of the order of the orbital period. The deviations come from second-order terms neglected in the linearization (7.34).

Let us assess these deviations in the case of a vehicle in a circular orbit of radius r, shifted by an initial angle $\alpha$ from the target as shown in Figure 7.11b.

The target and vehicle position and velocity in the Earth frame RG are

$$\begin{cases} \vec{r}_c = r_c \begin{pmatrix} \cos\theta_c \\ \sin\theta_c \end{pmatrix}, & \vec{v}_c = v_c \begin{pmatrix} -\sin\theta_c \\ \cos\theta_c \end{pmatrix}, & \theta_c = \omega_c t, & \omega_c = \sqrt{\frac{\mu}{r_c^3}}, & v_c = r_c\omega_c \\ \vec{r} = r \begin{pmatrix} \cos\theta \\ \sin\theta \end{pmatrix}, & \vec{v} = v \begin{pmatrix} -\sin\theta \\ \cos\theta \end{pmatrix}, & \theta = \omega t + \alpha, & \omega = \sqrt{\frac{\mu}{r^3}}, & v = r\omega \end{cases} \tag{7.44}$$

The coordinates of the vehicle in the rotating frame RC linked to the target are (taking into account the drive velocity $\vec{\omega}_c \wedge \vec{r}$ at the position $\vec{r}$)

$$\begin{cases} x = r\sin(\theta_c - \theta) \\ y = r\cos(\theta_c - \theta) - r_c \end{cases}, \quad \begin{cases} \dot{x} = \ \ (\omega_c - \omega)\, r\cos(\theta_c - \theta) \\ \dot{y} = -(\omega_c - \omega)\, r\sin(\theta_c - \theta) \end{cases} \tag{7.45}$$

The formulas (7.45) represent the "true" trajectory of the vehicle in RC.

By applying these exact formulas with the initial shift angle $\theta - \theta_c = \alpha$, we obtain the initial conditions

$$\begin{cases} x_0 = -r\sin\alpha \\ y_0 = r\cos\alpha - r_c \end{cases}, \quad \begin{cases} \dot{x}_0 = (\omega_c - \omega)\, r\cos\alpha \\ \dot{y}_0 = (\omega_c - \omega)\, r\sin\alpha \end{cases} \tag{7.46}$$

from which the equations (7.42) and (7.43) give the analytical solution HCW.

The HCW analytical solution is compared to the true trajectory (7.45) for a target in a circular orbit at 400 km altitude. The vehicle is in a circular orbit at the altitude of 399 km and it is shifted by an initial angle $\alpha = -0.1$ deg from the target.

The vehicle is thus 1 km below and 12 km behind the target at the initial time.

The deviations in x and y between the true trajectory and the HCW analytical solution are plotted in Figure 7.12. The error in x is 400 m after a period (92 minutes).

Let us now consider another situation where the vehicle and the target are on the same orbit ($r = r_c$). Their mean motions are then equal.

The initial conditions (7.46) become

$$\begin{cases} x_0 = -r_c\sin\alpha \\ y_0 = r_c(\cos\alpha - 1) \end{cases}, \quad \begin{cases} \dot{x}_0 = 0 \\ \dot{y}_0 = 0 \end{cases} \tag{7.47}$$

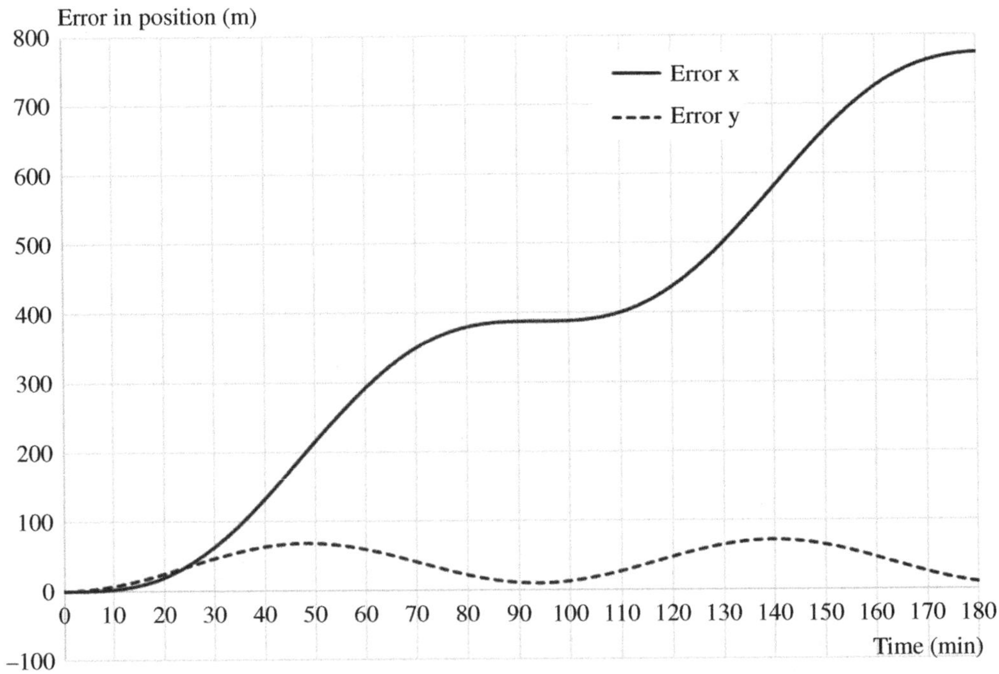


**Figure 7.12** Error on the analytical solution (target at 400 km).

and the true trajectory (7.45) is a stationary point in RC.

$$\begin{cases} x = x_0 \\ y = y_0 \end{cases}, \quad \begin{cases} \dot{x} = 0 \\ \dot{y} = 0 \end{cases} \tag{7.48}$$

On the other hand, the same initial conditions (7.47) substituted in (7.43) yield a non-null drift and a non-zero amplitude: $\dot{x}_e = 6\omega y_0, A = 3y_0$.

The analytical solution (7.42) departs from the true trajectory (7.48) in proportion to the initial angle $\alpha$. To retrieve the true trajectory, it would be necessary to impose $y_0 = 0$. These examples show that a scenario built on analytical solutions must be checked by numerical simulations, even in the case of a Keplerian motion.

#### 7.3.2.4 Initial Conditions

Equations (7.42) and (7.43) allow us to analyze the influence of the initial conditions on the shape of the relative trajectory. Figure 7.13 shows the effect of different initial conditions in the (xy) plane. The target orbit is circular at 400 km altitude.

- A positive value of $y_0$ or a or negative value of $\dot{x}_0$ increases the semi-major axis and thus the orbital period of the vehicle. The motion therefore exhibits a backward drift ($x > 0$) with respect to the target.

  The relative trajectory is drawn in green with round marks in Figure 7.13 for initial conditions $y_0 = 100$ or 200 m.
- A negative value of $\dot{y}_0$ increases the orthoradial velocity, which means that the initial point is the perigee of the vehicle orbit. The vehicle will first fly over the target ($y > 0$), then lose velocity while raising to the apogee and get caught up by the target ($x > 0$). After an orbit, it comes back to the altitude of the target ($y = 0$) behind it (with a backward shift $x > 0$).

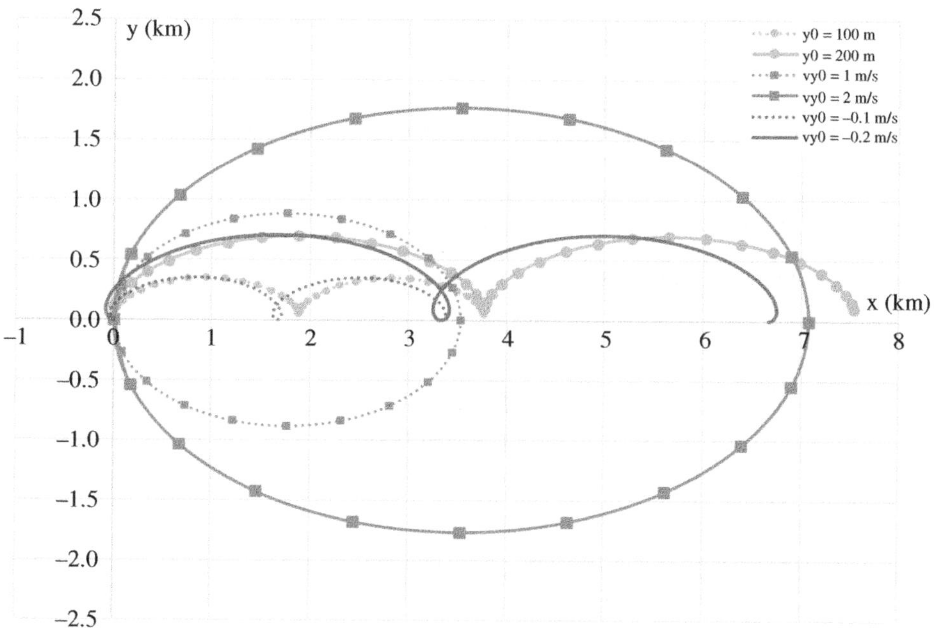


**Figure 7.13** Sensitivity to initial conditions (target at 400 km).

The relative trajectory is drawn in red without marks in Figure 7.13 for initial conditions $\dot{x}_0 = -0.1$ or $-0.2$ m/s.

- A positive value of $\dot{y}_0$ increases the radial velocity without changing the semi-major axis and the orbital period. The vehicle first goes upward ($y > 0$) and is caught up by the target ($x > 0$) during a half orbit; then it comes back to the initial point ($x = 0$).

  The relative trajectory is plotted in blue with square marks in Figure 7.13 for initial conditions $\dot{y}_0 = 1$ or $2$ m/s.

An appropriate choice of initial conditions allows for specific forms of relative motion that are analyzed in the next section.

### 7.3.3 Maneuvers

We examine in turn the use of HCW equations to define impulsive and continuous maneuvers.

#### 7.3.3.1 Impulsive Maneuvers

Equations (7.42) and (7.43) are useful to establish a rendezvous scenario based on impulsive maneuvers.

The initial position at the time $t_0$ is $(x_0, y_0, z_0)$ and the final position to be reached at the time $t_f$ is $(x_f, y_f, z_f)$. The goal is to find the required initial velocity $(\dot{x}_0, \dot{y}_0, \dot{z}_0)$ to achieve the transfer in the allocated duration $\Delta t = t_f - t_0$.

Equations (7.42) written at the initial time $t_0$ and at the final time $t_f$ give

$$\begin{cases} x_0 = x_e - 2A\cos\varphi \\ y_0 = y_e + A\sin\varphi \\ z_0 = B\cos\psi \end{cases} \quad \text{and} \quad \begin{cases} x_f = x_e + \dot{x}_e\Delta t - 2A\cos(\omega\Delta t + \varphi) \\ y_f = y_e + A\sin(\omega\Delta t + \phi) \\ z_f = B\cos(\omega\Delta t + \psi) \end{cases} \tag{7.49}$$

The conditions at $t_0$ give $A\cos\varphi$ and $A\sin\varphi$ as functions of $(x_0, y_0, x_e, y_e)$, and the relations (7.43) give $(x_e, y_e, \dot{x}_e)$ as functions of $(x_0, y_0)$. By substituting into the conditions at $t_f$, we obtain a system in $(\dot{x}_0, \dot{y}_0)$.

$$\begin{cases} x_0 = x_e - 2A\cos\varphi \\ y_0 = y_e + A\sin\varphi \\ z_0 = B\cos\psi \end{cases} \quad \text{and} \quad \begin{cases} x_f = x_e + \dot{x}_e\Delta t - 2A\cos(\omega\Delta t + \varphi) \\ y_f = y_e + A\sin(\omega\Delta t + \varphi) \\ z_f = B\cos(\omega\Delta t + \psi) \end{cases} \tag{7.50}$$

with the notations: $\begin{cases} \Delta x = x_f - x_0 \\ \Delta y = y_f - y_0 \end{cases}$ and $\begin{cases} C = \cos(\omega\Delta t) \\ S = \sin(\omega\Delta t) \end{cases}$.

The solution of this system gives the initial velocity required to perform the transfer in the allocated duration $\Delta t = t_f - t_0$.

$$\begin{cases} \dot{x}_0 = \omega \dfrac{S\Delta x - 2(1-C)\Delta y + 6y_0[2(1-C) - S\omega\Delta t]}{8(1-C) - 3\omega S\Delta t} \\[2ex] \dot{y}_0 = \omega \dfrac{2(1-C)\Delta x + (4S - 3\omega\Delta t)\Delta y - 3y_0(1-C)\omega\Delta t}{8(1-C) - 3\omega S\Delta t} \end{cases} \tag{7.51}$$

The rendezvous scenario is defined by a series of intermediate points $(x_k, y_k, z_k)$ at the times $t_k$. The formulas (7.51) can be used to quickly evaluate the maneuvers required and the associated propellant consumption.

The intermediate points are chosen to carry out the approach in a progressive way, controlling the proper functioning of the vehicle after each phase. If one of the planned maneuvers cannot be carried out, the free motion equations (7.42) make it possible to calculate the subsequent trajectory of the vehicle and to check that it is not likely to collide with the target.

The maneuvers (7.51) depend on their allocated duration $\Delta t$. The setting of the intermediate times allows to reduce the consumption of propellants, while satisfying constraints imposed on the scenario, such as lighting or communications with the ground.

#### 7.3.3.2 Continuous Maneuvers

The HCW equations (7.40) can be used to calculate the acceleration required to follow a relative trajectory of a given shape. Let us assume that a specific trajectory to be followed is defined by its coordinates $(x_r(t), y_r(t), z_r(t))_{t_0 \le t \le t_f}$ in the frame RC. The required accelerations are

$$\begin{cases} a_x(t) = \ddot{x}_r(t) - 2\omega\dot{y}_r(t) \\ a_y(t) = \ddot{y}_r(t) + 2\omega\dot{x}_r(t) - 3\omega^2 y_r(t) \\ a_z(t) = \ddot{z}_r(t) + \omega^2 z_r(t) \end{cases} \tag{7.52}$$

The total impulse required to follow the imposed relative trajectory is the integral of the acceleration: $\Delta v = \int_{t_0}^{t_f} \|\vec{a}(t)\|\, dt$.

Let us assess the needs for the usual relative motions, which are hold points, translations or circles in the target orbital plane.

- Hold point: $\begin{cases} x_r(t) = x_0 \\ y_r(t) = y_0 \end{cases} \Rightarrow \begin{cases} a_x(t) = 0 \\ a_y(t) = -3\omega^2 y_0 \end{cases}$

The impulse required for a hold point of duration $\Delta t$ is: $\Delta v = 3\omega^2 y_0 \Delta t$.

A hold point without thrust is possible only if $y_0 = 0$. The vehicle is then in the same circular orbit as the target.

- Translation: $\begin{cases} x_r(t) = x_0 + \dot{x}_0 t \\ y_r(t) = y_0 + \dot{y}_0 t \end{cases} \Rightarrow \begin{cases} a_x(t) = -2\omega\dot{y}_0 \\ a_y(t) = 2\omega\dot{x}_0 - 3\omega^2(y_0 + \dot{y}_0 t) \end{cases}$

The impulse required for a translation $\Delta x$ on the x axis $(y = 0)$ is: $\Delta v = 2\omega\Delta x$.

A translation without thrust is possible if $\dot{y}_0 = 0$ and $2\dot{x}_0 = 3\omega y_0$. These initial conditions cancel the amplitude A of the ellipse (7.43). The translation then follows the center of the ellipse along the x axis and with the drift velocity $\dot{x}_e$.

- Circle: $\begin{cases} x_r(t) = \rho\cos(\alpha t) \\ y_r(t) = \rho\sin(\alpha t) \end{cases} \Rightarrow \begin{cases} a_x(t) = -\alpha(\alpha + 2\omega)x_r \\ a_y(t) = -\alpha(\alpha + 2\omega)y_r - 3\omega^2 y_r \end{cases}$

In order to minimize the required acceleration, the pulsation $\alpha$ on the circle should be: $\alpha = -2\omega$. This pulsation is twice that of the target and the minus sign means that the circle is traversed in the opposite sense to the target orbit.

The required acceleration is thus reduced to: $a_y(t) = -3\omega^2 y_r$.

The impulse required for such a circle over one target orbit is: $\Delta v = 6\omega\rho$.

It is not possible to follow a circle without thrusting. However, it is possible to follow an ellipse centered in $(x_e, y_e = 0)$ if $\dot{x}_0 = 2\omega y_0$. This initial condition cancels the drift $\dot{x}_e$, and allows it to fly around the target without thrusting. The vehicle is then in an orbit having the same period as the target.

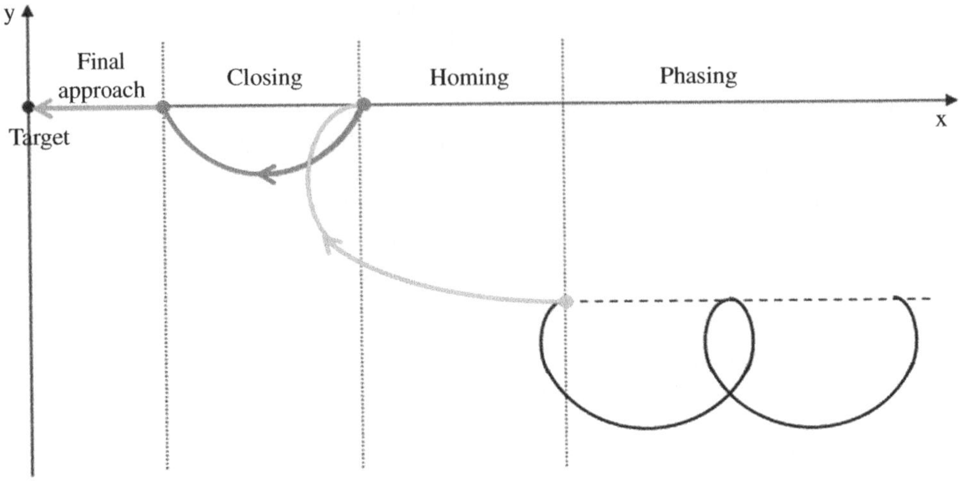


**Figure 7.14** Approach scenario.

### 7.3.4 Approach Scenario

The rendezvous is carried out in successive phases, represented in the target frame in Figure 7.14. These approach phases are separated by waiting phases, in order to control the progress of operations and wait for favorable sunlight and/or visibility conditions from the ground.

Homing (third part from the left in Figure 7.14) is a Hohmann transfer starting from the phasing orbit and arriving on the target orbit a few kilometers away. Closing (second part from the left in Figure 7.14) is a two-impulse transfer arriving within a few hundred meters of the target. These two coast sequences are controlled in absolute navigation (Earth reference frame) by the initial and final impulse. In the event of failure, the ballistic trajectory must pass at a sufficient distance from the target to avoid any risk of collision.

The final approach (on the left in Figure 7.14) to docking is achieved by forced translation along the x axis toward the target. Continuous thrust enables precise control of the relative position and velocity reduction. This phase takes place in relative navigation with respect to the target.

## 7.4 Control Laws

An orbital rendezvous requires precise control of the vehicle trajectory and its real-time updating in response to any perturbation encountered. This section presents the formulation of the optimal control problem and the main control laws adapted to onboard guidance.

### 7.4.1 Optimum Control

#### 7.4.1.1 Minimum Consumption

The vehicle is modeled as a material point of position $\vec{r}$ and velocity $\vec{v}$. It is subject to the gravity $\vec{g}(\vec{r})$ and to the thrust acceleration $\vec{a}$ whose norm can be controlled between 0 and a maximum level $a_m$. The equations of motion are

$$\begin{cases} \dot{\vec{r}} = \vec{v} \\ \dot{\vec{v}} = \vec{g}(\vec{r}) + \vec{a} \end{cases} \tag{7.53}$$

The initial conditions $(\vec{r}_0, \vec{v}_0)$ are given at a time $t_0$. The objective is to find the control law $\vec{a}(t), t_0 \leq t \leq t_f$ to reach the prescribed final conditions $(\vec{r}_f, \vec{v}_f)$ at a given time $t_f$, while minimizing the propellant consumption, or equivalently the total impulse delivered.

This optimal control problem is formulated as:

$$\min_{\vec{a}} J = \int_{t_0}^{t_f} \|\vec{a}\| dt \quad \text{s.t.} \quad \begin{cases} \dot{\vec{r}} = \vec{v} \\ \dot{\vec{v}} = \vec{g} + \vec{a} \end{cases}, \quad \begin{cases} \vec{r}(t_f) = \vec{r}_f \\ \vec{v}(t_f) = \vec{v}_f \end{cases} \tag{7.54}$$

Solving this minimum-consumption problem poses a number of difficulties.

- The numerical solution methods are complex to implement. They can be used at ground to prepare the mission, but they are not suited to onboard guidance.
- The solution structure is of "bang" type, composed of maximum thrust arcs and coast arcs. This type of control may be sensitive to in-flight perturbations.
- Taking into account intermediate or state constraints is difficult, both theoretically and numerically.

These issues are overcome by changing the consumption cost function into an energy cost function and relaxing the constraints by penalization.

#### 7.4.1.2 Minimum Energy

The problem is reformulated by minimizing the integral of $\|\vec{a}\|^2$ (associated with the energy) instead of the integral $\|\vec{a}\|$ (associated with the consumption).

The solution is no longer optimal in terms of consumption, but the resulting quadratic problem can be solved analytically and the control law has a continuous form suitable for onboard guidance.

The **minimum-energy** rendezvous problem is formulated as

$$\min_{\vec{a}} J = \int_{t_0}^{t_f} \|\vec{a}\|^2 dt \quad \text{s.t.} \quad \begin{cases} \dot{\vec{r}} = \vec{v} \\ \dot{\vec{v}} = \vec{g} + \vec{a} \end{cases}, \quad \begin{cases} \vec{r}(t_f) = \vec{r}_f \\ \vec{v}(t_f) = \vec{v}_f \end{cases} \tag{7.55}$$

This optimal control problem is solved by applying Pontryaguin Maximum Principle (PMP) [6]. The Hamiltonian of the problem is

$$H = \frac{1}{2}a^2 + \vec{p}_r \cdot \vec{v} + \vec{p}_v \cdot (\vec{g} + \vec{a}) \tag{7.56}$$

The costate vector has components $(\vec{p}_r, \vec{p}_v)$ associated with the state components in position and velocity $(\vec{r}, \vec{v})$. It follows the costate differential equations

$$\begin{cases} \dot{\vec{p}}_r = -G\vec{p}_v \\ \dot{\vec{p}}_v = -\vec{p}_r \end{cases} \quad \text{with} \quad G = \frac{\partial \vec{g}}{\partial \vec{r}} \tag{7.57}$$

The matrix G is the gravity gradient. For a short trajectory arc, the gravity field $\vec{g}$ can be considered as constant and its gradient is zero: $G = 0$. Owing to this simplification, the costate equations (7.57) integrate to give a linear costate.

$$\begin{cases} \vec{p}_r(t) = \vec{p}_{rf} \\ \vec{p}_v(t) = \vec{p}_{vf} + (t_f - t)\vec{p}_{rf} \end{cases} \tag{7.58}$$

According to the PMP, the optimal control minimizes the Hamiltonian (7.56).

$$\min_{\vec{a}} H \Rightarrow \vec{a} = -\vec{p}_v \tag{7.59}$$

The optimal acceleration is thus a linear function of time depending on two constants noted $(\vec{\alpha}_1, \vec{\alpha}_2)$.

$$\vec{a} = (t - t_f)\vec{\alpha}_1 + \vec{\alpha}_2 \quad \text{with} \quad \vec{\alpha}_1 = \vec{p}_{rf}, \vec{\alpha}_2 = -\vec{p}_{vf} \tag{7.60}$$

With this control law, the equations of motion (7.55) integrate analytically with the assumption of a constant gravity $\vec{g}$.

$$\begin{cases} \vec{v} = \frac{1}{2}(t - t_f)^2\vec{\alpha}_1 + (t - t_f)\,(\vec{\alpha}_2 + \vec{g}) + \vec{v}_f \\ \vec{r} = \frac{1}{6}(t - t_f)^3\vec{\alpha}_1 + \frac{1}{2}(t - t_f)^2(\vec{\alpha}_2 + \vec{g}) + (t - t_f)\vec{v}_f + \vec{r}_f \end{cases} \tag{7.61}$$

The constants $(\vec{\alpha}_1, \vec{\alpha}_2)$ are calculated from the initial conditions $(\vec{r}_0, \vec{v}_0)$ at $t_0$.
After calculation and substitution in (7.60), we obtain the acceleration $\vec{a}_0$ at $t_0$.

$$\vec{a}_0 = \frac{4}{t_f - t_0}(\vec{v}_f - \vec{v}_0) + \frac{6}{(t_f - t_0)^2}[\vec{r}_f - \vec{r}_0 - (t_f - t_0)\vec{v}_f] - \vec{g} \tag{7.62}$$

This **feedback control** calculated at the initial time applies in fact at any time t.

It only depends on the current conditions of position and velocity $(\vec{r}, \vec{v})$ and on the remaining time (called time-to-go) noted $\tau = t_f - t$.

The targeted final time can be varied during the trajectory (and the targeted final position and velocity should vary accordingly) so that acceleration remains below the maximum available bound $a_m$. To reach the target as soon as possible, one can choose at any time t the value of $t_f$ giving $\|\vec{a}\| = a_m$. When the time-to-go approaches zero, formula (7.62) becomes unstable. A possible strategy is to finish the rendezvous by fixing the value of $\tau$ when it falls below a threshold.

The control (7.62) is energy-optimal under the assumption of a constant gravity. It can nevertheless be applied using a variable gravity $\vec{g}(\vec{r}_0)$ in formula (7.62), although it is then no longer optimal for problem (7.55). Even if this control law does not minimize the propellant consumption, it provides a good compromise between performance and robustness. It is particularly suitable when the final conditions have to be reached precisely, as is the case for a landing.

#### 7.4.1.3 Linear Quadratic Formulation

The formulation (7.55) imposes strict final conditions on position and velocity. A more flexible formulation consists of relaxing the constraints by weighting them with the cost function. The weights can be adjusted to achieve a compromise between constraint accuracy and cost minimization.

Another option is to express the dynamics (7.53) in terms of deviations on position $\delta\vec{r}$ and velocity $\delta\vec{v}$ between the vehicle and the target. The position deviation is assumed to be small, which allows the expansion of the gravity to order one with the gradient matrix G.

The differential equations for the deviations (called variational equations) are

$$\begin{cases} \delta\dot{\vec{r}} = \delta\vec{v} \\ \delta\dot{\vec{v}} = G\delta\vec{r} + \vec{a} \end{cases} \tag{7.63}$$

With these models, the rendezvous is recast as a linear quadratic (LQ) problem without final constraints.

$$\min_{u,t_f} J = \frac{1}{2}(x^t S x)_f + \frac{1}{2}\int_{t_0}^{t_f} (x^t Q x + u^t R u)dt \quad \text{s.t.} \quad \begin{cases} \dot{x} = Ax + Bu \\ x(t_0) = x_0 \end{cases} \tag{7.64}$$

- The control vector $u = (\vec{a})$ of dimension three is the vehicle acceleration.
- The state vector $x = (\delta\vec{r}, \delta\vec{v})$ of dimension six contains the position and velocity deviations.
- The dynamics is linear with matrices A $(6\times 6)$ and B $(6\times 3)$ derived from (7.63). In the general case, these matrices are time dependent.

$$\dot{x} = Ax + Bu \quad \text{with} \quad A = \begin{pmatrix} O & I \\ G & O \end{pmatrix}, \quad B = \begin{pmatrix} O \\ I \end{pmatrix} \tag{7.65}$$

The weighting matrices S $(6\times 6)$, Q $(6\times 6)$, R $(3\times 3)$ are set to control respectively the norm of the final deviation $x(t_f)$, the norm of the current deviation $x(t)$ and the norm of the control $u(t)$ along the trajectory. The matrix R is positive definite, whereas the matrices S and Q are positive semidefinite. A usual choice is to set maximum allowed values on the components of $x(t_f)$, $x(t)$, and $u(t)$, and to define diagonal weightings matrices $S = \text{diag}(s_i)$, $Q = \text{diag}(q_i)$, $R = \text{diag}(r_i)$ by

$$s_i = \frac{1}{x_{i\,max}^2(t_f)}, \quad q_i = \frac{1}{(t_f - t_0)x_{i\,max}^2(t)}, \quad r_i = \frac{1}{(t_f - t_0)u_{i\,max}^2(t)} \tag{7.66}$$

The quadratic cost J defined in (7.64) takes into account all the rendezvous objectives in a single function. The zero final deviation constraints are not strictly imposed. The matrices S, Q, and R can be chosen to be constant or time-dependent, depending on the form of solution required. In particular, the Q matrix can be used to indirectly manage state constraints along the trajectory.

Applying the PMP to the linear quadratic problem (7.64) gives as Hamiltonian

$$H = \frac{1}{2}(x^t Q x + u^t R u) + p^t(Ax + Bu) \tag{7.67}$$

The costate vector $p(t)$, of the same dimension as the state vector $x(t)$, follows the costate equations and satisfies the transversality conditions at the final time.

$$\dot{p} = -Qx - A^t p, \quad p(t_f) = S(t_f)\,x(t_f) \tag{7.68}$$

According to the PMP, the optimal control minimizes the Hamiltonian (7.67).

$$\min_u H \quad \Rightarrow \quad u = -R^{-1}B^t p \tag{7.69}$$

Substituting this control into the dynamic equation (7.65), we obtain a linear differential system with conditions at both endpoints on the state and costate.

$$\begin{pmatrix} \dot{x} \\ \dot{p} \end{pmatrix} = \begin{pmatrix} A & -BR^{-1}B^t \\ -Q & -A^t \end{pmatrix} \begin{pmatrix} x \\ p \end{pmatrix} \quad \text{with} \quad \begin{cases} x(t_0) = x_0 \\ p(t_f) = S(t_f)\,x(t_f) \end{cases} \tag{7.70}$$

It can be shown [R9] that the costate depends linearly on the state in the form $p(t) = S(t)x(t)$, with the matrix $S(t)$ following the Riccati differential equation

$$\dot{S} = -SA - A^t S + SBR^{-1}B^t S - Q \quad \text{with} \quad S(t_f) = S_f \tag{7.71}$$

The final matrix $S_f$ is the one chosen in (7.64) to weight the final deviation. By backward integration of Riccati equation (7.71), we obtain the history of the matrix $S(t)$ and of the costate vector $p(t) = S(t)\,x(t)$ along the trajectory.

This allows to express the control (7.69) in state feedback with a gain matrix K.

$$u = Kx \quad \text{with} \quad K = -R^{-1}B^tS \tag{7.72}$$

This form of control is suitable for onboard guidance, as it takes into account the actual state of the vehicle when subject to perturbations. The gain matrix is either precomputed or updated during the trajectory by modulating the S, Q, and R weights.

### 7.4.2 Specific Controls

Optimal control solutions are demanding in terms of onboard memory storage and onboard computing capabilities. Furthermore, they may lack robustness in the presence of large perturbations. The specific control laws [R37] such as those presented in this section sacrifice optimality for greater simplicity and robustness.

#### 7.4.2.1 Proportional Navigation

Proportional navigation is a simplified case of the linear quadratic formulation, where the control is obtained explicitly, without computing the matrix S(t) defined by the differential equation (7.71).

The rendezvous problem is formulated in minimum energy with positive weights noted $k_r$ and $k_v$ on the final position and velocity deviations.

$$J = \frac{1}{2}k_r\delta r_f^2 + \frac{1}{2}k_v\delta v_f^2 + \frac{1}{2}\int_{t_0}^{t_f} a^2 dt \quad \text{s.t.} \quad \begin{cases} \delta\dot{\vec{r}} = \delta\vec{v} \\ \delta\dot{\vec{v}} = G\delta\vec{r} + \vec{a} \end{cases} \tag{7.73}$$

The application of the PMP is similar to (7.56)–(7.59) by assuming a null gravity gradient. In addition, the final costate must satisfy the following transversality conditions derived from the cost derivatives with respect to the final state.

$$\begin{cases} \vec{p}_r(t_f) = k_r\delta\vec{r}_f \\ \vec{p}_v(t_f) = k_v\delta\vec{v}_f \end{cases} \tag{7.74}$$

The constants $\vec{\alpha}_1 = \vec{p}_{rf}, \vec{\alpha}_2 = -\vec{p}_{vf}$ (7.60) are thus directly determined and the acceleration control is expressed as a function of the time-to-go $\tau = t_f - t$.

$$\vec{a} = -\tau k_r\, \delta\vec{r}_f - k_v\, \delta\vec{v}_f \tag{7.75}$$

The deviation equations (7.63) are integrated analytically with the assumption of a constant gravity $\vec{g}$.

$$\begin{cases} \delta\vec{v} = \frac{1}{2}\tau^2 k_r\delta\vec{r}_f + (1+\tau k_v)\delta\vec{v}_f \\ \delta\vec{r} = \left(1 - \frac{1}{6}\tau^3 k_r\right)\delta\vec{r}_f - \left(\tau + \frac{1}{2}\tau^2 k_v\right)\delta\vec{v}_f \end{cases} \tag{7.76}$$

To obtain a state feedback control, we have to express $(\delta\vec{r}_f, \delta\vec{v}_f)$ as functions of $(\delta\vec{r}, \delta\vec{v})$ from (7.76), then substitute in (7.75). After tedious calculations [R9], we obtain the control law under the form

$$\vec{a} = -\frac{\tau k_r\left(1 + \frac{1}{2}\tau k_v\right)\delta\vec{r} + \left(k_v + \tau^2 k_r + \frac{1}{3}\tau^3 k_r k_v\right)\delta\vec{v}}{1 + \tau k_v + \frac{1}{3}\tau^3 k_r + \frac{1}{12}\tau^4 k_r k_v} \tag{7.77}$$

The coefficients $k_r$ and $k_v$ weight the final deviations in position and velocity and the propellant consumption. This general control law can be simplified in the following three special cases, where the consumption is not taken into account.

- For a ballistic injection, the aim is to reach the targeted final velocity: $\delta v_f = 0$ without accounting for the final position.
  The control is of the "velocity-to-gain" form.

$$k_r = 0 \quad \text{and} \quad k_v = \infty \quad \Rightarrow \quad \vec{a} = -\frac{1}{\tau}\delta\vec{v} \tag{7.78}$$

- For an interception, the aim is to reach the targeted final position: $\delta r_f = 0$, without accounting for the final velocity.
  The control is of the "proportional navigation" form.

$$k_r = \infty \quad \text{and} \quad k_v = 0 \quad \Rightarrow \quad \vec{a} = -\frac{3}{\tau^2}(\delta\vec{r} + \tau\delta\vec{v}) \tag{7.79}$$

- For a rendezvous, the aim is to reach both the targeted final position: $\delta r_f = 0$ and the targeted final velocity $\delta v_f = 0$.
  The control is of the "proportional navigation" form.

$$k_r = \infty \quad \text{and} \quad k_v = \infty \quad \Rightarrow \quad \vec{a} = -\frac{1}{\tau^2}(6\delta\vec{r} + 4\tau\delta\vec{v}) \tag{7.80}$$

The proportional navigation control can be interpreted geometrically as follows.

The reference trajectory is a straight-line D traveled at constant velocity $v_d$. The remaining distance to the target C is $r_d = v_d\tau$. The acceleration $\vec{a}$ is normal to D and it acts on the transverse deviations in position $\delta r_n$ and velocity $\delta v_n$.

These variables are depicted in Figure 7.15.

The transverse deviations are functions of the sight angle $\sigma$ assumed to be small.

$$\begin{cases} \delta r_n = r_d\sigma \\ \delta v_n = \delta\dot{r}_n = r_d\dot{\sigma} + \dot{r}_d\sigma = r_d\dot{\sigma} - v_d\sigma \end{cases} \tag{7.81}$$

By substituting these deviations into formulas (7.79) and (7.80), we obtain the proportional navigation control as a function of the sight angle $\sigma$ and its rate $\dot{\sigma}$.

$$\begin{aligned} a_n &= -3v_d\dot{\sigma} && \text{for an interception} \\ a_n &= -v_d\left(2\frac{\sigma}{\tau} + 4\dot{\sigma}\right) && \text{for a rendezvous} \end{aligned} \tag{7.82}$$

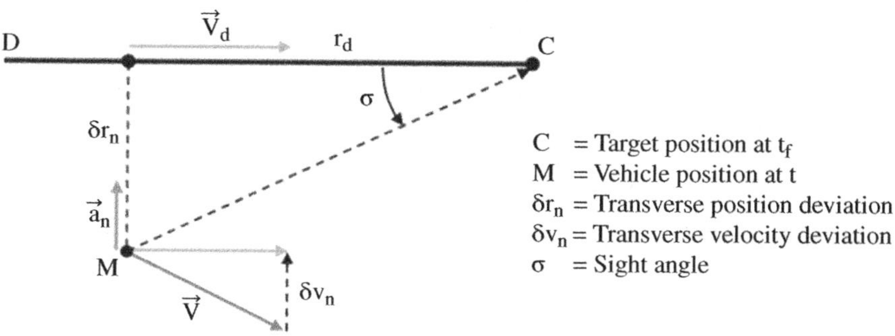


**Figure 7.15** Proportional navigation and line of sight.

This control only acts on transverse motion and is based on a highly simplified dynamic model assuming straight-line motions at constant velocities for the vehicle and the target.

It can be extended to more complex cases, either by adjusting the coefficients in the formulas (7.82) or by adding a term for controlling the longitudinal motion (augmented proportional navigation).

#### 7.4.2.2 Velocity-to-Gain

Velocity-to-gain control is a guidance method developed in the 1960s for ballistic missiles and used for the Apollo program. This method does not require the solution of an optimization problem.

The vehicle is modeled as a material point M of position $\vec{r}$ and velocity $\vec{v}$ subject to the gravity and the thrust. The gravity $\vec{g}(\vec{r})$ is a function of position and the thrust acceleration $\vec{a}$ is controlled. The objective is to reach a target C at a given final time $t_f$.

At the time $t_0$, the vehicle is in the point $M_0$ of position $\vec{r}_0$ and velocity $\vec{v}_0$. The required velocity $\vec{v}_{r0}$ in the point $M_0$ is the velocity that would enable the target to be reached following a purely ballistic trajectory, that is, subject only to gravity.

The calculation of the required velocity depends on the definition of the target.

- For a target defined by its position $\vec{r}_c(t)$ as a function of time, the required velocity $\vec{v}_r$ is calculated by solving Lambert's problem (see Section 6.4) between the vehicle's current position $\vec{r}(t_0)$ and the target position at the final time $\vec{r}_c(t_f)$. The final time can be updated during the trajectory to minimize the velocity-to-gain.
- For an injection in orbit, the required velocity is given by $\vec{v}_r = \frac{\vec{h}}{p} \wedge \left(\vec{e} + \frac{\vec{r}}{r}\right)$, where $\vec{h}, \vec{e}$ are the angular momentum and eccentricity vectors and p is the orbit parameter (see Equation (1.37)). The velocity-to-gain strategy does not control the final position, which will depend on the time chosen to start thrusting.
- For an escape maneuver, the required velocity is given by (see Section 14.2):

$\vec{v} = \frac{v_\infty}{2}\left[(s-1)\frac{\vec{r}}{r} + (s+1)\frac{\vec{v}_\infty}{v_\infty}\right]$ with $s = \sqrt{1 + \frac{4\mu}{r v_\infty^2 (1+\cos\psi)}}$, where $\vec{v}_\infty$ is the target velocity at infinity and $\psi = (\vec{r}, \vec{v}_\infty)$ is the transfer angle.

The required velocity is assumed to be a function $\vec{v}_r(\vec{r}, t)$ that can be calculated in any point $M(\vec{r}, t)$ along the vehicle trajectory. The velocity-to-gain $\vec{v}_g$ in M is the difference between the required velocity $\vec{v}_r$ and the current velocity $\vec{v}$.

$$\vec{v}_g \underset{\text{def}}{=} \vec{v}_r - \vec{v} \tag{7.83}$$

The aim is to bring the velocity-to-gain to zero in order to inject the vehicle on a ballistic rendezvous trajectory. To do this, we try to express the evolution of the velocity-to-gain as a function of the thrust acceleration applied.

At the time $t_1 = t_0 + dt$, the vehicle is in the point $M_1(\vec{r}_1, t_1)$ with the velocity $\vec{v}_1$ and the required velocity is $\vec{v}_{r1}$. The variation in velocity-to-gain between points $M_0$ and $M_1$ is

$$\vec{v}_{g1} - \vec{v}_{g0} = (\vec{v}_{r1} - \vec{v}_1) - (\vec{v}_{r0} - \vec{v}_0) = (\vec{v}_{r1} - \vec{v}_{r0}) - (\vec{v}_1 - \vec{v}_0) \tag{7.84}$$

The variation of the vehicle velocity is due to the action of gravity and thrust.

$$\vec{v}_1 - \vec{v}_0 = (\vec{g} + \vec{a})\, dt \tag{7.85}$$

Let us assume that the vehicle already has the required velocity $\vec{v}_{r0} = \vec{v}_r(\vec{r}_0, t_0)$ in the point $M_0$ at time $t_0$ and that it follows a ballistic trajectory from that time.

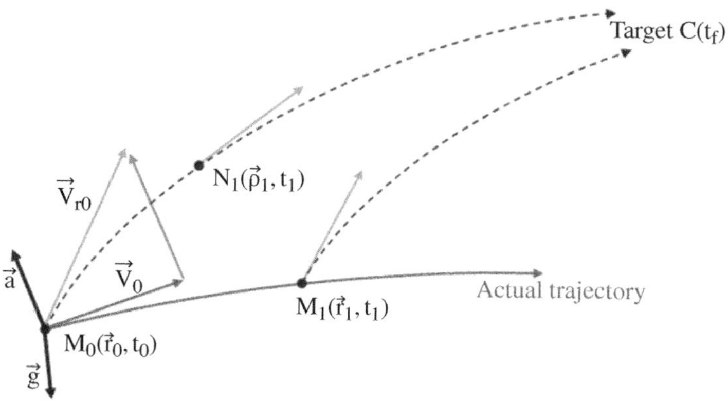


**Figure 7.16** Velocity-to-gain.

At time $t_1$, the vehicle will be in the point $N_1$ at a position noted $\vec{\rho}_1$ (Figure 7.16). It will then naturally have the required velocity $\vec{v}_r(\vec{\rho}_1, t_1)$ in this point since it is already on the ballistic trajectory reaching the target. The variation of the required velocity between $M_0$ and $N_1$ is therefore due only to the gravity.

$$\vec{v}_r(\vec{\rho}, t_1) - \vec{v}_r(\vec{r}_0, t_0) = \vec{g}\,dt \tag{7.86}$$

The difference in position between the points $M_1$ (on the actual vehicle trajectory) and $N_1$ (on the ballistic trajectory starting in $M_0$) at the time $t_1$ is

$$d\vec{r} = \vec{r}_1 - \vec{\rho} = (\vec{r}_0 + \vec{v}_0 dt) - (\vec{r}_0 + \vec{v}_{r0} dt) = -\vec{v}_{g0} dt \tag{7.87}$$

As both points $M_1$ and $N_1$ are considered at the same time $t_1$, the difference in their required velocities is due only to their position.

$$\vec{v}_r(\vec{r}_1, t_1) - \vec{v}_r(\vec{\rho}, t_1) = \frac{\partial \vec{v}_r}{\partial \vec{r}} d\vec{r} \tag{7.88}$$

Combining the formulas (7.86)–(7.88), we obtain the variation of the required velocity between $M_0$ and $M_1$.

$$\vec{v}_r(\vec{r}_1, t_1) - \vec{v}_r(\vec{r}_0, t_0) = \vec{g}dt - \frac{\partial \vec{v}_r}{\partial \vec{r}} \vec{v}_{g0} dt \tag{7.89}$$

Substituting (7.85) and (7.89) into (7.84) and taking the limit $dt \to 0$, we obtain the equation for the evolution of the velocity-to-gain.

$$\frac{d\vec{v}_g}{dt} = -\frac{\partial \vec{v}_r}{\partial \vec{r}} \vec{v}_g - \vec{a} \tag{7.90}$$

From this vector equation, we can assess the evolution of the modulus of the velocity-to-gain. By differentiating its squared modulus using (7.90), we obtain

$$\frac{dv_g^2}{dt} = 2\vec{v}_g \cdot \frac{d\vec{v}_g}{dt} = -2\vec{v}_g \cdot \left(\frac{\partial \vec{v}_r}{\partial \vec{r}} \vec{v}_g\right) - 2\vec{v}_g \cdot \vec{a} \tag{7.91}$$

Our aim is to reduce the velocity-to-gain as fast as possible. When the velocity-to-gain is null, the thrust can be stopped and the vehicle is on a ballistic trajectory reaching the target.

From equation (7.91), we observe that the instantaneous thrust direction that reduces at most the modulus of $\vec{v}_g$ is that of $\vec{v}_g$. The velocity-to-gain control consists thus to orientate at any time the thrust acceleration $\vec{a}$ along $\vec{v}_g$. This result obtained by the calculations (7.84)–(7.91) is somewhat intuitive.

The velocity-to-gain control is well-suited to onboard guidance, since it only needs the current flight conditions to assess the thrust direction. But it remains a "locally optimal" control, unlike linear quadratic control based on a "global optimization" of the command law along the remaining trajectory.

## 7.5 Conclusion

### 7.5.1 The Key Points

- A space rendezvous involves the injection on an orbit close to the target, the phasing in RAAN and anomaly, the transfer to the target orbit, and the final approach maneuvers until the docking.
- The equations of Hill–Clohessy–Wiltshire, obtained by linearization, represent the relative motion near a circular orbit. They allow us to define a preliminary rendezvous scenario based on impulsive maneuvers.
- The free motion near a circular orbit is an elliptical epicycloid (ellipse in translation) at the orbital pulsation and whose major axis along the horizontal is twice the minor axis along the vertical.
- Onboard control for rendezvous uses simplified command laws such as linear quadratic control, proportional navigation, or velocity-to-gain control.

### 7.5.2 To Go Further

- BATTIN R. – *An Introduction to the Mathematics and Methods of Astrodynamics*. AIAA (1999).

Chapter 11 presents in a very clear manner the derivation of specific control laws for interception and rendezvous with their practical application.

- CHOBOTOV V. – *Orbital Mechanics Third edition*. AIAA (2002).

Chapter 7 devoted to rendezvous derives the equations of relative motion, their application to preliminary scenarios, and also many other aspects related to the phasing and transfer strategies.

- VALLADO D. – *Fundamentals of Astrodynamics and Applications*. Microcosm Press, Springer (2007).

Chapter 6 presents the equations of Hill-Clohessy-Wiltshire and their use for assessing proximity maneuvers.

# 8

# Optimal Thrust Level

## 8.1 Introduction

The problem of finding the optimal thrust level of a rocket engine has been studied intensively since Goddard's pioneering work at the beginning of the 20th century. The classical formulation assumes that the thrust level varies freely between zero and a given maximum value. The optimal solution is then made up of maximum thrust arcs and coast arcs.

This chapter presents an alternative hypothesis consisting of a parametric thrust profile, either linear or bilevel. This modeling is representative of the operation of most engines. The objective is to simultaneously optimize the thrust profile and the trajectory of the vehicle to reach the targeted orbit while minimizing the propellant consumption. The initial conditions are given and can result from the flight of the launcher lower stages. The targeted orbit is an ellipse of a given apogee and perigee. The trajectory is assumed to be planar, which is the case in most practical applications dealing with launcher upper stages.

This hybrid optimum control problem including thrust profile is of great practical importance when designing new engines. The study of the optimal conditions shows that this problem admits an analytical solution in state feedback control, the optimal thrust direction being expressed as a function of the current kinematic conditions (position and velocity). The study also shows that the optimal injection point is the perigee of the targeted orbit and that the orbit is reached by a downward phase. This theoretical result explains the a priori surprising shape of many launch trajectories, first ascending and then descending. It turns out that this shape is actually associated with a choice of thrust level close to the optimum. These properties reduce the optimal control problem (OCP) to a nonlinear two-variable problem whose numerical resolution is simple.

The theoretical study also provides an analytical expression of the costate vector as a function of position and velocity conditions. The initialization of the costate vector is a hard point for the optimization of space trajectories. This analytical initialization, obtained when the thrust profile is optimized, is found to be effective in solving various problems, in particular when the engine thrust level is set.

The first section of this chapter presents the problem formulation and the second section demonstrates different properties of the optimal solution. These results are illustrated by numerical examples in the third section.

*Space Trajectories: Basic and Advanced Topics*, First Edition. Max Cerf.

## 8.2 Problem Formulation

This section formulates the hybrid OCP dealing simultaneously with the trajectory and the thrust profile. The optimal conditions are then expressed by applying the Pontryaguin Maximum Principle (PMP).

### 8.2.1 Optimal Control Problem

The minimum-consumption trajectory of a launcher upper stage is sought to reach an orbit from given initial conditions. The vehicle is considered as a material point and the trajectory takes place outside the atmosphere.

The dynamic equations in an inertial geocentric frame are as follows.

$$\begin{cases} \dot{\vec{r}}(t) = \vec{v}(t) \\ \dot{\vec{v}}(t) = \vec{g}[\vec{r}(t)] + \dfrac{T(t)}{m(t)}\vec{u}(t) \end{cases} \quad \text{and} \quad \dot{m}(t) = -\frac{T(t)}{v_e} \tag{8.1}$$

where $\vec{r}(t)$, $\vec{v}(t)$, and $m(t)$ denote the position, the velocity, and the mass of the vehicle at time t, $\vec{g}(\vec{r}) = -\frac{\mu}{r^3}\vec{r}$ is the acceleration due to the Earth's gravity in the position $\vec{r}$, $\mu$ is the Earth's gravitational constant, $T(t)$ is the thrust level modulus, $\vec{u}(t)$ is the unit vector along the thrust direction, and $v_e$ is the exhaust velocity of burnt gases.

All forces other than the thrust and the central gravity (aerodynamics, lunar or solar attraction, gravitational perturbations, and radiation pressure) are neglected.

The engine is ignited at the time $t_0$ with given initial conditions of position $\vec{r}_0$, velocity $\vec{v}_0$ and mass $m_0$. It burns continuously (without thrust-off phases) until the targeted orbit is reached at some free time $t_f$. The trajectory is assumed to be planar, which corresponds to most practical cases of launcher upper stage flight.

The targeted orbit is an ellipse, which can be defined in different ways: either by the altitudes of apogee and perigee or by the semi-major axis and eccentricity, or by the energy and angular momentum modulus per unit mass. All these definitions are strictly equivalent in the case of a planar transfer. The orientation of the orbit (perigee argument) is free.

In the problem formulation, we will consider constraints on the energy w and the angular momentum modulus h defined by

$$\begin{cases} w = \dfrac{v^2}{2} - \dfrac{\mu}{r} = -\dfrac{\mu}{2a} \\ \vec{h} = \vec{r} \wedge \vec{v}, \quad h = \sqrt{\mu a(1 - e^2)} \end{cases} \tag{8.2}$$

where a and e denote the semi-major axis and the eccentricity. The targeted values of w and h are noted $w_f$ and $h_f$, respectively.

Two parametric models are considered for the thrust level profile: a linear model and a bilevel model, respectively.

- The **linear profile** is defined by

$$T(t) = T_1 + (t - t_0)T_2 \quad \text{for } t_0 \leq t \leq t_f \tag{8.3}$$

- The **bilevel profile** is defined by

$$T(t) = \begin{cases} T_1 & \text{if } t_0 \leq t < t_1 \\ T_2 & \text{if } t_1 \leq t < t_f \end{cases} \tag{8.4}$$

The thrust level T thus depends on two parameters $T_1$ and $T_2$, and also on the switching time $t_1$ for the bilevel profile. It is assumed that the thrust remains strictly positive on $[t_0; t_f]$.

The OCP consists of finding the parameters $T_1, T_2, t_1$, the final time $t_f$, and the thrust direction history $\vec{u}(t)_{t_0 \le t \le t_f}$ along the trajectory in order to maximize the final mass injected into the targeted orbit. This problem differs from the conventional rocket problem, which considers the thrust level to be any function T(t) varying freely between zero and a maximum value. In the problem studied here, the thrust profile has an imposed shape.

Optimal Control Problem (OCP)

$$\min_{\substack{T_1,T_2,t_1,t_f \\ \vec{u}(t)}} \quad J = -m(t_f) \text{ s.t. } \begin{cases} \dot{\vec{r}} = \vec{v} \\ \dot{\vec{v}} = \vec{g} + \dfrac{T}{m}\vec{u} \end{cases}, \quad \dot{m} = -\frac{T}{v_e}$$

$$\text{with} \begin{cases} \vec{r}(t_0) = \vec{r}_0 \\ \vec{v}(t_0) = \vec{v}_0 \\ m(t_0) = m_0 \end{cases} \quad \text{fixed initial state}$$

$$\text{and} \begin{cases} w(t_f) = w_f \\ h(t_f) = h_f \end{cases} \quad \text{final constraints} \tag{8.5}$$

The Hamiltonian of this OCP is

$$H = \vec{p}_r \cdot \dot{\vec{r}} + \vec{p}_v \cdot \dot{\vec{v}} + p_m \dot{m} = \vec{p}_r \cdot \vec{v} + \vec{p}_v \cdot \vec{g} + T\left(\frac{\vec{p}_v \cdot \vec{u}}{m} - \frac{p_m}{v_e}\right) \tag{8.6}$$

where $\vec{p}_r(t)$, $\vec{p}_v(t)$, and $p_m(t)$ denote the costate vectors associated with the position, the velocity, and the mass, respectively. The velocity costate $\vec{p}_v$ is also called "primer vector" since Lawden's works [R25]. These three vectors cannot cancel simultaneously over any interval of $[t_0; t_f]$ and they are defined to within one multiplicative factor.

### 8.2.2 Conditions for Optimality

PMP (see [R9], [R38]) gives first-order necessary conditions that must be met along an optimal trajectory.

- The Hamiltonian minimization determines the thrust direction $\vec{u}(t)$. This unit vector is collinear and opposed to the velocity costate.

$$\min_{\vec{u}} H \quad \Rightarrow \quad \vec{u} = -\frac{\vec{p}_v}{p_v} \quad \text{with} \quad p_v = \|\vec{p}_v\| \tag{8.7}$$

By substituting $\vec{u}(t)$ into (8.6), the Hamiltonian takes the form

$$H = H_0 - T\,H_T \quad \text{with} \quad H_0 \underset{\text{def}}{=} \vec{p}_r \cdot \vec{v} + \vec{p}_v \cdot \vec{g} \quad \text{and} \quad H_T \underset{\text{def}}{=} \frac{p_v}{m} + \frac{p_m}{v_e} \tag{8.8}$$

- The costate evolves according to the following differential equations:

$$\begin{cases} \dot{\vec{p}}_r = -\dfrac{\partial H}{\partial \vec{r}} = -\dfrac{\partial \vec{g}}{\partial \vec{r}}\vec{p}_v \\ \dot{\vec{p}}_v = -\dfrac{\partial H}{\partial \vec{v}} = -\vec{p}_r \\ \dot{p}_m = -\dfrac{\partial H}{\partial m} = \dfrac{T}{m^2}\vec{p}_v \cdot \vec{u} = -\dfrac{T}{m^2}p_v \quad \text{since } \vec{u} = -\dfrac{\vec{p}_v}{p_v} \text{ given by (8.7)} \end{cases} \tag{8.9}$$

- The final costate satisfies the conditions of transversality. These conditions written in (8.10) involve the gradients of the final constraints (on w and h) with respect to the state $(\vec{r}, \vec{v}, m)$, their multipliers $\nu_1$ and $\nu_2$, and the gradient of the final cost $J = -m(t_f)$ with a positive multiplier $p_0$ (this sign convention is associated with the Hamiltonian minimization condition).

$$\begin{cases} \vec{p}_r(t_f) = \nu_1 \dfrac{dw}{d\vec{r}}(t_f) + \nu_2 \dfrac{dh}{d\vec{r}}(t_f) \\ \vec{p}_v(t_f) = \nu_1 \dfrac{dw}{d\vec{v}}(t_f) + \nu_2 \dfrac{dh}{d\vec{v}}(t_f) \\ p_m(t_f) = -p_0 \end{cases} \tag{8.10}$$

- The Hamiltonian is constant (autonomous problem) and null (final free time).

$$H(t) = H(t_f) = 0, \quad \forall t \in [t_0, t_f] \tag{8.11}$$

- The parameters $T_1$ and $T_2$ satisfy the first-order conditions (see [R9], [R21])

$$\begin{cases} \displaystyle\int_{t_0}^{t_f} \frac{\partial H}{\partial T_1} dt = 0 \quad \Rightarrow \quad \int_{t_0}^{t_f} \frac{\partial T}{\partial T_1} H_T \, dt = 0 \\ \displaystyle\int_{t_0}^{t_f} \frac{\partial H}{\partial T_2} dt = 0 \quad \Rightarrow \quad \int_{t_0}^{t_f} \frac{\partial T}{\partial T_2} H_T \, dt = 0 \end{cases} \quad \text{with} \quad H = H_0 - T H_T \ \text{from (8.8)} \tag{8.12}$$

**Remark 8.1** ***(discontinuity)*** *The time $t_1$ defines a dynamic change in an interior point. If a constraint were applied in such a point, it would generate discontinuities on the Hamiltonian and the costates (see* [R6], [R9], [R21]*). In the present case, there is no interior constraint. The costates and the Hamiltonian are thus continuous in $t_1$, justifying (8.9) and (8.11).*

**Remark 8.2** **(sign convention)** *The original formulation of the PMP (1962) assumes a negative multiplier $p_0$, which leads to the maximization of the Hamiltonian. Here we have chosen the opposite convention ($p_0 \geq 0$) leading to the minimization of the Hamiltonian.*

These conditions are now used to derive a property of the velocity costate along the optimal trajectory.

### 8.2.3 Property of the Velocity Costate

Property 1

The modulus of the velocity costate $\vec{p}_v$ is constant along an optimal trajectory and its value is given by

$$p_v(t) = \frac{m(t_f)}{v_e}, \quad \forall t \in [t_0, t_f] \tag{8.13}$$

---

**Demonstration**

The demonstration starts from the integral conditions (8.12), which depend on the function $H_T$ and on the derivatives of the thrust level with respect to the two parameters $T_1$ and $T_2$.

The function $H_T$ in (8.8) is first rewritten as:

$$\begin{aligned} H_T &= \frac{p_v}{m} + \frac{p_m}{v_e} = -\frac{m\dot{p}_m}{T} + \frac{p_m}{v_e} && \text{since } \dot{p}_m = -\frac{T}{m^2}p_v \text{ from (8.9)} \\ &= -\frac{m\dot{p}_m}{T} - \frac{\dot{m}p_m}{T} && \text{since } \dot{m} = -\frac{T}{v_e} \quad \text{from (8.1)} \\ &= -\frac{\dot{\Psi}}{T} && \text{by noting } \Psi \underset{def}{=} mp_m \end{aligned} \tag{8.14}$$

with the assumption that the thrust does not cancel on the interval $[t_0; t_f]$.

The derivatives of the thrust level in (8.12) depend on the type of thrust profile. The linear profile and the bilevel profile are considered successively.

- Linear thrust profile

For the linear thrust profile (8.3), substituting $H_T$ (8.14) into the conditions (8.12), and taking the derivatives of (8.3) with respect to $T_1$ and $T_2$, we obtain

$$\int_{t_0}^{t_f} \frac{\dot{\Psi}}{T}\,dt = 0 \quad \text{and} \quad \int_{t_0}^{t_f} \frac{\dot{\Psi}}{T}(t - t_0)\,dt = 0 \tag{8.15}$$

Integrating by parts the first condition (8.12) yields

$$\int_{t_0}^{t_f} \frac{\dot{\Psi}}{T}\,dt = \left[\frac{\Psi}{T}\right]_{t_0}^{t_f} + \int_{t_0}^{t_f} \Psi\frac{\dot{T}}{T^2}\,dt = \left[\frac{\Psi}{T}\right]_{t_0}^{t_f} + \int_{t_0}^{t_f} \Psi\frac{T_2}{T^2}\,dt = 0 \tag{8.16}$$

whereas integrating by parts the second condition (8.12) yields

$$\begin{aligned} \int_{t_0}^{t_f} \frac{\dot{\Psi}}{T}(t - t_0)\,dt &= \left[\frac{\Psi}{T}(t - t_0)\right]_{t_0}^{t_f} - \int_{t_0}^{t_f} \Psi\frac{T - (t - t_0)\dot{T}}{T^2}\,dt \\ &= \left[\frac{\Psi}{T}(t - t_0)\right]_{t_0}^{t_f} - \int_{t_0}^{t_f} \Psi\frac{T_1}{T^2}\,dt = 0 \end{aligned} \tag{8.17}$$

Adding (8.16) multiplied by $T_1$ and (8.17) multiplied by $T_2$ eliminates the integral term. After simplification, the equation reduces to

$$[\Psi]_{t_0}^{t_f} = \Psi(t_f) - \Psi(t_0) = 0 \tag{8.18}$$

By substituting $\Psi = mp_m$ defined in (8.14) and $p_m(t_f)$ given by (8.10), we get

$$m(t_0)p_m(t_0) = -m(t_f)p_0 \tag{8.19}$$

- Bilevel thrust profile

For the bilevel thrust profile (8.4), substituting $H_T$ (8.14) into the conditions (8.12), and taking the derivatives of (8.4) with respect to $T_1$ and $T_2$, we obtain

$$\int_{t_0}^{t_1} \frac{\dot{\Psi}}{T}dt = 0 \quad \text{and} \quad \int_{t_1}^{t_f} \frac{\dot{\Psi}}{T}dt = 0 \tag{8.20}$$

Since T is constant by pieces over the intervals $[t_0; t_1]$ and $[t_1; t_f]$, we obtain

$$\begin{cases} [\Psi]_{t_0}^{t_1} = \Psi\left(t_1^-\right) - \Psi(t_0) = 0 \\ [\Psi]_{t_1}^{t_f} = \Psi(t_f) - \Psi\left(t_1^+\right) = 0 \end{cases} \tag{8.21}$$

Since there is no interior constraint in $t_1$, the costate $p_m$ has no discontinuity and the function $\Psi = mp_m$ is continuous.

$$m(t_0)p_m(t_0) = m(t_1)p_m(t_1) = m(t_f)p_m(t_f) = -m(t_f)p_0 \tag{8.22}$$

We have obtained the same condition (8.19) or (8.22) for both thrust profiles.

*Now let us show that this condition on $p_m$ is actually satisfied at any time t along an optimal trajectory, and not only at the initial time $t_0$.*

First, note that the dynamic (8.1) is unchanged if the mass m(t) and the thrust level T(t) are multiplied by a same constant factor $\alpha$. Suppose that the optimal trajectory associated with the final constraints is known and yields a final time $t_f$ and a final mass $m(t_f)$.

The costates along this optimal path are equal to the partial derivatives of the cost $J = -m(t_f)$ with respect to the state, to within the multiplier $p_0 \geq 0$ ([R6] and [R9]).

For the mass costate, we have thus

$$p_m(t) = -p_0 \frac{\partial m(t_f)}{\partial m(t)}, \quad \forall t \in [t_0; t_f] \tag{8.23}$$

According to Bellman's principle of optimality [R6], a sub-trajectory from any time $t \in [t_0; t_f]$ to $t_f$ is also optimal. Consider a mass perturbation dm applied at some time t on the optimal trajectory.

This perturbation is equivalent to a multiplicative factor $\alpha$ calculated by

$$m(t) + dm = \left(1 + \frac{dm}{m(t)}\right) m(t) \underset{\text{def}}{=} \alpha m(t) \quad \Rightarrow \quad dm = (\alpha - 1)m(t) \tag{8.24}$$

According to the previous observation on the invariability of dynamic equations with respect to a multiplying factor, the optimal sub-trajectory from t to $t_f$ can still be followed by applying the original control law $\vec{u}(t)$, provided that the thrust level T(t)is multiplied by the same factor $\alpha$ applied to the mass. The final time will be unchanged and the final mass will be multiplied by the same factor $\alpha$.

$$m(t_f) + dm_f = \alpha m(t_f) \quad \Rightarrow \quad dm_f = (\alpha - 1)m(t_f) \tag{8.25}$$

Substituting into (8.23) dm and $dm_f$ given by (8.24) and (8.25) gives

$$p_m(t) = -p_0 \frac{\partial m(t_f)}{\partial m(t)} = -p_0 \frac{(\alpha - 1)m(t_f)}{(\alpha - 1)m(t)} = -p_0 \frac{m(t_f)}{m(t)} \tag{8.26}$$

The condition (8.19) therefore applies to any time t along the optimal trajectory, as stated in italics above.

$$m(t)p_m(t) = -m(t_f)p_0, \quad \forall t \in [t_0, t_f] \tag{8.27}$$

Deriving (8.27) and substituting $\dot{p}_m$ given by (8.9) and $\dot{m}$ given by (8.1), we get

$$p_v(t) = \frac{-m(t)p_m(t)}{v_e} = \frac{m(t_f)p_0}{v_e}, \quad \forall t \in [t_0, t_f] \tag{8.28}$$

The modulus $p_v$ of the velocity costate is therefore constant along the optimal trajectory. It cannot be null; otherwise equations (8.9) would give all costates $\vec{p}_r(t)$, $\vec{p}_v(t)$, and $p_m(t)$ simultaneously null, which would contradict the PMP. The multiplier $p_0$ is therefore non-null (normal solution) and we can take the usual normalization $p_0 = 1$ in the sequel.

This ends the demonstration of property 1.

It can be noted that the same property of constant modulus for the velocity costate holds for intermediate thrust arcs in the classical rocket problem (with the thrust level freely optimized between given bounds). Indeed, the function $H_T$ defined in (8.8) represents the switching function for the classical problem and it cancels out on singular arcs, which are associated with an intermediate thrust level. This property was established by Lawden [R25].

## 8.3 Analytical Solution

This section establishes a state feedback formula for the optimal thrust direction and shows that the optimal injection point is an apsis of the targeted orbit. These properties allow the OCP to be reduced to a nonlinear problem.

### 8.3.1 Direction of Thrust

Figure 8.1 depicts the vectors and angles used for the calculations. The trajectory takes place in a plane (Oxy) with the origin O at the Earth's center and arbitrary inertial axes Ox and Oy.

$\varphi$ is the longitude (or polar angle) measured from the Ox axis. $\theta$ and $\gamma$ are the local pitch angle of the vehicle and the local slope of the velocity (also called flight path angle), respectively. By convention, $\theta$ and $\gamma$ measured from the local horizontal are positive upwards. The pitch angle $\theta$ defines the thrust direction $\vec{u}$, and $\vec{n}$ is the vector normal to $\vec{u}$ in the direct sense.

$\omega$ is the angular rate of the thrust direction $\vec{u}$ in the inertial frame (Oxy).

Property 2

The optimal pitch angle $\theta$ satisfies at any time the equation

$$\sin(\theta - \gamma) = \pm \frac{v_c}{v} \frac{\sin\theta}{\sqrt{1 - 3\sin^2\theta}} \quad \text{with} \quad v_c = \sqrt{\frac{\mu}{r}} \tag{8.29}$$

where $v_c$ denotes the circular orbital velocity at radius r.

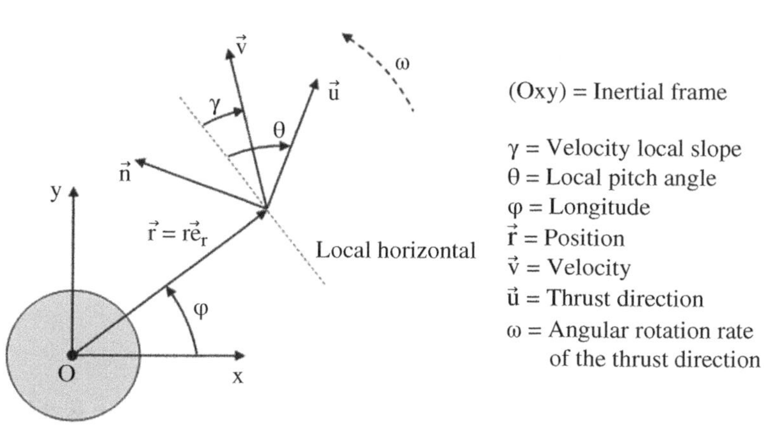


**Figure 8.1** Velocity and thrust orientation.

**Demonstration**

The costate equations (8.9) give a differential equation of order two for $\vec{p}_v$.

$$\dot{\vec{p}}_r = -\frac{\partial \vec{g}}{\partial \vec{r}}\vec{p}_v \text{ and } \dot{\vec{p}}_v = -\vec{p}_r \Rightarrow \ddot{\vec{p}}_v = \frac{\partial \vec{g}}{\partial \vec{r}}\vec{p}_v \tag{8.30}$$

The gravity gradient matrix noted G depends on the position $\vec{r}$. Its expression has been established in (2.86) as

$$\vec{g}(\vec{r}) = -\frac{\mu}{r^3}\vec{r} \Rightarrow G = \frac{\partial \vec{g}}{\partial \vec{r}} = \frac{\mu}{r^5}(3\vec{r}\vec{r}^t - r^2 I) = \frac{\mu}{r^3}\left(3\vec{e}_r\vec{e}_r^t - I\right) \tag{8.31}$$

where I is the identity matrix and $\vec{e}_r$ denotes the unit radial vector ($\vec{r} = r\vec{e}_r$).

The vector $\vec{p}_v$ rotates at some angular rate $\omega$ in the inertial frame (Oxy).

Since this vector is of constant modulus (Property 1), its derivative $\dot{\vec{p}}_v$ is a vector normal to $\vec{p}_v$ and of modulus $\omega p_v$. We have

$$\vec{p}_v = -p_v\vec{u} \Rightarrow \dot{\vec{p}}_v = -p_v\omega\vec{n} \Rightarrow \ddot{\vec{p}}_v = p_v(\omega^2\vec{u} - \dot{\omega}\vec{n}) \text{ since } \begin{cases} \dot{\vec{u}} = \omega\vec{n} \\ \dot{\vec{n}} = -\omega\vec{u} \end{cases} \tag{8.32}$$

Substituting into (8.30) G given by (8.31) and $\ddot{\vec{p}}_v$ given by (8.32), then dividing by the non-null constant $p_v$, we get

$$\dot{\omega}\vec{n} - \omega^2\vec{u} = \frac{\mu}{r^3}\left(3\vec{e}_r\vec{e}_r^t - I\right)\vec{u} = \frac{3\mu}{r^3}(\vec{e}_r \cdot \vec{u})\vec{e}_r - \frac{\mu}{r^3}\vec{u} \tag{8.33}$$

By substituting $\vec{e}_r = \vec{u}\sin\theta - \vec{n}\cos\theta$, this equation becomes

$$\vec{u}\left[\omega^2 + \frac{\mu}{r^3}(3\sin^2\theta - 1)\right] - \vec{n}\left[\dot{\omega} + \frac{3\mu}{r^3}\sin\theta\cos\theta\right] = \vec{0} \tag{8.34}$$

This vector identity gives two relations between the pitch angle $\theta$ and the angular rate $\omega$ in the inertial frame.

$$\begin{cases} \omega^2 = \frac{\mu}{r^3}(1 - 3\sin^2\theta) \\ \dot{\omega} = -\frac{3\mu}{r^3}\sin\theta\cos\theta \end{cases} \tag{8.35}$$

The first relation bounds the pitch angle $\theta$, because $(1 - 3\sin^2\theta)$ must be positive.

$$\sin^2\theta \le \frac{1}{3} \Rightarrow \begin{cases} -35.26 \text{ deg} \le \theta \le 35.26 \text{ deg} \\ \text{or } 144.73 \text{ deg} \le \theta \le 215.26 \text{ deg} \end{cases} \tag{8.36}$$

A third relation between $\theta$ and $\omega$ can be obtained from the Hamiltonian (8.8), since this one must be constantly null (8.11).

$$H = H_0 - TH_T = 0 \text{ with } \begin{cases} H_0 = \vec{p}_r \cdot \vec{v} + \vec{p}_v \cdot \vec{g} \\ H_T = \frac{p_v}{m} + \frac{p_m}{v_e} \end{cases} \tag{8.37}$$

From (8.27) and (8.28), we have $p_v v_e = -mp_m$. The term $H_T$ in the Hamiltonian is thus constantly null and so is the term $H_0$. This term $H_0$ is expressed by

$$H_0 = \vec{p}_r \cdot \vec{v} + \vec{p}_v \cdot \vec{g} = 0$$

$$\text{with} \quad \begin{cases} \dot{\vec{p}}_v = -p_v\vec{u}, \quad \vec{p}_r = -\dot{\vec{p}}_v = \omega p_v\vec{n} \\ \vec{v} = v[\vec{u}\cos(\theta-\gamma) + \vec{n}\sin(\theta-\gamma)] \\ \vec{g} = -g\vec{e}_r = -g(\vec{u}\sin\theta - \vec{n}\cos\theta) \end{cases} \quad \text{and } g = \frac{\mu}{r^2} \tag{8.38}$$

After substituting all vectors in $H_0$, the result is

$$\omega v \sin(\theta-\gamma) + \frac{\mu}{r^2}\sin\theta = 0 \tag{8.39}$$

By substituting into this relation $\omega$ given by (8.35), we obtain

$$\sin(\theta-\gamma) = \pm\frac{v_c}{v}\frac{\sin\theta}{\sqrt{1-3\sin^2\theta}} \quad \text{with} \quad v_c = \sqrt{\frac{\mu}{r}} \tag{8.40}$$

This ends the demonstration of property 2.

---

Equation (8.29) implicitly defines the pitch angle $\theta$ as a function of the current kinematic conditions $(r, v, \gamma)$. For the solution of (8.29) to exist, the second term must be comprised between $-1$ and $+1$. This imposes further bounds on $\theta$ that are more restrictive than (8.36).

$$\begin{cases} -\theta_m \le \theta \le \theta_m \\ \text{or} \\ \pi - \theta_m \le \theta \le \pi + \theta_m \end{cases} \quad \text{with} \quad \theta_m = \text{Arcsin}\frac{1}{\sqrt{3+\left(\frac{v_c}{v}\right)^2}}, \quad 0 < \theta_m < \frac{\pi}{2} \tag{8.41}$$

We observe that if $\theta$ is a solution of (8.29), then so is $\theta + \pi$. We can thus restrict the search interval to $[-\theta_m, \theta_m]$, and then deduce all possible solutions.

Equation (8.29) has no analytical solution and it must be solved numerically.

For this, $\gamma$ is expressed in terms of $\theta$ by taking the Arcsin of both members.

$$\gamma = \theta \pm \text{Arcsin}\left(\frac{v_c}{v}\frac{\sin\theta}{\sqrt{1-3\sin^2\theta}}\right) + k\pi, \quad k \text{ integer} \tag{8.42}$$

Since the angle $\theta$ is defined modulo $\pi$ (see previous remark on the solutions), it is sufficient to solve (8.42) for $k = 0$. Depending on the sign chosen for Arcsin, we are left with two possible equations to solve in the interval $[-\theta_m, \theta_m]$.

$$\gamma = F^+(\theta) \quad \text{or} \quad \gamma = F^-(\theta) \tag{8.43}$$

The functions $F^+(\theta)$ and $F^-(\theta)$ denote the second term of (8.42), either with a plus or a minus sign. These functions depend on the velocity ratio $\frac{v_c}{v}$, which ranges in practice from 0.7 (escape velocity) to 1.5 (low initial velocity).

They are plotted in Figures 8.2 and 8.3, respectively, for three typical values of this velocity ratio.

Let us examine the solution of (8.43) for a given slope value $\gamma$, depending on the function chosen $F^+$ or $F^-$.

- The function $F^+$ is monotonous in $[-\theta_m, \theta_m]$ whatever the velocity ratio, and the equation $F^+(\theta) = \gamma$ admits a unique solution whatever the slope $\gamma$.
- The function $F^-$ is not monotonous for all velocity ratio values, and the equation $F^-(\theta) = \gamma$ does not admit a solution for all values of $\gamma$.

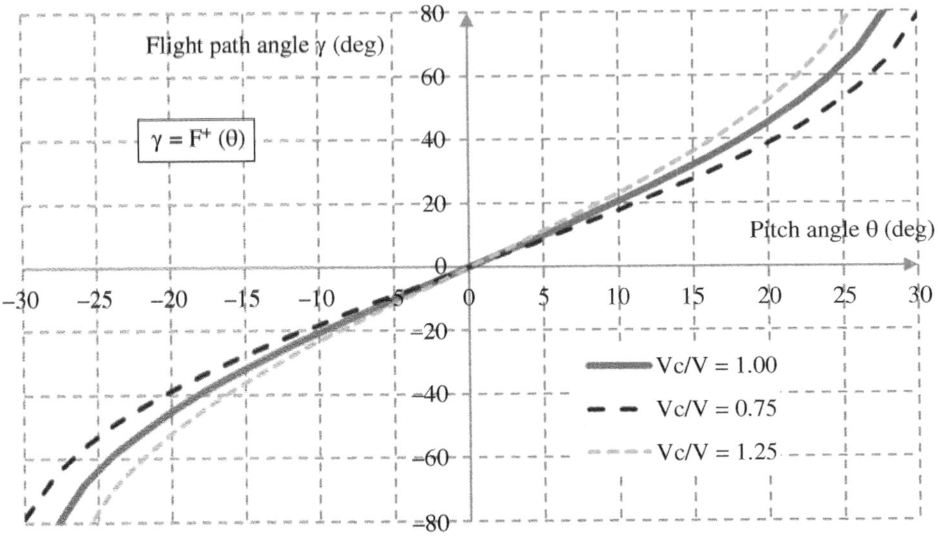


**Figure 8.2** Function $F^{+}(\theta)$ for three values of the velocity ratio.

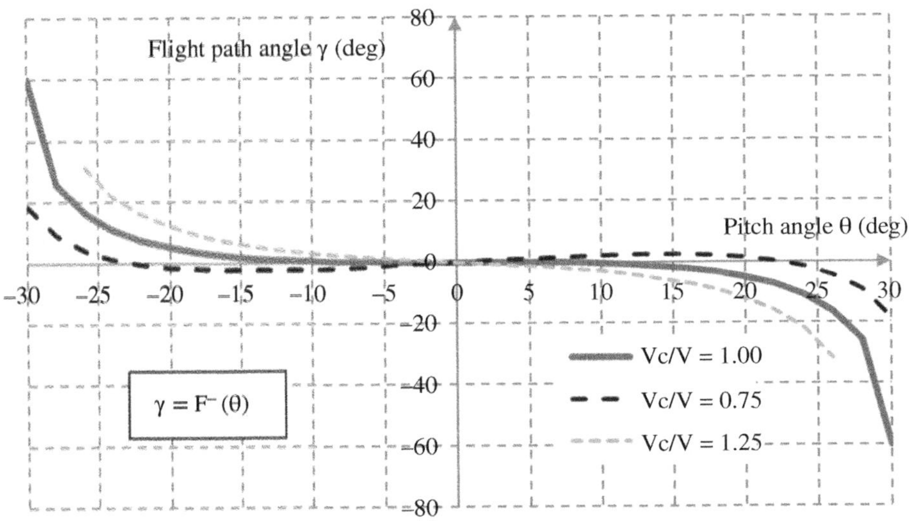


**Figure 8.3** Function $F^{-}(\theta)$ for three values of the velocity ratio.

Furthermore, the solution $\theta^{+}$ associated with the first equation $F^{+}(\theta) = \gamma$ is more physical, because it yields a thrust direction closer to the velocity direction. This solution is therefore the only one considered in the sequel.

At any time t with current kinematic conditions (r, v, γ), the thrust direction is thus obtained by solving the following algebraic equation:

$$\gamma = \theta + \text{Arcsin}\left(\frac{v_c}{v}\frac{\sin\theta}{\sqrt{1-3\sin^2\theta}}\right)$$
$$\text{with } -\theta_m \le \theta \le \theta_m,\ \theta_m = \text{Arcsin}\frac{1}{\sqrt{3+\left(\frac{v_c}{v}\right)^2}} \tag{8.44}$$

This equation is solved numerically by any classical method (Newton, dichotomy, and fixed point). Starting from the initial guess $\theta = \gamma$, the convergence is ensured in a few iterations. This results in a state feedback control, which makes it unnecessary to calculate the costate.

Although the costate is no longer useful to solve the optimal problem control (8.5), the following section derives an analytical expression that will prove useful for the applications presented in Chapters 9 and 12.

### 8.3.2 Costate Vector

Property 3

The costate along the optimal trajectory has the expression

$$\vec{p}_r = \omega \begin{pmatrix} -\cos(\theta - \varphi) \\ \sin(\theta - \varphi) \end{pmatrix}, \quad \vec{p}_v = -\begin{pmatrix} \sin(\theta - \varphi) \\ \cos(\theta - \varphi) \end{pmatrix}, \quad p_m = -\frac{v_e}{m} \tag{8.45}$$

where $\varphi$ is the longitude, $\theta$ is the local pitch angle solution of (8.44) and $\omega$ is the angular rate in the inertial frame (Figure 8.1) given by (8.35).

---

**Demonstration**

The final (unknown) mass is noted $m_f$. The mass costate is given by (8.27).

$$p_m = -\frac{m_f}{m} \tag{8.46}$$

The costates of position $\vec{p}_r$ and velocity $\vec{p}_v$ depend on $p_v$, $\omega$, $\vec{u}$ and $\vec{n}$ by

$$\begin{cases} \vec{p}_v = -p_v \vec{u} \\ \vec{p}_r = -\dot{\vec{p}}_v = -p_v \dot{\vec{u}} = p_v \omega \vec{n} \end{cases} \tag{8.47}$$

The modulus $p_v$ is constant, given by (8.13): $p_v = \frac{m_f}{v_e}$.

The components of $\vec{u}$ and $\vec{n}$ in the inertial frame are (Figure 8.1)

$$\vec{u} = \begin{pmatrix} \sin(\theta - \varphi) \\ \cos(\theta - \varphi) \end{pmatrix} \quad \text{and} \quad \vec{n} = \begin{pmatrix} -\cos(\theta - \varphi) \\ \sin(\theta - \varphi) \end{pmatrix} \tag{8.48}$$

Substituting these vectors into (8.47), the costate $\vec{p}_r$ and $\vec{p}_v$ are expressed as:

$$\vec{p}_r = \omega \frac{m_f}{v_e} \begin{pmatrix} -\cos(\theta - \varphi) \\ \sin(\theta - \varphi) \end{pmatrix} \quad \text{and} \quad \vec{p}_v = -\frac{m_f}{v_e} \begin{pmatrix} \sin(\theta - \varphi) \\ \cos(\theta - \varphi) \end{pmatrix} \tag{8.49}$$

Since the costates $(\vec{p}_r, \vec{p}_v, p_m)$ are defined to within one positive factor, they can be multiplied by $\frac{v_e}{m_f}$. This eliminates the unknown final mass $m_f$ and we finally obtain from (8.46) and (8.49)

$$\vec{p}_r = \omega \begin{pmatrix} -\cos(\theta - \varphi) \\ \sin(\theta - \varphi) \end{pmatrix}, \quad \vec{p}_v = -\begin{pmatrix} \sin(\theta - \varphi) \\ \cos(\theta - \varphi) \end{pmatrix}, \quad p_m = -\frac{v_e}{m} \tag{8.50}$$

This ends the demonstration of property 3.

---

With the initial state $(\vec{r}_0, \vec{v}_0, m_0)$ given, the initial costate is calculated by (8.45) after solving equation (8.44) giving $\theta_0$ at the initial point and with the angular velocity $\omega$ assessed from (8.39).

This costate is not useful to solve the OCP (8.5), since the optimal pitch angle can be determined at any time via equation (8.44). However, it turns out that the formulas (8.45) provide a good initial guess for other OCPs. This is in particular the case for the classical rocket problem at fixed thrust level, as will be illustrated in Section 8.4.

### 8.3.3 Injection Point and Direction

Property 4

The optimal injection point is an apsis of the targeted orbit. The orbit is reached either from above at the perigee, or from below at the apogee. For a circular orbit, the orbit is reached from above.

---

**Demonstration**

The demonstration is divided into three stages: injection point, sense of injection for an elliptical orbit, and sense of injection for a circular orbit.

- First part: injection point

The demonstration starts from the transversality conditions (8.10) linking the final costates to the derivatives of the energy and the angular momentum modulus.

The derivatives of the energy w are calculated using the fact that the gravity is derived from the potential.

$$w = \frac{v^2}{2} - \frac{\mu}{r} \quad \text{with} \quad \vec{g} = -\frac{\mu}{r^3}\vec{r} \quad \Rightarrow \quad \frac{\partial w}{\partial \vec{r}} = -\vec{g}, \ \frac{\partial w}{\partial \vec{v}} = \vec{v} \tag{8.51}$$

For the derivatives of the modulus h of the angular momentum, we consider its half-square noted k. The constraints on k and h are equivalent and the derivatives of k are easier to calculate.

$$k \underset{\text{def}}{=} \frac{1}{2}h^2 = \frac{1}{2}(\vec{r} \wedge \vec{v}) \cdot (\vec{r} \wedge \vec{v}) \quad \Rightarrow \quad \frac{\partial k}{\partial \vec{r}} = -\vec{h} \wedge \vec{v}, \ \frac{\partial k}{\partial \vec{v}} = \vec{h} \wedge \vec{r} \tag{8.52}$$

Substituting these derivatives into the transversality equations (8.10), we obtain two relations at the final time $t_f$.

$$\begin{cases} \vec{p}_r = -\nu_1 \vec{g} - \nu_2 \vec{h} \wedge \vec{v} \\ \vec{p}_v = \ \ \nu_1 \vec{v} + \nu_2 \vec{h} \wedge \vec{r} \end{cases} \tag{8.53}$$

By taking the scalar products of these two relations with $\vec{r}$ and $\vec{v}$, respectively, and using the equalities $\begin{cases} (\vec{h} \wedge \vec{v}) \cdot \vec{r} = \vec{h} \cdot (\vec{v} \wedge \vec{r}) = -h^2 \\ (\vec{h} \wedge \vec{r}) \cdot \vec{v} = \vec{h} \cdot (\vec{r} \wedge \vec{v}) = h^2 \end{cases}$, we get

$$\begin{cases} \vec{p}_r \cdot \vec{r} = \nu_1 v_c^2 + \nu_2 h^2 \\ \vec{p}_v \cdot \vec{v} = \nu_1 v^2 + \nu_2 h^2 \\ \vec{p}_v \cdot \vec{r} = \nu_1 \vec{r} \cdot \vec{v} \end{cases} \quad \text{with} \quad v_c = \sqrt{\frac{\mu}{r}} \tag{8.54}$$

The multipliers $\nu_1$ and $\nu_2$ are eliminated by linear combination, resulting in

$$(\vec{p}_r \cdot \vec{r} - \vec{p}_v \cdot \vec{v})(\vec{r} \cdot \vec{v}) = (\vec{p}_v \cdot \vec{r})\left(v_c^2 - v^2\right) \tag{8.55}$$

The position and velocity components in the frame (Oxy) are (Figure 8.1)

$$\vec{r} = r\begin{pmatrix}\cos\varphi \\ \sin\varphi\end{pmatrix} \quad \text{and} \quad \vec{v} = v\begin{pmatrix}\sin(\gamma-\varphi) \\ \cos(\gamma-\varphi)\end{pmatrix} \tag{8.56}$$

By substituting into (8.55) the state (8.56) and the costate (8.45), we obtain

$$v^2\cos\gamma\sin(\gamma-\theta) = \omega rv\sin\gamma\cos\theta - v_c^2\sin\theta \tag{8.57}$$

Combining this equation with (8.39) written in the form

$$\omega rv\sin(\theta-\gamma) = -v_c^2\sin\theta \quad \text{since } v_c^2 = \frac{\mu}{r} = rg \tag{8.58}$$

leads to

$$\omega rv\sin\theta\cos\gamma = \omega rv\sin\gamma\cos\theta - v_c^2\sin\theta \tag{8.59}$$

which is substituted in the second member of (8.57).

Dividing then by $v\cos\gamma$ (which cannot be null at the final time, corresponding to the injection on the targeted orbit), we have

$$v\sin(\gamma-\theta) = \omega r\sin\theta \tag{8.60}$$

Then, by substitution in (8.58), we finally arrive at

$$\left(v_c^2 + \omega^2 r^2\right)\sin\theta = 0 \tag{8.61}$$

Recall that these calculations derived from the transversality conditions (8.53) apply to the final time.

Condition (8.61) implies that the final thrust direction is horizontal: $\theta(t_f) = 0$.

We deduce from (8.44) that the final slope is zero.

$$\sin\theta(t_f) = 0 \quad \Rightarrow \quad \gamma(t_f) = 0 \tag{8.62}$$

Therefore, the injection takes necessarily place at the perigee or at the apogee of the targeted orbit. This demonstrates the first part of Property 4.

- Second part: injection direction (elliptical orbit)

To demonstrate the second part of Property 4 concerning the direction of injection, the inertial angular rate $\omega$ is expressed as a function of the derivatives of $\varphi$ and $\theta$ (Figure 8.1).

$$\omega = \dot{\varphi} - \dot{\theta} \quad \text{with} \quad \dot{\varphi} = \frac{v\cos\gamma}{r} \tag{8.63}$$

The angular rate $\omega$ is given by (8.35): $\omega^2 = \frac{\mu}{r^3}(1 - 3\sin^2\theta)$. Its sign is deduced from (8.39): $\omega v \sin(\gamma - \theta) = g \sin\theta$ by observing from (8.44) that $\gamma - \theta$ and $\theta$ must have the same sign. The angular rate is therefore always positive and its value in the final point ($\theta_f = 0$) is given by (8.35): $\omega = \sqrt{\frac{\mu}{r^3}} = \frac{v_c}{r}$.

The derivative $\dot{\theta}$ at the final time $t_f$ is then deduced from (8.63)

$$\dot{\theta} = \dot{\varphi} - \omega = \frac{v - v_c}{r} \tag{8.64}$$

In the case of the injection at the perigee of an elliptical orbit, the velocity v is greater than the circular velocity $v_c$, giving a positive derivative $\dot{\theta}$. The angles $\theta$ and $\gamma$ linked by (8.44) have the same sense of variation, as shown by the plot of the function $\gamma = F^+(\theta)$ in Figure 8.2. The derivative $\dot{\gamma}$ is therefore also positive and the final condition $\gamma_f = 0$ is reached by increasing values, which means that the slope just before injection is negative. The trajectory ends with a descending arc and the perigee of the elliptical orbit is reached from above.

By a similar reasoning, an injection at the apogee is achieved from below by an ascending arc. This demonstrates the second part of Property 4.

- Third part: injection direction (circular orbit)

In the case of a circular orbit, the relation (8.64) gives $\dot{\theta} = 0$, which does not define the sense of injection. The second derivative must then be examined.

$$\ddot{\theta} = \ddot{\varphi} - \dot{\omega} \quad \text{with} \quad \dot{\omega} = -\frac{3\mu}{r^3}\sin\theta\cos\theta \quad \text{given by (8.35)} \tag{8.65}$$

The derivative $\ddot{\varphi}$ is calculated from the dynamic equations:

$$\dot{\varphi} = \frac{v\cos\gamma}{r} \quad \text{with} \quad \begin{cases} \dot{r} = v\sin\gamma \\ \dot{v} = \frac{T}{m}\cos(\theta - \gamma) - g\sin\gamma \end{cases}$$

For $\theta_f = \gamma_f = 0$, most terms cancel out and there remains

$$\ddot{\theta} = \frac{r\dot{v} - \dot{r}v}{r^2} = \frac{T}{mr} \tag{8.66}$$

This second derivative is strictly positive (as T). The pitch and slope just before injection are therefore negative. The trajectory ends with a descending arc and the circular orbit is reached from above.

This ends the demonstration of property 4.

---

In practice, the targeted orbits are usually either circular or at high apogee, like the geostationary transfer orbit (GTO). In such cases, the optimal injection always occurs from above, at the perigee for an elliptical orbit.

Property 4 explains the ascending, then descending, shape that can be frequently observed on launch trajectories. This shape is associated with a thrust level close to the optimal one for the launcher last stage.

### 8.3.4 Reduced Problem

The above properties allow a significant reduction of the OCP.

- Property 2 defines the optimal thrust direction in feedback form by (8.44). Solving this equation "on-line" gives the pitch angle $\theta$ as a function of the current kinematic conditions (r, v, γ). The costate equations (8.9) are no longer needed and only the state equations (8.1) are numerically integrated.

- Property 4 indicates that the injection takes place at the perigee reached from above. The integration of the trajectory can therefore be stopped as soon as the slope reaches zero by increasing values. The final time is discarded from the problem unknowns.

The remaining unknowns are the thrust profile parameters. The control problem reduces to a nonlinear optimization problem with two variables ($T_1$, $T_2$) for a linear profile (8.3) or three variables ($T_1$, $T_2$, $t_1$) for a bilevel profile (8.4). If we choose to sweep the switching time $t_1$ for the bilevel model, there remain only two unknowns ($T_1$, $T_2$) for both models.

The two constraints are the apogee and perigee altitudes of the targeted orbit.

We thus end up with a system of nonlinear equations of dimension two. This system is easily solved by Newton's method from an approximate guess of the unknowns ($T_1$, $T_2$).

### 8.3.5 Performance Estimate

The feedback control (8.44) can be used to estimate the velocity losses along the optimal trajectory. The evolution of the velocity modulus is given by

$$\dot{v} = \frac{T}{m}\cos(\theta - \gamma) - g\sin\gamma \tag{8.67}$$

The velocity losses are caused by the gravity and by the thrust not aligned with the velocity (incidence losses). These terms noted $\Delta v_G$ and $\Delta v_T$ respectively are assessed by integration of (8.67).

$$\Delta v_G = \int_{t_0}^{t_f} g\sin\gamma\, dt, \quad \Delta v_T = \int_{t_0}^{t_f} \frac{T}{m}[1 - \cos(\theta - \gamma)]\, dt \tag{8.68}$$

For small values of γ and θ (this necessarily happens as the injection at apsis approaches) and considering that v and $v_c$ have the same order of magnitude, equation (8.44) can be simplified to give

$$\sin(\gamma - \theta) \approx \sin\theta \quad \Rightarrow \quad 2\theta \approx \gamma \tag{8.69}$$

With this approximation, equation (8.35) gives for $\dot{\omega}$

$$\dot{\omega} = -\frac{3\mu}{2r^3}\sin 2\theta \approx -\frac{3}{2r}g\sin\gamma \quad \text{with} \quad g = \frac{\mu}{r^2} \tag{8.70}$$

Replacing this expression of $g\sin\gamma$ in the gravity loss integral (8.68) yields

$$\Delta v_G = \int_{t_0}^{t_f} g\sin\gamma\, dt \approx -\frac{2}{3}\int_{t_0}^{t_f} r\dot{\omega}\, dt \tag{8.71}$$

The radius varies little along the propelled flight of an upper stage (typically from 6500 to 6800 km for a targeted perigee of 400 km).

Assuming a constant radius equal to $r_0$, the integral (8.71) can be computed.

$$\Delta v_G \approx \frac{2}{3}r_0(\omega_0 - \omega_f) \tag{8.72}$$

The angular rate is given by (8.35): $\omega = \sqrt{\frac{\mu}{r^3}(1 - 3\sin^2\theta)}$.

With the previous approximations ($r \approx r_0$, $2\theta \approx \gamma$), we get at the first order

$$\Delta v_G \approx \frac{1}{4} v_c \gamma_0^2 \quad \text{with} \quad v_c = \sqrt{\frac{\mu}{r_0}} \tag{8.73}$$

The term $\Delta v_G$ is the main contributor to the velocity losses. Indeed, the incidence losses $\Delta v_T$ are low because the angles $\gamma$ and $\theta$ satisfying (8.44) have close values along the optimal trajectory. The total velocity impulse $\Delta v$ required for the injection is the sum of the actual velocity increment $v_f - v_0$ and the losses. The final velocity is known because the injection occurs at the perigee of the targeted orbit: $v_f = v_P$. The final mass can then be estimated by Tsiolkovski's formula.

$$\Delta v = v_e \ln \frac{m_0}{m_f} \quad \Rightarrow \quad m_f \approx m_0\, e^{-\frac{v_P - v_0 + \Delta v_G}{v_e}} \tag{8.74}$$

## 8.4 Application

This section illustrates the application of the previous theoretical results. The first example concerns the optimization of a linear or bilevel thrust profile. The second example shows the resolution of a classical problem at a fixed thrust level by using the analytical costate as an initial guess.

### 8.4.1 Optimized Thrust Profile

This first example deals with a transfer at optimized thrust profile, considering either the linear or the bilevel profile. The initial mass is 10 000 kg and the engine specific impulse is 300 s corresponding to an exhaust velocity of 2942 m/s.

The initial conditions are an altitude of 150 km, a velocity of 5000 m/s and a local slope of 30 deg. These conditions correspond to altitudes of apogee at 661.7 km and of perigee at −5209.4 km. The targeted orbit is a GTO with an apogee at 36 000 km and a perigee at 300 km.

Formulas (8.73) and (8.74) give a preliminary estimate of the performance.

$$\begin{cases} m_0 = 10000 \text{ kg} \\ h_0 = 150 \text{ km} \\ v_0 = 5000 \text{ m/s} \\ \gamma_0 = 30 \text{ deg} \end{cases} \quad \text{and} \quad \begin{cases} h_p = 300 \text{ km} \\ v_p = 10155 \text{ m/s} \end{cases}$$

$$\Rightarrow \Delta V_G \approx \frac{1}{4} v_c \gamma_0^2 = 536 \text{ m/s} \quad \Rightarrow \quad m_f \approx 1445 \text{ kg} \tag{8.75}$$

The reduced problem (Section 8.3.4) is then solved, firstly for a linear thrust profile and secondly for a bilevel thrust profile by fixing three different values of the switching time $t_1$, respectively 250, 500, 750 s.

This reduced problem is a nonlinear system of two equations (final apogee and perigee) with two unknowns ($T_1$, $T_2$). For each set of values ($T_1$, $T_2$), the state equations (8.1) are integrated using the feedback control $\theta$ assessed from (8.44), until a null slope is reached. This yields the final apogee and perigee that must match the targeted values of the GTO. This system is solved using Newton's method.

Table 8.1 compares the solutions and performances obtained with the different thrust profiles.

The optimized thrust profiles are shown in Figure 8.4.

Figure 8.5 shows the altitude evolution along the optimal trajectory (for the linear profile and for the bilevel profile with a switching time of 500 s).

We can observe in Table 8.1 that the final mass is almost identical in all cases and very close to the preliminary estimate (8.75). The fact that the switching time has no influence on performance could perhaps be demonstrated analytically.

The main difference between the different solutions is the final time and the angular range between the initial position and the perigee injection. In the case of a GTO orbit, the angular range must be controlled to place the apogee above the equator (perigee argument of 0 or 180 deg). This is required to allow the GEO circularization maneuver in a single impulse (as in Figure 6.6). This example shows that the perigee argument constraint could to some extent be managed without loss of performance through an appropriate setting of the switching time.

**Table 8.1** Performances with optimized thrust profiles.

| Variable | Linear model | Bilevel model | | |
|---|---|---|---|---|
| Switching time, $t_1$ | — | 250 s | 500 s | 750 s |
| Thrust parameter, $T_1$ | 26.467 kN | 37.942 kN | 26.339 kN | 23.378 kN |
| Thrust parameter, $T_2$ | −10.976 N/s | 12.842 kN | 13.799 kN | 14.015 kN |
| Final mass, $m_f$ | **1422.5 kg** | **1422.7 kg** | **1422.5 kg** | **1422.5 kg** |
| Final time, $t_f$ | 1308.5 s | 1476.3 s | 1374.4 s | 1299.5 s |
| Angular range, $\varphi_f$ | 69.7 deg | 79.2 deg | 73.6 deg | 69.3 deg |
| Gravity losses, $\Delta v_G$ | 555 m/s | 545 m/s | 555 m/s | 558 m/s |
| Incidence losses, $\Delta v_T$ | 27 m/s | 37 m/s | 28 m/s | 25 m/s |

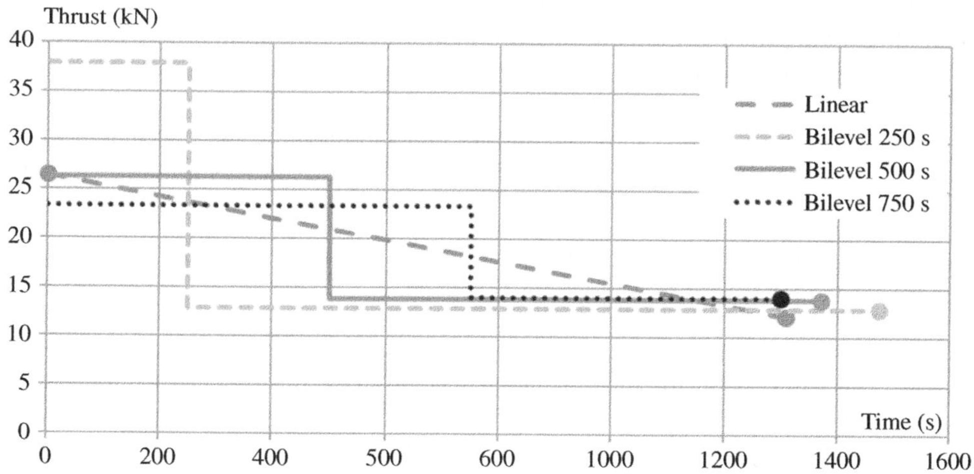


**Figure 8.4** Optimized thrust profiles.

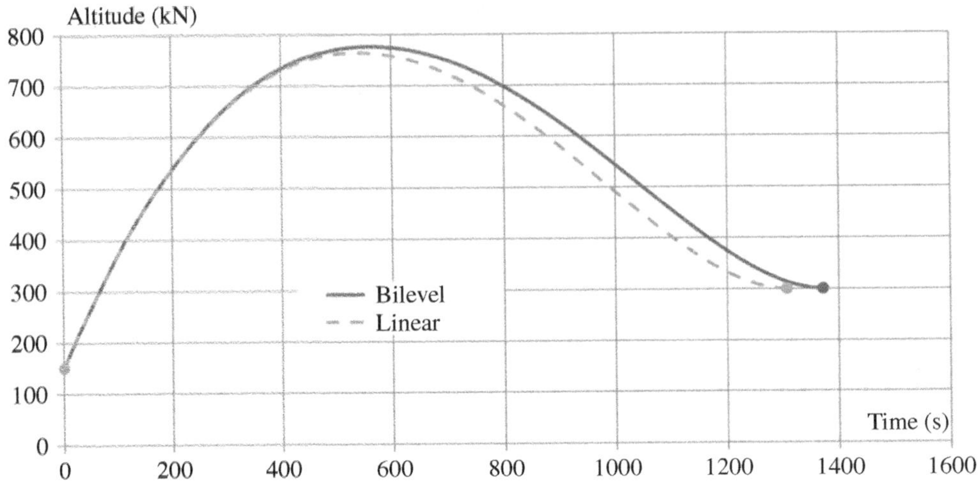


**Figure 8.5** Altitude along trajectories at optimized thrust profile.

### 8.4.2 Fixed Thrust Level

This second example deals with the classical transfer problem at fixed thrust level. The optimality conditions resulting from the PMP lead to a system of nonlinear equations (called shooting problem) of dimension eight:

- the eight unknowns are the initial costates $(\vec{p}_r, \vec{p}_v, p_m)_{t_0}$, the final time $t_f$ and the multipliers $\nu_1$ and $\nu_2$ (recalling that the trajectory is planar);
- the eight equations are the two final constraints (apogee and perigee), the five transversality conditions (8.10), and the Hamiltonian value (8.11).

The main issue is to find a good initialization of the costate. Indeed, this shooting problem is extremely sensitive to the initial costate value and Newton's method is very likely to fail if the initial guess is not sufficiently close to the solution.

The strategy applied here is to use the analytical formula (8.45) obtained in the case of an optimized thrust profile. The procedure consists of first solving the optimized linear profile problem as described in Section 8.3.4 to obtain the optimal values of the two unknowns $T_1$ and $T_2$. These values are noted $T_{1opt}$ and $T_{2opt}$.

Suppose that we now want to solve the transfer problem for some given constant thrust level $T_{max}$. A continuation method is applied, which consists in varying progressively the parameters $T_1$ and $T_2$:

- the value of $T_1$ progressively changes from $T_{1opt}$ to $T_{max}$;
- the value of $T_2$ progressively changes from $T_{2opt}$ to 0.

These variations are processed simultaneously, with a predefined step.

For each new pair $(T_1, T_2)$, the shooting problem is solved by Newton's method initialized with the previous solution (costates, final time, and multipliers). If the variations of $T_1$ and $T_2$ from one problem to the next are small enough, the initial costates allow the convergence of the new problem. In this way, the transfer problem can be solved automatically for the prescribed thrust level $T_{max}$, without any tedious search for a satisfactory initial costate. This procedure is illustrated below with an example.

Numerical Data

The initial mass is 40 000 kg and the specific impulse is 350 s, corresponding to an exhaust velocity of 3432 m/s. The apogee and perigee altitudes at the engine ignition are 400 and −5000 km, respectively, and the initial anomaly is 169 deg. These orbital parameters correspond to an altitude of 164.5 km, a velocity of 4909.8 m/s, and a slope of 19.84 deg. The Ox axis of the inertial frame is defined by the initial perigee direction. The initial longitude is thus 169 deg. The targeted orbit is a GTO with an apogee at 36 000 km and a perigee at 250 km.

We wish to solve the transfer problem for different constant thrust levels ranging from 100 to 230 kN.

The first step consists of assessing the analytical costate by (8.45). The initial pitch angle $\theta_0$ is assessed from (8.44) and the angular rate $\omega_0$ is given by (8.35).

$$\begin{cases} r_0 = 6542.632 \text{ km} \\ v_0 = 4909.82 \text{ m/s} \\ \gamma_0 = 19.8414 \text{ deg} \end{cases} \Rightarrow \theta_0 = 7.5153 \text{ deg}, \quad \omega_0 = 0.06658 \text{ deg/s} \tag{8.76}$$

The initial costate components in the inertial frame (Oxy) are assessed by (8.45).

$$\begin{cases} \vec{p}_r(t_0) = \begin{pmatrix} 1.1018 \\ -0.3691 \end{pmatrix} \text{kg/km}, \quad \vec{p}_v(t_0) = \begin{pmatrix} 0.3176 \\ 0.9482 \end{pmatrix} \text{kg/(m/s)} \\ p_m(t_0) = -0.0858 \text{ kg/kg} \end{cases} \tag{8.77}$$

The shooting method is initialized with this costate guess. The series of transfer problems is then solved for thrust levels varying from 100 to 230 kN.

Figures 8.6 and 8.7 show the final mass and velocity losses as a function of the thrust level. The best performance ($m_f = 7899$ kg) is obtained with a thrust level of 131 kN, for which the analytical costate is the solution to the shooting problem.

Figure 8.8 shows the two Hamiltonian terms $H_0$ and $TH_T$ defined in (8.37) and Figure 8.9 shows the components of the initial costate for thrust levels ranging from 100 to 230 kN. Compared to the values in (8.77), these costates have been multiplied by $-\frac{m_f}{v_e}$, in order to retrieve the usual normalization $p_0 = 1$.

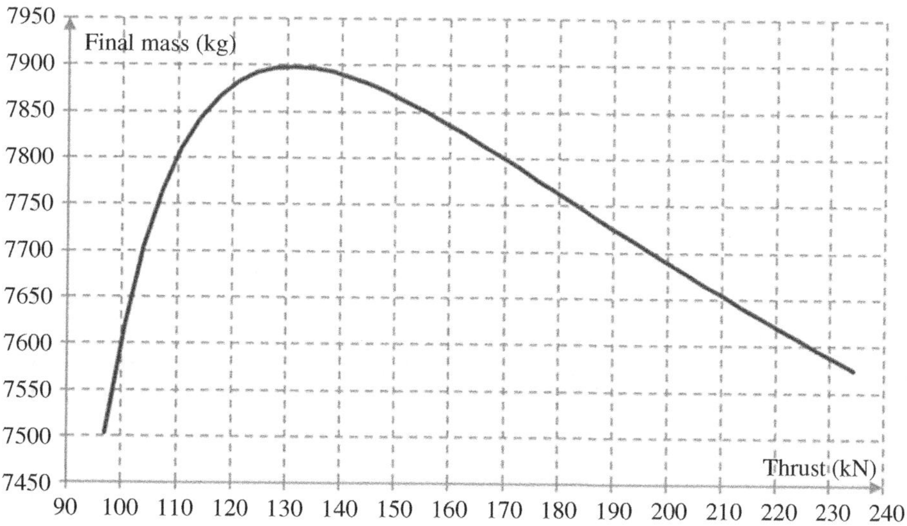


**Figure 8.6** Performance as a function of thrust level.

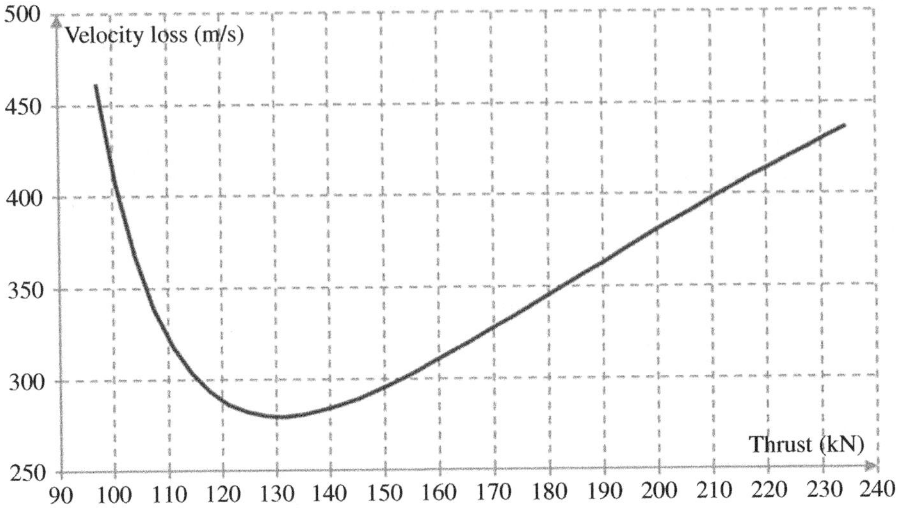


**Figure 8.7** Velocity losses as a function of thrust level.

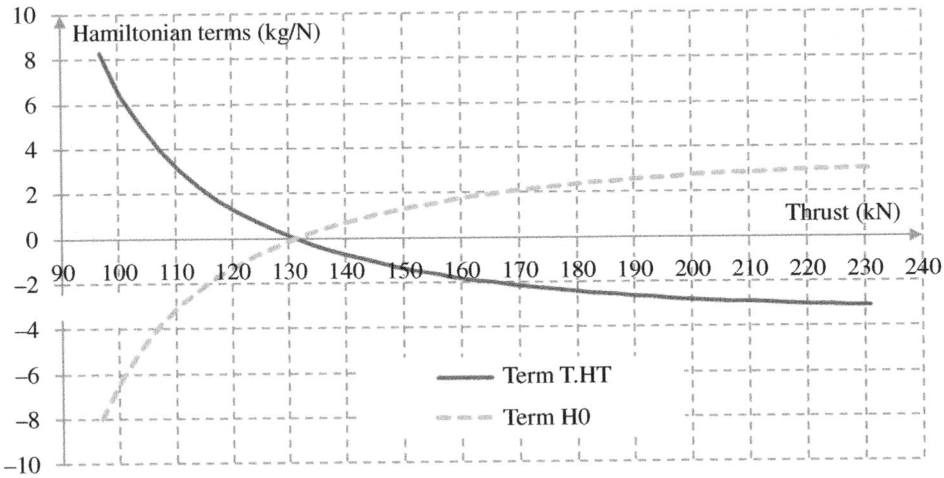


**Figure 8.8** Hamiltonian components as a function of the thrust level.

As expected, it is observed that the total Hamiltonian is zero for each thrust level and that both terms $H_0$ and $TH_T$ cancel simultaneously for the optimal thrust level of 131 kN. The components of the costate change smoothly with the thrust level, which explains that the initialization with the analytical costate (associated with the optimal thrust level of 131 kN) works well and allows the convergence of Newton's method.

Figures 8.10 and 8.11 show the evolution of the altitude and the velocity along the trajectory for the optimal thrust level of 131 kN. The altitude increases up to 450 km before descending for an injection at the perigee.

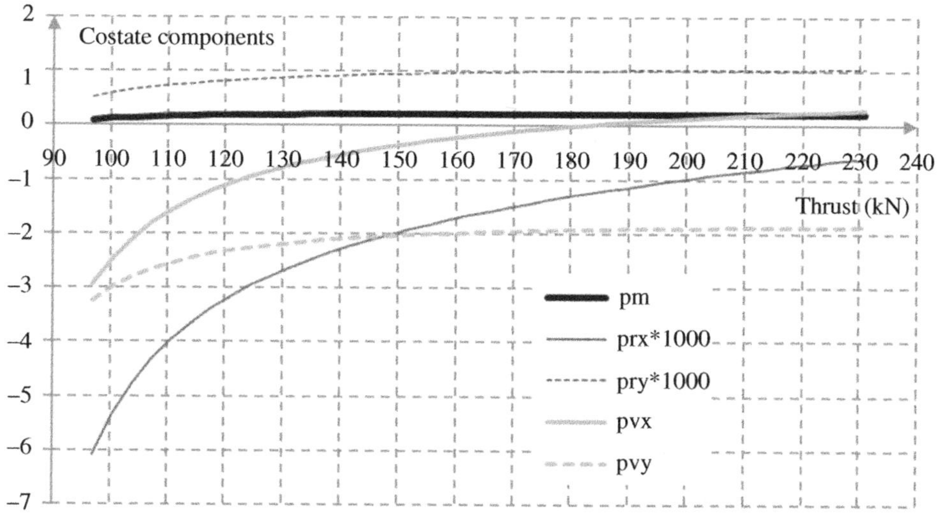


**Figure 8.9** Initial costate components as a function of the thrust level.

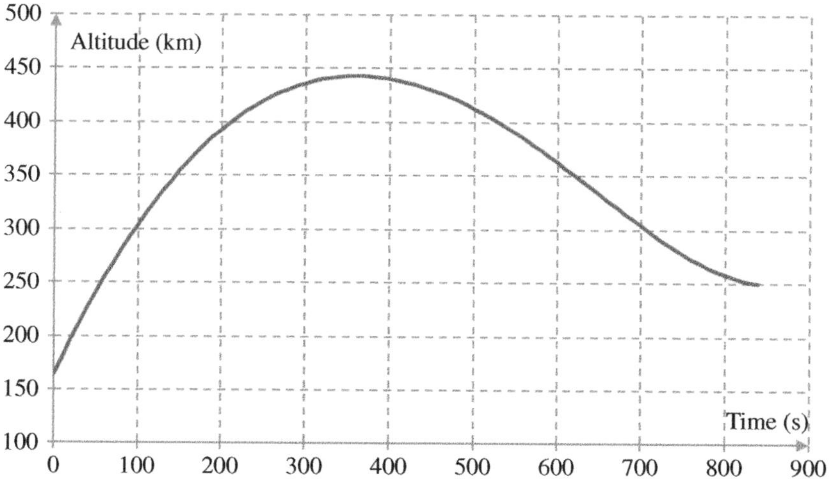


**Figure 8.10** Altitude evolution for optimal thrust level.

Figure 8.12 shows the evolution of the pitch angle and the flight path angle along the trajectory for the optimal thrust level of 131 kN. These angles cancel out the first time around 350 s, when reaching the local apogee visible in Figure 8.10. They become then negative before canceling a second time at the perigee injection.

Figure 8.13 shows the evolution of the osculating apogee and the perigee along the trajectory for the optimal thrust level of 131 kN. There is a first phase increasing the perigee, followed by a second phase increasing the apogee. The change occurs around 600 s when the velocity slope is minimal.

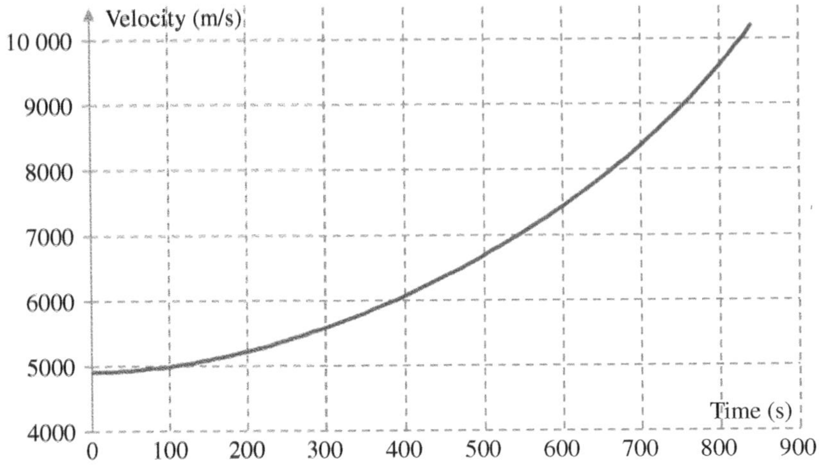


**Figure 8.11** Velocity evolution for optimal thrust level.

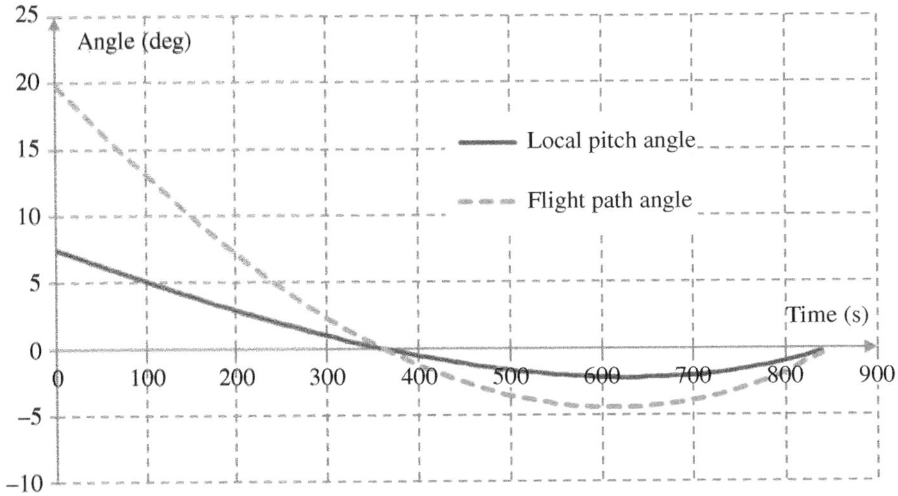


**Figure 8.12** Pitch angle and flight path angle for optimal thrust level.

## 8.5 Conclusion

### 8.5.1 The Key Points

These key points hold under the assumption of an optimized thrust profile.

- The optimal thrust direction admits a solution in state feedback control.
- The optimum injection point is the perigee of the targeted orbit reached by a descending phase.
- The costate vector has an analytical expression depending on the state.
- The analytical costate provides a good initial guess for solving other space trajectory problems.

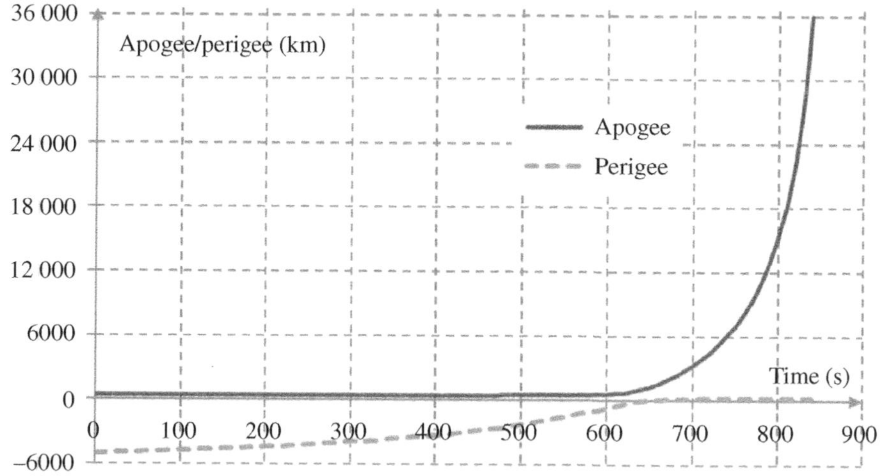


**Figure 8.13** Osculating apogee and perigee for optimal thrust level.

### 8.5.2 To Go Further

- AZIMOV D. M. – *Extremal Space Trajectories*. Butterworth-Heinemann (2018).

This recent book resumes Lawden's work (see next reference) in a much more complete way, using the tools of optimal control theory developed during the second half of the 20th century. The various kinds of rocket problems are examined (variable or constant thrust, variable or constant specific impulse) as well as the possible theoretical solutions (coast, intermediate thrust, or maximum thrust arcs).

- LAWDEN D. F. – *Optimal Trajectories for Space Navigation*. Butterworths (1963)

This book is one of the first extensive studies of optimal rockets transfers. The Pontryaguin Maximum Principle was not published yet, but many important results are demonstrated by the calculus of variations. Of great interest is Chapter 5, which is dedicated to the properties of the primer vector, which is the name given to the velocity costate. In particular, the study of singular arcs leads to properties quite similar to those established here in the case of an optimized parametric thrust profile.

- MAREC J.-P. – *Optimal Space Trajectories*. Elsevier (1979)

This book, originating from a French course in engineering schools, provides an extensive study of optimal rockets transfers.

Authorizations

- Examples in Section 8.4 are reprinted by permission of Elsevier Ltd. from "Optimal injection point for launch trajectories with parametric thrust profile" published in Acta Astronautica 151, 752–760 (2018).

# 9

# Low-Thrust Transfer

## 9.1 Introduction

Electric propulsion is becoming widespread on satellites of the 21st century. Its principle is to accelerate ionized particles by an electrostatic or electromagnetic field in order to eject them at very high velocities. The specific impulses (1000–8000 s) are significantly better than those of chemical propellants based on the combustion of propellants (300–450 s), but the low mass flow rates usually lead to thrust levels below 1 N. Electric propulsion is therefore reserved for the orbital maintenance or repositioning phases. It allows significant mass savings at the cost of more complex and longer maneuvers. The engine also requires a high level of electrical power, unlike a chemical propulsion system. Since this power is generated by solar panels, the maneuvers must take eclipses into account, during which propulsion is cut off.

This chapter deals with the transfer of an electrically propelled satellite from an injection orbit reached by a launcher. The first section formulates the optimization problem in the general case and presents some solution methods.

The second section concerns the deployment of satellites in high orbit. These satellites for television, meteorology or navigation are to be placed in circular orbits such as GEO (Geostationary Earth Orbit) equatorial at 35 786 km altitude or MEO (Medium Earth Orbit) at about 55 deg inclination and 23 000 km altitude. The launcher typically injects the satellite into an elliptical low perigee orbit such as the GTO (Geostationary Transfer Orbit) or on a LEO (Low Earth Orbit) circular orbit. These satellites represent very expensive investments and must be made operational as soon as possible. Since the transfer may take several weeks or months, the main objective is to minimize the duration of the transfer taking into account the repetition of eclipses.

The third section concerns the deployment of satellite constellations in LEO. These satellites for communications or Earth observation are to be placed in circular orbits such as LEO or SSO (Sun Synchronous Orbit) distributed on different planes covering the Earth's surface. The constellation is deployed by launching several tens or hundreds of satellites into an initial low orbit. Each satellite must then go to its final orbit, which requires an expensive plane change. The main objective is then to minimize the consumption by using the natural precession due to the Earth's flattening.

## 9.2 Problem Formulation

An orbital transfer is defined by the initial orbit, the targeted orbit, and the engine characteristics (thrust level, specific impulse and power). The goal is to achieve the transfer by minimizing either the time or the propellant consumption. This section presents the problem formulation in the general case.

### 9.2.1 Dynamics

The dynamic equations are written by considering the vehicle as a material point subject to the gravity of the main attracting body, to the engine thrust, and possibly to other perturbations such as gravity higher terms, attraction of other bodies, atmospheric friction and radiation pressure.

#### 9.2.1.1 Propulsion System

The thrust is obtained by mass ejection. Figure 9.1 shows a vehicle of mass m at rest (left), then after ejection at velocity $v_e$ of a mass dm burnt in a time dt (right).

By conservation of the total momentum, we have

$$\underset{\text{before ejection}}{0} = \underset{\text{after ejection}}{(m - dm)dv + dm(-v_e)} \tag{9.1}$$

where the exhaust velocity is directed backward.

The velocity increment received by the remaining mass m – dm in a time dt is dv. Neglecting the second-order terms, dividing by dt, and noting q as the mass flow rate, we obtain

$$m\frac{dv}{dt} = qv_e \quad \text{with } q = \frac{dm}{dt} \tag{9.2}$$

The second member is the increase in momentum due to mass ejection in vacuum.

It is called the "thrust force" since it plays the same role in the dynamic equation as the real physical forces such as weight or aerodynamics.

The thrust force in vacuum noted T is thus defined by

$$T \underset{\text{def}}{=} qv_e \tag{9.3}$$

By integrating (9.2) between $t_0$ and $t_f$, we obtain Tsiolkovski's formula (1905) linking the velocity increment to the mass consumption.

$$dv = \frac{dm}{m}v_e \;\Rightarrow\; v_f - v_0 = v_e \ln\frac{m_0}{m_f} \tag{9.4}$$

The power produced by the thrust is the kinetic energy ejected per unit time.

$$P = \frac{dEc}{dt} = \frac{1}{2}qv_e^2 = \frac{1}{2}Tv_e \tag{9.5}$$

This power is obtained from a chemical or electrical energy source.

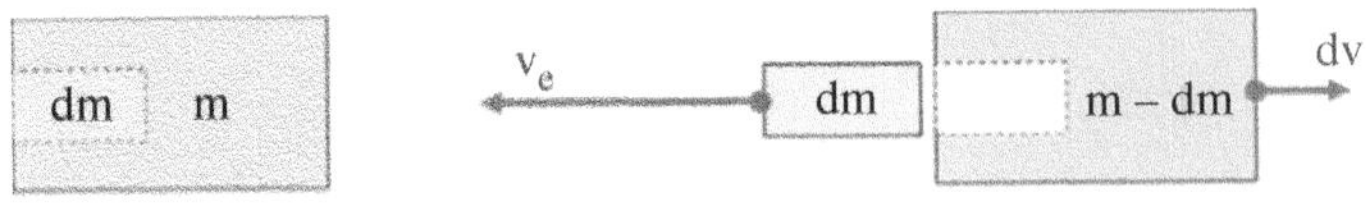


**Figure 9.1** Thrust of a rocket engine.

- A chemical propulsion engine uses the energy released by the combustion of propellants. Depending on the type of propellant (solid, storable, cryogenic), the exhaust velocities range from 2.5 to 4.5 km/s. The mass flow rate and the thrust are high, generating accelerations of about 1 g. The thrusting durations are short relative to the orbital period. The main objective when using a chemical engine is to minimize the propellant consumption.
- An electrical propulsion engine uses the energy produced by solar panels to ionize a plasma and accelerate charged particles (electrons or atoms). Depending on the system (electrostatic or electromagnetic), the exhaust velocities range from 10 to 40 km/s. The mass flow rate and the thrust are low, generating accelerations generally comprised between 0.001 and 0.01 g. An orbital transfer using electrical propulsion can then take several months. The main objective when using an electric engine is to minimize the transfer time.

#### 9.2.1.2 Motion Equations

The vehicle state can be defined either by position and velocity or by osculating orbital parameters. Let us detail both formulations.

The equations in position and velocity are derived from Newton's second law.

$$\begin{cases} \dot{\vec{r}} = \vec{v} \\ \dot{\vec{v}} = \vec{g} + \dfrac{T}{m}\vec{u} + \vec{a}_p \end{cases} \tag{9.6}$$

where $\vec{u}$ is the unit vector orientating the thrust, $\vec{g}$ is the gravity central term, and $\vec{a}_p$ represents the perturbative acceleration due to all forces other than central gravity and thrust.

These equations are based directly on the position $\vec{r}$ and velocity $\vec{v}$ vectors in Cartesian coordinates. They have a very simple form, which eases both the analysis of the optimal control problem and the implementation of numerical simulation. This motivates their use for the transfer problem addressed in Section 9.3.

On the other hand, these equations require sufficiently small time steps due to their nonlinearity. This may raise numerical issues for low-thrust transfers lasting several months. In such cases, the formulation in orbital parameters is preferable.

The equations in orbital parameters have been derived in Section 2.3.3. They express the variation of the osculating orbital parameters as functions of the radial, orthoradial, and normal components of the perturbative acceleration $\vec{a}_p$. Gauss equations (2.45) take the form.

$$\begin{cases} \dfrac{da}{dt} = \dfrac{2a^2}{h}[a_r e \sin\theta + a_\theta(1 + e\cos\theta)] \\ \dfrac{de}{dt} = \dfrac{p}{h}\left[a_r \sin\theta + a_\theta\left(\cos\theta + \dfrac{e + \cos\theta}{1 + e\cos\theta}\right)\right] \\ \dfrac{di}{dt} = a_h \dfrac{r\cos\nu}{h} \\ \dfrac{d\Omega}{dt} = a_h \dfrac{r\sin\nu}{h\sin i} \\ \dfrac{d\omega}{dt} = \dfrac{r}{he}[-a_r\cos\theta(1 + e\cos\theta) + a_\theta\sin\theta(2 + e\cos\theta)] - a_h\dfrac{r\sin\nu}{h\tan i} \\ \dfrac{dM}{dt} = n + \dfrac{b}{aeh}[a_r(p\cos\theta - 2re) - a_\theta(p + r)\sin\theta] \end{cases} \tag{9.7}$$

When the orbit is close to circular ($e = 0$) or equatorial ($i = 0$), the equinoctial orbital parameters shall be used to avoid singularities. The associated differential equations are given in (2.58).

For a low-thrust transfer, the perturbative acceleration is mainly due to the thrust and its magnitude is generally less than 0.1 m/s$^2$. The second members of Gauss equations are then close to zero, which allows a numerical integration with large time steps (up to one-tenth of the orbital period). The trajectory calculation is thus faster and more accurate than in Cartesian coordinates. The drawback is a more complex formulation and implementation of the optimal control problem.

### 9.2.2 Optimal Control Problem

The goal of an orbital transfer is to reach the targeted orbit by minimizing either the time or the propellant consumption. This formulates as an optimal control problem, whose standard form is

$$\min_{\vec{u}(t),T(t),t_f} J = \varphi(x_f, t_f) + \int_{t_0}^{t_f} L(x, u, t)dt \quad \text{s.t.} \begin{cases} \dot{x} = f(x, u, t) \\ x(t_0) = x_0 \\ \psi(x_f, t_f) = 0 \end{cases} \tag{9.8}$$

The state vector x(t) is of dimension seven. The first six components are either the position and velocity vectors, or the osculating orbital parameters, whereas the seventh component is the mass. The initial state $x(t_0)$ is given. The function f(x, u, t) represents the dynamics, either in Cartesian coordinates (9.6) or in orbital parameters (9.7). The function $\psi(x_f, t_f)$ defines the final constraints associated with the targeted orbit. The cost function J to minimize in the sum of a final cost $\varphi(x_f, t_f)$ depending only on the final state, and an integral cost depending on the path followed. The trajectory is controlled by the thrust direction $\vec{u}(t)_{t_0 \le t \le t_f}$ (unit vector), the thrust magnitude $T(t)_{t_0 \le t \le t_f}$ (bounded between 0 and $T_{max}$), and possibly the final time $t_f$ if not prescribed.

#### 9.2.2.1 Optimality Conditions

Pontryaguin Maximum Principle (PMP) (see [R9], [R38]) gives first-order necessary conditions that must be met along an optimal trajectory. These conditions are summarized hereafter. For more details, see Section 11.4.2.

The Hamiltonian of the optimal control problem (9.8) is

$$H(x, u, p, t) = L(x, u, t) + p^t f(x, u, t) \tag{9.9}$$

where p(t) is the costate vector having the same dimension as the state vector x(t).

The Hamiltonian minimization determines the optimal control.

$$\min_u H(x, u, p, t) \quad \Rightarrow \quad u_{opt}(x, p, t) \tag{9.10}$$

The costate evolves according to the differential system

$$\dot{p} = -\frac{\partial H}{\partial x}, \quad p(t_f) = \frac{\partial \varphi}{\partial x_f} + \frac{\partial \psi}{\partial x_f}\nu \tag{9.11}$$

where the final conditions (called transversality conditions) depend on the final cost and final constraint derivatives, with multipliers $\nu$.

If the final time is optimized, an additional transversality condition applies.

$$H(t_f) = -\frac{\partial \varphi}{\partial t_f} - \frac{\partial \psi}{\partial t_f}\nu \tag{9.12}$$

These conditions result in a two-point boundary value problem (initial state given, initial costate unknown, constraints on the final state and costate):

- The unknowns are the initial costate $p(t_0)$, the constraint multipliers $\nu$ and the final time $t_f$ (if free).

- The equations to satisfy are the transversality conditions (9.11) on the final costate $p(t_f)$, the final constraints $\psi(x_f, t_f) = 0$ and the final Hamiltonian value (9.12) (if the final time is free).

This nonlinear system is called a shooting problem. It will be further detailed in Section 11.4.3 with the associated numerical solution methods.

For the orbital transfer problem, a preliminary analysis is nevertheless possible by considering the formulation in Cartesian coordinates as explained in the subsequent text.

#### 9.2.2.2 Cartesian Coordinate Formulation

Let us consider the state vector defined by $x = (\vec{r}, \vec{v}, m)$. The motion equations take the form (9.6) with the mass equation derived from (9.2). For simplicity, the perturbations $\vec{a}_p$ are considered as negligible and discarded from (9.6).

$$\dot{\vec{r}} = \vec{v}, \quad \dot{\vec{v}} = \vec{g} + \frac{T}{m}\vec{u}\,, \quad \dot{m} = -\frac{T}{v_e} \tag{9.13}$$

Two possible cost functions will be considered:

- for a minimum-time transfer, the final time is free, and the cost function is $J = t_f$;
- for a minimum-consumption transfer, the final time is fixed, and the cost function is $J = -m_f$.

In both cases, the integral cost L is null and the Hamiltonian (9.9) takes the form

$$H = \vec{p}_r \cdot \dot{\vec{r}} + \vec{p}_v \cdot \dot{\vec{v}} + p_m \dot{m} = \vec{p}_r \cdot \vec{v} + \vec{p}_v \cdot \vec{g} + T\left(\frac{\vec{p}_v \cdot \vec{u}}{m} - \frac{p_m}{v_e}\right) \tag{9.14}$$

where $\vec{p}_r(t)$, $\vec{p}_v(t)$ and $p_m(t)$ denote the costate components associated with the position, the velocity and the mass, respectively.

The thrust direction $\vec{u}$ is determined by the Hamiltonian minimization

$$\min_{\vec{u}} H \quad \Rightarrow \quad \vec{u} = -\frac{\vec{p}_v}{p_v} \quad \text{with } p_v = \|\vec{p}_v\| \tag{9.15}$$

By substituting $\vec{u}$ into (9.14), the Hamiltonian becomes

$$H = \vec{p}_r \cdot \vec{v} + \vec{p}_v \cdot \vec{g} - TS \quad \text{with } S \underset{\text{def}}{=} \frac{p_v}{m} + \frac{p_m}{v_e} \tag{9.16}$$

The thrust magnitude T is determined also by the Hamiltonian minimization.

$$\min_{T} H \quad \Rightarrow \quad \begin{cases} T = T_{max} & \text{if } S > 0 \\ T = 0 & \text{if } S < 0 \end{cases} \tag{9.17}$$

The sign of the switching function S determines the thrust level, either maximal or minimal. The case of a null switching function on a time interval leads to a singular arc, whose study is more complex.

With the Hamiltonian (9.16), the costate equations (9.11) take the form

$$\begin{cases} \dot{\vec{p}}_r = -\dfrac{\partial H}{\partial \vec{r}} = -\dfrac{\partial \vec{g}}{\partial \vec{r}}\vec{p}_v \\ \dot{\vec{p}}_v = -\dfrac{\partial H}{\partial \vec{v}} = -\vec{p}_r \\ \dot{p}_m = -\dfrac{\partial H}{\partial m} = -\dfrac{T}{m^2}p_v \end{cases} \tag{9.18}$$

We observe that the mass costate $p_m$ is decreasing. Its final value is given by the transversality condition (9.11). For an orbital transfer, the final constraints $\psi(x_f, t_f)$ define the targeted orbit and they do not involve the final mass.

The final mass costate is thus

$$p_m(t_f) = \frac{\partial \phi}{\partial m_f} = \begin{cases} 0 & \text{for minimum time} \\ -1 & \text{for minimum consumption} \end{cases} \tag{9.19}$$

For a minimum-time transfer, the mass costate is always positive and so is the switching function (9.16). The thrust is therefore constantly maximal throughout the trajectory.

For a minimum-consumption transfer, the final mass costate is negative. The solution structure depends on the evolution of the switching function, and the trajectory can be composed of maximum thrust arcs and coast arcs. The number of such arcs is not known a priori. In the simplified case of a constant gravity field, it is possible to pursue the theoretical analysis and completely solve the optimal control problem, as will be done in Section 13.5.

#### 9.2.2.3 Minimum-Energy Transfer

Low-thrust transfer problems are difficult to solve because of the high sensitivity of the Hamiltonian system (9.11) coupled with large integration times. For a minimum-consumption transfer, an additional difficulty comes from the unknown solution structure with bang and coast arcs. A strategy to discover the solution structure consists of first solving the minimum-energy problem, and then passing to the minimum-consumption problem by a homotopy procedure.

The energy is the integral of the power. For an electrical engine at fixed flow rate, the exhaust velocity depends on the power by (9.5), which can be rewritten as:

$$P = \frac{1}{2} q v_e^2 = \frac{T^2}{2q} \tag{9.20}$$

and the cost function for the minimum-energy transfer is

$$J = \int_{t_0}^{t_f} P dt = \int_{t_0}^{t_f} \frac{T^2}{2q} dt \tag{9.21}$$

where the control is the thrust level T and the final time is fixed.

The Hamiltonian for the minimum-energy problem is

$$H = \frac{T^2}{2q} + \vec{p}_r \cdot \vec{v} + \vec{p}_v \cdot \vec{g} + T \left( \frac{\vec{p}_v \cdot \vec{u}}{m} - \frac{p_m}{v_e} \right) \tag{9.22}$$

The thrust direction $\vec{u}$ is still defined by (9.15). The minimization with respect to the thrust magnitude gives

$$\min_T H = \frac{T^2}{2q} + \vec{p}_r \cdot \vec{v} + \vec{p}_v \cdot \vec{g} - TS \quad \Rightarrow \quad T_{opt} = \frac{S}{q} \quad \text{with } S \underset{\text{def}}{=} \frac{p_v}{m} + \frac{p_m}{v_e} \tag{9.23}$$

This solution $T_{opt}$ is then projected within the bounds $[0; T_{max}]$.

Owing to the squared term in the energy cost function, the thrust magnitude becomes a smooth function instead of a bang function, which eases the numerical solution of the optimal control problem. This motivates the development of a homotopy procedure based on the following parameterized cost function

$$J = \int_{t_0}^{t_f} \frac{(1-\alpha)T^2 + \alpha T}{v_e} dt \tag{9.24}$$

When the parameter $\alpha$ passes from 0 to 1, the problem (9.24) passes from the minimum-energy problem (easier to solve) to the minimum-consumption problem (with unknown solution structure).

The costate equations are still (9.18) and the final mass costate is $p_m(t_f) = 0$ since the cost function (9.24) is purely integral, without final term $\varphi(x_f, t_f)$.

The optimal thrust level for the cost function (9.24) is

$$\min_T H = \frac{(1-\alpha)T^2 + \alpha T}{v_e} + \vec{p}_r \cdot \vec{v} + \vec{p}_v \cdot \vec{g} - T\left(\frac{p_v}{m} + \frac{p_m}{v_e}\right)$$
$$\Rightarrow T_\alpha = \frac{v_e}{2(1-\alpha)} S_\alpha \quad \text{with } S_\alpha \underset{\text{def}}{=} \frac{p_v}{m} + \frac{p_m - \alpha}{v_e} \tag{9.25}$$

This solution $T_\alpha$ is then projected within the bounds $[0; T_{max}]$.

The initial costate solution for a value $\alpha$ is used as initial guess to solve the next problem for the value $\alpha + \delta\alpha$. If the successive increments $\delta\alpha$ are sufficiently small, the procedure can end with the solution of the minimum-consumption problem with only bang arcs. The solution structure appears progressively as $\alpha$ passes from 0 to 1.

When $\alpha$ approaches 1, the optimal thrust level (9.25) becomes either $+\infty$ or $-\infty$ depending on the sign of the function $S_\alpha$. The projection on the bounds $[0; T_{max}]$ gives then either $T_{max}$ or 0. For $\alpha = 1$, we retrieve the switching function S defined in (9.16) for the minimum-consumption transfer, with the variable change $p'_m = p_m + 1$ on the mass costate, corresponding to the formulation with an integral cost (with $p_m(t_f) = 0$) instead of a final cost (with $p_m(t_f) = -1$).

### 9.2.3 Local Control Laws

Since the optimal control problem is hard to solve, it is interesting for preliminary studies to consider sub-optimal control laws for faster performance assessments.

From Gauss equations (9.7), it is possible to find the instantaneous thrust direction that maximizes the variation of an orbital parameter. We consider successively two approaches based on this principle: the first one deals only with apogee and perigee variations, whereas the second one due to Petropoulos deals with all orbital parameters simultaneously.

#### 9.2.3.1 Apogee and Perigee Local Control

The radius at an apsis is $r_{aps} = a(1 + se)$, with $s = \pm 1$ (apogee or perigee), and the variation is

$$\frac{dr_{aps}}{dt} = (1 + se)\frac{da}{dt} + sa\frac{de}{dt} \tag{9.26}$$

The thrust acceleration is noted f. It is directed by the angle $\beta$ with the orbital plane and the angle $\alpha$ with the orthoradial direction (normal to the vector radius in the orbital plane). The components of the thrust acceleration in the radial, orthoradial and normal directions are noted $(f_r, f_t, f_n)$.

$$f_r = f\cos\beta\sin\alpha\ ,\ f_t = f\cos\beta\cos\alpha\ ,\ f_n = f\sin\beta \tag{9.27}$$

Substituting into (9.26) the derivatives of a and e given by Gauss equations (9.7) with the components (9.27) yields the apsis variation depending on $\alpha$ and $\beta$.

$$\frac{dr_{aps}}{dt}(\alpha, \beta) = \frac{a}{h} f\cos\beta(A\sin\alpha + B\cos\alpha)$$
$$\text{with } \begin{cases} A = [2ae(1+se) + sp]\sin\theta \\ B = 2a(1+se)(1+e\cos\theta) + s[(p+r)\cos\theta + re] \end{cases} \tag{9.28}$$

The value of the in-plane angle $\alpha$ maximizing the instantaneous change in $r_{aps}$ is obtained by canceling the derivative of (9.28). This yields

$$\tan\alpha = \frac{A}{B} = \frac{(1+e\cos\theta)(e+s)\sin\theta}{(1+s\cos\theta)[2+e(\cos\theta+s)+se^2(\cos\theta-s)]} \tag{9.29}$$

This formula defines the in-plane thrust orientation depending on the current state (eccentricity and anomaly). It can be used as a state feedback control to propagate the trajectory. The optimal control problem is then simplified as follows.

The trajectory is divided into thrust arcs applying the in-plane control angle $\alpha$ given by (9.29) with $s = +1$ around the perigee and $s = -1$ around the apogee.

The switching anomaly and the out-of-plane angle $\beta$ (if there is an inclination change) are defined by polynomials depending on the time (or another variable). The coefficients of these polynomials become the optimization variables. The optimal control problem is thus reduced to a small-size optimization problem. A good initial guess of the polynomials can be easily found, and the problem is solved by standard nonlinear programming software. Although a sub-optimal control is applied, very good quality solutions can be obtained quickly, with polynomials of degree two or three at most.

#### 9.2.3.2 Petropoulos Method

The approach proposed by Petropoulos (See paper reference in Section 9.5.2) is based on a Lyapunov function, called proximity quotient. This function Q measures the remaining distance from the current osculating orbit to the targeted final orbit. The thrust direction along the trajectory is derived from a local minimization of the function Q with respect to the thrust angles. The resulting control law is called **Q-law**.

Let us define the state vector $X = (a, e, i, \Omega, \omega) = (x_k)_{k=1,\cdots,5}$, whose components are the current osculating orbital parameters on the trajectory and the state vector $X_f = (a_f, e_f, i_f, \Omega_f, \omega_f) = (x_{f,k})_{k=1,\cdots,5}$ giving the targeted orbital parameters.

The distance from the current state X to the targeted state $X_f$ is defined by

$$d(x_k, x_{f,k}) = \begin{cases} x_k - x_{f,k} & \text{for } k = 1,2,3 \quad \rightarrow \quad a, e, i \\ \text{acos}[\cos(x_k - x_{f,k})] & \text{for } k = 4,5 \quad \rightarrow \quad \Omega, \omega \end{cases} \tag{9.30}$$

The evolution of the orbital parameters under the action of the thrust is given by Gauss equations (9.7). The thrust acceleration has radial, orthoradial and normal components ($f_r$, $f_\theta$, $f_h$). By noting $f_{max}$ the acceleration modulus, $\alpha$ the in-plane angle with the radial direction and $\beta$ the out-of-plane angle (which is an alternative projection to (9.27)), these components are

$$f = \begin{pmatrix} f_r \\ f_\theta \\ f_h \end{pmatrix} = f_{max} \begin{pmatrix} \cos\beta\cos\alpha \\ \cos\beta\sin\alpha \\ \sin\beta \end{pmatrix} \tag{9.31}$$

Gauss equations (9.7) for the orbital parameters $(a, e, i, \Omega, \omega)$ take the form

$$\dot{x}_k = g_k(X, \theta, \alpha, \beta) \tag{9.32}$$

where $\theta$ is the true anomaly.

The maximal instantaneous variation of an orbital parameter $x_k$ that can be achieved at the current anomaly $\theta$ is

$$\dot{x}_{k,max} \underset{def}{=} \max_{\alpha,\beta} \dot{x}_k = \max_{\alpha,\beta} g_k(X, \theta, \alpha, \beta) \tag{9.33}$$

This maximization problem is solved analytically from the Gauss equation for each orbital parameter. The resulting formulas are given in Petropoulos's paper (See paper reference in Section 9.5.2) and are not reproduced here. These formulas yield the maximal achievable variations $\dot{x}_{k,max}$ as functions of the true anomaly. Each maximal variation is associated with a specific thrust direction. In order to account for the simultaneous variation of all orbital parameters, the function Q is defined as the following weighted sum.

$$Q(X, \theta, \alpha, \beta) = (1 + w_p S_p) \sum_{k=1}^{5} w_k S_k \left( \frac{d(x_k, x_{f,k})}{\dot{x}_{k,max}} \right)^2 \tag{9.34}$$

The scalar weights $w_k$, to be chosen freely, are associated with the orbital parameters $x_k$. The penalty functions $S_k$ are defined by

$$S_1 = \sqrt{1 + \left( \frac{a - a_f}{3a_f} \right)^4}, \quad S_{2,3,4,5} = 1 \tag{9.35}$$

The purpose of the penalty $S_1$ is to prevent an excessive increase of the semi-major when minimizing the Q function. Indeed, large semi-major axis values would increase the variations $\dot{x}_{k,max}$ resulting from Gauss equations and help minimizing the function Q, to the detriment of reducing the distance to the targeted orbit.

An additional weight $w_p$ and a penalty function $S_p$ are defined for the perigee, in order to prevent too low altitudes on intermediate orbits

$$S_p = e^{c\left(1 - \frac{r_p}{r_{min}}\right)} \tag{9.36}$$

where $r_{min}$ is the minimal allowed perigee radius and c is a constant.

The function Q measuring the distance to the target is used as Lyapunov function to define a locally optimal control. The principle is to select the control (here the thrust direction) that minimizes the derivative $\dot{Q}(X, \theta, \alpha, \beta)$ at the current state X and anomaly θ.

$$\min_{\alpha,\beta} \dot{Q}(X, \theta, \alpha, \beta) \quad \rightarrow \quad (\alpha, \beta)_{opt} \tag{9.37}$$

In this way, the Q function is driven to zero as quickly as possible, in order to reach the targeted orbit as soon as possible.

To solve the minimization problem (9.37), we retrieve Gauss equations (9.7)

$$\begin{pmatrix} \frac{da}{dt} \\ \frac{de}{dt} \\ \frac{di}{dt} \\ \frac{d\Omega}{dt} \\ \frac{d\omega}{dt} \end{pmatrix} = \begin{pmatrix} \frac{2a^2}{h} e \sin\theta & \frac{2a^2}{h}(1 + e\cos\theta) & 0 \\ \frac{p}{h}\sin\theta & \frac{p}{h}\left(\cos\theta + \frac{e + \cos\theta}{1 + e\cos\theta}\right) & 0 \\ 0 & 0 & \frac{r\cos\nu}{h} \\ 0 & 0 & \frac{r\sin\nu}{h\sin i} \\ -\frac{r}{he}\cos\theta(1 + e\cos\theta) & \frac{r}{he}\sin\theta(2 + e\cos\theta) & -\frac{r^2\sin\nu}{h^2 e\tan i} \end{pmatrix} f_n \begin{pmatrix} \cos\beta\cos\alpha \\ \cos\beta\sin\alpha \\ \sin\beta \end{pmatrix} \tag{9.38}$$

and rewrite them in matrix form as

$$\dot{X} = Gf \tag{9.39}$$

where the $5 \times 3$ matrix G depends only on X and θ, and f is the thrust acceleration vector defined in (9.31), which depends on the angles α and β.

The time derivative $\dot{Q}(X, \theta, \alpha, \beta)$ is then expanded as:

$$\dot{Q}(X, \theta, \alpha, \beta) = \left(\frac{dQ}{dX}\right)^T \dot{X} = Q_X^T G f = (G^T Q_X)^T f \tag{9.40}$$

where the $5 \times 1$ vector $\frac{dQ}{dX} \underset{noted}{=} Q_X$ is the gradient of Q with respect to the state X.

The derivative $\dot{Q}(X, \theta, \alpha, \beta)$ is thus expressed as a scalar product of the thrust acceleration vector f with a vector $G^T Q_X$ that does not depend on $\alpha$ and $\beta$.

Since f is of a given modulus $f_{max}$, the solution of (9.37) consists in orientating f opposite to the vector $G^T Q_X$.

$$\min_{\alpha,\beta} \dot{Q} = (G^T Q_X)^T f \quad \Rightarrow \quad f_{opt} = -f_{max} \frac{G^T Q_X}{\|G^T Q_X\|} \tag{9.41}$$

This formula yields the optimal thrust direction at the current state and anomaly. The analytical expression is quite lengthy, since it involves the gradient of Q with respect to the state X, with the maximal variations $\dot{x}_{k,max}$ assessed by (9.33) depending themselves on the state and anomaly. These expressions can be derived by symbolic calculus.

Applying the optimal thrust direction given by (9.41) yields the minimum derivative of Q noted $\dot{Q}_m$.

$$\dot{Q}_m = (G^T Q_X)^T f_{opt} = -f_{max} \|G^T Q_X\| \tag{9.42}$$

This minimum value depends on the current anomaly $\theta$. However, constantly thrusting throughout the trajectory is not efficient, and it is interesting to cut off the engine on orbit parts where thrusting is inefficient. For that purpose, the minimum value $\dot{Q}_m(\theta)$ is assessed over one revolution for anomalies ranging from 0 to 360 deg assuming that the orbital parameters X keep their current values. This minimum value noted $\dot{Q}_{min}$ depends only on the orbital parameters X.

$$\dot{Q}_{min} = \min_{\theta} \dot{Q}_m(\theta) \tag{9.43}$$

The thrust efficiency at the current anomaly is measured by the ratio $\eta(\theta)$.

$$\eta(\theta) = \frac{\dot{Q}_m(\theta)}{\dot{Q}_{min}} \tag{9.44}$$

This ratio ranges from 0 to 1 and reaches its maximum when the anomaly is the solution of (9.43). By fixing a minimum efficiency threshold $\eta_{min}$, it is possible to define the thrust strategy along the trajectory as follows.

$$f = \begin{cases} f_{max} & \text{if } \eta(\theta) \geq \eta_{min} \\ 0 & \text{if } \eta(\theta) < \eta_{min} \end{cases} \tag{9.45}$$

The value of $\eta_{min}$ is chosen according to the transfer objective.

- Setting $\eta_{min} = 0$ leads to constant thrusting, yielding a minimum-time transfer.
- Setting $\eta_{min} = 1$ leads to impulsive thrusting with one impulse per revolution, since it allows to thrust only when the anomaly matches the solution of (9.43).
  This yields a minimum-consumption transfer.
- Intermediate values of $\eta_{min}$ yield a compromise between the transfer time and the total consumption.

Since the Q-law method calculates the control within the trajectory simulation, no optimization is required and values of $\eta_{min}$ can be swept to select the best solution.

This method has proved quite efficient and the results match those obtained by optimal control methods. The main settings are the weights $w_k$ on the targeted orbital parameters and the cut-off threshold $\eta_{min}$. These settings can be optimized by particle swarm or genetic algorithms. They can also be defined as time-varying (instead of constant) in order to improve the transfer performance.

### 9.2.4 Edelbaum's Solution

The analytical solution of Edelbaum (1962) gives the minimum-time transfer between circular noncoplanar orbits [R12]. This solution is based on a simplified dynamic model. The orbit is constantly circular, the satellite is subject to the central gravity in $\mu/r^2$, and the thrust acceleration noted f is assumed to be constant.

The in-plane angle $\alpha$ (between the velocity and the thrust direction) is constantly null. The out-of-plane angle $\beta$ (between the thrust direction and the orbital plane is constant over each half-revolution with a sign change at the antinodes, which are the points of extreme latitudes located at an angular distance $\nu = \omega + \theta = \pm\frac{\pi}{2}$ from the nodes. The value of $\beta$ changes at each revolution.

The average circular orbit is defined by its velocity V and its inclination I. The evolution of V and I are obtained by averaging Gauss equations over one period.

For a circular orbit of radius a, the relations $v = \sqrt{\frac{\mu}{a}}$ and $n = \frac{d\nu}{dt} = \sqrt{\frac{\mu}{a^3}}$ allow to reduce Gauss equations (9.7) to the form

$$\frac{dv}{dt} = -f\cos\beta, \quad \frac{di}{d\nu} = \frac{\cos\nu}{nv} f\sin\beta \tag{9.46}$$

Averaging over one period with a sign change of $\beta$ when $\nu = \pm\frac{\pi}{2}$ yields

$$\dot{V} = -f\cos\beta, \quad \dot{I} = \frac{2f}{\pi V}\sin\beta \tag{9.47}$$

The variables in uppercase (V, I) represent the averaged values.

Edelbaum's problem consists of finding the control law $\beta(t)$ minimizing the transfer time $t_f$ from the initial orbit $(V_0, I_0)$ to the final orbit $(V_f, I_f)$.

The optimal control problem is formulated as:

$$\min_{\beta(t),t_f} J = t_f \quad \text{s.t.} \quad \begin{cases} \dot{V} = -f\cos\beta, & V(t_f) = V_f \\ \dot{I} = \dfrac{2f}{\pi V}\sin\beta, & I(t_f) = I_f \end{cases} \tag{9.48}$$

Since this optimal control problem is autonomous, the Hamiltonian is constant and equal to its final value given by (9.12).

$$H = -p_V f\cos\beta + p_I \frac{2f}{\pi V}\sin\beta = -1 \tag{9.49}$$

The maximization of H with respect to $\beta$ (unbounded) gives the condition

$$\frac{\partial H}{\partial \beta} = p_V f\sin\beta + p_I \frac{2f}{\pi V}\cos\beta = 0 \tag{9.50}$$

Combining (9.49) and (9.50) gives the costates $p_V$ and $p_I$.

$$p_V = \frac{\cos\beta}{f}, \quad p_I = -\frac{\pi V\sin\beta}{2f} \tag{9.51}$$

The costate equations associated with the Hamiltonian (9.49) are

$$\dot{p}_V = p_I \frac{2f \sin \beta}{\pi V^2}, \quad \dot{p}_I = 0 \tag{9.52}$$

The costate $p_I$ is constant, and its expression (9.51) gives the relation

$$V \sin \beta = \text{Cte} = V_0 \sin \beta_0 \tag{9.53}$$

Deriving this relation gives the differential equation followed by β

$$\dot{\beta} = \frac{f \sin^2 \beta}{V_0 \sin \beta_0} \tag{9.54}$$

which is integrated into

$$\tan \beta = \frac{V_0 \sin \beta_0}{V_0 \cos \beta_0 - ft} \tag{9.55}$$

The expression of V is obtained from (9.53) by substituting β given by (9.55).

$$V = \sqrt{V_0^2 + f^2 t^2 - 2\, ft\, V_0 \cos \beta_0} \tag{9.56}$$

The expression of I is obtained by integrating its differential equation (9.48) with V given by (9.53) and $\dot{\beta}$ by (9.54).

$$\dot{I} = \frac{2}{\pi} \frac{f \sin^2 \beta}{V_0 \sin \beta_0} = \frac{2}{\pi} \dot{\beta} \quad \Rightarrow \quad I - I_0 = \frac{2}{\pi}(\beta - \beta_0) \tag{9.57}$$

The formulas for β (9.55), V (9.56), and I (9.57) depend on the initial value $\beta_0$.

This value is obtained by substituting into (9.55) the time t calculated from (9.56) and β calculated from (9.57). This yields

$$\tan \beta_0 = \frac{\sin\left(\frac{\pi}{2}(I - I_0)\right)}{\frac{V_0}{V} - \cos\left(\frac{\pi}{2}(I - I_0)\right)} \tag{9.58}$$

This formula gives $\beta_0$ as a function of the initial orbit $(V_0, I_0)$ and the final orbit $(V_f, I_f)$ of the transfer. The trajectory and control are then fully determined by (9.55)–(9.57). The final time $t_f$ is obtained from (9.55) and (9.57) for $I = I_f$.

$$t_f = \frac{V_0}{f}\left[\cos \beta_0 - \sin \beta_0 / \tan\left(\frac{\pi}{2}(I_f - I_0) + \beta_0\right)\right] \tag{9.59}$$

Edelbaum's solution makes it possible to evaluate analytically a minimum-time transfer between circular orbits, assuming constant acceleration. However, a real engine operates at a constant thrust level instead of a constant acceleration level.

In order to obtain a more realistic performance assessment, Edelbaum's solution is calculated twice.

- Edelbaum's first solution considers the acceleration at the engine ignition, obtained by the ratio of the engine thrust by the vehicle initial mass.

  This yields a first estimation of the transfer time $t_f$, which allows to calculate successively the resulting final mass $m_f$, the total velocity impulse $\Delta v$ and the actual mean acceleration $f_{mean}$ during the transfer.

$$m_f = m_0 - \frac{T}{v_e} t_f \quad \rightarrow \quad \Delta v = v_e \ln \frac{m_0}{m_f} \quad \rightarrow \quad f_{mean} = \frac{\Delta v}{t_f} \tag{9.60}$$

- Edelbaum's second solution considers the mean acceleration $f_{mean}$. This updated solution is more representative of the engine's actual performance.
  If necessary, the process is repeated until the assumed acceleration matches the calculated mean acceleration. Generally, one iteration is sufficient.

With this two-step method, Edelbaum's formulas provide good-quality solutions that can be used for preliminary studies. This analytical solution will be used in Section 9.4 as an initial guess for the optimal control problem and in Chapter 10 for the planning of debris removal missions.

## 9.3 Transfers to the Geostationary Orbit

This section deals with transfers in minimum time toward the geostationary orbit taking into account eclipses.

### 9.3.1 Dynamic Model

#### 9.3.1.1 Reference Frame and Sun Motion

We place ourselves in an inertial geocentric frame (OXYZ) called ECI (Earth Centered Inertial) with axes X and Z pointing respectively to the vernal point V and to the North Pole. The relative motion of the Sun around the Earth is a circular orbit in the ecliptic plane, whose inclination (called obliquity) is $i_s = 23.45$ deg.

The ecliptic and equatorial planes intersect along OX (Figure 9.2), which defines the vernal point V. The Sun anomaly $\alpha_s$ is measured from the X axis. The values $\alpha_s = 0, 90, 180$, and 270 deg correspond respectively to the spring equinox, the summer solstice, the autumn equinox, and the winter solstice in the northern hemisphere. The initial anomaly $\alpha_0$ depends on the start day of the transfer.

The Sun anomaly $\alpha_s$ evolves linearly with the angular rate $\omega_s = 360$ deg/year.

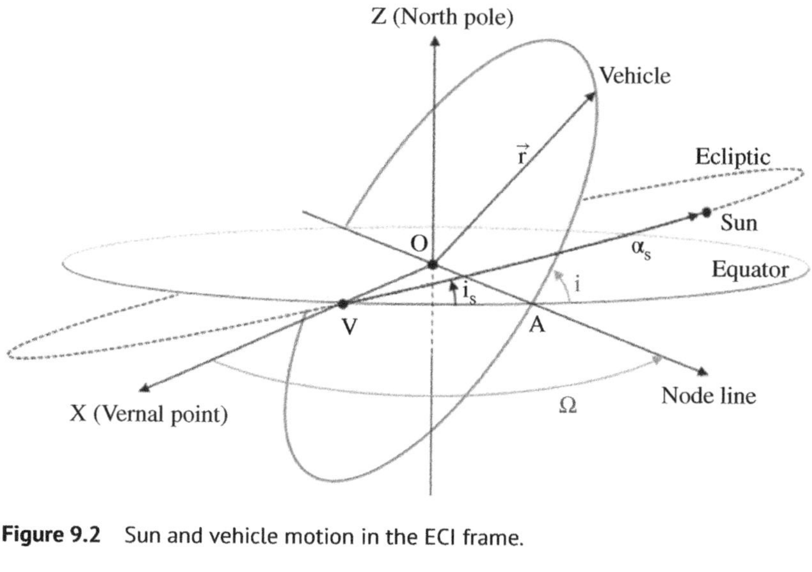


**Figure 9.2** Sun and vehicle motion in the ECI frame.

The direction of the Sun in ECI is given by the unit vector $\vec{e}_s(t)$.

$$\vec{e}_s(t) = \begin{pmatrix} \cos\alpha_s(t) \\ \sin\alpha_s(t)\cos i_s \\ \sin\alpha_s(t)\sin i_s \end{pmatrix} \quad \text{with} \quad \begin{cases} \alpha_s(t) = \alpha_0 + \omega_s t \\ i_s = 23.45 \text{ deg} \\ \omega_s = 0.986 \text{ deg/day} \end{cases} \tag{9.61}$$

The direction of the Sun determines the shadow zone of the Earth.

#### 9.3.1.2 Vehicle Motion

The vehicle is a material point of position $\vec{r}(t)$, velocity $\vec{v}(t)$ and mass $m(t)$, subject to the Earth central gravity $\vec{g}(\vec{r})$ and to the engine thrust $\vec{T}(t)$. All other forces (gravitational perturbations, Sun and Moon attraction, radiation pressure and drag) are neglected.

The thrust is directed along to the unit vector $\vec{u}(t)$, which can be freely oriented.

The thrust level T and the exhaust velocity $v_e$ are constant. The mass flow rate is the ratio $T/v_e$. During the eclipses, the engine is switched off to spare the battery power.

The motion equations in ECI are as follows (the time dependency is omitted to simplify the notation).

$$\begin{cases} \dot{\vec{r}} = \vec{v} \\ \dot{\vec{v}} = \vec{g}(\vec{r}) + \dfrac{\varepsilon T}{m}\vec{u} \\ \dot{m} = -\dfrac{\varepsilon T}{v_e} \end{cases} \quad \text{with} \quad \begin{cases} \varepsilon = 1 \text{ in light} \\ \varepsilon = 0 \text{ in shadow} \end{cases} \tag{9.62}$$

A dynamic discontinuity occurs at each eclipse entry or exit, leading to a control discontinuity for the optimal control problem formulated in Section 9.3.2.

#### 9.3.1.3 Eclipses

The eclipses depend on the position of the vehicle $\vec{r}(t)$ and the direction of the Sun $\vec{e}_s(t)$ (Figure 9.3). The Earth shadow is modelled as a cylinder having the radius of the Earth $R_E$.

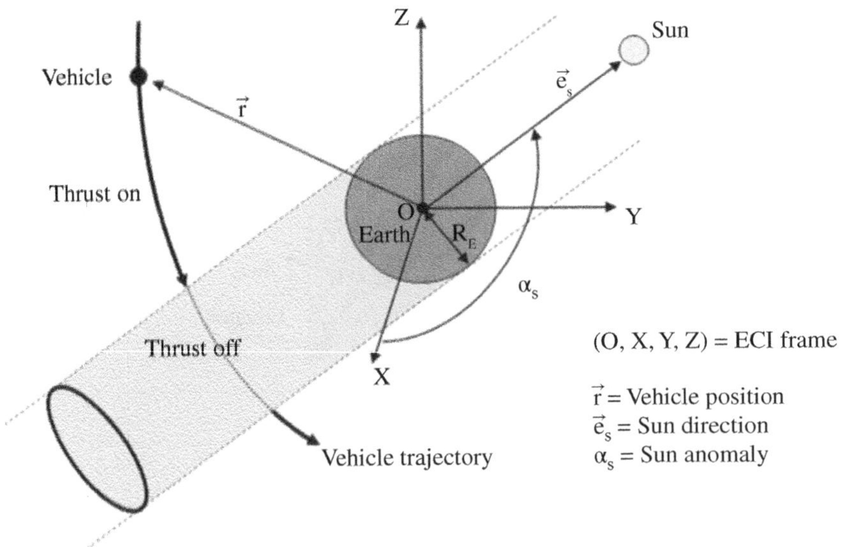


**Figure 9.3** Eclipse conditions.

A more accurate model would be a cone centered on the Sun. The difference between these two models (angle deviation of about $2.10^{-3}$ deg) is not significant for the low-thrust transfer problem.

An eclipse occurs whenever the following two conditions are met.

$$\|\vec{e}_s(t) \wedge \vec{r}(t)\| \leq R_E \text{ and } \vec{e}_s(t) \cdot \vec{r}(t) \leq 0 \tag{9.63}$$

The first condition expresses that the vehicle is inside the cylinder, the second condition verifies that the vehicle and the Sun are opposite the Earth.

### 9.3.2 Optimal Control Problem

The problem is formulated by taking as state variables $\vec{r}(t)$, $\vec{v}(t)$ and $m(t)$ and as control variables the thrust direction $\vec{u}(t)$ and the final time $t_f$. The initial state and the final orbit are fixed. For a transfer with many revolutions, the arrival anomaly on the final orbit has a negligible influence on the transfer time. By fixing it (for example, to zero), the final position and velocity become fully determined. This removes the associated transversality conditions, which will significantly simplify the resolution of the optimal control problem.

The minimum-time transfer problem is then formulated as

$$\min_{\vec{u}(t),t_f} J = t_f \text{ with } \begin{cases} \dot{\vec{r}} = \vec{v} \\ \dot{\vec{v}} = \vec{g}(\vec{r}) + \dfrac{\varepsilon T}{m}\vec{u} \\ \dot{m} = -\dfrac{\varepsilon T}{v_e} \end{cases} \text{ and } \begin{cases} \vec{r}(t_0) = \vec{r}_0 \\ \vec{v}(t_0) = \vec{v}_0 \\ m(t_0) = m_0 \\ \begin{cases} \vec{r}(t_f) = \vec{r}_f \\ \vec{v}(t_f) = \vec{v}_f \end{cases} \end{cases} \tag{9.64}$$

The dynamics is discontinuous at each eclipse entry and exit. The number of such discontinuities is unknown a priori because it depends on the number of revolutions made during the transfer and on the Sun's position at each revolution. The problem (9.64) is not autonomous, as it depends on the initial date defining the Sun anomaly $\alpha_0$, as well as the orientation of the initial orbit in the ECI frame.

#### 9.3.2.1 Optimality Conditions

To solve the optimal control problem (9.64), the PMP (see [R9], [R38]) is applied by introducing the costates $\vec{p}_r(t)$, $\vec{p}_v(t)$ and $p_m(t)$ associated respectively with the position, the velocity and the mass.

The Hamiltonian is defined by

$$H = \vec{p}_r \cdot \dot{\vec{r}} + \vec{p}_v \cdot \dot{\vec{v}} + p_m\dot{m} = \vec{p}_r \cdot \vec{v} + \vec{p}_v \cdot \vec{g} + \varepsilon T \left( \frac{\vec{p}_v \cdot \vec{u}}{m} - \frac{p_m}{v_e} \right) \tag{9.65}$$

The PMP provides the following set of first-order necessary conditions that must be met by any optimal trajectory.

- Minimizing the Hamiltonian with respect to the control $\vec{u}(t)$ gives the thrust direction opposite to the velocity costate.

$$\min_{\vec{u}} H \Rightarrow \vec{u} = -\frac{\vec{p}_v}{p_v} \text{ with } p_v = \|\vec{p}_v\| \tag{9.66}$$

With this control, the Hamiltonian (9.65) becomes

$$H = \vec{p}_r \cdot \vec{v} + \vec{p}_v \cdot \vec{g} - \varepsilon T S \text{ with } S \underset{\text{def}}{=} \frac{p_v}{m} + \frac{p_m}{v_e} \tag{9.67}$$

- The costates follow the following differential equations with transversality conditions at the final time. With the final position and velocity imposed, the corresponding final costates are free.

$$\begin{cases} \dot{\vec{p}}_r = -\dfrac{\partial H}{\partial \vec{r}} = -\dfrac{\partial \vec{g}}{\partial \vec{r}}\vec{p}_v, & \vec{p}_r(t_f) \text{ free} \\ \dot{\vec{p}}_v = -\dfrac{\partial H}{\partial \vec{v}} = -\vec{p}_r, & \vec{p}_v(t_f) \text{ free} \\ \dot{p}_m = -\dfrac{\partial H}{\partial m} = -\dfrac{\varepsilon T}{m^2}p_v, & p_m(t_f) = \dfrac{\partial J}{\partial m_m} = 0 \end{cases} \tag{9.68}$$

- The final Hamiltonian satisfies the transversality condition

$$H(t_f) = -\frac{\partial J}{\partial t_f} = -1 \tag{9.69}$$

- The dynamic discontinuities are associated with the lightening function $\psi_s$ defined by

$$\psi_s(\vec{r}, t) = \|\vec{e}_s(t) \wedge \vec{r}(t)\|^2 - R_E^2 \tag{9.70}$$

This function, defined from (9.63), changes sign at the boundary of the shadow cylinder. A discontinuity occurs whenever $\psi_s$ cancels with the vehicle and the Sun located in opposite directions with respect to the Earth. Each time $t_d$ of eclipse entry or exit thus satisfies the following interior point constraint.

$$\psi_s[\vec{r}(t_d), t_d] = \|\vec{e}_s(t_d) \wedge \vec{r}(t_d)\|^2 - R_E^2 = 0 \tag{9.71}$$

- The costate and Hamiltonian discontinuities at $t_d$ depend on the constraint derivatives and an associated unknown multiplier $\nu$ (see [R6], [R9], [R21]).

$$\begin{cases} \Delta\vec{p}_r(t_d) = -\nu\dfrac{\partial \psi_s}{\partial \vec{r}} & \\ \Delta\vec{p}_v(t_d) = -\nu\dfrac{\partial \psi_s}{\partial \vec{v}} & \Rightarrow \Delta\vec{p}_v(t_d) = \vec{0} \\ \Delta p_m(t_d) = -\nu\dfrac{\partial \psi_s}{\partial m} & \Rightarrow \Delta p_m(t_d) = 0 \\ \Delta H(t_d) = \nu\dfrac{\partial \psi_s}{\partial t} & \end{cases} \tag{9.72}$$

#### 9.3.2.2 Multiplier Calculation

The multiplier $\nu$ is part of the problem unknowns, as well as the initial costates and the final time. In the present case, an explicit expression can be derived that will simplify the problem resolution.

For that purpose, the Hamiltonian (9.67) is expressed just before and just after the discontinuity, at times noted $t_d^-$ and $t_d^+$. The state is continuous and the only discontinuities concern the coefficient $\varepsilon$, which passes from 1 to 0 (eclipse entry) or from 0 to 1 (eclipse exit), and the position costate $\vec{p}_r$. Indeed, the constraint $\psi_s$ (9.71) does not depend on the velocity or the mass, so that the velocity and mass costate discontinuities in (9.72) are null.

We have thus

$$\begin{cases} H(t_d^-) = \vec{p}_r(t_d^-) \cdot \vec{v} + \vec{p}_v \cdot \vec{g} - \varepsilon(t_d^-)\,TS \\ H(t_d^+) = \vec{p}_r(t_d^+) \cdot \vec{v} + \vec{p}_v \cdot \vec{g} - \varepsilon(t_d^+)\,TS \end{cases} \tag{9.73}$$

The member-to-member subtraction of these two equations gives a relation between the discontinuities of the Hamiltonian, of the costate $\vec{p}_r$ and of $\varepsilon$.

$$\Delta H(t_d) = \Delta\vec{p}_r(t_d)\cdot\vec{v} - \Delta\varepsilon(t_d)TS \tag{9.74}$$

Using (9.72) to substitute $\Delta H$ and $\Delta\vec{p}_r$ gives the expression of the multiplier $\nu$.

$$\nu = -\frac{TS\Delta\varepsilon}{\dot{\psi}_s} \quad \text{with } \dot{\psi}_s = \frac{\partial\psi_s}{\partial t} + \frac{\partial\psi_s}{\partial\vec{r}}\cdot\vec{v} \tag{9.75}$$

where $\dot{\psi}_s$ is the total derivative of the lightening function $\psi_s$ along the vehicle trajectory and $\Delta\varepsilon = -1$ (respectively $+1$) at the shadow entry (respectively exit).

The total derivative $\dot{\psi}_s$ is calculated from (9.70) by applying the derivation rules of the vector product and the scalar product.

$$\dot{\psi}_s = \frac{d}{dt}\|\vec{e}_s \wedge \vec{r}\|^2 = 2(\vec{e}_s \wedge \vec{r})\cdot(\dot{\vec{e}}_s \wedge \vec{r} + \vec{e}_s \wedge \vec{v}) \tag{9.76}$$

In this expression, $\vec{r}$ and $\vec{v}$ are the position and velocity of the vehicle at time $t_d$ and $\vec{e}_s$ is the direction of the Sun whose derivative is obtained from (9.61).

$$\dot{\vec{e}}_s(t) = \omega_s \begin{pmatrix} -\sin\alpha_s \\ \cos\alpha_s \cos i_s \\ \cos\alpha_s \sin i_s \end{pmatrix} \tag{9.77}$$

The formulas (9.75)–(9.77) allow a direct calculation of the costate discontinuities (9.72) during the propagation of the differential equations (9.68). This necessitates that the eclipse entries and exits are detected during the numerical integration. The optimal trajectory is then the solution of the following shooting problem.

#### 9.3.2.3 Shooting Problem

The shooting problem is formulated as follows: find the initial costates $\vec{p}_r(t_0)$, $\vec{p}_v(t_0)$, and $p_m(t_0)$ and the final time $t_f$ such that the final constraints (9.64) on $(\vec{r}_f, \vec{v}_f)$ and the transversality conditions (9.68) and (9.69) on $p_m(t_f)$ and $H(t_f)$ are satisfied.

This problem is a nonlinear system of dimension eight. The relation between the unknowns associated with the initial time and the equations expressed at the final time requires the numerical integration of the differential equations of the state (9.64) and the costate (9.68) with the optimal control given by (9.66). This system has the intrinsic characteristic of being extremely sensitive to the initial costate.

### 9.3.3 Solution Method

Many numerical methods can be used to solve the optimal control problem (9.64) [R3], [R6], [R9], [R13]. Direct methods discretize the control and the state to formulate a large finite-dimension optimization problem. These methods that do not use the PMP have the advantage of manipulating only "physical" variables (control, state). They are easy to initialize and handle, but their convergence can be slow and imprecise. Indirect methods exploit the PMP by trying to solve the shooting problem. These methods are fast and accurate, but the high sensitivity to the initial costate makes them very difficult to apply.

For a transfer with eclipses, it is also necessary to accurately detect the eclipse entries and exits during the numerical integration and possibly regulate the discontinuities to make them differentiable. These numerical difficulties are accentuated by propagation times ranging from several days to several months.

The method presented here bypasses these numerical difficulties by relying on three elements: a parametric costate guess, the reformulation into a constraint-free minimization problem, and the use of a gradient-free algorithm.

#### 9.3.3.1 Costate Parametric Guess

The first element of the solution method is the guess of an initial costate sufficiently close to the solution. The analytical solution obtained in Chapter 8 for coplanar transfers at an optimized thrust profile is used for this purpose. This analytical costate (8.45) is of the form

$$\vec{p}_r(t_0) = \omega_0 \begin{pmatrix} -\cos\theta_0 \\ \sin\theta_0 \end{pmatrix}, \quad \vec{p}_v(t_0) = -\begin{pmatrix} \sin\theta_0 \\ \cos\theta_0 \end{pmatrix} \tag{9.78}$$

In these formulas, $\theta_0$ is the pitch angle (angle between the local horizontal and the thrust direction) and $\omega_0$ is the angular rate of the thrust direction in ECI.

These variables depend on the initial conditions (radius $r_0$, velocity $v_0$, and local slope $\gamma_0$) by equations (8.29) and (8.35) recalled here.

$$\sin(\theta_0 - \gamma_0) = \frac{v_{c0}}{v_0} \frac{\sin\theta_0}{\sqrt{1 - 3\sin^2\theta_0}} \quad \text{with } v_{c0} = \sqrt{\frac{\mu}{r_0}}$$

$$\text{and } \omega_0 = \sqrt{\frac{\mu}{r_0^3}(1 - 3\sin^2\theta_0)} \tag{9.79}$$

These formulas (9.78) and (9.79) have been obtained for a planar transfer.

For a three-dimensional transfer, an additional out-of-plane component with an angle $\beta$ is introduced as follows.

The costates of position $\vec{p}_r(t_0)$ and velocity $\vec{p}_v(t_0)$ are parameterized by their respective angles $(\theta_r, \beta_r)$ and $(\theta_v, \beta_v)$.

- The in-plane angle $\theta$ is the angle between the initial velocity and the vector projection on the initial orbital plane defined by $(\vec{r}_0, \vec{v}_0)$.
- The out-of-plane angle $\beta$ is the angle between the vector and the initial orbital plane.

The costates thus take the following form

$$\vec{p}_r(t_0) = \omega_v \begin{pmatrix} -\cos\theta_r \cos\beta_r \\ \sin\theta_r \cos\beta_r \\ \sin\beta_r \end{pmatrix}, \quad \vec{p}_v(t_0) = -\begin{pmatrix} -\sin\theta_v \cos\beta_v \\ \cos\theta_v \cos\beta_v \\ \sin\beta_v \end{pmatrix}$$

$$\text{with } \omega_v = \sqrt{\frac{\mu}{r_0^3}(1 - 3\sin^2\theta_v)} \tag{9.80}$$

For a low-thrust transfer starting at the perigee of the initial orbit, the thrust should be nearly aligned with the initial velocity. It is then expected that the angle values $(\theta_r, \theta_v, \beta_r, \beta_v)$ are close to zero. This allows us to start the optimization from a null guess and reduce the search space on $(\theta_r, \theta_v, \beta_r, \beta_v)$ to a small interval around zero.

The evolution of the costates $\vec{p}_r(t)$ and $\vec{p}_v(t)$ is assessed by the equations (9.68) with the control $\vec{u}(t)$ defined by (9.66). The mass costate is not used.

#### 9.3.3.2 Problem Reformulation

The second element of the solution method is to reformulate the optimal control problem into an unconstrained minimization problem. Most transfers consist in bringing the vehicle into a higher orbit such as the geostationary orbit. In such cases, the perigee altitude increases constantly along the trajectory. This property makes it possible to eliminate the final time of the problem unknowns, by stopping the numerical integration as soon as the perigee reaches the targeted altitude.

The problem unknowns are reduced to the four angles $(\theta_r, \theta_v, \beta_r, \beta_v)$. Their values must be found so that the targeted position $\vec{r}_f$ and velocity $\vec{v}_f$ are reached at the trajectory end. This problem is recast as an unconstrained minimization problem.

$$\min_{\theta_r,\theta_v,\beta_r,\beta_v} \frac{\|\vec{r}(t_f) - \vec{r}_f\|}{\|\vec{r}_f\|} + \frac{\|\vec{v}(t_f) - \vec{v}_f\|}{\|\vec{v}_f\|} \tag{9.81}$$

The cost function is the scaled sum of the deviations observed on the final positions and velocities. The final orbit is reached when this cost function is zero. It is not guaranteed that the formulation (9.81) gives the optimal trajectory in minimum time. There could indeed exist different sets of values $(\theta_r, \theta_v, \beta_r, \beta_v)$ allowing to reach the final orbit with different final times. In practice, since the search interval for these angles is narrow around zero, it can be observed that the trajectories present the expected shapes and that the transfer times are very close to those obtained by other methods.

#### 9.3.3.3 Gradient Free Algorithm

The third element of the solution method is the choice of the minimization algorithm applied to the problem (9.81). The propagation of the differential equations of state (9.64) and costate (9.68) over long durations is numerically very sensitive. In addition, the entries and exits of the shadow region must be detected precisely to ensure smooth integration. These numerical effects are detrimental to gradient-based algorithms. An algorithm without gradient is less sensitive to the numerical "noise" of integration. An attractive choice for the small size problem (9.81) is the DIRECT (DIviding RECTangles) algorithm, which explores the search space by cutting it into hyperrectangles. Unlike metaheuristics (particle swarms, simulated annealing, genetic algorithms, …), which have a random part and a risk of missing the global minimum, this dichotomy-like approach ensures that the solution can be located under reasonable assumptions (Lipschitz function) and with a controlled number of cost function evaluations.

For the problem (9.81) the variables are the four angles $(\theta_r, \theta_v, \beta_r, \beta_v)$ defining the initial costate (9.80). The search intervals are set respectively to:

- $[-10\,\text{deg}; +10\,\text{deg}]$ for the in-plane angles $(\theta_r, \theta_v)$;
- $[-1\,\text{deg}; +1\,\text{deg}]$ for the out-of-plane angles $(\beta_r, \beta_v)$.

With this gradient-free algorithm, a precise numerical integration is not required and the Cartesian coordinates can be used. This choice is not the best numerically (a formulation in orbital parameters is preferable), but it gives very simple equations for the state (9.64), the costate (9.68), and the eclipse conditions (9.71).

In particular, the gradient of gravity in the costate equations (9.68) is expressed as $\vec{g}(\vec{r}) = -\frac{\mu}{r^3}\vec{r} \Rightarrow \frac{\partial \vec{g}}{\partial \vec{r}} = \frac{\mu}{r^5}(3\vec{r}\vec{r}^t - r^2 I)$, as was established in (2.86).

The integration is carried out by the eighth-order Runge–Kutta method (DOP853) with an adaptive time step of 1000 s. The time step is furthermore corrected by dichotomy to pass at the eclipse entry and exit dates with an accuracy of 10 s. The costate discontinuities (9.72) are applied during integration with the multiplier calculated by (9.75)–(9.77). The convergence requires about 1000 simulations and takes a few minutes on a standard Linux computer.

### 9.3.4 Application Cases

The solution method is illustrated on transfers to the geostationnary orbit. The targeted GEO orbit is circular equatorial at an altitude of 35 786 km. The Earth is spherical with a radius of 6378.137 km and a central gravity field of gravitational constant $\mu = 3.986\,005 \cdot 10^{14}\ \mathrm{m^3/s^2}$.

#### 9.3.4.1 Initial Orbit GTO

We first consider an initial GTO orbit with an apogee at 35 786 km, a perigee at 500 km, an inclination of 10 deg and a perigee argument of 0 deg (the major axis of the orbit is in the equatorial plane).

The vehicle's initial mass is 1000 kg, with an engine thrust of 1 N and a specific impulse of 1500 s, equivalent to an exhaust velocity $v_e = g_0$ Isv = 147 10 m/s with $g_0 = 9.806\,65\ \mathrm{m/s^2}$.

Figure 9.4 shows the trajectories projected in the equatorial plane (OXY) for transfers beginning at the equinox or solstice and for initial local times at the perigee of 0, 6, or 12 h.

The thrust direction is indicated by arrows during the first and last revolutions. Arrows orthogonal to the trajectory indicate an out-of-plane thrust direction associated with an inclination correction.

The shadow zone (in the dark on the figures) rotates with the apparent motion of the Sun at a rate of 0.986 deg/day. With a thrust of 1 N, the transfer takes 25 days and the direction of the Sun rotates of about 30 deg. Eclipses occur early in the transfer and they decrease as the vehicle gains altitude. A transfer at the equinox undergoes long eclipse durations because the Sun is in the equatorial plane in which the targeted GEO orbit is located. Conversely, a transfer at the solstice has shorter eclipse durations because the Sun is at a latitude of 23.45 deg, outside the targeted equatorial plane. In this configuration, the final GEO orbit is fully lightened and the last portion of the transfer does not enter the shadow region. For a local time of 12 h at the perigee, there is even no eclipse at all because the low-altitude portion of the orbit is on the lightened side of the Earth.

Figure 9.5 shows the same GTO-GEO transfer at the equinox with a thrust of 0.1 N instead of 1 N. The transfer period is increased from 25 days to 8 months. Eclipses (white areas) may then occur at the beginning and at the end of the transfer due to the rotation of the Sun direction. The cumulative duration of the eclipse is plotted below the figures.

Figure 9.6 shows the evolution of the osculating orbital parameters (apogee, perigee, inclination) and of the thrust angles (angle $\theta$ in the plane between velocity and thrust, angle $\beta$ out of the plane) for a transfer at the equinox with a local time at perigee of 0 h and a thrust of 1 N.

The transfer shows two distinct phases.

- During the first phase of about 8 days, the out-of-plane angle $\beta$ increases up to 90 deg, while the in-plane angle $\theta$ remains below 40 deg. This first phase increases the apogee altitude beyond that of the GEO up to 45 000 km. The out-of-plane thrust phases are accentuated toward the apogee so as to change the inclination by benefiting from a lower velocity.
- During the second phase, the out-of-plane angle $\beta$ decreases, while the in-plane angle $\theta$ oscillates between 0 deg at the apogee (acceleration to raise the perigee) and 180 deg at the perigee (braking to lower the apogee to 35 786 km).

This two-phase behavior is systematically observed for low-thrust transfers starting from a highly elliptical orbit. The apogee passes above the targeted value in order to achieve the change of inclination at a lower cost.

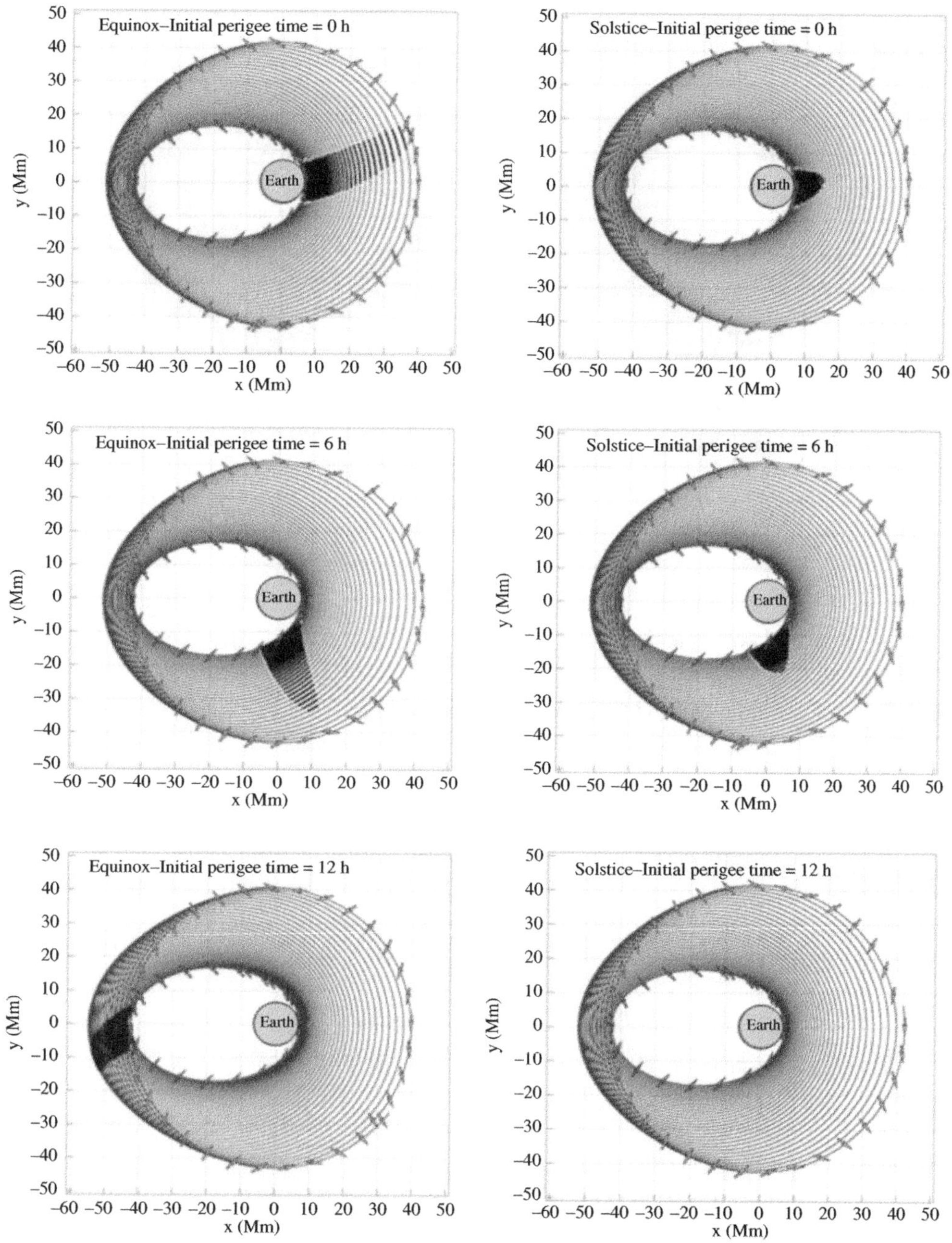


**Figure 9.4** Transfer from GTO to GEO (thrust = 1 N).

The numerical results are presented in Table 9.1 for the different configurations (thrust, initial date, local time at perigee).

The four optimized angles defining the initial costate (9.80) have the following orders of magnitude: $\theta_r \approx 0.08$ deg, $\theta_v \approx -4.2$ deg, $\beta_r \approx 10^{-4}$ deg, $\beta_v \approx -0.05$ deg.

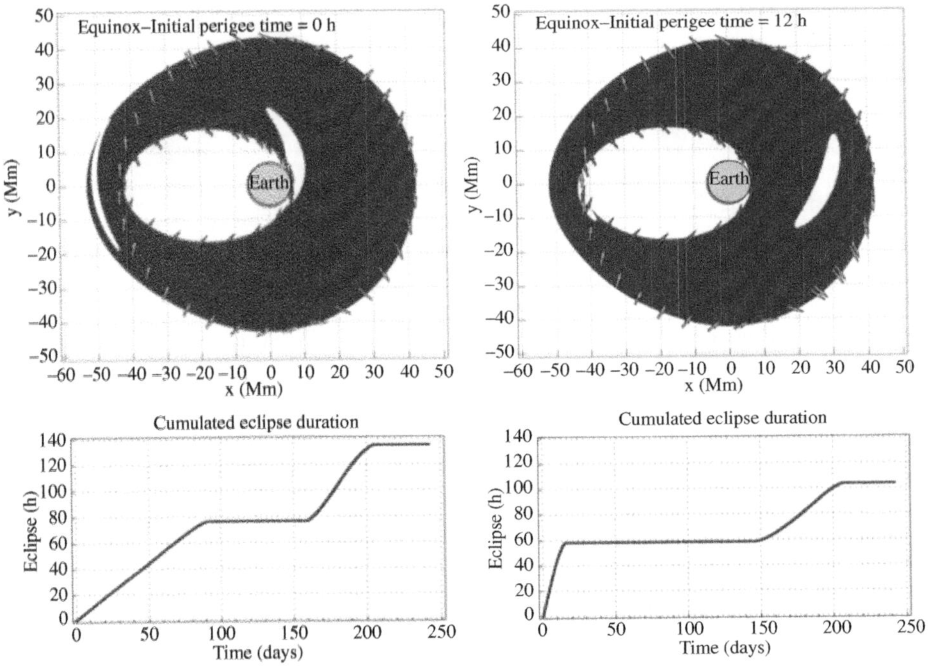


**Figure 9.5** Transfer from GTO to GEO at equinox (thrust = 0.1 N).

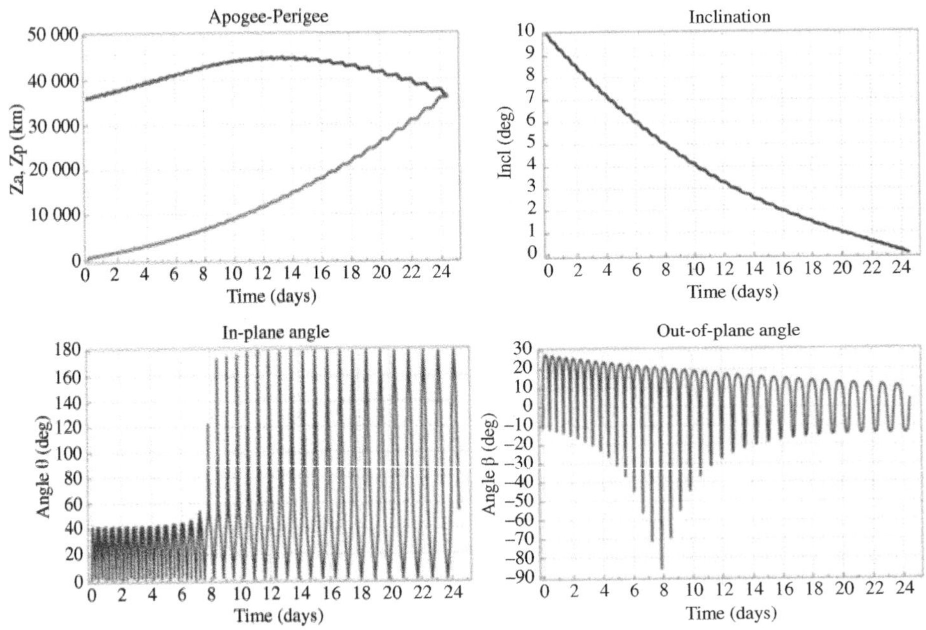


**Figure 9.6** Orbital parameters and thrust direction (GTO to GEO with thrust = 1 N).

**Table 9.1** Performances for GTO-GEO transfers.

| **Thrust level** | **1 N** | | | | | | **0.1 N** | |
|---|---|---|---|---|---|---|---|---|
| **Initial date** | **Equinox** | **Solstice** | **Equinox** | **Solstice** | **Equinox** | **Solstice** | **Equinox** | **Solstice** |
| Perigee local time | 0 h | 0 h | 6 h | 6 h | 12 h | 12 h | 0 h | 0 h |
| Transfer duration (d) | 24.04 | 23.99 | 24.38 | 23.99 | 26.11 | 23.89 | 242.18 | 242.24 |
| Eclipse duration (h) | 19.4 | 6.1 | 17.8 | 9.3 | 44.3 | 0.0 | 134.9 | 103.4 |
| Number of eclipses | 34 | 18 | 29 | 18 | 26 | 0 | 219 | 104 |
| Final mass (kg) | 863.6 | 860.6 | 861.1 | 861.4 | 857.5 | 859.7 | 861.1 | 860.2 |
| Total impulse (m/s) | 2157.8 | 2208.2 | 2199.2 | 2195.3 | 2261.3 | 2224.0 | 2200.4 | 2214.4 |

#### 9.3.4.2 Initial Orbit LEO

The solution method is then applied to a transfer to the GEO from a circular LEO at an altitude of 500 km and an inclination of 20 deg. The transfer is calculated at the equinox for a thrust of 1 N. The local time at perigee is set at 0 h, but it has no influence when the initial orbit is circular.

The trajectory in the equatorial plane is plotted in Figure 9.7, with the orbital parameter evolution and the eclipse duration. It is observed that the orbit remains almost circular throughout the transfer. The total duration is 62.1 days with 400 eclipses for a cumulative duration of 269.5 h.

#### 9.3.4.3 Solution Method Recap

The solution method presented proves robust for a wide variety of initial conditions (initial orbit, transfer start date, perigee local time) and vehicle features (mass, thrust, specific

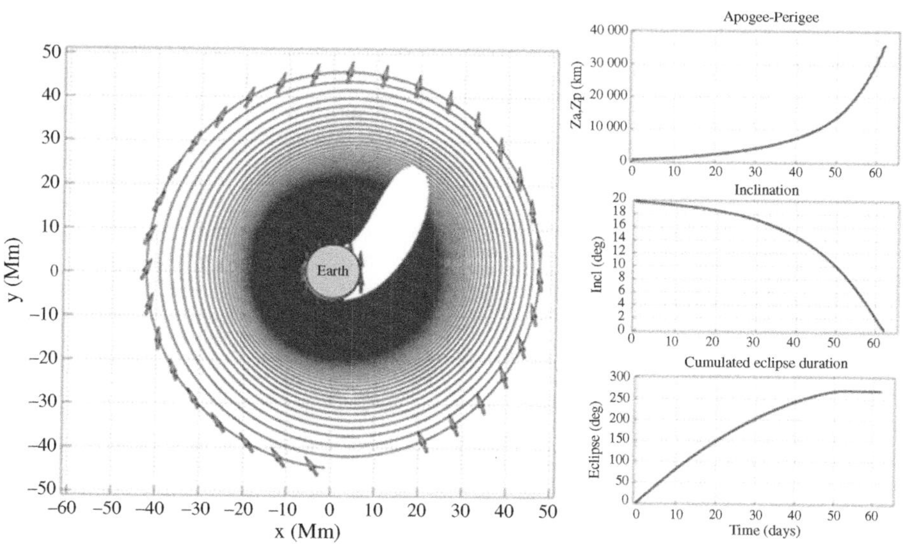


**Figure 9.7** Transfer from LEO to GEO at equinox (thrust = 1 N).

impulse). The main advantages compared to usual optimal control methods are the simplicity of the equations in Cartesian coordinates and the ability to find the solution without a specific initialization task. The solutions are obtained in a very short time without human help. The numerical results match closely those obtained by other optimization methods.

## 9.4 Transfers Between Circular Orbits

This section deals with transfers at minimum consumption between circular orbits using the natural precession due to the Earth flattening.

### 9.4.1 Dynamic Model

In this section, we consider low-thrust transfers between circular orbits. We assume furthermore that all intermediate orbits during the transfer remain circular. This assumption may seem restrictive, but it is observed that it yields results close to a general strategy with free intermediate orbit shapes during the transfer. This is observed for example in Figure 9.7 for the LEO–GEO transfer. This assumption simplifies the resolution of the minimum-consumption problem.

We place ourselves in the inertial geocentric frame (OXYZ) already described in Section 9.3.1.1. The orbital plane is defined by the inclination I and the right ascension of the ascending node Ω (RAAN). The inclination I is the angle of the orbital plane with the equatorial plane (OXY). The intersection of the orbital plane with the equator is the line of nodes. The RAAN Ω is the angle between the X axis and the direction of the ascending node AN as shown in Figure 9.8.

The shape of a circular orbit is defined in an equivalent way by its radius a or by its constant orbital velocity V. A circular orbit is thus completely defined in the Earth frame by the three parameters (V, I, Ω).

By restricting us to circular orbits, we can use the model of Edelbaum presented in Section 9.2.4. This model is based on an averaging of Gauss equations (9.7) with the following assumptions:

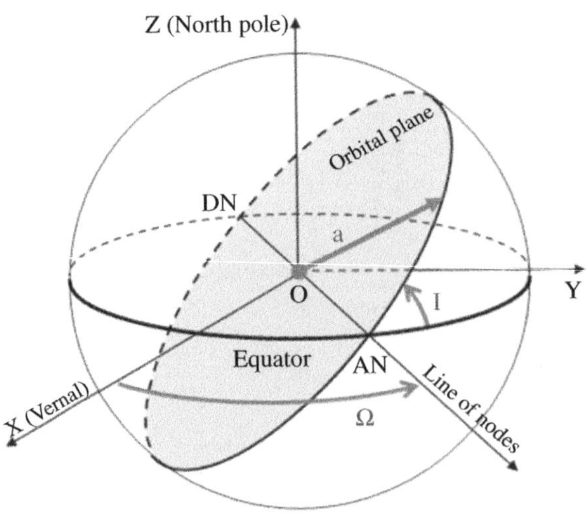


Figure 9.8 Orbital parameters for a circular orbit.

- the variations of the orbital parameters are averaged over the orbital period;
- the mean orbit is constantly circular throughout the transfer;
- the acceleration level noted f is constant throughout the transfer;
- the thrust direction is orthoradial (orthogonal to the vector radius) and makes a constant angle $\beta$ with the orbital plane over an orbital period, with a sign change at the antinodes (points located at a latitude argument of $\pm 90$ deg). This change in the sign of $\beta$ maximizes the variation in inclination over an orbital period.

Subsequently, the notations $(V, I, \Omega)$ denote the mean orbital parameters.

Edelbaum's problem is to minimize the transfer time to pass from an initial circular orbit $(V_0, I_0)$ to a final circular orbit $(V_f, I_f)$. The control variable is the mean out-of-plane angle $\beta$, which may vary at each orbital period.

The analytical solution of Edelbaum's problem is detailed in Section 9.2.4. For the problem studied here, there are additional RAAN constraints on the initial and the final orbit. These constraints will be managed using the natural precession due to the Earth flattening (see Section 2.4.4).

The Earth's equatorial bulge creates a nodal precession rate, which depends on the orbital parameters by the formula (2.103). For a circular orbit of radius a and inclination I, this rate is

$$\dot{\Omega} = -\frac{3}{2} J_2 \sqrt{\mu} R_E^2 a^{-\frac{7}{2}} \cos I = -k\left(\frac{\mu}{a}\right)^{\frac{7}{2}} \cos I = -kV^7 \cos I \tag{9.82}$$

The constants of the Earth's gravitational model are:

- equatorial radius $R_E = 6378.137$ km
- gravitational constant $\mu = 3.986\,005 \cdot 10^{14}\ m^3/s^2$
- coefficient of the first zonal term of the potential $J_2 = 1.082\,66 \cdot 10^{-3}$

which gives the constant $k = \frac{3J_2R_E^2}{2\mu^3} = 1.0425 \times 10^{-33}\ s^6m^{-7}$ in equation (9.82).

The Earth flattening (term $J_2$) does not change the shape and inclination of the mean orbit. In the absence of thrust, this mean orbit remains circular.

The evolution of the mean parameters $(V, I, \Omega)$ is thus given by the equations

$$\begin{cases} \dot{V} = -f \cos \beta \\ \dot{I} = \dfrac{2}{\pi V} f \sin \beta \\ \dot{\Omega} = -kV^7 \cos I \end{cases} \tag{9.83}$$

The control variables are the acceleration f and the out-of-plane angle $\beta$ over each orbital period. This model is analogous to that of Edelbaum supplemented with the evolution of the RAAN under the effect of the $J_2$ term.

### 9.4.2 Optimal Control Problem

The goal is to bring the vehicle from an initial circular orbit to a final circular orbit in a given time while minimizing the propellant consumption. The consumption is directly related to the velocity impulse $\Delta V = \int_{t_0}^{t_f} f dt$.

The minimum-consumption transfer problem is then formulated as:

$$\min_{f,\beta} J = \int_{t_0}^{t_f} f dt \text{ with } \begin{cases} \dot{V} = -f \cos \beta \\ \dot{I} = \dfrac{2}{\pi V} f \sin \beta \\ \dot{\Omega} = -kV^7 \cos I \end{cases} \tag{9.84}$$

The initial conditions are fixed at $t_0$: $V(t_0) = V_0$, $I(t_0) = I_0$, $\Omega(t_0) = \Omega_0$.

The final conditions are fixed at $t_f$: $V(t_f) = V_f$, $I(t_f) = I_f$, $\Omega(t_f) = \Omega_f$.

The acceleration f is bounded between 0 and $f_{max}$. The angle β is free.

To solve this optimal control problem, the PMP is applied by introducing the costates $p_V(t)$, $p_I(t)$, $p_\Omega(t)$ associated with the states V(t), I(t), Ω(t), respectively and the positive or zero cost multiplier $p_0$.

The Hamiltonian has the following expression:

$$H = p_0 f + p_V \dot{V} + p_I \dot{I} + p_\Omega \dot{\Omega} = fS + p_\Omega \dot{\Omega} \tag{9.85}$$

by noting S the function defined by

$$S = p_0 - p_V \cos\beta + p_I \frac{2}{\pi V} \sin\beta \tag{9.86}$$

The PMP provides the following set of first-order necessary conditions that must be met by any optimal trajectory.

- The costates $p_V(t)$, $p_I(t)$, $p_\Omega(t)$ do not cancel simultaneously over an interval of $[t_0; t_f]$ and they follow the differential system

$$\begin{cases} \dot{p}_V = -\dfrac{\partial H}{\partial V} = 7p_\Omega k V^6 \cos I + \dfrac{2}{\pi V^2} p_I f \sin\beta \\ \dot{p}_I = -\dfrac{\partial H}{\partial I} = -p_\Omega k V^7 \sin I \\ \dot{p}_\Omega = -\dfrac{\partial H}{\partial \Omega} = 0 \quad \Rightarrow \quad p_\Omega = Cte \end{cases} \tag{9.87}$$

Since the initial and final conditions are completely fixed (time and state), there are no transversality conditions on the costates and on the Hamiltonian.

- The control (f, β) minimizes the Hamiltonian $H = fS + p_\Omega \dot{\Omega}$.

The minimization with respect to f gives

$$\begin{cases} f = f_{max} & \text{if } S < 0 \\ f = 0 & \text{if } S > 0 \end{cases} \tag{9.88}$$

The sign of S (called the switching function) determines the acceleration level. A singular arc occurs if S cancels on an interval. The acceleration level is then no longer determined by the PMP and a specific analysis is required.

The minimization with respect to β gives

$$\begin{cases} \dfrac{\partial H}{\partial \beta} = 0 \quad \Rightarrow \quad p_V \sin\beta + p_I \dfrac{2}{\pi V} \cos\beta = 0 \\ \dfrac{\partial^2 H}{\partial \beta^2} \geq 0 \quad \Rightarrow \quad p_V \cos\beta - p_I \dfrac{2}{\pi V} \sin\beta \geq 0 \end{cases} \tag{9.89}$$

The costates $p_V$ and $p_I$ cannot cancel simultaneously. Otherwise (9.87) would also lead to $p_\Omega = 0$, in contradiction with PMP (null costates). The system (9.89) thus defines the angle β in terms of the costates $p_V$ and $p_I$.

- The Hamiltonian is constant because the problem (9.84) is autonomous (dynamic and cost not explicitly dependent on the time).

$$H = fS + p_\Omega \dot{\Omega} = Cte \tag{9.90}$$

**Shooting Problem**

The PMP optimality conditions lead to a two-point boundary value problem called shooting problem. This problem is a system of equations of dimension three:

- the three unknowns are the initial costates: $p_V(t_0)$, $p_I(t_0)$, $p_\Omega(t_0)$;
- the three equations define the final state: $V(t_f) = V_f$, $I(t_f) = I_f$, $\Omega(t_f) = \Omega_f$;
- the final state is calculated by integrating the state equations (9.83) and the costate equations (9.87) with the control $(f, \beta)$ defined in (9.88) and (9.89).

This problem is solved by Newton's method. The resolution is difficult without a very good initialization of the costate. For that purpose, a theoretical analysis of the solution properties is presented in Section 9.4.3. This analysis makes it possible to develop an initialization of the costate presented in Section 9.4.4.

## 9.4.3 Form of Optimal Trajectories

Any solution satisfying the necessary conditions of the PMP is called an extremal.

- An extremal is said normal if $p_0 > 0$ and abnormal if $p_0 = 0$.
- An extremal is said regular if $S \neq 0$ and singular if $S = 0$.

The optimal trajectory may be either normal or abnormal, and it may also be composed of several regular and singular arcs. The options are discussed in the subsequent text.

### 9.4.3.1 Singular Solution: S = 0

For a singular solution, the costates $p_V$ and $p_I$ are calculated by

$$\begin{cases} S = 0 \\ \dfrac{\partial H}{\partial \beta} = 0 \end{cases} \Rightarrow \begin{cases} p_0 - p_V \cos\beta + p_I \dfrac{2}{\pi V} \sin\beta = 0 \\ p_V \sin\beta + p_I \dfrac{2}{\pi V} \cos\beta = 0 \end{cases} \Rightarrow \begin{cases} p_V = p_0 \cos\beta \\ p_I = -\dfrac{\pi V}{2} p_0 \sin\beta \end{cases} \tag{9.91}$$

A singular solution cannot be abnormal ($p_0 = 0$), otherwise we would have $p_V = p_I = 0$, and with (9.87) $p_\Omega = 0$. All costates would be null, which is ruled out by the PMP. To find the optimal acceleration level $f_{opt}$ along the singular arc, one must derive the switching function S until f appears. Since this solution is not of practical interest, the (long) calculations are not detailed here.

### 9.4.3.2 Abnormal Solution: $p_0 = 0$

The inequality in (9.89) gives an upper bound on the switching function S (9.86).

$$S = p_0 - p_V \cos\beta + p_I \frac{2}{\pi V} \sin\beta \le p_0 \tag{9.92}$$

For an abnormal solution ($p_0 = 0$), the switching function S is negative or zero and the trajectory consists of maximum thrust arcs and singular arcs. However, it has been shown just above that a singular solution cannot be abnormal. Therefore, the trajectory has a single arc at maximum thrust and the problem (9.84) is reduced to Edelbaum's minimum-time problem. This situation corresponds to the very special case where the minimum-time trajectory undergoes a precession matching exactly the required deviation between the initial and final RAAN.

#### 9.4.3.3 Specific Solution: $p_\Omega = 0$

If $p_\Omega = 0$, the Hamiltonian (9.90) is expressed as $H = fS \leq 0$ according to (9.88).

- If $H = 0$, then the solution consists of zero-thrust arcs and singular arcs. This form of solution is not useful in practice and is not further discussed here.
- If $H < 0$, then $S < 0$ and the trajectory has a single arc with maximum thrust. As in the abnormal case, we retrieve Edelbaum's minimum-time problem. This situation corresponds to very specific values of RAAN.

In the sequel, we restrict ourselves to regular normal solutions such that $p_\Omega \neq 0$.

These solutions, composed of maximum thrust arcs and coast arcs, form the general case and they are the most useful in practice.

#### 9.4.3.4 Normal Regular Solution with $p_\Omega \neq 0$

Since the costate $p_\Omega$ is constant and nonzero, one can express the Hamiltonian in (9.90) by introducing a constant noted $\dot{\Omega}_d$.

$$H(t) = fS + p_\Omega \dot{\Omega} \underset{\text{def}}{=} p_\Omega \dot{\Omega}_d = \text{Cte} \tag{9.93}$$

This constant represents the precession rate on a coast arc ($f = 0$). It is thus observed all coast arcs of an optimal trajectory necessarily have the same precession rate $\dot{\Omega}_d$. Using (9.88) which gives $fS \leq 0$, we obtain an inequality satisfied over the whole trajectory

$$p_\Omega \dot{\Omega} \geq p_\Omega \dot{\Omega}_d \tag{9.94}$$

This inequality shows that the precession rate reaches its maximum (if $p_\Omega < 0$) or its minimum (if $p_\Omega > 0$) on coast arcs.

During a coast arc, the velocity and inclination remain constant while the RAAN evolves linearly with the precession rate $\dot{\Omega}_d$. Figure 9.9 illustrates a possible evolution of $p_\Omega \dot{\Omega}$ along the trajectory with three thrust arcs ($f = f_{max}$) separated by two coast arcs. This scenario is called scenario 1.

The following relations apply between the beginning and the end of coast arcs.

$$\begin{cases} V_2 = V_1 \\ I_2 = I_1 \\ \Omega_2 = \Omega_1 + \dot{\Omega}_d \cdot (t_2 - t_1) \end{cases} \quad \text{and} \quad \begin{cases} V_4 = V_3 \\ I_4 = I_3 \\ \Omega_4 = \Omega_3 + \dot{\Omega}_d \cdot (t_4 - t_3) \end{cases} \tag{9.95}$$

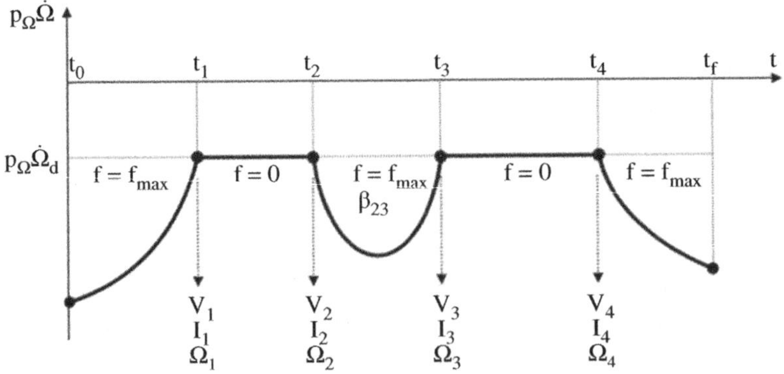


**Figure 9.9** Scenario 1.

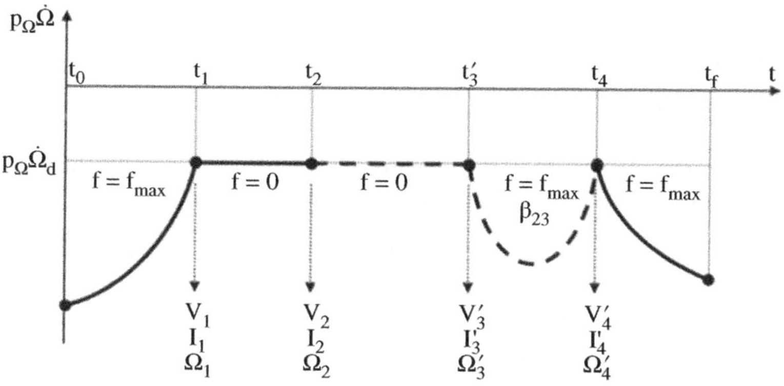


**Figure 9.10** Scenario 2.

The arc from $t_2$ to $t_3$ is controlled with the maximum acceleration $f_{max}$ and with a thrust direction law noted $\beta_{23}$. This control transfers the state from $(V_2, I_2, \Omega_2)$ to $(V_3, I_3, \Omega_3)$ in a duration $\Delta t_{23} = t_3 - t_2$. The change in RAAN during this propelled arc is noted $\Delta\Omega_{23} = \Omega_3 - \Omega_2$.

Let us now consider the scenario obtained by swapping the third and fourth arc, each arc keeping its duration and its control law. The times $t_2$ and $t_4$ are unchanged, whereas the intermediate time $t_3$ is shifted to $t_3'$. This new scenario is called scenario 2 and it is shown in Figure 9.10.

This scenario is identical to scenario 1 until $t_2$. The ballistic arc from $t_2$ to $t_3'$ has the duration $\Delta t_{34} = t_4 - t_3$ and the precession rate $\dot{\Omega}_d$. It ends at $t_3'$ with the state

$$\begin{cases} V_3' = V_2 \\ I_3' = I_2 \\ \Omega_3' = \Omega_2 + \dot{\Omega}_d \cdot \Delta t_{34} \end{cases} \tag{9.96}$$

The control $(f_{max}, \beta_{23})$ of scenario 1 is then applied from $t_3'$ to $t_4$. Compared to the arc from $t_2$ to $t_3$ in scenario 1, this arc from $t_3'$ to $t_4$ has a different start time ($t_3'$ instead of $t_2$), the same duration $(\Delta t_{23})$, and the same initial state $\left(V_3' = V_2, I_3' = I_2\right)$ except for RAAN ($\Omega_3'$ instead of $\Omega_2$). The initial time shift does not change the trajectory because the problem (9.84) is autonomous. The initial RAAN shift does not affect the evolution of (V, I) given by (9.83): this shift is simply reported on the final RAAN value. Applying the control $(f_{max}, \beta_{23})$ from the state $\left(V_2, I_2, \Omega_3'\right)$ at time $t_3'$ until time $t_4$ gives an arc identical to scenario 1 with a change in RAAN of $\Delta\Omega_{23}$. The final state at time $t_4$ is thus

$$\begin{cases} V_4' = V_3 \\ I_4' = I_3 \\ \Omega_4' = \Omega_3' + \Delta\Omega_{23} = \Omega_2 + \dot{\Omega}_d \cdot \Delta t_{34} + \Delta\Omega_{23} = \Omega_3 + \dot{\Omega}_d \cdot \Delta t_{34} \end{cases} \tag{9.97}$$

This final state at $t_4$ is identical to that of scenario 1 (9.95). Scenario 2 reaches the same end state as scenario 1 with the same cost because the thrust arcs have the same duration. Thus, a solution with several coast arc can be replaced by an equivalent solution with a single coast arc. We can therefore restrict ourselves to solutions in three arcs of the form $(f_{max} - 0 - f_{max})$.

### 9.4.4 Solution Method

The goal is to solve the shooting problem formulated in Section 9.4.2. The main issue is to find a good initialization of the costates $p_V(t_0)$, $p_I(t_0)$, $p_\Omega(t_0)$ allowing the convergence of Newton's method.

For this purpose, we use the structure of the solution $(f_{max} - 0 - f_{max})$ established in Section 9.4.3.4, which gives the transfer strategy illustrated in Figure 9.11.

- The first thrust arc transfers from the known initial orbit $(V_0, I_0)$ at time $t_0$ to a drift orbit $(V_d, I_d)$. This orbit (to be determined) is reached at time $t_1$.
- The coast arc on the drift orbit $(V_d, I_d)$ lasts from time $t_1$ to time $t_2$.
- The second thrust arc transfers from the drift orbit $(V_d, I_d)$ at time $t_2$ to the known final orbit $(V_f, I_f)$ at time $t_f$.

#### 9.4.4.1 Simplified Problem

An approximate analytical solution can be constructed for the strategy depicted in Figure 9.11, based on the following ideas.

- The goal of the thrust arcs is to minimize the transfer time, regardless of the RAAN. These arcs are thus assessed by Edelbaum's formulas recalled below.
- Most of the variation of RAAN is carried out during the coast arc on the drift orbit.

This simplified problem has for unknowns $(V_d, I_d)$ and it is formulated as

$$\min_{V_d, I_d} \Delta V \text{ with } \Omega(t_f) = \Omega_f \tag{9.98}$$

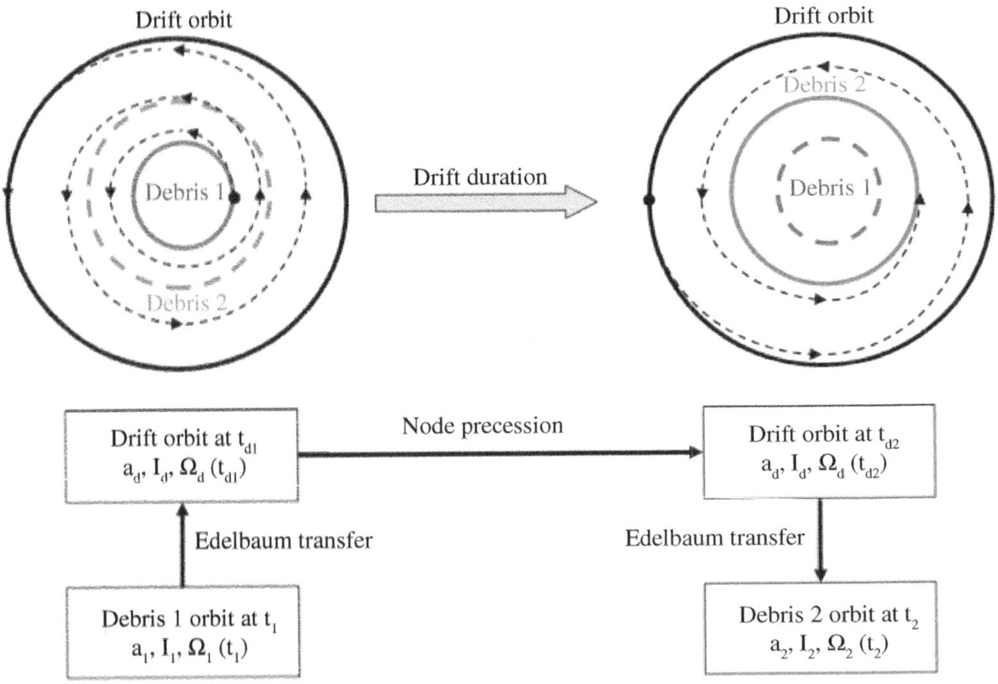


**Figure 9.11** Transfer strategy.

The cost $\Delta V$ is the sum of the velocity impulses of the two propelled transfers. These impulses calculated by Edelbaum's formulas are noted $\Delta V_{e1}$ and $\Delta V_{e2}$.

$$\Delta V = \Delta V_{e1} + \Delta V_{e2} \tag{9.99}$$

#### 9.4.4.2 Edelbaum's Solution

Edelbaum's problem deals with the transfer in minimum time from a circular orbit $(V_1, I_1)$ to a circular orbit $(V_2, I_2)$ with the dynamic model described in Section 9.4.1. The analytical solution is given by the following formulas (see Section 9.2.4).

- Initial angle $\beta_1$: $\tan\beta_1 = \dfrac{V_2 \sin\delta I}{V_1 - V_2\cos\delta I}$ by noting $\delta I = \dfrac{\pi}{2}(I_2 - I_1)$
- Duration and impulse: $\Delta t = \dfrac{V_1}{f}\left[\cos\beta_1 - \dfrac{\sin\beta_1}{\tan(\delta I + \beta_1)}\right]$, $\Delta V_e = f\Delta t$
- State and costate: $\begin{cases} V = \sqrt{V_1^2 + f^2t^2 - 2ftV_1\cos\beta_1} \\ I = I_1 + \dfrac{2}{\pi}(\beta - \beta_1) \\ \tan\beta = \dfrac{V_1\sin\beta_1}{V_1\cos\beta_1 - ft} \end{cases}$, $\begin{cases} p_{eV} = \dfrac{1}{f}\cos\beta \\ p_{eI} = -\dfrac{\pi V}{2f}\sin\beta \end{cases}$

Edelbaum's formulas give the durations of the two propelled arcs from $(V_0, I_0)$ to $(V_d, I_d)$ and from $(V_d, I_d)$ to $(V_f, I_f)$. Since the transfer initial time $t_0$ and final time $t_f$ are fixed, the coast arc times $t_1$ and $t_2$ can be assessed. The resulting RAAN variation on the drift orbit $(V_d, I_d)$ is

$$\Omega(t_2) = \Omega(t_1) - kV_d^7\cos I_d(t_2 - t_1) \tag{9.100}$$

Assuming that the propelled durations are negligible compared to the coast duration, the RAAN variation assessed by (9.84) must match the required RAAN difference between the initial and the final orbit.

$$\Omega_f = \Omega_0 - kV_d^7\cos I_d(t_f - t_0) \tag{9.101}$$

Since the times $(t_0, t_f)$ and the RAAN $(\Omega_0, \Omega_f)$ are given, this gives a relation between the drift orbit parameters $V_d$ and $I_d$.

By expressing $I_d$ as a function of $V_d$, the cost function (9.99) thus depends only on the variable $V_d$. Its gradient is expressed analytically as:

$$\begin{aligned} \frac{d\Delta V}{dV_d} &= \frac{\partial\Delta V}{\partial V_d} + \frac{\partial\Delta V}{\partial I_d}\frac{dI_d}{dV_d} \\ &= [p_{eV1}(t_1) + p_{eV2}(t_2)] + [p_{eI1}(t_1) + p_{eI2}(t_2)]\frac{7}{V_d\tan I_d} \end{aligned} \tag{9.102}$$

This formula is established by using on the one hand the relation (9.101) between $I_d$ and $V_d$, and on the other hand the following relations between the costates and the cost derivatives with respect to the state [R12].

$$\begin{cases} \dfrac{\partial(\Delta V_{e1})}{\partial V_d} = p_{eV1}(t_1) \\ \dfrac{\partial(\Delta V_{e1})}{\partial I_d} = p_{eI1}(t_1) \end{cases} \text{ and } \begin{cases} \dfrac{\partial(\Delta V_{e2})}{\partial V_d} = p_{eV2}(t_2) \\ \dfrac{\partial(\Delta V_{e2})}{\partial I_d} = p_{eI2}(t_2) \end{cases} \tag{9.103}$$

The costates ($p_{eV}$, $p_{eI}$) are those of Edelbaum's analytical solution. The gradient in (9.102) is thus a function of the single variable $V_d$. A zero of this gradient is easily found by a numerical method (dichotomy, Newton, or secant). We then deduce the parameters ($V_d$, $I_d$) of the drift orbit, the coast times ($t_1$, $t_2$), and the total transfer cost $\Delta V$ of the simplified problem (9.98).

#### 9.4.4.3 Costate Initial Guess

Let us now try to estimate the initial costates $p_V(t_0)$, $p_I(t_0)$, $p_\Omega(t_0)$ associated with the problem (9.84). This is done by solving the problem (9.98) a first time for the reference initial conditions, and then by varying the initial state components and the initial time one by one. This gives the cost partial derivatives with respect to the initial state and the initial time. These cost derivatives calculated for the simplified problem (9.98) yield approximations of the costates and Hamiltonian of the problem (9.84).

Let us indeed recall that for an optimal control problem of state X, costate P, Hamiltonian H, and whose optimal cost is J*, we have (see [R6], [R9])

$$P(t_0) = \frac{\partial J^*}{\partial X(t_0)} \quad , \quad H(t_0) = \frac{\partial J^*}{\partial t_0} \tag{9.104}$$

**Remark 9.1** ***(local minima)*** *Since the RAAN difference $\Omega_f - \Omega_0$ is defined modulo $2\pi$, the problem (9.84) admits several local minima. Usually, the best solution can be guessed a priori depending on the orbit configuration. In some cases, it is nevertheless advisable to compare different solutions to ensure that the global minimum is found.*

**Remark 9.2** ***(transfer duration)*** *The solution of problem (9.98) is close to that of problem (9.84) as long as the propelled phase durations are small compared to the coast duration. Most of the RAAN deviation is then carried out on the coast arc. If the total time allocated to the transfer is too short, this assumption is no longer correct and the costates given by (9.104) may become inappropriate. This issue can be alleviated by a continuation method, which consists in firstly solving the problem (9.84) for a long transfer time and progressively reducing it until the desired time is reached.*

### 9.4.5 Application Case

The solution method is illustrated by an example representative of the deployment of a constellation of satellites in near sun-synchronous orbits. The initial orbit is at an altitude of 800 km, inclined at 98 deg with a RAAN of 0 deg. The final orbit is at an altitude of 900 km, inclined at 99 deg with a RAAN of 30 deg.

The maximum acceleration level is $3.5 \cdot 10^{-3}$ m/s$^2$ and the duration of the transfer is set to 100 days. The precession rate of the final orbit is of 0.982 deg/day, so that its RAAN at the end of the transfer will be 128.2 deg (=30 + 0.982 × 100). This corresponds to the targeted value for the transfer problem.

The initial and final orbital parameters are given in Table 9.2.

The RAAN deviation of 30 deg must be achieved within 100 days. Therefore, the expected precession rate of the drift orbit should be about 0.3 deg/day higher than that of the targeted orbit.

The procedure with the successive steps presented in Section 9.4.4 is applied.

**Table 9.2** Initial and final orbits.

| | Time (d) | Altitude (km) | Velocity (m/s) | Inclination (deg) | Precession (deg/d) | RAAN at $t_0$ (deg) | RAAN at $t_f$ (deg) |
|---|---|---|---|---|---|---|---|
| Initial | 0 | 800 | 7450.0 | 98.0 | 0.917 | 0.0 | 91.7 |
| Final | 100 | 900 | 7398.6 | 99.0 | 0.982 | 30.0 | 128.2 |

**Table 9.3** Simplified problem solution.

| | Time (d) | Altitude (km) | Velocity (m/s) | Inclination (deg) | RAAN (deg) | Precession (deg/d) | Impulse (m/s) |
|---|---|---|---|---|---|---|---|
| $t_0$ | 0.000 | 800.0 | 7450.0 | 98.00 | 0.00 | 0.917 | 0.0 |
| $t_1$ | 1.075 | 404.7 | 7664.0 | 99.20 | 1.18 | 1.284 | 328.7 |
| $t_2$ | 99.137 | 404.7 | 7664.0 | 99.20 | 127.26 | 1.284 | 328.7 |
| $t_f$ | 100.000 | 900.0 | 7398.6 | 99.00 | 128.20 | 0.982 | 598.1 |

#### 9.4.5.1 First Step

The first step is to solve the simplified problem (9.98). The unknowns $(V_d, I_d)$ are initialized by canceling the cost gradient (9.102) using Newton's method. This yields $V_d = 7564$ m/s (altitude 585.2 km) and $I_d = 100.08$ deg. This initialization allows us to solve the problem (9.98).

A drift orbit is found at an altitude of 404.7 km and an inclination of 99.20 deg.

This orbit is reached at $t_1 = 1.075$ days and it is followed until $t_2 = 99.137$ days.

The total impulse is 598.1 m/s. Table 9.3 summarizes the transfer sequences and the cumulated impulse.

The drift precession (1.284 deg/day) is close to the expected value ($\approx 0.982 + 0.3$). It is achieved by a decrease in altitude and an increase in inclination.

#### 9.4.5.2 Second Step

The second step consists in estimating the costates of the shooting problem formulated in Section 9.4.2. The estimation is given by the cost derivatives of the problem (9.98) with respect to the initial state. These derivatives calculated by centered finite differences are presented in Table 9.4. The velocity variation is equivalent to an altitude variation of ±50 km. The derivative with respect to the final time is also evaluated.

These partial derivatives provide estimates of the initial costates. The derivatives with respect to the angles (inclination, RAAN) are converted to radians.

$$\begin{cases} p_V(t_0) \approx -0.630 \\ p_I(t_0) \approx -8852.2 \\ p_\Omega(t_0) \approx -760.31 \end{cases} \tag{9.105}$$

The Hamiltonian is estimated from the partial derivative with respect to the final time, taking into account the induced variation in RAAN.

$$H = -\frac{\partial J^*}{\partial t_f} = -\frac{dJ^*}{dt_f} + \frac{\partial J^*}{\partial \Omega(t_f)} \times \frac{d\Omega(t_f)}{dt_f} = -\frac{dJ^*}{dt_f} + p_\Omega \dot{\Omega}_f$$
$$\Rightarrow \quad H = 3.695 + (-13.27) \times 0.982 = -9.336 \tag{9.106}$$

**Table 9.4** Cost derivatives.

| | Variation | Cost ΔV (m/s) | Centered derivative |
|---|---|---|---|
| Variable | 0 | 598.1 | — |
| Initial velocity (equivalent altitude) | −25.9 m/s (+50 km) | 615.5 | −0.630 (m/s)/(m/s) |
| | +26.0 m/s (−50 km) | 582.8 | |
| Initial inclination | +0.1 deg | 583.2 | −154.5 (m/s)/deg |
| | −0.1 deg | 614.1 | |
| Initial RAAN | +5 deg | 539.3 | −13.27 (m/s)/deg |
| | −5 deg | 672.0 | |
| Final time | +5 days | 582.3 | −3.695 (m/s)/day |
| | −5 days | 619.3 | |

#### 9.4.5.3 Third Step

The third step is to solve the shooting problem by Newton's method starting from the initialization given by (9.105). As Table 9.5 shows, this initialization is close to the solution and Newton's method converges well.

Table 9.6 details the sequences of the optimal transfer and the cumulated impulse.

In this example, the optimal solution is very close to the approximate solution. This is the case if a sufficient duration is allocated to the transfer, since most of the RAAN change takes place during the drift phase. If the time allocated to the transfer becomes too

**Table 9.5** Shooting problem solution.

| | Initialization | | Solution | |
|---|---|---|---|---|
| | Unknowns | Constraints | Unknowns | Constraints |
| Velocity (m/s) | $p_V(t_0) = -0.630$ | $V(t_f) - V_f = 5.7$ | $p_V(t_0) = -0.644$ | $V(t_f) - V_f = 10^{-5}$ |
| Inclination (deg) | $p_I(t_0) = -8852.2$ | $I(t_f) - I_f = -0.02$ | $p_I(t_0) = -9215.9$ | $I(t_f) - I_f = 10^{-8}$ |
| RAAN (deg) | $p_\Omega(t_0) = -760.31$ | $\Omega(t_f) - \Omega_f = -1.59$ | $p_\Omega(t_0) = -816.97$ | $\Omega(t_f) - \Omega_f = 10^{-6}$ |
| Drift start (d) | $t_1 = 1.075$ | $S(t_1) = 0.032$ | $t_1 = 1.092$ | $S(t_1) = 10^{-8}$ |
| Drift end (d) | $t_2 = 99.137$ | $S(t_2) = 0.119$ | $t_2 = 99.114$ | $S(t_2) = 10^{-8}$ |

**Table 9.6** Optimal transfer sequences and orbits.

| | Time (d) | Altitude (km) | Velocity (m/s) | Inclination (deg) | RAAN (deg) | Precession (deg/d) | Impulse (m/s) |
|---|---|---|---|---|---|---|---|
| $t_0$ | 0.000 | 800.0 | 7450.0 | 98.00 | 0.00 | 0.917 | 0.0 |
| $t_1$ | 1.092 | 407.1 | 7664.6 | 99.22 | 1.19 | 1.287 | 330.2 |
| $t_2$ | 99.114 | 407.1 | 7664.6 | 99.22 | 127.20 | 1.287 | 330.2 |
| $t_f$ | 100.000 | 900.0 | 7398.6 | 99.00 | 128.20 | 0.982 | 598.1 |

short, Edelbaum's control strategy (constant out-of-plane angle $\beta$ with sign change at the antinodes) is no longer suitable for the mission. In such cases, the optimal control problem is much more complex. It shall be processed directly without averaging the dynamic equations and the optimum thrust law shall aim at simultaneously changing the inclination and the RAAN.

## 9.5 Conclusion

### 9.5.1 The Key Points

- Edelbaum's analytical solution applies to constant-acceleration minimum-time transfers between circular mutually inclined orbits.
- A transfer in minimum time accounting for eclipses can be reformulated as an unconstrained nonlinear programming problem.
- The minimum-time transfer is solved by a gradient-free optimization method using an analytical costate guess.
- The strategy for a minimum-fuel transfer between circular orbits consists of going on a drift orbit to benefit from the natural precession due to the Earth flattening.
- The minimum-fuel solution is obtained from a costate guess derived from Edelbaum's analytical solution.

### 9.5.2 To Go Further

- CHOBOTOV V. – *Orbital Mechanics Third Edition.* AIAA (2002).

Chapter 14 is dedicated to low-thrust transfers. It provides in particular the complete derivation of Edelbaum's solution with further applications.

- CONWAY B. A. – *Spacecraft Trajectory Optimization.* Cambridge University Press (2010).

Chapter 6 written by Kechichian J. A. studies different types of low-thrust transfers based on Edelbaum's dynamics model. Some papers were written later by the same author still on that topic.

- GERGAUD J., HABERKORN T. – *Homotopy Method for Minimum Consumption Orbit Transfer Problem.* ESAIM: Control, Optimisation and Calculus of Variations, Tome 12 (2006) no. 2, pp. 294–310.

This paper presents a homotopic procedure to pass progressively from the minimum-energy problem to the minimum-consumption problem.

- PETROPOULOS A.E. – *Simple Control Laws for Low-Thrust Orbit Transfers.* AAS/AIAA Astrodynamics Specialists Conference (August 2003).

This paper presents the method proposed by Petropoulos, based on a local minimization of a Lyapunov function. This method allows to define the thrust direction along the trajectory, with an efficiency ratio to define the thrust and coast legs. Very good quality solutions are obtained without the need for an optimization algorithm.

### 9.5.3 Authorizations

- Examples in Section 9.3 are reprinted by permission of SAGE from "Fast solution of minimum-time low-thrust transfer with eclipses" published in Proc IMechE Part G: J Aerospace Engineering Volume 233(7), 2699–2714 (2019).
- Examples in Section 9.4 are reprinted by permission of the American Institute of Aeronautics and Astronautics, Inc., from "Low-Thrust Transfer Between Circular Orbits Using Natural Precession" published in Journal of Guidance, Control and Dynamics Volume 39, Number 10, October 2016. Copyright c AIAA.

# 10

# Space Debris Cleaning

## 10.1 Introduction

The near-Earth space is crowded by thousands of debris of all sizes. The latest estimates indicate about one million objects of size 1–10 cm and about 36 000 objects of size greater than 10 cm. These objects moving at an orbital velocity of 7–8 km/s pose a constant threat to operational satellites and to the space station. They require daily monitoring and precise trajectography in order to anticipate the risk of collisions and, if necessary, to carry out evasive maneuvers.

The debris come from old satellites and launch vehicle stages that have been abandoned in orbit since the beginning of the space age. The erosion of these vehicles (mainly by collisions with particles) constantly generates new debris, which in turn lead to new collisions. To stop this exponential growth, known as Kessler's syndrome, it is necessary to avoid leaving new vehicles in orbit and also to remove the largest current debris. Studies have concluded that the disposal of at least five large debris per year (satellites or launch stages), in addition to strict compliance with regulations, is necessary to stabilize the debris population and not jeopardize the use of space in the coming decades. A critical region is that of sun-synchronous orbits (SSO) and Polar Earth orbits (PEO) in the altitude range of 700–900 km. These orbits, which are well suited to Earth observation, contain a large number of satellites and therefore debris.

The cleaning program consists of launching a series of identical vehicles, each tasked with capturing and deorbiting five pieces of debris. The goal is to design the cheapest vehicle compliant with the cleaning program specifications. This leads to a mixed continuous/combinatorial time-dependent problem, where the debris choice, the capture times and the transfer maneuver must be simultaneously optimized.

This chapter presents a systematic solution method in order to plan the successive missions and carry out the cleaning program at minimum cost.

The first section formulates the optimization problem by coupling a path problem and embedded transfer problems.

A generic transfer strategy adapted to high or low thrust is defined in the second section, which simplifies the transfer problem and provides a nearly analytical solution.

The third section describes a three-step optimization procedure based on the simulated annealing method and a grid of precomputed cost matrices. The methodology presented takes into account the deorbiting strategy (by vehicle or autonomous kits) and the priorities assigned to debris. A detailed application case is presented in the fourth section, considering either high or low thrust.

*Space Trajectories: Basic and Advanced Topics*, First Edition. Max Cerf.

## 10.2 Problem Formulation

This section defines the characteristics of the debris orbits. The objectives of the cleaning program are specified and the optimization problem is formulated.

### 10.2.1 Debris Orbits

The debris to be removed as a priority are located in the LEO region below an altitude of 2000 km and they evolve on near-circular orbits. We will therefore only consider circular orbits defined by their radius a and the angles $(I, \Omega)$ orientating the orbital plane in the inertial geocentric reference frame (Figure 10.1).

The inclination I is the angle between the orbital plane and the equatorial plane. The intersection of the orbital plane with the equator is the line of nodes. The right ascension of the ascending node $\Omega$ (RAAN) is the angle between the X axis of the inertial geocentric frame and the direction of the ascending node (intersection with the equator, crossed from south to north).

The main orbital perturbation is due to the Earth flattening and the corresponding first zonal term of coefficient $J_2$. It creates a nodal precession rate depending on the orbital parameters by the formula (2.103). For a circular orbit of radius a and inclination I, this rate $\dot{\Omega}$ is

$$\dot{\Omega} = -\frac{3}{2} J_2 \sqrt{\mu} R_E^2 a^{-\frac{7}{2}} \cos I \tag{10.1}$$

The constants of the Earth's gravitational model are:

- equatorial radius $R_E = 6378.137$ km;
- gravitational constant $\mu = 3.986\,005 \cdot 10^{14}\ m^3/s^2$;
- coefficient of the first zonal term of the potential $J_2 = 1.082\,66 \cdot 10^{-3}$.

The effect of nodal precession is depicted in Figure 10.2.

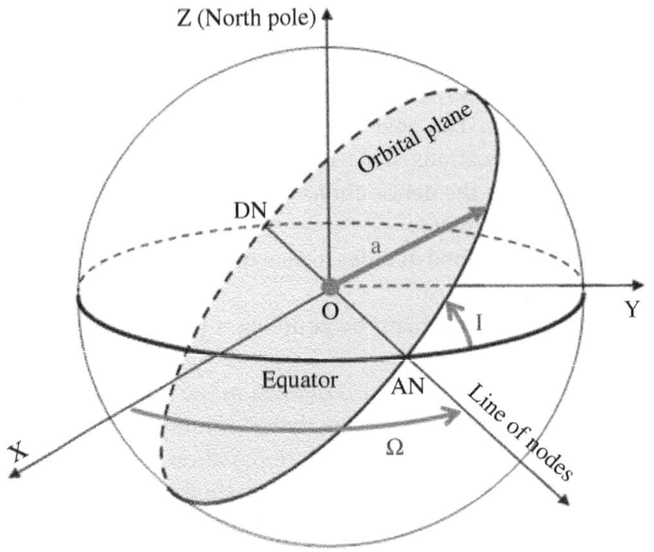


**Figure 10.1** Circular orbit parameters.

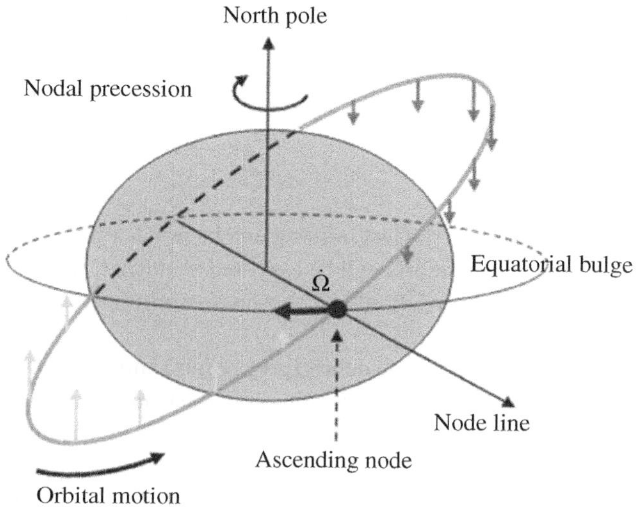


**Figure 10.2** Nodal precession due to the Earth flattening.

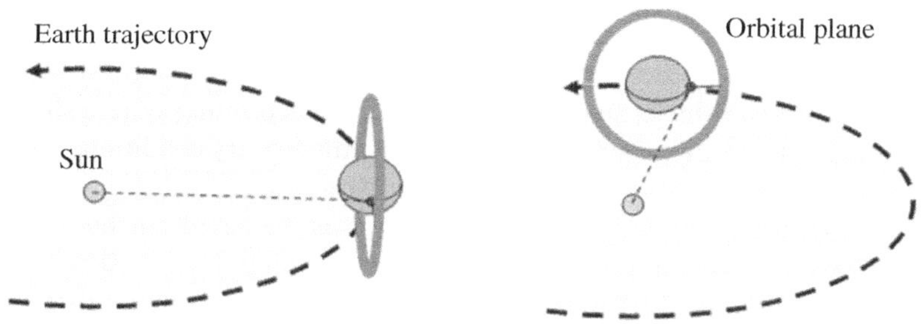


**Figure 10.3** Sun-synchronous orbit.

On the other hand, the $J_2$ perturbation does not generate secular variations of the semi-major axis, eccentricity and inclination. The average orbit remains circular with radius a and inclination I. The nodal precession rate $\dot{\Omega}$ is therefore constant and the RAAN angle changes linearly with time.

$$\Omega(t) = \Omega(t_0) + \dot{\Omega} \cdot (t - t_0) \tag{10.2}$$

For a sun-synchronous orbit, the precession rate matches the Earth's revolution rate around the Sun (360 deg/year). The orbital plane thus maintains a constant angle with the Sun direction as shown in Figure 10.3. This property is crucial for Earth observation because a region of a given latitude is always flown over at the same local solar time.

### 10.2.2 Cleaning Program

The cleaning program consists of launching a series of dedicated vehicles, each tasked with removing several selected debris. The disposal of a debris consists of capturing it and then deorbiting it by a propelled maneuver before moving on to the next debris. A series of cleaning vehicles will have to be launched to gradually clean the LEO area (below 2000 km altitude) of the most dangerous debris.

In order to limit the overall cost of the cleaning program, the vehicles used for the successive missions should have the same design.

The following notations are used to formulate the optimization problem:

- N is the total number of candidate debris;
- n is the number of debris to be collected per mission;
- m is the number of missions planned.

The total number of debris removed at the end of the last mission will be $m \times n$ among the N candidates of the initial list. The $m \times n$ selected debris will be captured at times $t_1, t_2, \ldots, t_{m \times n}$. The order of the debris and the times of capture should be chosen to minimize the cost of the missions.

Figure 10.4 illustrates a situation with twenty candidate debris (grey points) and three successive missions eliminating five debris each. At the end of each mission, the vehicle is either destroyed by a controlled reentry or it is refueled into orbit for the next mission. In the end, 15 out of 20 debris have been deorbited and five debris remain in orbit.

The duration of each mission is limited (e.g. one year for five debris) and the cost of a single mission depends on the vehicle mass. This mass includes the propellants consumed by the maneuvers (orbital transfers from one debris to another, deorbitation of the debris if carried out by the vehicle itself) and the subsystems used for debris operations (rendezvous, capture, deorbitation if carried out by autonomous kits to be attached to the debris).

High or low-thrust architectures need to be compared in order to choose the most economical solution. For a low-thrust vehicle, an additional high-thrust deorbiting system shall nevertheless be implemented to ensure a sufficient accuracy on the reentry conditions and the impact zone.

As the vehicles are identical for each mission, their design must be based on the most demanding mission. The optimization problem consists of planning the missions to minimize the cost of the most demanding mission. The evaluation of this sizing mission is illustrated in Figure 10.5. The 15 debris selected out of 20 candidates are captured at

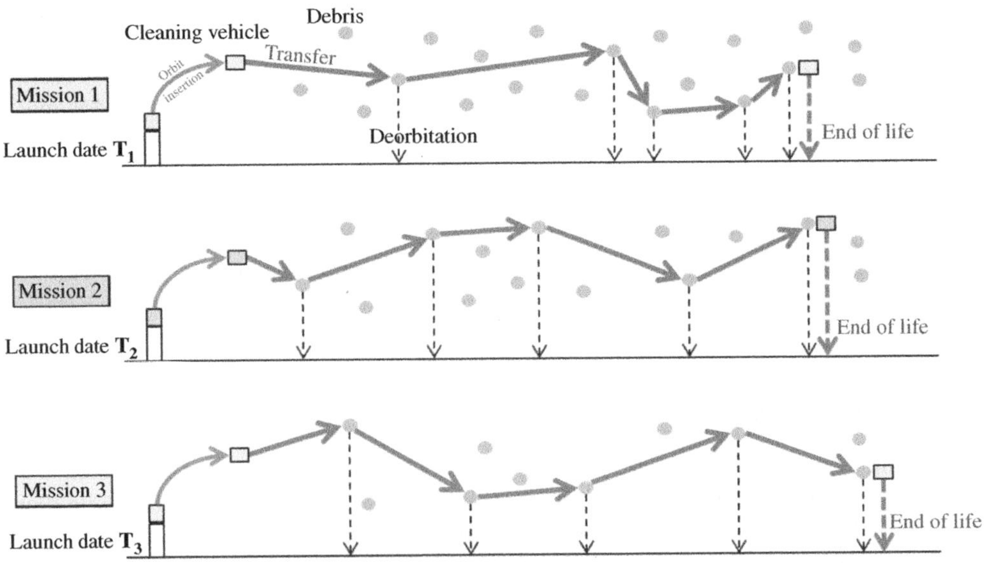


**Figure 10.4** Program of three missions deorbiting five debris each.

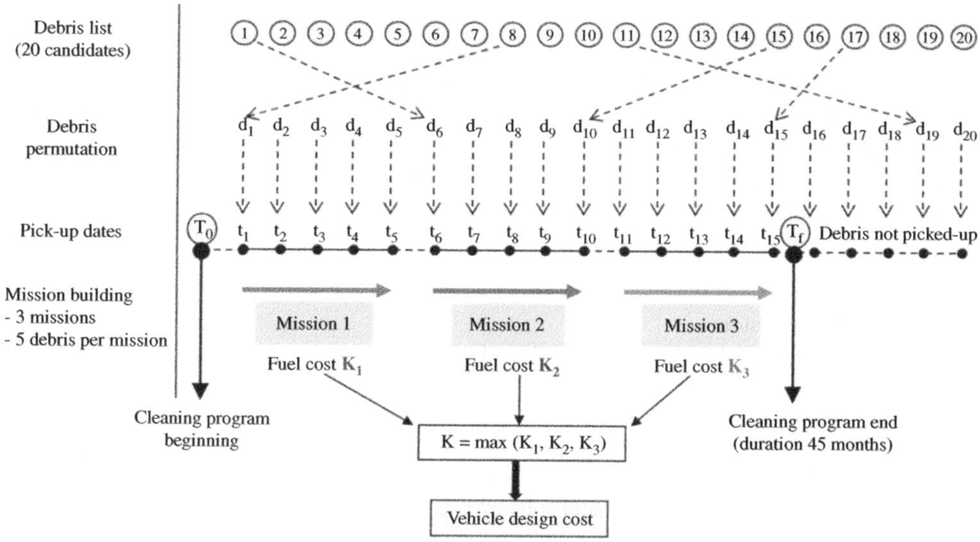


**Figure 10.5** Debris and date choice – sizing mission cost.

successive times $t_1 < t_2 < \ldots < t_{15}$. The costs of the three successive missions are $K_1$, $K_2$ and $K_3$, respectively. The vehicle is sized for the most expensive mission. The initial time $T_0$ is set and the final time $T_f$ is set accordingly to achieve the targeted average disposal rate of five debris per year.

### 10.2.3 Optimization Problem

The formulation of the problem involves integer variables (debris numbers), real variables (capture times) and control laws (orbital transfer maneuvers).

Two embedded problems of different nature need to be addressed.

- Choosing the debris and their capture times constitutes a **path problem** in a time-dependent graph (because the debris orbits move due to the precession).
- Finding the control to move from one debris to another at minimum consumption constitutes an optimal control problem, that will be referred to as **transfer problem**.

#### 10.2.3.1 Path Problem

The path problem is a combinatorial problem that can be addressed in three different ways.

- The explicit enumeration of all possible combinations ensures that the optimal path is found. For a problem with m binary variables, the total number of combinations is $2^m$. For $m = 50$, the enumeration would take a year with a computer estimating one billion solutions per second. The explicit enumeration is therefore limited to problems of small size (a few dozen variables at most).
- Implicit enumeration explores the tree of combinations with branch elimination strategies (separation/evaluation methods, dynamic programming methods ...). Elimination is based on an estimate of the best potential solution contained in a branch, which is compared to the best overall solution already found. The efficiency depends on the exploration strategy, but computation times increase exponentially with the problem size. The implicit enumeration is applicable to medium-sized problems (a few hundred variables).

- Approximate solutions can be found by metaheuristics such as genetic algorithms, simulated annealing, swarms or ant colonies. These methods limit a priori the number of combinations evaluated and thus the calculation time. The search space is explored by a random oriented strategy. If the best solution found is considered unsatisfactory, the search can be resumed with new settings that improve the exploration strategy. Unlike continuous problems, there are usually no optimality conditions for discrete problems, so there is no guarantee of the quality of the solution. Approximate methods are the only ones applicable for large problems (more than a thousand variables).

#### 10.2.3.2 Transfer Problem

The transfer problem is an optimal control problem that can be addressed in two different ways.

- Direct methods discretize the control law with time steps. The functional problem is transformed into a large finite-dimension problem for which nonlinear programming (NLP) algorithms can be applied. Direct methods are easy to handle and they can be applied to any optimization problem without specific implementation. But their convergence can be slow and imprecise for large size problems.
- Indirect methods solve the control problem using Pontryaguin Maximum Principle (PMP). The optimality conditions involve a costate vector and lead to a boundary value problem (BVP) that can be solved by a shooting method. Indirect methods converge quickly and accurately, but they require problem-specific implementation and raise issues of initialization and sensitivity.

This review of algorithms shows that a direct solution to the debris-cleaning problem is not feasible. To reduce the overall complexity, a simplified transfer strategy is defined in the next section.

## 10.3 Transfer Problem

The transfer problem consists of finding the minimum-consumption trajectory between two debris. This optimal control problem is difficult in the general case. To simplify it, a generic transfer strategy is defined, which is suited to the debris orbits and to the propulsion system. This transfer strategy reduces the optimal control problem to a small-size nonlinear optimization problem that is easier to solve.

### 10.3.1 Generic Transfer Strategy

The simplifications are based on the mission following specificities:

- The debris orbits are almost circular, since most of the debris originate from old observation satellites in circular orbits.
- The targeted removal rate (five debris per year) gives sufficient time to achieve the RAAN changes by using the natural precession due to the $J_2$ term.

The transfer strategy between two debris consists of putting the cleaning vehicle in a circular drift orbit and waiting for the RAAN change to occur naturally. Once the desired RAAN change is completed, the vehicle goes to the orbit of the targeted debris to capture it.

The choice of a circular drift orbit facilitates the operations (tracking and control from ground stations, energy management during eclipses and calculation of maneuvers).

**Remark 10.1** *This transfer strategy using the RAAN natural precession is near-optimal as long as a sufficient time is allocated to the transfer. For short durations, this strategy would no longer be applicable and the RAAN change would require expensive propelled maneuvers. In the context of debris removal, where time is not an issue, such maneuvers are not envisioned and the drift strategy is the most interesting.*

**Remark 10.2** *The deorbitation of the debris is carried out either by the vehicle itself or by a kit to be attached to the debris. The associated durations and consumptions are evaluated independently of the transfer. They are taken into account in the mission global optimization according to the procedure described in Section* 10.4.2.

The transfer from debris 1 to debris 2 is divided into three phases:

- a propelled transfer from the orbit of debris 1 to the drift orbit;
- a waiting time on the drift orbit;
- a propelled transfer from the drift orbit to the debris orbit 2.

The transfer begins on a time $t_1$ and ends on a time $t_2$. The orbital parameters of the successive orbits are shown in Table 10.1.

The maneuver calculation depends on the propulsion system. The following sections present the dynamic models adapted for high-thrust or low-thrust engines.

### 10.3.2 High-Thrust Propulsion

In the case of a high-thrust engine, the propelled transfers are modeled by impulse maneuvers. Each transfer (debris 1 to drift orbit, then drift orbit to debris 2) is achieved by a two-impulse Hohmann transfer with a partial inclination change at each impulse. The inclination of the intermediate elliptic orbit is given by Lisowski approximation, which minimizes the sum of the squares of the impulses (instead of the sum of the norms). This near-optimal solution is given by (6.20).

**Table 10.1** Successive orbits during the transfer.

| | Propelled transfer 1 | Drift | | Propelled transfer 2 |
|---|---|---|---|---|
| Orbit | Debris 1 | Drift start | Drift end | Debris 2 |
| Time | $t_1$ | $t_{d1}$ | $t_{d2}$ | $t_2$ |
| Radius | $a_1$ | $a_d$ | $a_d$ | $a_2$ |
| Inclination | $I_1$ | $I_d$ | $I_d$ | $I_2$ |
| RAAN debris 1 | $\Omega_1(t_1)$ | $\Omega_1(t_{d1})$ | $\Omega_1(t_{d2})$ | $\Omega_1(t_2)$ |
| RAAN debris 2 | $\Omega_2(t_1)$ | $\Omega_2(t_{d1})$ | $\Omega_2(t_{d2})$ | $\Omega_2(t_2)$ |
| RAAN vehicle | $\Omega_v(t_1) = \Omega_1(t_1)$ | $\Omega_v(t_{d1})$ | $\Omega_v(t_{d2})$ | $\Omega_v(t_2) = \Omega_2(t_2)$ |

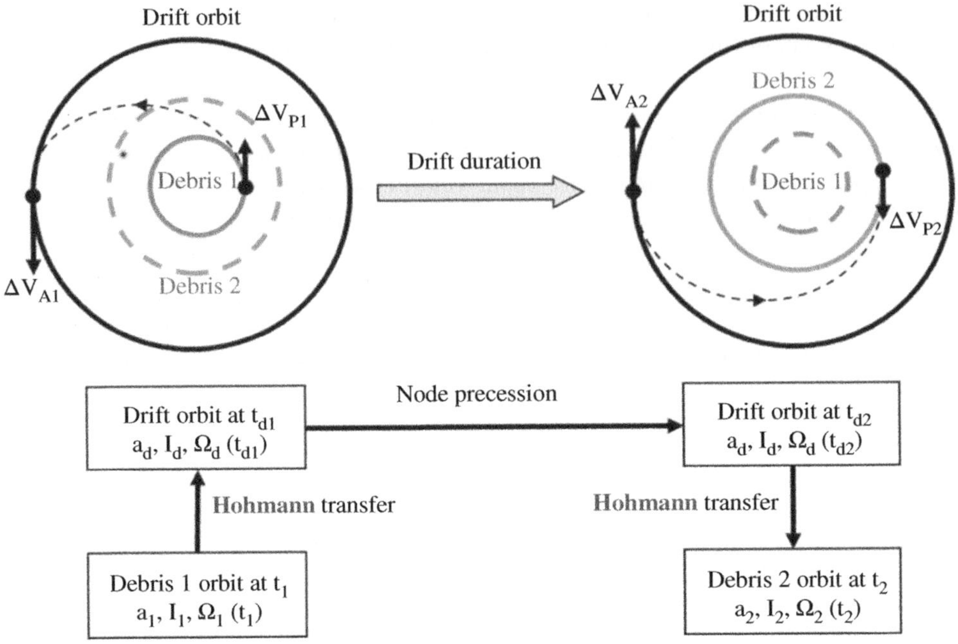


**Figure 10.6** High-thrust transfer strategy.

The transfer strategy is illustrated in Figure 10.6 with the four successive impulses required by the Hohmann transfers. In this figure, the drift orbit is higher than the targeted orbit of debris 2.

#### 10.3.2.1 RAAN Constraint

Hohmann transfers are of negligible duration (a few hours) compared to the drift time (several days or weeks). The RAAN precession can be neglected during these transfers, so that we have

$$\begin{cases} t_{d1} \approx t_1 & \Rightarrow \quad \Omega_v(t_{d1}) \approx \Omega_v(t_1) \\ t_2 \approx t_{d2} & \Rightarrow \quad \Omega_v(t_2) \approx \Omega_v(t_{d2}) \end{cases} \tag{10.3}$$

The RAAN constraint is thus explicitly linked to the drift phase parameters.

#### 10.3.2.2 Cost Assessment

The total cost of the transfer is the sum of the four velocity impulses. This impulsive cost does not depend on the thrust level of the vehicle, so that it can be directly used when assessing the mission total cost. The mass costs induced by the debris deorbitation will also be accounted as described in Section 10.4.2.

### 10.3.3 Low-Thrust Propulsion

In the case of a low-thrust engine, the propelled transfers are modeled by Edelbaum's dynamics supplemented by the nodal precession due to the $J_2$ term.

This dynamic model has been studied in Section 9.2.4. It assumes that the average orbit is constantly circular and varies at each revolution following the differential equations (9.83).

The vehicle state is defined by the velocity V (equivalent to the radius a for a circular orbit), the inclination I and the RAAN Ω. The transfer is controlled by the acceleration level f and by the out-of-plane thrust angle β.

The optimal control problem is formulated with fixed endpoint conditions corresponding to the pick-up of successive debris.

$$\min_{f,\beta} J = \int_{t_0}^{t_f} f dt \quad \text{with} \quad \begin{cases} \dot{V} = -f \cos \beta \\ \dot{I} = \dfrac{2}{\pi V} f \sin \beta \\ \dot{\Omega} = -kV^7 \cos I \end{cases} \tag{10.4}$$

- The initial conditions are fixed at $t_0$: $V(t_0) = V_0$, $I(t_0) = I_0$, $\Omega(t_0) = \Omega_0$.
- The final conditions are fixed at $t_f$: $V(t_f) = V_f$, $I(t_f) = I_f$, $\Omega(t_f) = \Omega_f$.
- The acceleration f is bounded between 0 and $f_{max}$. The angle β is free.

#### 10.3.3.1 Problem Solution

It has been shown in Section 9.4.3 that the solution structure is ($f_{max} - 0 - f_{max}$) and a solution method has been presented in order to obtain a reliable costate guess.

In this chapter dealing with space debris cleaning, it is assumed that a sufficiently long time is allocated for each transfer from a debris to another. This allows using the approximate analytical solution derived in Section 9.4.4. The main simplification consists of assessing separately each propelled transfer by Edelbaum's formulas, which gives the minimum-time solution. This solution is suboptimal in terms of propellant consumption, as it does not account for the global optimization problem (10.4). Nevertheless, the difference should remain small as long as the allocated time allows a drift orbit close to the debris orbits and consequently short propelled transfers.

Since Edelbaum's solution assumes a constant acceleration level, it is necessary to proceed in two steps, as explained in Section 9.2.4. The solution is first calculated with the initial acceleration ($T/M_0$), which gives a first value of the duration and mean acceleration during the transfer. Edelbaum's solution is then recalculated with this new mean acceleration to obtain a more representative solution in terms of duration and consumption.

The transfer strategy for the low-thrust case is depicted in Figure 10.7 with the spiral trajectories associated with Edelbaum's transfers.

#### 10.3.3.2 RAAN Constraint

Unlike high-thrust transfers, low-thrust transfers can lead to significant durations during which the RAAN varies due to the $J_2$ effect. The approximation (10.3) is no longer valid.

Nevertheless, the RAAN variations can be calculated from Edelbaum's solution, which gives the evolution of the velocity V(t) and the inclination I(t). We can indeed deduce the evolution of the precession rate $\dot{\Omega} = -kV^7 \cos I$, then the evolution of the RAAN by numerical quadrature.

$$\Omega_v(t_{d1}) = \Omega_v(t_1) + \int_{t_1}^{t_{d1}} \dot{\Omega}_v(t) dt \quad \text{during the first propelled transfer}$$

$$\Omega_v(t_2) = \Omega_v(t_{d2}) + \int_{t_{d2}}^{t_2} \dot{\Omega}_v(t) dt \quad \text{during the second propelled transfer} \tag{10.5}$$

#### 10.3.3.3 Cost Assessment

The transfer total cost is the sum of the velocity impulses delivered during the two propelled transfers: $\Delta V = f \cdot (t_{d1} - t_1) + f \cdot (t_2 - t_{d2})$. Unlike high-thrust transfers, this cost depends on

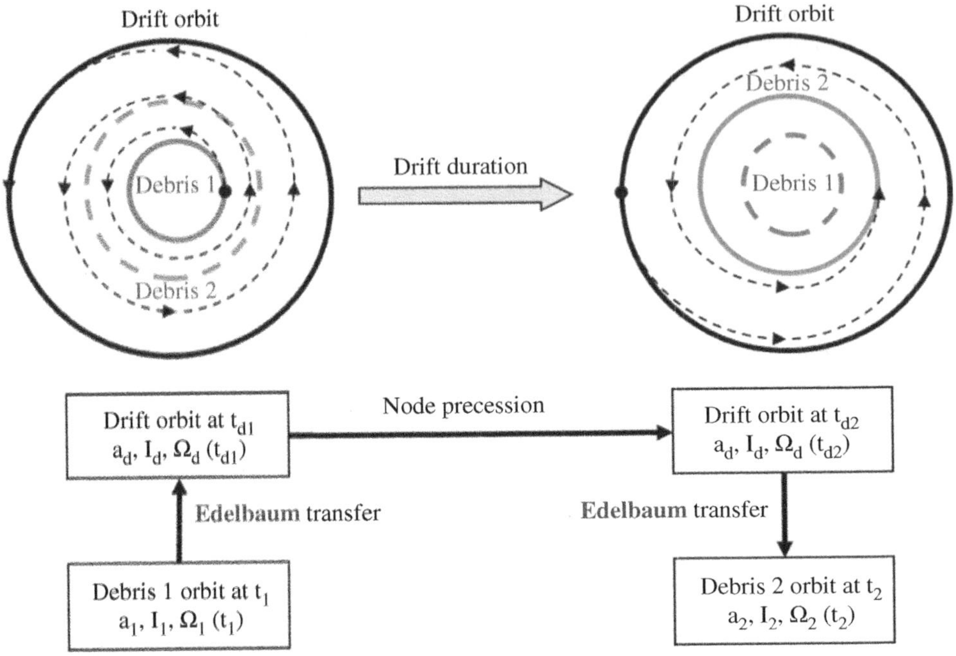


**Figure 10.7** Low-thrust transfer strategy.

the acceleration level, which itself depends on the engine thrust and on the vehicle mass at the transfer beginning. The mass evolution will therefore be defined when assessing the mission's total cost, accounting for the debris deorbitation. This cost assessment procedure is described in Section 10.4.2.

**Remark 10.3** *A low-thrust engine is not suitable for a deorbitation maneuver, which requires a sufficient accuracy to control the reentry and impact areas. It is necessary either to implement an additional high-thrust engine dedicated to these maneuvers, or to carry out the debris deorbitation by autonomous kits to attach to the debris.*

### 10.3.4 Reduced Formulation

With the previous models adapted to high-thrust or low-thrust engines, the transfer problem is reduced to finding the radius $a_d$ and the inclination $I_d$ of the drift orbit to reach the RAAN of debris 2, while minimizing the total consumption.

The vehicle RAAN noted $\Omega_V(t)$ must pass from $\Omega_1(t_1)$ (RAAN of the debris 1 at the time $t_1$) to $\Omega_2(t_2)$ (RAAN of the debris 2 at the time $t_2$).

The RAAN constraint is expressed by

$$\Omega_V(t_2) = \Omega_2(t_2)$$
$$\text{with} \begin{cases} \Omega_V(t_2) = \Omega_1(t_1) + \int_{t_1}^{t_{d1}} \dot{\Omega}_V(t)dt + \dot{\Omega}_d \cdot (t_{d2} - t_{d1}) + \int_{t_{d2}}^{t_2} \dot{\Omega}_V(t)dt \\ \Omega_2(t_2) = \Omega_2(t_1) + \dot{\Omega}_2 \cdot (t_2 - t_1) \end{cases} \tag{10.6}$$

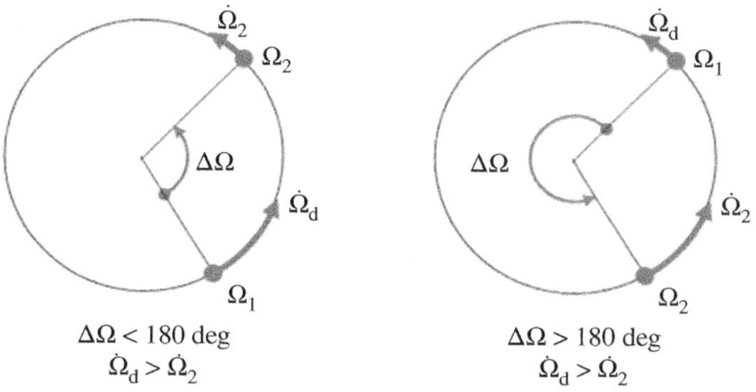


**Figure 10.8** Drift sense depending on RAAN difference.

The transfer problem from a debris to another is formulated as:

$$\min_{a_d, I_d} \Delta V \text{ s.t. } \Omega_v(t_2) = \Omega_2(t_2) \tag{10.7}$$

**Remark 10.4** *Solving the problem (10.7) does not pose any difficulty as long as some care is taken for the initialization. Indeed, there are two local minima depending on the drift sense chosen for carrying out the RAAN deviation, as shown in Figure 10.8. The optimal drift sense depends on the RAAN deviation between debris 1 and 2.*

*The initialization of the drift parameters $a_d$ and $I_d$ should account for the sense chosen, whether the drift velocity $\dot{\Omega}_d$ is higher or lower than that of the target debris $\dot{\Omega}_2$. This initialization must also take into account drift altitude bounds (typically greater than 400 km to limit the atmospheric braking).*

The problem (10.7) is formulated with fixed initial and final conditions and times. It must be solved for each new trial solution, defined by a sequence of debris and times. Although this problem is very simple, it is not possible to solve it "on-line" within the path problem. The solution method presented in the next section shows how to embed this transfer problem into the global optimization problem.

## 10.4 Solution Method

For the planning of several successive missions with a few dozen candidate debris, the combinatorial problem cannot be solved exactly while keeping a reasonable calculation time. We must be satisfied with approximate solutions obtained by metaheuristics, as described in Section 10.2.3. This section presents a resolution procedure based on the simulated annealing method.

### 10.4.1 Simulated Annealing

Simulated annealing (SA) is an optimization algorithm inspired by the metallurgical process of annealing. It is very effective on difficult combinatorial problems such as the Travelling Salesman Problem and it can be customized to the planning of debris cleaning missions.

This algorithm applies to the minimization of a function f of several variables x by applying the following principles.

- The current solution $x_0$ has a cost $f_0$.
- A random perturbation on $x_0$ gives a trial solution x of cost f.
- The solution x is accepted with the probability P defined by

$$P = \begin{cases} 1 & \text{if } f \leq f_0 \\ e^{\frac{f_0 - f}{T}} & \text{if } f > f_0 \end{cases} \quad \text{where T is a “temperature” parameter.}$$

The variables x are interpreted as an arrangement of atoms having an energy level defined by f(x). Moves of atoms reducing the energy are always accepted. On the other hand, moves increasing the energy may be accepted with a high probability at high temperature (metal melting) and with a low probability at low temperature (metal solidifying). This mechanism avoids to be trapped in local minima by potentially accepting cost-increasing solutions. Starting from a high and slowly decreasing temperature, one can explore the space of variables extensively and hopefully converge toward a good minimum.

The main settings of the algorithm are:

- the initial temperature $T_0$ and the decrease factor;
- the number of trials per temperature step and the total number of trials;
- the definition of random moves to apply for each trial.

A too-fast temperature decrease may trap the solution into a local minimum, whereas a too-slow temperature decrease may result in long calculation times.

For each application case, it is necessary to try different settings to achieve a satisfactory convergence.

Four elementary moves are considered for the mission planning problem: insertion, swap, permutation and time shift. The insertion, swap and permutation change the order of the debris on the path. The time shift only changes the pick-up time while keeping the debris order. These moves are illustrated in Figure 10.9.

An evaluation (or trial) involves the following steps:

- Random drawing of an elementary move (insertion, swap or permutation)
- Random drawing of the debris on which the move is applied
- Application of the move to get a trial path

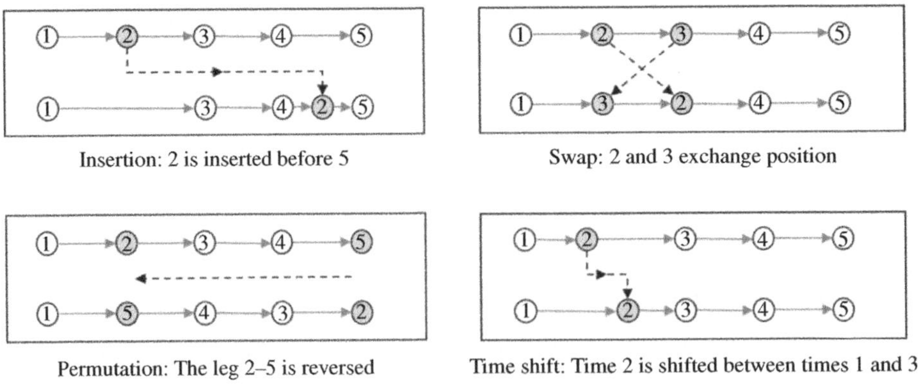


**Figure 10.9** Elementary moves for simulated annealing.

- Random drawing of a debris number on the trial path
- Random drawing of the pick-up time between the previous and next times on the path
- Evaluation of the cost of the trial path
- Uniform drawing of a random number between 0 and 1, comparison to the probability P as a function of the current temperature and the path cost, and acceptation or not of the trial path as a new reference solution

An iteration of the simulated annealing consists of lowering the temperature by a fixed factor (typically 0.999) after a given number of trials. The algorithm is started with a solution built by a greedy insertion method: each debris is inserted successively into the path to the position minimizing the cost.

The initial temperature $T_0$ is set so that an elementary random move is accepted with a probability of 90%. This high level allows for large variations during the first iterations. When the solution stagnates after several temperature steps, a local search is carried out by systematic exploration of the vicinity of the solution (defined by single elementary moves). If this search succeeds, the iterations start again from the local solution found; otherwise the algorithm is stopped.

Evaluating the cost of a trial solution requires solving a series of transfer problems between successive debris. Each transfer problem (10.7) consists of finding the drift orbit parameters $a_d$ and $I_d$ for given endpoint conditions and times, corresponding to successive pairs of debris along the trial path.

Although this small-size problem can be solved quickly by a standard nonlinear optimization algorithm, an "on-line" solution cannot be envisaged. Indeed, the simulated annealing algorithm typically requires millions of trials before reaching a satisfactory solution, so that an instantaneous cost assessment is compulsory. The next section describes a procedure based on interpolation in cost matrices.

### 10.4.2 Cost Function

The optimization of the mission planning must take into account:

- the propellant consumptions associated with the transfers between debris;
- the debris operations (capture and deorbitation);
- the priority given to the most critical debris, either for the space environment or for the risks in case of an uncontrolled reentry and fallout.

This section presents the formulation of a cost function combining these different criteria in a form that can be quickly assessed for the simulated annealing method.

#### 10.4.2.1 Transfer Costs

The cost of the transfer from one debris to another is measured by the required propellant consumption. It depends on the start time and duration of the transfer.

For a given start time $\tau$ and a given transfer time $\Delta\tau$, the transfer costs between all pairs of the N candidate debris can be assessed. This requires $N \times (N-1)$ resolutions of the elementary problem (10.7). The results are stored in an $N \times N$ cost matrix depicted in Figure 10.10.

A series of cost matrices is evaluated for a grid of start times and transfer durations covering the total duration of the cleaning program. We note:

- $T_0$ the start time of the cleaning program;
- $\Delta T$ the total duration of the cleaning program;

| | | | |
|---|---|---|---|
| Time $\tau$ Duration $\Delta\tau$ | | Arrival debris k ($1 \le k \le N$) | |
| | | | |
| Departure debris j ($1 \le j \le N$) | | Cost $C(\tau, \Delta\tau, j, k)$ | |
| | | | |

**Figure 10.10** Cost matrix for start time $\tau$ and duration $\Delta\tau$.

- $n_t$ the number of discretized transfer start times;
- $n_d$ the number of discretized transfer durations;
- $\tau_i$ the start time number i in the grid ($1 \le i \le n_t$);
- $\Delta\tau_d$ the transfer duration number d in the grid ($1 \le d \le n_d$).

For a start time $\tau_i$, a transfer duration $\Delta\tau_d$ and a pair of debris j and $k \neq j$, we denote $C(i, d, j, k)$ the transfer cost from debris j at time $\tau_i$ to debris k at time $\tau_i + \Delta\tau_d$.

The submatrix $C(i, d, 1{:}N, 1{:}N)$ of size $N \times N$ corresponds to that of Figure 10.10 for the time $\tau_i$ and the duration $\Delta\tau_d$. Some transfers may be unfeasible within the prescribed duration because the drift altitude is lower bounded to 400 km (to limit the atmospheric braking). In this case, the element of the matrix is set to an arbitrary high value, so that it will not be selected on the optimal path.

The total number of $N \times N$ submatrices to compute is $n_t \times n_d$ for a grid of $n_t$ initial times and $n_d$ transfer durations. Submatrices with an end time $(\tau_i + \Delta\tau_d)$ beyond the end of the cleaning program $(T_0 + \Delta T)$ are unnecessary. They are filled with high costs indicating the impossibility of such transfers. Figure 10.11 illustrates the grid of $n_t \times n_d$ cost submatrices covering the duration of the cleaning program.

The discretization step is the result of a compromise between the interpolation accuracy and the total computation time. The following empirical choices are based on average

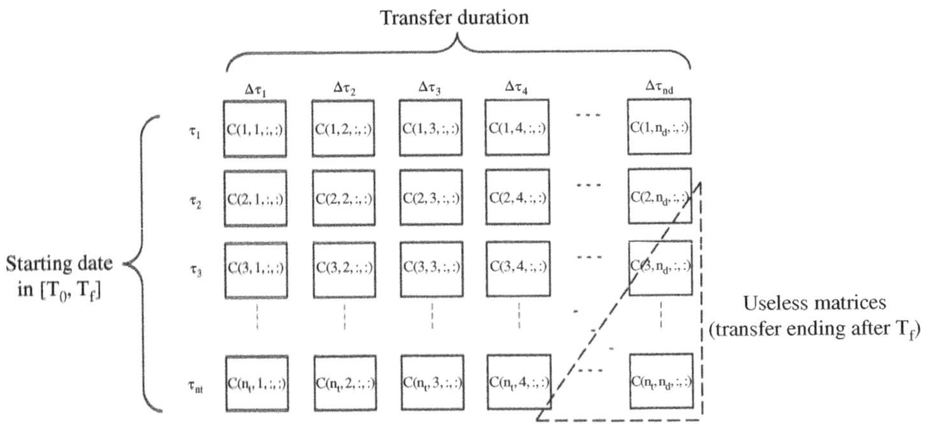


**Figure 10.11** Grid of cost matrices.

durations per mission and average durations per transfer. They provide a suitable compromise for practical applications.

- $n_t = n \times m$ (as many initial times as debris to be selected)
- $\tau_1 = T_0$ (start time of the cleaning program)
- $\tau_{nt} = T_0 + \Delta T$ (end time of the cleaning program)
- $n_d = n$ (as many transfer times as debris per mission)
- $\Delta\tau_1 = \frac{\Delta T}{2\,m\,n}$ (minimum duration = half the average duration of a transfer)
- $\Delta\tau_{nd} = \frac{\Delta T}{2\,m}$ (maximum duration = half the average duration of a mission)

A total of $n_t \times n_d \times N \times (N-1)$ optimizations is performed to fill the cost matrix grid. This grid is used in the simulated annealing algorithm to estimate the cost of a trial path by bilinear interpolation on the initial time and on the transfer duration.

A simulated annealing trial solution is a path defined by:

- the numbers of the successive debris noted $(d_1, d_2, \ldots, d_N)$.
  The set $(d_1, d_2, \ldots, d_N)$ is a permutation of $(1, 2, \ldots, N)$.
- the successive pick-up times noted $(t_1, t_2, \ldots, t_N)$;
- These times increase strictly along the path: $t_1 < t_2 < \cdots < t_N$.

For each transfer of the trial solution, the cost is evaluated by bilinear interpolation on the transfer initial time and duration. The complete path is then subdivided into m sub-paths of n debris each, representing the successive missions.

The m costs of the subpaths are noted $(K_i)_{1 \le i \le m}$. The global cost of the solution is that of the most expensive subpath $K = \max(K_1, K_2, \ldots, K_m)$.

This calculation process is depicted in Figure 10.5 in Section 10.2.2.

#### 10.4.2.2 Capture and Deorbitation Costs

In addition to the propellant consumption associated with the transfers between debris, the mission optimization must take into account the debris operations (approach, capture, deorbitation), both in terms of durations and masses.

A fixed amount of time is allocated to the debris operations (typically a few hours or days). This duration noted $\Delta t_{oper}$ is taken into account in the construction of the cost matrices by including a last waiting sequence of $\Delta t_{oper}$ duration in the transfer modeling. A transfer from debris 1 at time $t_1$ to debris 2 at time $t_2$ is thus completed at time $t_2 - \Delta t_{oper}$. This procedure allows the duration of operations on the debris to be implicitly taken into account within the optimization of the path.

The approach, capture and deorbitation maneuvers induce propellant consumption and additional equipment masses, which may depend on the characteristics of the debris to be removed. These operational costs, noted $C_{oper}$, are stored on the diagonal of each cost matrix C(i, d, j, k=j), and they are thus taken into account in the global optimization of the mission. Two deorbitation options are considered.

- Deorbitation performed by the vehicle itself

In this case, the velocity impulse required for the deorbitation depends on the debris altitude and it can be assessed a priori. The deorbitation impulses of the N debris are stored on the diagonals of the cost matrix. For a transfer from debris j to debris k, the deorbitation $\Delta V$ of debris k is added to the interpolated $\Delta V$ of the transfer, so that it is accounted for in the propellant consumption.

- Autonomous deorbitation by a "kit" brought by the vehicle

The kit is attached to the debris, then the kit + debris assembly is released to perform the deorbitation maneuver. In this option, the masses of the N kits designed for the N debris (standard size, or specific size for some debris) are stored on the diagonals of the cost matrix. At the end of the transfer arriving on debris k, the total mass of the vehicle is simply reduced by the mass of the kit released on debris k.

#### 10.4.2.3 Priority of Debris

It is also possible to consider weights on debris for prioritization based on their dangerousness to the space environment and the risks they would create in the event of uncontrolled fallout. The weight $w_k$ associated with debris k acts as a multiplier on column k of the cost matrix.

The cost of the transfer from the debris j at time $t_p$ to the debris k at time $t_p+\Delta t_p$, taking into account the operations on the debris k and its weight $w_k$, is thus given by $w_k \times [C_{transfer}(t_p, \Delta t_p, j, k) + C_{oper}(k)]$.

The cost $K_i$ of the mission number i is obtained by summing the selected debris.

### 10.4.3 Solution Process

The transfer cost obtained by solving problem (10.7) is a nonlinear function of the start and end times. A bilinear interpolation may therefore differ significantly from the actual value, depending on the discretization of the grid.

In order to obtain a reliable assessment of the costs, the missions are reoptimized a posteriori with simulated transfers and maneuvers. For each mission, the order of the debris is determined by the simulated annealing solution. Only the capture dates and drift orbits are reoptimized. This precise optimization with simulation of the trajectories gives the actual propellant masses to use for the vehicle design.

During this step, the various maneuvers (propelled transfer, deorbitation and possible reorbitation) are precisely simulated and it is possible to take into account some operational constraints regarding the tracking by a network of ground stations and the reentry and fallout in the targeted impact area.

If the precise solution gives a cost too different from that obtained by the simulated annealing, the optimization of the path is resumed with a finer discretization grid.

This resolution process is schematized in Figure 10.12 with the optimization algorithms used for the three steps.

## 10.5 Application Case

A list of 21 debris in circular orbits is considered. The altitudes range from 700 to 900 km, the inclinations range from 97 to 99 deg and the initial RAAN range from 0 to 360 deg. The altitudes, inclinations and RAAN are evenly distributed within their respective ranges. Table 10.2 gives the orbital parameters of the debris, with their nodal precession rate in the last column.

The objective is to plan three successive missions deorbiting five debris each, while minimizing the impulse of the most expensive mission. The total duration is limited to 45 months, giving an average of 3 months per transfer. A fixed period of five days between each transfer is reserved for the debris operations (capture and deorbitation). The drift orbit altitudes are

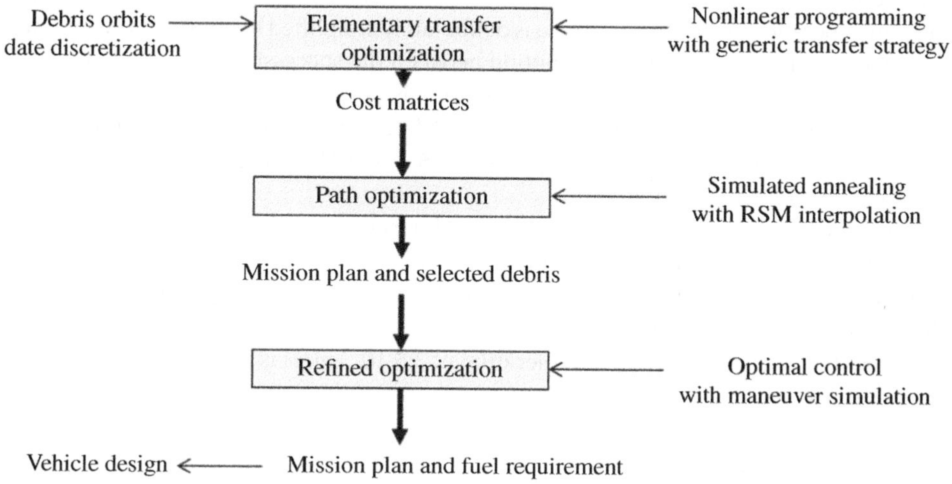


**Figure 10.12** Three-step solution process.

**Table 10.2** List of 21 candidate debris.

| Debris number | Altitude (km) | Inclination (deg) | Initial RAAN (deg) | Precession rate (deg/d) |
|---|---|---|---|---|
| 1 | 700 | 97.0 | 0. | 0.8429 |
| 2 | 710 | 97.3 | 90. | 0.8745 |
| 3 | 720 | 97.6 | 180. | 0.9058 |
| 4 | 730 | 97.9 | 270. | 0.9367 |
| 5 | 740 | 98.2 | 018. | 0.9672 |
| 6 | 750 | 98.5 | 108. | 0.9975 |
| 7 | 760 | 98.8 | 198. | 1.0273 |
| 8 | 770 | 97.1 | 288. | 0.8260 |
| 9 | 780 | 97.4 | 36. | 0.8565 |
| 10 | 790 | 97.7 | 126. | 0.8866 |
| 11 | 800 | 98.0 | 216. | 0.9165 |
| 12 | 810 | 98.3 | 306. | 0.9460 |
| 13 | 820 | 98.6 | 54. | 0.9752 |
| 14 | 830 | 98.9 | 144. | 1.0040 |
| 15 | 840 | 97.2 | 234. | 0.8094 |
| 16 | 850 | 97.5 | 324. | 0.8389 |
| 17 | 860 | 97.8 | 72. | 0.8681 |
| 18 | 870 | 98.1 | 162. | 0.8969 |
| 19 | 880 | 98.4 | 252. | 0.9254 |
| 20 | 890 | 98.7 | 342. | 0.9536 |
| 21 | 900 | 99.0 | 360. | 0.9815 |

bounded between 400 and 2000 km. It is assumed here that all debris have the same priority and the same operation costs. This example is deliberately simplified to highlight the optimal solution features in terms of debris distribution between the successive missions. For actual applications, reference lists of debris and internationally agreed priority criteria are used.

The interpolation grid is constructed with 16 initial dates (from 0 to 45 months) and 6 transfer durations (from 20 to 200 days). This requires $16 \times 6 \times 20 \times 21 = 40\,320$ elementary transfer optimizations (about 4 h of computation).

The simulated annealing is applied using the cost interpolation grid to find a mission planning. It converges in 10 minutes with 200 million trials.

The maneuvers are then refined by precise optimization with trajectory simulation. Cases with high or low trust are treated successively.

For the low-thrust case, an average acceleration of $3.5 \cdot 10^{-3}$ m/s$^2$ is considered to calculate Edelbaum's transfers.

### 10.5.1 High-Thrust Case

Table 10.3 shows a sampling of the simulated annealing iterations, with the debris numbers selected and the total impulse required. It can be seen from the last iterations that there are several paths giving very similar costs.

Table 10.4 presents the simulated annealing mission planning for the high-thrust case. It is observed that the three missions have close impulses, between 820 and 838 m/s, indicating a good convergence of the simulated annealing (minimization of the most expensive mission). The $\Delta V$ shown here are associated with transfers between debris. For real applications, the optimization takes into account the deorbitation $\Delta V$ (if carried out by the vehicle) or the masses of the kits to be attached to the debris according to the procedure described in Section 10.4.2.

Table 10.5 presents the mission planning obtained after the reoptimization of the times and maneuvers (debris order fixed). The $\Delta V$ per mission is shown in the last column, to be compared with the interpolated $\Delta V$ given in Table 10.4 and recalled in parentheses. The comparison with the simulated annealing results shows an improvement on all missions, thanks to the precise optimization of the drift orbit parameters. The main changes are marked by the orange cells.

- The improvement in mission 2 is due to the advance of the second transfer time (−52 days), reducing the $\Delta V$ by 84 m/s. The transfer duration is increased to 218 days. This solution was not found by interpolation, as the mesh size was defined with a maximum transfer time of 200 days. In this case, it may be useful to resume the process with a finer discretization of the durations.
- The improvement in mission 3 is due to the advance of the time of the second transfer (−38 days). The $\Delta V$ of this transfer increases by about 50 m/s, but this is compensated by significant gains on the previous and next transfer. A more precise discretization could also help to detect these nonlinearities of the cost function during the simulated annealing step.

Another interesting observation concerns the RAAN of the debris selected for each mission. Table 10.6 shows the RAAN of the debris at the start time of the mission. It can be noted that the debris are rather grouped by increasing initial RAAN values, with fairly regular spacing. This is not an absolute rule (especially for mission 3), as the optimal order also depends on the other orbital parameters (radius and inclination). Where large differences in

**Table 10.3** Iterations of simulated annealing.

| ΔV (m/s) | Debris order | | | | | | | | | | | | | | | | | | | | |
|---|---|---|---|---|---|---|---|---|---|---|---|---|---|---|---|---|---|---|---|---|---|
| 28353.3 | 15 | 14 | 13 | 12 | 4 | 10 | 9 | 8 | 7 | 6 | 5 | 11 | 3 | 1 | 2 | 16 | 17 | 18 | 19 | 20 | 21 |
| 22706.3 | 13 | 12 | 4 | 14 | 15 | 6 | 7 | 9 | 19 | 10 | 11 | 3 | 16 | 2 | 1 | 17 | 18 | 20 | 8 | 5 | 21 |
| 20859.4 | 13 | 12 | 7 | 14 | 15 | 4 | 9 | 10 | 19 | 17 | 11 | 3 | 16 | 2 | 1 | 18 | 20 | 8 | 5 | 6 | 21 |
| 20810.4 | 1 | 16 | 5 | 15 | 18 | 17 | 19 | 8 | 9 | 10 | 3 | 12 | 4 | 7 | 20 | 14 | 6 | 11 | 21 | 2 | 13 |
| 20644.6 | 1 | 16 | 8 | 15 | 18 | 9 | 5 | 19 | 17 | 10 | 3 | 12 | 4 | 7 | 20 | 14 | 6 | 11 | 21 | 2 | 13 |
| 18544.8 | 3 | 8 | 20 | 19 | 14 | 5 | 2 | 4 | 11 | 6 | 16 | 1 | 7 | 15 | 13 | 21 | 10 | 9 | 12 | 17 | 18 |
| 16611.0 | 3 | 7 | 20 | 19 | 14 | 5 | 8 | 4 | 1 | 6 | 11 | 16 | 2 | 15 | 13 | 21 | 10 | 9 | 12 | 17 | 18 |
| 16459.2 | 3 | 7 | 20 | 19 | 16 | 5 | 8 | 4 | 1 | 6 | 11 | 14 | 2 | 15 | 13 | 21 | 10 | 9 | 12 | 17 | 18 |
| 14129.7 | 18 | 7 | 20 | 8 | 1 | 12 | 2 | 6 | 10 | 16 | 19 | 14 | 4 | 15 | 13 | 21 | 9 | 5 | 17 | 3 | 11 |
| 13434.5 | 18 | 5 | 1 | 13 | 7 | 12 | 4 | 21 | 20 | 9 | 2 | 15 | 10 | 14 | 19 | 6 | 17 | 3 | 16 | 8 | 11 |
| 1769.6 | 16 | 12 | 1 | 20 | 9 | 4 | 7 | 19 | 11 | 8 | 13 | 18 | 15 | 5 | 10 | 14 | 6 | 2 | 17 | 21 | 3 |
| 1685.1 | 16 | 12 | 1 | 20 | 9 | 4 | 7 | 19 | 11 | 8 | 13 | 18 | 15 | 5 | 10 | 14 | 6 | 2 | 17 | 21 | 3 |
| 1604.2 | 10 | 14 | 18 | 6 | 15 | 4 | 16 | 7 | 8 | 11 | 2 | 5 | 21 | 20 | 17 | 12 | 19 | 13 | 1 | 9 | 3 |
| 1553.8 | 20 | 21 | 9 | 17 | 5 | 3 | 15 | 14 | 8 | 11 | 4 | 1 | 19 | 7 | 12 | 16 | 2 | 13 | 6 | 18 | 10 |
| 1514.0 | 10 | 6 | 17 | 5 | 21 | 16 | 4 | 1 | 7 | 12 | 15 | 13 | 18 | 3 | 8 | 2 | 19 | 14 | 9 | 11 | 20 |
| 1470.4 | 16 | 8 | 4 | 19 | 7 | 14 | 15 | 18 | 3 | 6 | 20 | 17 | 2 | 5 | 21 | 11 | 9 | 13 | 10 | 12 | 1 |
| 1374.6 | 12 | 16 | 8 | 19 | 7 | 18 | 6 | 3 | 11 | 14 | 17 | 20 | 2 | 5 | 21 | 15 | 1 | 10 | 9 | 4 | 13 |
| 1326.3 | 2 | 6 | 18 | 3 | 14 | 16 | 19 | 7 | 1 | 4 | 13 | 15 | 21 | 5 | 10 | 8 | 9 | 20 | 17 | 11 | 12 |
| 1301.8 | 2 | 6 | 18 | 3 | 14 | 19 | 16 | 7 | 1 | 4 | 13 | 15 | 21 | 10 | 5 | 8 | 9 | 20 | 17 | 11 | 12 |
| 1268.6 | 19 | 8 | 4 | 12 | 1 | 15 | 3 | 6 | 11 | 14 | 20 | 17 | 21 | 5 | 10 | 9 | 2 | 13 | 18 | 16 | 7 |
| 1251.0 | 3 | 7 | 19 | 4 | 16 | 15 | 14 | 11 | 8 | 6 | 17 | 20 | 21 | 5 | 10 | 9 | 12 | 13 | 18 | 2 | 1 |
| 1239.8 | 11 | 15 | 3 | 18 | 6 | 4 | 16 | 7 | 12 | 9 | 17 | 20 | 21 | 10 | 5 | 13 | 2 | 8 | 19 | 14 | 1 |
| 1231.6 | 10 | 6 | 14 | 3 | 11 | 4 | 16 | 19 | 9 | 12 | 18 | 13 | 15 | 5 | 21 | 17 | 20 | 8 | 7 | 2 | 1 |
| 1213.0 | 15 | 3 | 14 | 18 | 6 | 19 | 7 | 16 | 4 | 1 | 20 | 17 | 21 | 5 | 10 | 9 | 13 | 11 | 8 | 2 | 12 |
| 1210.4 | 15 | 3 | 14 | 18 | 6 | 19 | 7 | 16 | 4 | 1 | 17 | 20 | 21 | 5 | 10 | 9 | 13 | 11 | 8 | 2 | 12 |
| 1142.3 | 12 | 16 | 4 | 19 | 7 | 3 | 14 | 11 | 8 | 6 | 13 | 21 | 5 | 10 | 15 | 9 | 18 | 17 | 1 | 20 | 2 |
| 1128.2 | 17 | 13 | 6 | 18 | 3 | 16 | 4 | 7 | 12 | 9 | 2 | 21 | 5 | 10 | 15 | 19 | 14 | 20 | 11 | 1 | 8 |
| 1119.5 | 7 | 11 | 15 | 3 | 14 | 12 | 1 | 4 | 16 | 19 | 20 | 2 | 21 | 5 | 10 | 13 | 9 | 17 | 6 | 8 | 18 |
| 1105.9 | 12 | 16 | 4 | 7 | 19 | 18 | 14 | 11 | 8 | 6 | 13 | 15 | 10 | 21 | 5 | 9 | 2 | 1 | 17 | 3 | 20 |
| 1052.4 | 12 | 16 | 4 | 7 | 19 | 3 | 14 | 11 | 8 | 6 | 13 | 15 | 10 | 21 | 5 | 9 | 2 | 1 | 17 | 18 | 20 |
| 996.7 | 12 | 16 | 4 | 19 | 7 | 3 | 14 | 11 | 8 | 6 | 13 | 15 | 21 | 5 | 10 | 9 | 2 | 1 | 17 | 18 | 20 |
| 962.6 | 12 | 8 | 4 | 19 | 7 | 18 | 6 | 3 | 11 | 14 | 13 | 15 | 10 | 5 | 21 | 1 | 2 | 20 | 9 | 16 | 17 |
| 958.0 | 12 | 4 | 8 | 19 | 7 | 18 | 6 | 3 | 11 | 14 | 13 | 15 | 10 | 5 | 21 | 1 | 2 | 20 | 9 | 16 | 17 |
| 946.7 | 18 | 3 | 14 | 15 | 6 | 16 | 7 | 4 | 12 | 9 | 17 | 20 | 21 | 5 | 10 | 8 | 11 | 19 | 13 | 2 | 1 |
| 900.6 | 6 | 18 | 14 | 3 | 15 | 19 | 16 | 7 | 4 | 1 | 17 | 20 | 21 | 5 | 10 | 13 | 2 | 8 | 12 | 11 | 9 |
| 897.8 | 6 | 18 | 14 | 3 | 15 | 19 | 16 | 7 | 4 | 1 | 17 | 20 | 21 | 5 | 10 | 13 | 9 | 8 | 2 | 11 | 12 |

**Table 10.3** (Continued)

| ΔV (m/s) | Debris order | | | | | | | | | | | | | | | | | | | | |
|---|---|---|---|---|---|---|---|---|---|---|---|---|---|---|---|---|---|---|---|---|---|
| 891.2 | 6 | 18 | 14 | 3 | 15 | 19 | 16 | 7 | 4 | 1 | 20 | 17 | 21 | 5 | 10 | 13 | 9 | 8 | 11 | 2 | 12 |
| 889.2 | 2 | 13 | 17 | 5 | 21 | 15 | 3 | 14 | 11 | 8 | 1 | 4 | 9 | 7 | 12 | 19 | 16 | 18 | 6 | 10 | 20 |
| 884.0 | 2 | 13 | 17 | 5 | 21 | 15 | 3 | 14 | 11 | 8 | 1 | 4 | 9 | 7 | 12 | 19 | 16 | 18 | 6 | 10 | 20 |
| 881.7 | 2 | 13 | 17 | 5 | 21 | 15 | 3 | 14 | 11 | 8 | 1 | 4 | 9 | 7 | 12 | 19 | 16 | 18 | 6 | 10 | 20 |
| 879.3 | 1 | 21 | 5 | 9 | 20 | 15 | 3 | 14 | 11 | 8 | 16 | 19 | 4 | 7 | 12 | 18 | 6 | 13 | 2 | 17 | 10 |
| 871.8 | 1 | 21 | 5 | 9 | 20 | 15 | 3 | 14 | 11 | 8 | 16 | 19 | 4 | 7 | 12 | 18 | 6 | 13 | 2 | 17 | 10 |
| 838.0 | 16 | 20 | 21 | 5 | 17 | 15 | 3 | 14 | 11 | 8 | 1 | 4 | 9 | 7 | 12 | 19 | 2 | 13 | 18 | 6 | 10 |

The main results (total DV) are in bold.

**Table 10.4** Mission planning with high thrust (simulated annealing solution).

| | Dates (d) | Debris visited | Total ΔV (m/s) |
|---|---|---|---|
| Mission 1 | 0–545.3 | 16–20–21–5–17 | **820.0** |
| Mission 2 | 552.7–935.7 | 15–3–14–11–8 | **838.0** |
| Mission 3 | 942.1–1365.9 | 1–4–9–7–12 | **837.5** |

| Mission 1 | | | Mission 2 | | | Mission 3 | | |
|---|---|---|---|---|---|---|---|---|
| Debris number | Date (d) | ΔV (m/s) | Debris number | Date (d) | ΔV (m/s) | Debris number | Date (d) | ΔV (m/s) |
| **16** | 3.1 | 285.8 | **15** | 552.7 | 145.0 | **1** | 942.1 | 154.7 |
| **20** | 184.8 | 224.4 | **3** | 616.0 | 375.5 | **4** | 1014.6 | 362.8 |
| **21** | 375.0 | 216.0 | **14** | 771.5 | 140.4 | **9** | 1179.8 | 246.9 |
| **5** | 488.7 | 93.9 | **11** | 831.9 | 177.1 | **7** | 1268.0 | 73.2 |
| **17** | 545.3 | **820.0** | **8** | 935.7 | **838.0** | **12** | 1365.9 | **837.5** |

The main results (total DV) are in bold.

RAAN exist at the mission start time (e.g. for the third and fifth debris of mission 3), they are compensated by longer transfer times (up to 6 months for the second and fourth transfers).

### 10.5.2 Low-Thrust Case

Table 10.7 presents the simulated annealing mission planning for the low-thrust case. As with the high-thrust case, it is observed that all missions have close impulses, between 963 and 973 m/s, reflecting the good convergence of the simulated annealing.

Table 10.8 presents the mission planning obtained after the reoptimization of the times and maneuvers (debris order fixed). The ΔV per mission is shown in the last column, to be compared with the interpolated ΔV given in Table 10.7 and recalled in parentheses.

The comparison with the simulated annealing results shows an improvement on all missions, thanks to the precise optimization of the drift orbit parameters. The main changes are marked by the orange cells.

**Table 10.5** Mission planning with high thrust (after reoptimization).

| | Dates (d) | Debris visited | Total ΔV (m/s) |
|---|---|---|---|
| Mission 1 | 0–545.3 | 16–20–21–5–17 | **811.1** (820.0) |
| Mission 2 | 552.7–935.7 | 15–3–14–11–8 | **711.9** (838.0) |
| Mission 3 | 942.1–1365.9 | 1–4–9–7–12 | **785.1** (837.5) |

| Mission 1 | | | Mission 2 | | | Mission 3 | | |
|---|---|---|---|---|---|---|---|---|
| Debris number | Date (d) | ΔV (m/s) | Debris number | Date (d) | ΔV (m/s) | Debris number | Date (d) | ΔV (m/s) |
| 16 | 3.1 | 287.1 | 15 | 552.7 | 141.5 | 1 | 942.1 | 119.2 |
| 20 | 183.1 | 210.8 | 3 | 563.3 | 291.8 | 4 | 976.8 | 411.8 |
| 21 | 389.3 | 202.2 | 14 | 781.7 | 132.2 | 9 | 1143.4 | 183.5 |
| 5 | 513.6 | 111.0 | 11 | 823.0 | 146.4 | 7 | 1177.3 | 70.6 |
| 17 | 545.4 | **811.1** | 8 | 935.8 | **711.9** | 12 | 1365.8 | **785.1** |

The main results (total DV) are in bold.

**Table 10.6** RAAN of selected debris at mission beginning (high thrust).

| Mission 1 ($t_1$ = 3.1 d) | | Mission 2 ($t_6$ = 552.7 d) | | Mission 3 ($t_{11}$ = 942.1 d) | |
|---|---|---|---|---|---|
| Debris number | RAAN (deg) | Debris number | RAAN (deg) | Debris number | RAAN (deg) |
| 16 | −33.4 | 15 | −38.6 | 1 | 74.1 |
| 20 | −15.1 | 3 | −39.4 | 4 | 72.4 |
| 21 | 3.0 | 14 | −21.1 | 9 | 122.8 |
| 5 | 21.0 | 11 | 2.5 | 7 | 74.1 |
| 17 | 74.7 | 8 | 24.5 | 12 | 117.2 |

The main results (total DV) are in bold.

**Table 10.7** Mission planning with low thrust (simulated annealing solution).

| | Dates (d) | Debris visited | Total ΔV (m/s) |
|---|---|---|---|
| Mission 1 | 0–556.9 | 6–14–18–3–15 | **969.2** |
| Mission 2 | 563.8–978.7 | 19–7–16–4–1 | **963.7** |
| Mission 3 | 987.6–1361.0 | 17–20–21–5–10 | **972.4** |

| Mission 1 | | | Mission 2 | | | Mission 3 | | |
|---|---|---|---|---|---|---|---|---|
| Debris number | Date (d) | ΔV (m/s) | Debris number | Date (d) | ΔV (m/s) | Debris number | Date (d) | ΔV (m/s) |
| 6 | 0.7 | 377.7 | 19 | 563.8 | 240.7 | 17 | 987.6 | 220.2 |
| 14 | 174.1 | 184.9 | 7 | 621.0 | 276.6 | 20 | 1047.4 | 443.5 |
| 18 | 302.9 | 262.3 | 16 | 714.1 | 274.7 | 21 | 1252.3 | 188.1 |
| 3 | 483.8 | 144.4 | 4 | 886.3 | 171.6 | 5 | 1285.8 | 120.6 |
| 15 | 556.9 | **969.2** | 1 | 978.7 | **963.7** | 10 | 1361.0 | **972.4** |

The main results (total DV) are in bold.

**Table 10.8** Mission planning with low thrust (after reoptimization).

| | Dates (d) | Debris visited | Total ΔV (m/s) |
|---|---|---|---|
| Mission 1 | 0–558.1 | 6–14–18–3–15 | **903.6** (969.2) |
| Mission 2 | 563.8–977.5 | 19–7–16–4–1 | **921.2** (963.7) |
| Mission 3 | 987.6–1359.9 | 17–20–21–5–10 | **926.0** (972.4) |

| Mission 1 | | | Mission 2 | | | Mission 3 | | |
|---|---|---|---|---|---|---|---|---|
| **Debris number** | **Date (d)** | **ΔV (m/s)** | **Debris number** | **Date (d)** | **ΔV (m/s)** | **Debris number** | **Date (d)** | **ΔV (m/s)** |
| 6 | 0.7 | 369.8 | 19 | 563.8 | 266.1 | 17 | 987.6 | 197.8 |
| 14 | 178.8 | 194.3 | 7 | 622.3 | 247.4 | 20 | 1059.0 | 440.4 |
| 18 | 243.4 | 219.2 | 16 | 680.3 | 222.3 | 21 | 1262.9 | 182.3 |
| 3 | 486.6 | 120.3 | 4 | 954.3 | 185.4 | 5 | 1272.0 | 105.5 |
| 15 | 558.1 | **903.6** | 1 | 977.5 | **921.2** | 10 | 1359.9 | **926.0** |

The main results (total DV) are in bold.

**Table 10.9** RAAN of selected debris at mission beginning (low thrust).

| Mission 1 ($t_1$ = 0.7 d) | | Mission 2 ($t_6$ = 563.8 d) | | Mission 3 ($t_{11}$ = 987.6 d) | |
|---|---|---|---|---|---|
| **Debris number** | **RAAN (deg)** | **Debris number** | **RAAN (deg)** | **Debris number** | **RAAN (deg)** |
| 6 | 108.7 | 19 | 53.7 | 17 | 209.3 |
| 14 | 144.7 | 7 | 57.2 | 20 | 203.7 |
| 18 | 162.6 | 16 | 76.9 | 21 | 249.3 |
| 3 | 180.6 | 4 | 78.1 | 5 | 253.2 |
| 15 | 234.6 | 1 | 115.2 | 10 | 281.6 |

The main results (total DV) are in bold.

- The improvement in mission 1 is due to the advance of the third transfer time (−59 days), which reduces the ΔV by 43 m/s. The duration of the transfer is increased to 243 days. This solution was not found by interpolation, as the mesh size was defined with a maximum transfer time of 200 days.
- The improvement in mission 2 is due to the time shift of the fourth transfer (+68 days). The ΔV of the third transfer is reduced by about 50 m/s at the expense of a very long transfer time (274 days).

As in the high-thrust case, it can be observed in Table 10.9 that the debris are rather grouped by increasing values of RAAN at the beginning of the successive missions, with fairly regular spacing.

### 10.5.3 Comparison of High and Low-Thrust Solutions

Low-thrust transfers require a significant duration that is no longer available for the drift phase. Consequently, the drift orbit must be farther away from the targeted orbit in order to increase the precession rate deviation and achieve the RAAN change in a shorter time.

This results in higher impulses required for the low-thrust missions: the total $\Delta V$ of the three missions is 2751 m/s in the low-thrust case (Table 10.8) compared to 2308 m/s in the high-thrust case (Table 10.5).

The low-thrust transfer also results in velocity losses due to gravity during the propelled phases (these losses are zero in the impulsive modeling). The $\Delta V$ for low-thrust missions is therefore higher than for high-thrust missions. But this is largely compensated by the reduction in consumption due to a higher exhaust velocity (at least three times higher).

Despite these major differences, it is observed that some debris sequences remain common to both solutions: (5–17–20–21), (3–14–15) and (1–4–7).

## 10.6 Conclusion

### 10.6.1 The Key Points

- The transfer strategy consists of going on a drift orbit to achieve the RAAN change by natural precession. The same strategy applies to high- or low-thrust engines.
- The path optimization (debris order and capture times) is completed in a few minutes by a simulated annealing method.
- The transfer costs are first assessed by interpolation into a precomputed grid of cost matrices covering all possible times of transfers, then updated by trajectory simulation.
- The planning of the successive missions accounts for the targeted removal rate, the number of debris per mission, the debris priority, and the operations costs (capture and deorbitation).
- The result is the cheapest vehicle able to perform the successive missions.

### 10.6.2 To Go Further

- MADI M. – *Space Debris Peril: Pathways to Opportunities: Capacity Building in the New Space Era*. CRC Press (2020).
- BIESBROEK R. – *Active Debris Removal in Space: How to Clean the Earth's Environment from Space Debris*. (2015).
- VAIBHAV DATTA L. – *Introduction to Space Debris*. LAP LAMBERT Academic Publishing (2013).
- https://www.orbitaldebris.jsc.nasa.gov/reference-documents/
  The reference website of NASA on space debris.
- https://www.esa.int/Space_Safety/Space_Debris/Active_debris_removal/
  The reference website of ESA on space debris.

### 10.6.3 Authorizations

- Examples in Section 10.5 are reprinted by permission of Springer Nature "Multiple Space Debris Collecting Mission: Optimal Mission Planning" published in Journal of Optimization Theory and Applications Volume 167, 195–218 (2015)

# Part III

# Launch and Reentry

# 11

# Launch into Orbit

## 11.1 Introduction

The mission of a launch vehicle is to transport a satellite from the Earth's surface to a specified orbit. The lowest stable orbits are at altitudes of at least 200 km to limit the atmospheric friction and the orbital velocity required for the satellite injection is of the order of 8 km/s. Reaction propulsion is the only way to reach these kinematic conditions at the cost of large quantities of fuel. As most of the trajectory takes place in vacuum, it is not possible to use oxygen from the air, as is the case with aircrafts. All the fuel has to be taken on board at lift-off, resulting in a very high total mass, in the order of 100 times that of the satellite. Conventional launchers are consumable, meaning that no components are recovered for the next flight. The primary objective is to reduce the gross mass by optimizing both the configuration and the trajectory.

The configuration is optimized for a reference mission defined by the payload mass and the orbit to be reached. Design choices concern the number of stages and propulsion technologies. Qualification of the launcher and its subsystems requires several years of study and testing, and the qualified configuration will then be identical for all flights. Its overall performance over a range of missions enables us to compare it with competing launchers.

The ascent trajectory from ground to orbit lasts around 30 minutes, including 2 minutes spent in the lower atmosphere below 50 km. This part of the flight is the most demanding in terms of aerodynamic loads. The nominal trajectory is optimized to reach the orbit with the lowest possible fuel consumption, while respecting the launcher's mechanical and thermal capacities and limiting the risks in case of failure. In-flight guidance updates the command in response to any perturbations encountered, so as to reach the targeted orbit precisely. A propellant reserve guarantees successful flight with a probability of around 99%.

The first section of this chapter presents the modeling of the launcher motion in terms of trajectory and attitude.

The second section formulates the trajectory optimization problem taking into account all constraints imposed on the launch and presents the solution methods.

The third section gives an overview of the on-board guidance and control systems.

*Space Trajectories: Basic and Advanced Topics*, First Edition. Max Cerf.

## 11.2 Launcher Dynamics

The launcher is considered as a nondeformable solid of variable mass subject to internal and external pressure forces. This chapter presents the force models, the equations of motion, and the staging design.

### 11.2.1 Force Models

#### 11.2.1.1 Pressure Forces

Rocket propulsion is based on the principle of reaction. The propellants are mixed and burnt in a combustion chamber and the gases are expelled by the pressure toward the nozzle. The convergent–divergent nozzle shape accelerates the gases, firstly to a transonic velocity in the throat (section $A_c$), then to a supersonic velocity $v_s$ at exhaust (section $A_s$). The thrust is generated by the ejected momentum flow rate.

Figure 11.1 shows the outer surface of the launcher ($S_{ext}$), the combustion chamber wall ($S_{int}$) and the nozzle ($S_{noz}$). The outer surface and the nozzle are subject to the pressure $p_{ext}$ due to the external atmosphere. The chamber wall and the nozzle are subject to the pressure $p_{int}$ of the burnt gases. The pressure in the exit section is $p_s$.

The resultant on the launcher of the external and internal pressure forces is

$$\vec{R} = \int_{S_{ext}+S_{noz}} p_{ext}\vec{n}\,dS - \int_{S_{int}+S_{noz}} p_{int}\vec{n}\,dS \tag{11.1}$$

where $\vec{n}$ denotes the internal normal to surfaces.

The first integral in (11.1) noted $\vec{R}_{ext}$ is calculated by introducing the atmospheric pressure $p_a$ (static pressure at an infinite distance of the launcher) and the integral on the nozzle exit section $A_s$.

$$\vec{R}_{ext} = \int_{S_{ext}+S_{noz}} (p_{ext} - p_a)\vec{n}\,dS + \int_{S_{ext}+S_{noz}+A_s} p_a\vec{n}\,dS - \int_{A_s} p_a\vec{n}\,dS \tag{11.2}$$

Since the integral of $p_a$ on the closed surface bounded by $S_{ext}$, $S_{noz}$ and $A_s$ is zero, we obtain

$$\vec{R}_{ext} = \int_{S_{ext}+S_{noz}} (p_{ext} - p_a)\vec{n}\,dS - p_a A_s \vec{n}_s \tag{11.3}$$

The second integral in (11.1) noted $\vec{R}_{int}$ is calculated by applying Euler's theorem (assuming a perfect fluid) to the volume bounded by the surfaces $S_{int}$, $S_{noz}$ and $A_s$. Euler's theorem states that the pressure forces applied by the surfaces on the fluid balance the momentum flow rate through the section $A_s$.

By noting q, the ejected mass flow rate, Euler's theorem gives

$$\int_{S_{int}+S_{noz}+A_s} p_{int}\vec{n}\,dS = -qv_s\vec{n}_s \tag{11.4}$$

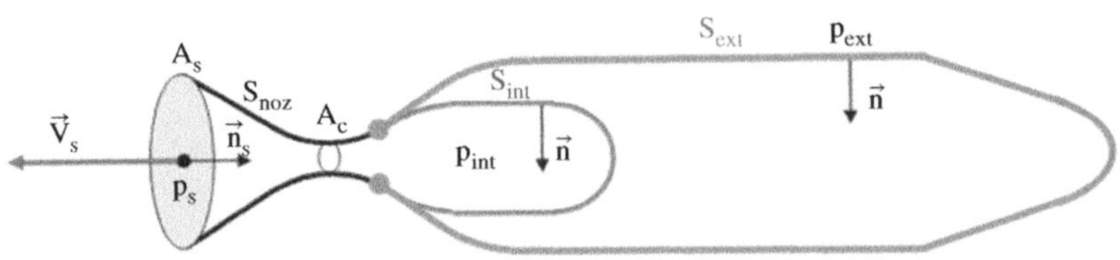


**Figure 11.1** External and internal pressure.

Assuming a uniform pressure $p_s$ in the exit section, we obtain for $\vec{R}_{int}$

$$\vec{R}_{int} = \int_{S_{int}+S_{noz}} p_{int}\vec{n}\, dS = -(qv_s + p_s A_s)\, \vec{n}_s \tag{11.5}$$

By substituting (11.3) and (11.5) into (11.1), the resultant of all pressure forces is

$$\vec{R} = [qv_s + (p_s - p_a)A_s]\, \vec{n}_s + \int_{S_{ext}+S_{noz}} (p_{ext} - p_a)\vec{n}\, dS \tag{11.6}$$

#### 11.2.1.2 Thrust and Aerodynamic Resultant

The first term in (11.6) depends on the conditions in the nozzle exit section. This term defines the thrust $\vec{T}$ directed along the normal $\vec{n}_s$.

$$\vec{T} = [qv_s + (p_s - p_a)A_s]\vec{n}_s \tag{11.7}$$

The velocity and pressure at the nozzle exit depend on the combustion conditions in the chamber, on the divergent shape and on the section ratio $\Sigma = A_s/A_c$.

The study of the flow in the nozzle shows that the velocity $v_s$ increases and the pressure $p_s$ decreases as $\Sigma$ increases, and that the thrust is maximal when the gases are expanded up to the atmospheric pressure $p_s = p_a$ (see [R15], [R26], and [R34]). The nozzle is then said to be adapted to the external pressure.

The specific impulse in vacuum noted Isv (in seconds) is defined by

$$T_{vacuum} = qv_s + p_s A_s \underset{def}{=} qg_0 Isv \quad \text{with} \quad g_0 = 9.80665\ \text{m/s}^2 \tag{11.8}$$

The actual thrust (11.7) in an atmosphere at the pressure $p_a$ is then given by

$$T = T_{vacuum} - p_a A_s = qg_0 Isv - p_a A_s \tag{11.9}$$

The specific impulse depends on the propellant type, the combustion conditions and the nozzle geometry. The typical ranges are 250–280 s for solid propellants, 300–350 s for storable propellants (liquids at ambient temperature) and 400–450 s for cryogenic propellants. The counterpart of a higher specific impulse is a heavier and more complex propulsion system (storage and cooling of liquid propellants during the flight). This is taken into account for preliminary design choices.

The second term in (11.6) depends on the external geometry and pressure, but not on the propulsion system. This term defines the aerodynamic resultant $\vec{R}_a$.

$$\vec{R}_a = \int_{S_{ext}+S_{noz}} (p_{ext} - p_a)\vec{n}\, dS \tag{11.10}$$

Calculating this integral requires computational fluid dynamics methods and it must be carried out for every possible flight condition. In the very simplified case of an axisymmetric incompressible perfect flow around the launcher, Bernoulli's theorem would give at the stagnation point: $p_{ext} - p_a = \frac{1}{2}\rho v_r^2$, where $\rho$ (kg/m$^3$) is the atmospheric density and $v_r$ is the velocity of the launcher in the atmosphere.

The numerical calculation introduces aerodynamic coefficients of drag $C_x$, lift $C_z$, and lateral force $C_y$ depending on the velocity (Mach number) and orientation in the flow (angle of attack, bank angle and skip angle). The aerodynamic resultant in the launcher axes is expressed by

$$\vec{R}_a = -QS_{ref} \begin{pmatrix} C_x \\ C_y \\ C_z \end{pmatrix} \quad \text{with} \quad Q = \frac{1}{2}\rho v_r^2 \tag{11.11}$$

where Q (Pa) is the **dynamic pressure** due to the relative motion in the atmosphere, and $S_{ref}$ ($m^2$) is the apparent cross section of the launcher in the ambient airflow.

#### 11.2.1.3 Gravity

The launcher is subject to the gravitational acceleration $\vec{g}$, which depends on the position $\vec{r}$. Considering in first approximation the Earth as a homogeneous sphere, the central gravity field is given by

$$\vec{g}(\vec{r}) = -\frac{\mu}{r^3}\vec{r} \tag{11.12}$$

where $\mu = 3.986007\ 10^{14}\ m^3/s^2$ is the Earth gravitational constant.

For a more accurate calculation, the first zonal term ($J_2$) must be taken into account, with the Cartesian components given in (2.111).

The gravity field is considered to be uniform on the launcher, whose dimensions are a few tens of meters at most. The center of gravity then coincides with the center of mass G to which the weight applies: $\vec{W} = m\vec{g}$. The position of G varies during the flight due to the consumption of propellants and the stage separations.

### 11.2.2 Motion Equations

The dynamic equations for a solid are split into the motion of the center of gravity (called trajectory) and the motion around the center of gravity (called attitude).

#### 11.2.2.1 Trajectory Equations

The trajectory equations are obtained by applying in the inertial geocentric frame the fundamental relation of dynamics (also called Newton's second law) to the system bounded by the surfaces $S_{ext}$, $S_{noz}$ and $A_s$ in Figure 11.1.

$$\frac{d\vec{P}}{dt} = \vec{F}_{ext} \tag{11.13}$$

where $\vec{P}$ is the momentum and $\vec{F}_{ext}$ is the sum of the external forces.

Figure 11.2 shows the system before and after mass ejection. The momentum before ejection is $\vec{P}^- = m\,\vec{v}$, where m is the total mass of the launcher and $\vec{v}$ its velocity in the inertial frame. The mass dm is ejected at velocity $\vec{v}_s$ and the velocity of the remaining mass (m − dm) is increased by $d\vec{v}$. The momentum after ejection is $\vec{P}^+ = dm\,(\vec{v} + \vec{v}_s) + (m - dm)(\vec{v} + d\vec{v})$, which gives the variation

$$d\vec{P} = \vec{P}^+ - \vec{P}^- = dm\,\vec{v}_s + md\vec{v} \tag{11.14}$$

The external forces are the pressure force $\vec{R}_{ext}$ (11.3) on the surfaces $S_{ext}$ and $S_{noz}$, the pressure force $p_s A_s \vec{n}_s$ on the surface $A_s$, and the weight $\vec{W}$.

By substituting into (11.13), we get

$$\frac{dm}{dt}\vec{v}_s + m\frac{d\vec{v}}{dt} = \vec{W} + (p_s - p_a)A_s\vec{n}_s + \int_{S_{ext}+S_{tuy}} (p_{ext} - p_a)\vec{n}\,dS \tag{11.15}$$

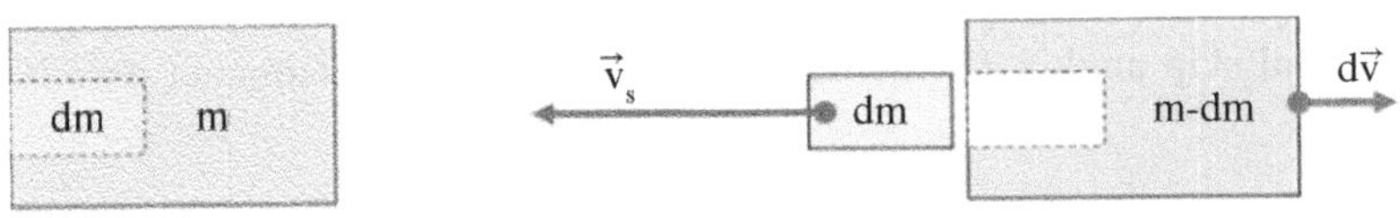


**Figure 11.2** Momentum variation.

Passing in the second member the term in $\vec{v}_s(= -v_s\vec{n}_s)$ and with $dm = q\,dt$, we retrieve the thrust and the aerodynamic resultant defined in (11.7) and (11.10).

$$m\frac{d\vec{v}}{dt} = \vec{W} + \vec{T} + \vec{R}_a \tag{11.16}$$

The trajectory of the launcher is obtained by integrating the differential system

$$\begin{cases} \dot{\vec{r}} = \vec{v} \\ \dot{\vec{v}} = \vec{g} + \vec{a} \quad \text{with } \vec{a} = \dfrac{\vec{T} + \vec{R}_a}{m} \\ \dot{m} = -q \end{cases} \tag{11.17}$$

The expression of the forces in the inertial geocentric frame involves frame changes depending on the attitude and the velocity of the launcher. The forces also depend on tabulated models defined from simulations and tests. For these reasons, the equations (11.17) cannot be analytically integrated and numerical methods are required. However, for the purposes of staging calculations, it is interesting to develop them in polar coordinates with the notations in Figure 11.3.

The local slope (also called flight path angle) $\gamma$ and the local pitch angle $\theta$ are the respective angles of the velocity and launcher axis with the local horizontal plane (normal to the Earth's surface). The incidence (or angle of attack) $\alpha$ is the angle between the velocity and the launcher axis. If the trajectory is not planar, the launcher axis makes the sideslip angle $\delta$ with the local vertical plane defined by the current position and velocity vectors $(\vec{r}, \vec{v})$.

The thrust $\vec{T}$ applies at point B of articulation of the nozzle and it makes the deflection angle $\beta$ with the launcher axis. The aerodynamic resultant $\vec{R}_a$ applies at the aerodynamic pressure center C and it is projected in drag $\vec{D}$ along the launcher axis and lift $\vec{L}$ normal to the launcher axis. The weight $\vec{W}$ applies at G.

The unit vectors along the velocity and the normal are noted $\vec{u}_v$ and $\vec{u}_n$.

By deriving the velocity: $\vec{v} = v\vec{u}_v \Rightarrow \dot{\vec{v}} = \dot{v}\vec{u}_v + v\dot{\gamma}\vec{u}_n$, then projecting the forces on $\vec{u}_v$ and $\vec{u}_n$, and noting $\varphi$ the polar angle of the position in the inertial frame, we obtain the following equations in polar coordinates.

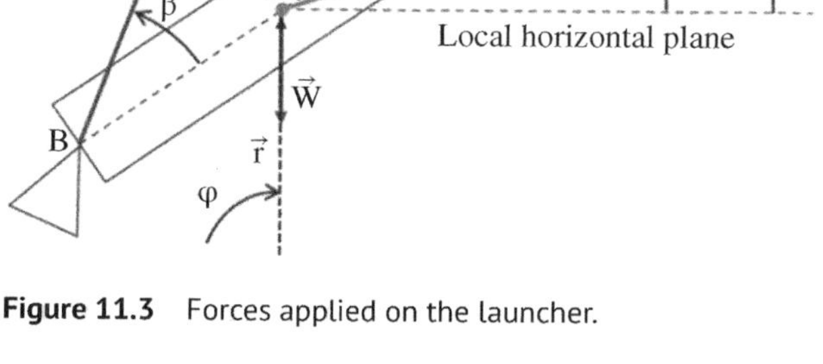


**Figure 11.3** Forces applied on the launcher.

$$\begin{cases} \dot{r} & = v \sin\gamma \\ r\dot{\varphi} & = v \cos\gamma \\ \dot{v} & = a_T \cos(\alpha+\beta) - a_D \cos\alpha - a_L \sin\alpha - g \sin\gamma \\ v(\dot{\gamma}-\dot{\varphi}) & = a_T \sin(\alpha+\beta) - a_D \sin\alpha + a_L \cos\alpha - g \cos\gamma \end{cases} \tag{11.18}$$

where $\vec{a}_T, \vec{a}_D, \vec{a}_L$ denote the respective accelerations of thrust, drag and lift.

The third equation will be used for the staging calculations in Section 11.3.1.

#### 11.2.2.2 Attitude Equations

The attitude equations are written in the inertial barycentric frame linked to the launcher. The origin of this frame noted RG is at the center of gravity G and the axes are inertial. The dynamic equation in rotation of the launcher is

$$\left.\frac{d\vec{H}_G}{dt}\right|_{RG} = \vec{M}_{ext/G} \tag{11.19}$$

where $\vec{M}_{ext/G}$ is the sum of the moments with respect to G of the external forces, $\vec{H}_G$ is the angular moment with respect to G of the launcher, and the derivative is taken in the inertial frame RG.

The launcher's angular moment is the moment of the momentum summed up over all mass elements of the launcher. A mass element dm at position $\vec{r}_m(x_m, y_m, z_m)$ in RG has a velocity $\vec{\Omega}_L \wedge \vec{r}_m$, where $\vec{\Omega}_L$ is the instantaneous angular velocity of the launcher in RG. The total angular moment is assessed by

$$\vec{H}_G = \iiint_{\text{launcher}} \vec{r}_m \wedge (\vec{\Omega}_L \wedge \vec{r}_m) dm = I_{RG} \vec{\Omega}_L \tag{11.20}$$

by introducing the inertia matrix $I_{RG} \underset{\text{def}}{=} \iiint_{\text{launcher}} \begin{pmatrix} y_m^2 + z_m^2 & -x_m y_m & -x_m z_m \\ -x_m y_m & x_m^2 + z_m^2 & -y_m z_m \\ -x_m z_m & -y_m z_m & x_m^2 + y_m^2 \end{pmatrix} dm$ in RG.

The inertia matrix is diagonalized in the principal frame of inertia of the launcher noted RL. The origin of this rotating frame linked to the launcher is G. The roll axis Gx is along the longitudinal axis, and the yaw axis Gy and the pitch axis Gz are respectively in the vertical plane and perpendicular to the plane of the trajectory.

By passing in the rotating frame RL and expressing the moments of the forces (Figure 11.3), the dynamic equation (11.19) becomes

$$\left.\frac{d\vec{H}_G}{dt}\right|_{RL} + \vec{\Omega}_L \wedge I_{RL} \vec{\Omega}_L = \overrightarrow{GB} \wedge \vec{T} + \overrightarrow{GC} \wedge \vec{L} \tag{11.21}$$

The projection of this vector equation in components (x,y,z) of the RL frame gives Euler's equations.

Since the roll velocity of a launcher is generally small and the launcher inertia varies slowly, the pitch equation is reduced to

$$I_z \ddot{\theta}_G = T.\overline{GB} \sin\beta + L.\overline{GC} \tag{11.22}$$

The pitch angle in the inertial frame RG is $\theta_G = \theta - \varphi = \alpha_r + \gamma_r - \varphi$ (Figure 11.3), where $\alpha_r$ and $\gamma_r$ are the angle of attack and the slope associated with the relative velocity $\vec{v}_r$ (launcher velocity in the atmosphere, which is rotating with the Earth), while the angles $\alpha$ and $\gamma$ are associated with the absolute velocity $\vec{v}$ (launcher velocity in the inertial frame RG).

The equation (11.22) is linearized by making the following approximations.

- The kinematic angles $\gamma_r$ and $\varphi$ change slowly (compared to the rotational dynamic), so that $\ddot{\theta}_G \approx \ddot{\alpha}_r$.
- The lift varies linearly with the angle of attack: $L \approx QS_{ref}C_{z\alpha}\alpha_r$, by noting $C_{z\alpha} > 0$ the derivative of $C_z$ with respect to the angle of attack. This approximation is correct for small angles of attacks, which is the nominal flight condition in the atmosphere (see the subsequent text and Section 11.4.1).
- The maximum available deflection is small (of the order of 5 deg), so that $\sin\beta \approx \beta$.

The resulting linearized equation (11.23) is used to analyze the controllability and the stability of the launcher.

$$I_z\ddot{\alpha}_r = A_\alpha\alpha_r + A_\beta\beta \quad \text{with} \quad \begin{cases} A_\alpha = QS_{ref}C_{z\alpha}\,.\,\overline{GC} \\ A_\beta = T\,.\,\overline{GB} \end{cases} \tag{11.23}$$

The launcher is said **controllable** if the available thrust moment is sufficient to balance the aerodynamic moment. For this purpose, the **static deflection** $\beta_s$ defined by $A_\alpha\alpha_r + A_\beta\beta_s = 0$ must not exceed the maximum available deflection.

A nominal trajectory at zero angle of attack allows one to fly at null static deflection, which keeps a maximum margin with respect to dispersions (especially wind). In addition, the **dynamic deflection** is calculated to follow the attitude control. Since the rotational dynamics are much faster than the center of gravity dynamics, the launch trajectory can be calculated (Section 11.4.3) by assuming a perfect (instantaneous) attitude control and a null dynamic deflection.

At null deflection, the attitude motion is stable if the aerodynamic pressure center is below the center of gravity ($A_\alpha < 0$). The stability depends on the geometry of the launcher and on its internal mass distribution. It evolves in flight at the same time as the aerodynamics changes and the center of gravity location moves. An unstable launcher is more efficient because it responds faster to rotation controls.

## 11.3 Launcher Configuration

Several propulsive stages are required to reach the orbital velocity. The optimization of the launcher staging is based on the concept of propulsive $\Delta V$.

### 11.3.1 Propulsive ΔV

The velocity modulus evolution $\dot{v}$ is given by the third equation in (11.18).

The velocity increment during a flight phase without dry mass jettisoning is calculated by integration.

$$v(t_2) - v(t_1) = \int_{t_1}^{t_2} \frac{T}{m}\cos\alpha\,dt - \int_{t_1}^{t_2} \frac{D}{m}\cos\alpha\,dt - \int_{t_1}^{t_2} \frac{L}{m}\sin\alpha\,dt - \int_{t_1}^{t_2} g\sin\gamma\,dt \tag{11.24}$$

Using (11.9) with the notation $v_e = g_0 Isv$ (exhaust velocity), the thrust term is put in the form

$$\int_{t_1}^{t_2} \frac{T}{m}\cos\alpha\,dt = v_e\int_{t_1}^{t_2} \frac{q}{m}dt - \int_{t_1}^{t_2} \frac{p_a A_s}{m}dt - \int_{t_1}^{t_2} \frac{T}{m}(1-\cos\alpha)dt \tag{11.25}$$

The first integral noted $\Delta V_{prop}$ is called the **propulsive velocity increment**. It can be calculated analytically because $q = -\dot{m}$. This gives the formula established by Tsiolkovski in 1905.

$$\Delta V_{prop} = v_e \ln \frac{m_1}{m_2} \tag{11.26}$$

If the propellant combustion were instantaneous, the actual velocity increment in (11.24) would be equal to the propulsive increment $\Delta V_{prop}$. In reality, the other integrals called **velocity losses** must be subtracted. These integrals depend on the trajectory and have the following orders of magnitude:

- drag losses (D term): 50–100 m/s depending on the launcher shape;
- lift losses (L term): ≈0 m/s (at null angle of attack atmospheric flight);
- gravity losses (g term): 700–1500 m/s depending on the altitude variation;
- pressure losses ($p_a$ term): 200–500 m/s depending on atmospheric flight;
- incidence losses (T term): very variable.

Incidence losses occur when the thrust is not aligned with the absolute velocity (angle $\alpha$). Part of the thrust rotates the velocity instead of increasing it. The loss is measured by the term in $1 - \cos\alpha$. Except for active constraint (Section 11.4.1), the optimal trajectory should have very low incidence losses in exo-atmospheric flight. During the atmospheric flight at a null angle of attack, the thrust is aligned with the *relative velocity* (velocity in the rotating Earth frame) and not with the absolute velocity. This results in incidence losses of the order of 500 m/s.

The objective of the launcher is to reach a specified orbit (velocity $v_f$) starting from the ground (velocity $v_0$). The velocities are defined in the inertial geocentric frame. The final velocity $v_f$ depends on the targeted orbit and on the injection point. It is calculated by equation (11.38). The initial velocity $v_0$ on the launch pad, which is rotating with the Earth, is $v_0 = \Omega_T R_T \cos\varphi$, where $R_T = 6\,378\,137$ m is the equatorial radius, $\Omega_T = 7.2921{\cdot}10^{-5}$ rad/s is the Earth rotation rate around the pole axis and $\varphi$ is the latitude of the launch pad. This "free" inertial velocity oriented eastward has a maximum value of 465 m/s for a launch pad located at the equator. It is fully beneficial for launches in equatorial orbits like the GEO (Geostationary Earth Orbit), but has no (or even adverse) effect for launches in polar or retrograde orbits, such as SSO (sun-synchronous orbits).

The propulsive $\Delta V$ required for the launcher includes the actual velocity increment to be provided and the losses incurred during the ascent trajectory.

$$\Delta V_{req} = v_f - v_0 - \Delta V_{loss} \tag{11.27}$$

For preliminary staging studies, this required propulsive $\Delta V$ is estimated by guessing the value of the losses (1500–2500 m/s depending on the mission) and the location of injection point (perigee). This $\Delta V$ guess is used to define the first staging of the launcher. It is later updated during the design loops with the actual value assessed by trajectory simulation.

### 11.3.2 Staging

A single-stage launcher consists of a propellant mass $m_p$ and a dry mass $m_d$.

It can deliver a propulsive $\Delta V$ equal to

$$\Delta V_{prop} = v_e \ln \frac{m_p + m_d}{m_d} = v_e \ln\left(1 + \frac{1}{k}\right) \quad \text{with } k \underset{def}{=} \frac{m_d}{m_p} \tag{11.28}$$

The ratio k between the dry mass and the propellant mass is called the constructive or **structural index** of the stage. It depends on the propulsion system (solid, storable or cryogenic). Even with the best technological assumptions (Isv = 430 s, k = 12% → $\Delta V = 9400$ m/s), a single-stage launcher can barely reach an orbital velocity.

Distributing the required ΔV over several propulsive stages allows to eliminate superfluous dry masses in flight by jettisoning the empty stages. The launcher's gross mass can thus be reduced at the expense of a more complex system (different engines, jettisoning of empty stages and engine ignition in flight). In practice, the launchers have two, three or four stages.

Figure 11.4 shows a three-stage linear launcher. The stage number j has a structural index $k_j$ and it is composed of a propellant mass $m_{p,j}$ and a dry mass $m_{d,j}$ separated at the end of combustion. The dry mass consists of a part depending on the propellant mass (tanks) and a fixed part (engines, equipment, skirts, ...). In first approximation, the relation between $m_p$ and $m_d$ is of the form: $m_d = a + bm_p^{2/3}$, where the power 2/3 comes from the ratio between the surface and the volume of the tanks. The coefficients a and b depend on the propulsion system.

The propulsive ΔV delivered by stage j is $\Delta V_j = v_{e,j}\frac{M_{i,j}}{M_{f,j}}$, where $M_{i,j}$ and $M_{f,j}$ are the launcher total masses at ignition and shutdown of stage j, and $v_{e,j} = g_0 Isp_j$ is the engine exhaust velocity.

The aim of staging optimization is to find the lightest launcher able to carry a given payload mass $m_u$ on the targeted orbit. The optimization variables are the propellant masses $m_{p,j}$ of the stages. They determine the total mass of the launcher and its propulsive ΔV, which must be equal to the ΔV required (11.27) to reach the orbit.

The constrained optimization problem is formulated as follows.

$$\min_{m_{p,j}} M_0 \text{ s.t. } \Delta V_{tot} = \Delta V_{req} \text{ with } \begin{cases} M_0 = m_u + \sum_j (1 + k_j) m_{p,j} \\ \Delta V_{tot} = \sum_j v_{e,j} \ln \frac{M_{i,j}}{M_{f,j}} \end{cases} \quad (11.29)$$

This problem is reformulated by:

- changing the cost function from $\min_{m_{p,j}} M_0$ to $\max_{m_{p,j}} \frac{m_u}{M_0}$ (mass ratio between the payload and the launcher);
- defining new variables $x_j = \frac{M_{i,j}}{M_{f,j}}$, with the convention $m_u = M_{i,n+1}$.

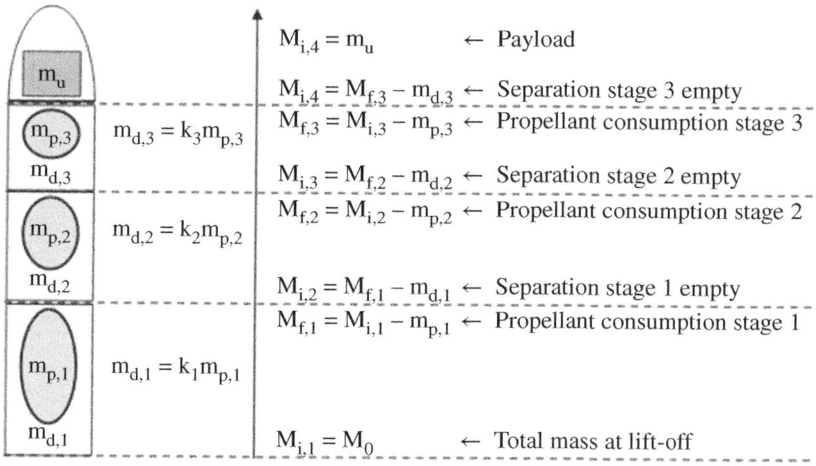


**Figure 11.4** Launcher staging.

The formulation in variables $(x_j)_{j=1,\cdots,n}$ instead of $(m_{p,j})_{j=1,\cdots,n}$ is

$$\max_{x_j} \prod_j \left( \frac{1+k_j}{x_j} - k_j \right) \quad \text{s.t.} \quad \sum_j v_{e,j} \ln x_j = \Delta V_{req} \tag{11.30}$$

The Karush, Kuhn, and Tucker (KKT) optimality conditions (assuming constant structural indexes $k_j$) allow to establish the relation (demonstrated in Chapter 12)

$$v_{e,j} \left( 1 - \frac{k_j}{1+k_j} x_j \right) = Cte \tag{11.31}$$

The constraint on the $\Delta V$ reduces to an equation with a single unknown (e.g. $x_1$), which is solved numerically. The launcher masses $M_{f,j}$, $M_{i,j}$, and the propellant masses $m_{p,j}$. are then obtained. Since this solution assumes constant structural indexes $k_j$, it is necessary to iterate with the updated values $k_j(m_{p,j})$.

This staging method is further developed in Section 12.3.2.

### 11.3.3 Preliminary Design

The purpose of the preliminary design of a launcher is to define the optimal configuration for a reference mission defined by a payload and a targeted orbit. The main choices concern the number of stages, the linear or parallel architecture (boosters), the propulsion systems (type of propellant and engines) and the geometry (diameters and heights).

The staging loop (Figure 11.5) consists of the following steps.

- Optimization of the propellant masses at given losses $\Delta V_{loss}$ assumption and fixed structural indexes k (Section 11.3.3).
- Definition of the configuration (tanks, stages, sizes) and assessment of the detailed mass breakdown at fixed thrust levels T.
- Update of the structural indexes k from the mass breakdown and iteration on the staging until convergence on the structural indexes.
- Configuration complete definition: masses, propulsion and aerodynamics.
- Optimization of the trajectory giving the mechanical loads and actual losses.
- Adaptation of the thrust levels T (to comply with the load constraints).
- Update of the losses $\Delta V_{loss}$ and iteration on the staging from the beginning.

The convergence is achieved by iterations for every architecture envisioned (defined by the number of stages, the type of propellants and the linear or parallel configuration), then the most promising configuration is selected based on the cost and the maturity of the propulsive technologies.

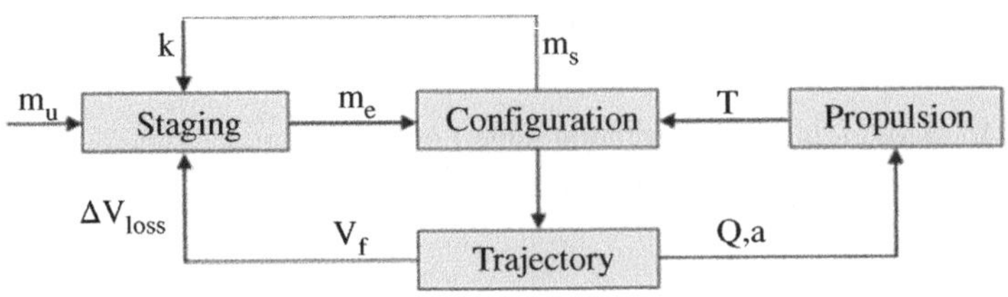


**Figure 11.5** Staging loop.

Multidisciplinary approaches based on evolutionary algorithms can be applied to optimize simultaneously the masses and the geometry (continuous variables) together with the architectural and technological choices (discrete variables).

The aspects of versatility (multiple missions), propellant reserves and thrust level setting are also considered during the preliminary staging studies. These aspects are briefly discussed hereafter and will be further investigated in Chapter 12.

### 11.3.4 Versatility

Since developing a launcher is a lengthy and expensive process, it is preferable to modify an existing configuration to adapt it to new missions. These developments are anticipated from the design stage through versatility studies. This may include the addition of boosters, modification of engines, or variation of propellant masses.

For a parallel staging, the parallel thrusters are summed to an equivalent thruster. The equivalent exhaust velocity is derived from the thrust and flow rate.

$$\begin{cases} T = T_1 + T_2 \\ q = q_1 + q_2 \\ A_s = A_{s1} + A_{s2} \end{cases} \Rightarrow v_e = \frac{q_1 v_{e1} + q_2 v_{e2}}{q_1 + q_2} \tag{11.32}$$

For multiple missions, the staging problem (11.29) is extended to deal simultaneously with several configurations and missions. For example, if two missions are to be performed with two configurations differing by one launcher element (such as the number of boosters or the last stage propellant mass), the problem is formulated with a weighted cost function of both configuration masses and two constraints defining the required $\Delta V$ of each configuration.

$$\min_{m_p} c_1 M_{0,1} + c_2 M_{0,2} \ \text{ s.t. } \begin{cases} \Delta V_{tot,1} = \Delta V_{req,1} \\ \Delta V_{tot,2} = \Delta V_{req,2} \end{cases} \tag{11.33}$$

#### 11.3.4.1 Reserves

The configuration definition is based on nominal propulsion and mass models. Since these models are subject to uncertainties, a propellant reserve is needed in the last stage to ensure that the orbit can be reached with a sufficient probability level. The propellant reserve mass $\delta m_p$ associated with a perturbation $\delta w$ of a parameter w (such as Isv, mass) can be estimated by considering that the total $\Delta V$ must remain equal to the $\Delta V$ required for the mission.

$$\Delta V(m_p + \delta m_p, w + \delta w) = \Delta V_{req} \Rightarrow \frac{\partial \Delta V}{\partial m_p} \delta m_p + \frac{\partial \Delta V}{\partial w} \delta w = 0 \tag{11.34}$$

The derivatives of the propulsive $\Delta V$ with respect to the masses and specific impulses are calculated analytically from the formula (11.29) giving the launcher total impulse.

The contributions of the different perturbations are considered as Gaussian and independent and they are summed quadratically. The reserve is then adjusted taking into account the trajectory and guidance of the launcher (Section 11.5).

#### 11.3.4.2 Engine Mixture Ratio

Liquid propulsion is based on the mixture of an oxidant (O) and a reductant (R) in a ratio noted $r_m$ (usually of the order of 5). This ratio is not perfectly known and it has a standard deviation $\sigma_m$. The total propellant mass $m_p$ of the stage is distributed between the oxidant mass $m_O$ and the reductant mass $m_R$ in a ratio $r_e$ that can be freely chosen. The mass of the remaining propellant (either O or R) at the end of combustion depends on $r_m$ as follows:

$$\begin{cases} \delta m_O(r_m) = \dfrac{r_e - r_m}{1 + r_e} m_p & \text{if } r_m < r_e \\ \delta m_R(r_m) = \dfrac{r_m - r_e}{(1 + r_e) r_m} m_p & \text{if } r_m > r_e \end{cases} \quad \text{with } r_m = \frac{q_O}{q_R},\ r_e = \frac{m_O}{m_R} \tag{11.35}$$

The choice $r_e = r_m$, which seems the most natural, would consume all propellants in nominal conditions. But in dispersed conditions, the unconsumed oxidant mass $\delta m_O$ is statistically higher than the unconsumed reductant mass $\delta m_R$, due to the mixture ratio $r_m$ greater than 1. For negative dispersions on $r_m$, which have the same probability as positive dispersions, a large mass of oxidant will be unconsumed, which is a waste of propellant. In order to balance the risk between the oxidant and the reductant, the value of $r_e$ is set to have a symmetrical risk at a level of 99% probability ($3\sigma$). The corresponding value of $r_e$ is given by

$$\delta m_O(r_m - 3\sigma_m) = \delta m_R(r_m + 3\sigma_m) \quad \Rightarrow \quad r_e = \frac{1 + r_m - 3\sigma_m}{1 + r_m + 3\sigma_m}(r_m + 3\sigma_m) \tag{11.36}$$

## 11.4 Trajectory Optimization

The launcher must bring the payload to the targeted orbit by minimizing the propellant consumption and satisfying several constraints shown in Figure 11.6. This section details the constraints, formulates the trajectory optimization problem and presents the solution methods.

### 11.4.1 Constraints

The constraints applied on the trajectory concern the targeted orbit, the in-flight loads and the risks associated with fallouts.

#### 11.4.1.1 Targeted Orbit

The orbit targeted by the launcher is either the operational orbit of the satellite or a transfer orbit from which the satellite will maneuver by its own propulsive means. The transfer orbit

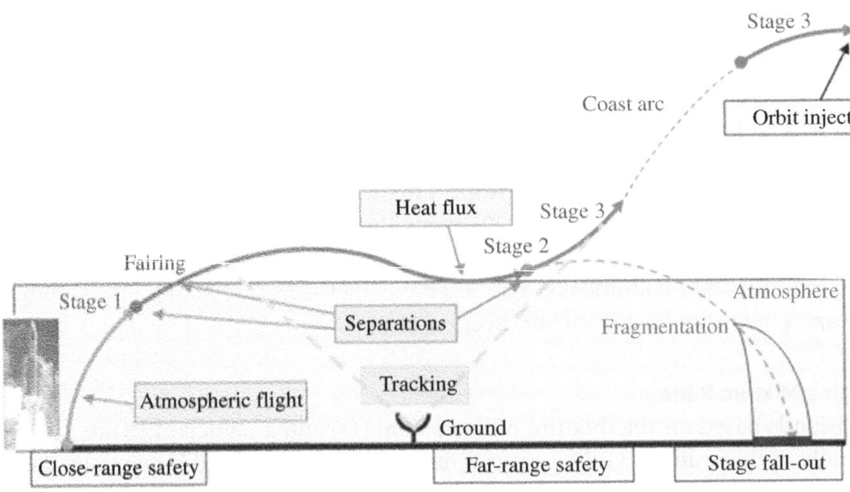


**Figure 11.6** Constraints on an ascent trajectory.

strategy is more efficient for high orbits: it avoids bringing the launcher's last stage to the final orbit, which is fuel-consuming.

The targeted orbit is defined by its Keplerian orbital parameters (a, e, i, Ω, ω, θ) in the inertial geocentric frame (OXYZ). The inclination i is the angle between the equator (XY) and the orbital plane. The right ascension of the ascending node Ω is the angle between X and the ascending node N, which is the point of the equator flown over northward. The argument of the perigee ω is the angle between the node N and the perigee P. The orbit shape can be defined either by the radii at perigee $r_P$ and apogee $r_A$, or by the semi-major axis a and the eccentricity e. The true anomaly θ locates the current position from the perigee (Figure 11.7).

The orbit can equivalently be defined by the angular momentum vector $\vec{h}$ and the eccentricity vector $\vec{e}$. These vectors are prime integrals of the Keplerian motion, as seen in Section 1.2.2.

$$\vec{h} = \vec{r} \wedge \vec{v}\,, \quad \vec{e} = \frac{\vec{v} \wedge \vec{h}}{\mu} - \frac{\vec{r}}{r} \tag{11.37}$$

The velocity v depends on the vector radius r by

$$v^2 = \mu\left(\frac{2}{r} - \frac{1}{a}\right) \tag{11.38}$$

This formula shows that an injection at a low altitude requires a higher velocity. The optimal injection point is not known in advance and it results from the trajectory optimization. Nevertheless, it is usually located in the vicinity of the perigee, as demonstrated in Chapter 8 in the case of an optimized thrust profile.

The targeted value of the right ascension of the ascending node Ω may come either from geometric considerations (for constellations of satellites to place in specific planes), or from solar time constraints (for sun-synchronous satellites), or from lightening constraints at the apogee (for geostationary transfer orbit) or from planetary ephemeris (for interplanetary missions discussed in Chapter 14).

The targeted orbit is fixed in the inertial geocentric frame, whereas the Earth rotates around the axis of the poles. To minimize the propellant consumption, the launcher must wait on the launch pad until it passes close to the targeted plane. The launch window (generally of the order of a few minutes) depends on the launcher's capability to perform a plane change in the event that the take-off does not occur at the optimal time.

Before separating the satellite, the launcher is oriented in a direction stabilized by spinning, in order to control the orientation of the satellite solar panels and antennas after separation. The deployment of the solar panels allows the satellite to recharge its batteries and operate autonomously. The deployment of the antennas ensures communications with the tracking

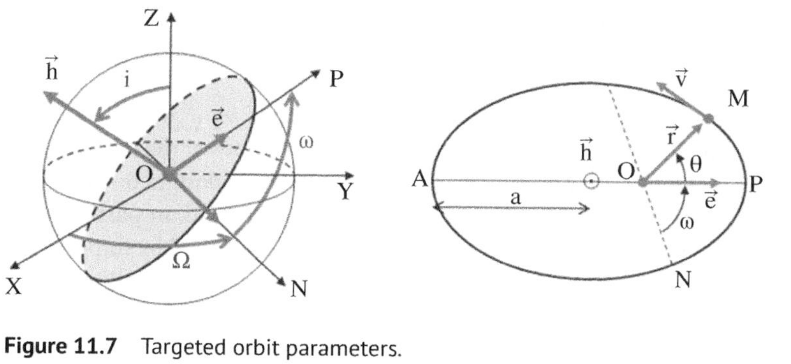


**Figure 11.7** Targeted orbit parameters.

ground stations and the reception of control orders. The precise positioning maneuvers will then be commanded from the ground control centers. After separation, the last stage is in an orbit that can pass through a protected area such as LEO (Low Earth Orbit) below 2000 km altitude or GEO (Geostationnary Earth Orbit) torus of radius 300 km. In such cases, the space laws impose that the area must be evacuated in less than 25 years. The evacuation may be achieved passively by atmospheric friction or actively by a deorbitation-propelled maneuver, as studied in Section 15.3.1. Before it is abandoned (either in space or after a deorbitation maneuver), the stage is passivated by emptying it of its residual propellants in order to avoid a subsequent explosion that could generate a lot of debris.

#### 11.4.1.2 In-Flight Loads

The mechanical, vibratory and thermal loads encountered during the flight must remain within the limits tolerated by the launcher and the satellite.

The nongravitational acceleration must remain below a threshold compliant with the launcher and the satellite structural design. A typical threshold value is 5 g. This limitation is taken into account right from the design of the launcher by the setting of the engine thrust levels, as discussed in Section 11.3.3. The aerodynamic loads are proportional to the dynamic pressure Q (11.11), which must remain below a threshold of typically 40 kPa. The transverse loads on the launcher structure are proportional to the product $Q\alpha_r$. The launcher has an elongated shape (for minimizing the aerodynamic drag), which is very sensitive to these transverse loads. These loads increase furthermore the need for attitude steering control, as seen in (11.23). They are minimized by forcing a nominal flight at a null angle of attack. The perturbations (noninstantaneous control, steady wind, wind gusts, launcher shape asymmetry) may generate a transient non-null angle of attack that must remain compatible with the structural and control capabilities of the launcher, as discussed in Section 11.2.2.2 (attitude equations).

After its shutdown and separation, the first stage (or the boosters) receives an impulse designed to move it away from the upper part of the launcher. In order to ensure a safe separation without uncontrolled rotation, it is necessary that the aerodynamic effects are sufficiently low. This is achieved by requiring a dynamic pressure limitation at the separation, typically of the order of 1 kPa. This constraint imposes a nearly vertical atmospheric flight in order to escape from the dense atmosphere before the end of combustion of the first stage.

Vibratory loads occur during the take-off (acoustic waves), the transonic passage (shock waves), the separations (pyrotechnic devices) and the engine ignitions (thrust increase phase). These transient phenomena are difficult to control because they concern the whole launcher in specific flight conditions, which cannot be fully simulated by ground tests. Resonances may also appear in flight between the liquid propellants and the structure (POGO), which may lead (in the absence of a specific damping system) to the destruction of the launcher.

Aerodynamic friction generates a convection heat flux proportional to $\sqrt{\rho v_r^3}$. The fairing protecting the satellite is separated as soon as the convective flux becomes of the order of the solar radiative flux in orbit (1300–1500 W/m$^2$) for which the satellite is designed. After the fairing has been jettisoned, the flux level shall not rise above this threshold, in particular if the launcher-optimized trajectory presents a downward leg. This is controlled by imposing a maximum heat flux level not to be exceeded after the fairing separation. If the satellite is carrying heat-sensitive instruments, the launcher attitude is further constrained from the fairing separation until the satellite injection in order to avoid exposing these instruments to the Sun.

### 11.4.2 Risk Mitigation

The risks of a launch are related to the fallouts of stages or debris. Such fallouts occurs either in abort cases due to engine failures or in-flight explosions, or on the nominal flight after the stage separations. The lower stages fall out naturally since they do not reach an orbital velocity, whereas the last stage is injected in the same orbit as the payload.

A propulsion failure places the launcher on a ballistic trajectory reentering the atmosphere. Aerodynamic forces and heat fluxes will lead to fragmentation and/or explosion. The impact zone of all possible debris is calculated at each time of the ascent trajectory. It depends on the current conditions (time, position, velocity and launcher configuration), on destruction models (fragmentation or explosion) and on debris features associated with elements of the launcher (mass, shape and resistance to loads). During the flight, the impact zone progresses along the trajectory ground track, which is the set of points flown over. Flight safety depends on the envelope of the impact zones, which should, whenever possible, be located in uninhabited areas like oceans or deserts.

At take-off, the impact zone of debris in the event of an explosion is centered on the launch pad. After a vertical ascent, the launcher makes a tilting maneuver of an angle $\gamma_0$ in a direction defined by the launch azimuth $\chi_0$ (Figure 11.8). The debris impact zones become ellipses oriented along the launch direction $\chi_0$ and they increase in size (because the launcher altitude and velocity increase with time). Whenever possible, launch pads are located by an ocean allowing a safe range of launch azimuths.

As the launcher gains altitude, the impact zone advances and finally reaches some land situated along the trajectory ground track. The risk is then measured by a casualty probability depending on the launcher failure probability, on the debris likely to reach the ground and on the population density in the impact area. This probability can be reduced by deflecting the trajectory to avoid flying over densely populated areas. An overall probability of $10^{-5}$ on the trajectory is generally acceptable, compared to the natural casualty probability of $10^{-7}$ associated with meteorites. It is also necessary to block sea and air corridors in the vicinity of the ground track and to check that the nominal launch trajectory will not pass too close to identified space objects (operational satellites or debris). If the risk of collision is too high, the launch must be postponed while staying within the launch window.

On the nominal flight, the debris originates from the fallouts of the lower stages after their separation. Their impact zones must be located in uninhabited areas. The size of the

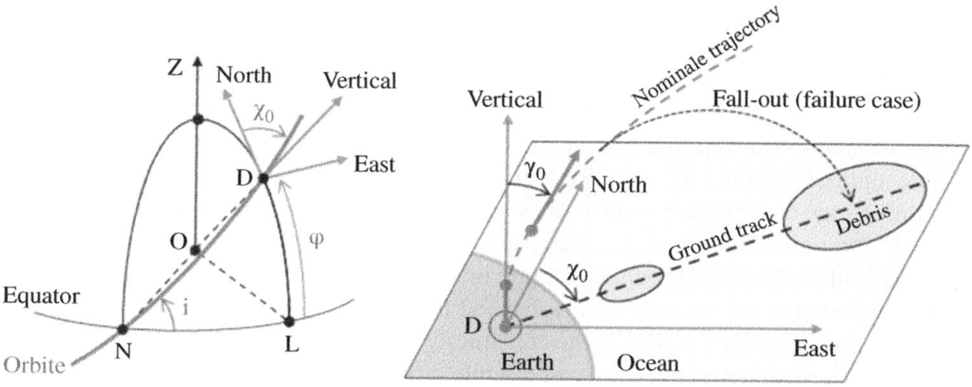


**Figure 11.8** Launch azimuth and trajectory ground track.

impact zone can be reduced by emptying the stage of its residual propellants (passivation) and constraining the atmospheric reentry conditions (flight path angle, angle of attack).

When several satellites are to be deployed, the launcher's last stage performs specific injection maneuvers to ensure some spacing between the satellites. The goal is to avoid possible collisions as long as the satellites are not controlled from the ground. The last floor then carries out its own spacing maneuver before a possible deorbitation targeting an uninhabited area.

The launcher flight is controlled automatically by the on-board GNC system (Guidance-Navigation-Control) described in Section 11.5.3. The only possible control from the ground is the neutralization, which stops the propulsion. The trajectory is tracked in real time by radar ground stations or by satellites. In the event of an abnormal deviation, the launcher becomes dangerous and it must be neutralized. The flight tracking requires several ground stations located along or near the trajectory ground track. In some cases, the optimal trajectory must be constrained to ensure the full flight visibility by the selected ground stations. The ground stations also collect the measurements recorded by on-board sensors. These measurements (acceleration, masses in tanks, combustion chamber pressure and temperature …) are analyzed after the flight to improve the launcher modeling and reduce the uncertainties for the preparation of subsequent missions.

### 11.4.3 Optimal Control Problem

The launch trajectory is controlled by the thrust. The goal is to reach the targeted orbit, while minimizing the propellant consumption and satisfying the design and operational constraints described in the previous section. This problem is formulated as an optimal control problem.

#### 11.4.3.1 Standard Formulation

The optimal control problem to solve is in the standard form

$$\min_{u,t_f,t_p} \varphi(x_f, t_f) \quad \text{s.t.} \quad \dot{x} = f(x, u, t) \quad \text{with} \quad \begin{cases} \psi_f(x_f, t_f) = 0 \\ \psi_p(x_p, t_p) \leq 0 \\ \psi_e(x, u, t) \leq 0 \end{cases} \tag{11.39}$$

The launcher is modeled as a material point and the state vector $x = (\vec{r}, \vec{v}, m)$ is defined by the position, the velocity and the total mass. The control vector $u = (\vec{T})$ is the thrust. The dynamics $f(x, u, t)$ is defined by the motion equations (11.17). The cost function is $\varphi = -m(t_f)$, since the final mass should be maximized.

The final constraints $\psi_f$ (targeted orbit) apply at the final time $t_f$, the interior constraints $\psi_p$ (separations and fallout conditions) apply at times $t_p$ and the path constraints $\psi_e$ (mechanical loads, heat flux and visibility from ground stations) apply on flight sequences. The state and the dynamics present discontinuities at the engine ignitions and shutdowns due to the change of thrust level and at the stages and fairing separations due to the masses jettisoned. The times $t_f$ and $t_p$ are either fixed or optimized, depending on the propulsion control capabilities (throttleable engine, liquid or solid propulsion)

A control law $u(t)_{t_0 \leq t \leq t_f}$ produces an associated trajectory $x(t)_{t_0 \leq t \leq t_f}$. The control and trajectory are admissible (or feasible) if the constraints are satisfied. The optimal control and trajectory are those that minimize the cost $\varphi$ while satisfying all constraints.

#### 11.4.3.2 Pontryaguin Maximum Principle

The Pontryaguin Maximum Principle (PMP) states necessary conditions of local optimality by introducing a costate vector p(t) of the same dimension as the state vector x(t), a costate vector q(t) associated with the path constraints $\psi_e$ of the same dimension as $\psi_e$, and the Hamiltonian H defined by

$$H(x, u, p, q, t) = p^t f(x, u, t) + q^t \psi_e(x, u, t) \tag{11.40}$$

The first-order conditions of the PMP are as follows.

- The optimal control minimizes the Hamiltonian.

$$\min_u H(x, u, p, q, t) \quad \rightarrow \quad u_{opt}(x, p, q, t) \tag{11.41}$$

- The state and the costate follow the Hamiltonian system.

$$\dot{x} = \frac{\partial H}{\partial p}, \quad \dot{p} = -\frac{\partial H}{\partial x} \tag{11.42}$$

- The final costate and the final Hamiltonian satisfy transversality conditions with multipliers $\nu_f$ for the constraints $\psi_f$ and a multiplier $\nu_0$ for the final cost $\varphi$, usually set to 1 (normal solution).

$$p_f = \frac{\partial \Phi}{\partial x_f}, \quad H_f = -\frac{\partial \Phi}{\partial t_f} \quad \text{with } \Phi = \nu_0 \varphi + \nu_f^t \psi_f \tag{11.43}$$

- The costate $q_j$ associated with the constraint $\psi_{e,j}$ is positive and it cancels when the constraint is inactive.

$$\begin{cases} q \geq 0 \ \text{ if } \ \psi_e = 0 \\ q = 0 \ \text{ if } \ \psi_e < 0 \end{cases} \tag{11.44}$$

- The costate and the Hamiltonian may have discontinuities when an interior inequality constraint $\psi_p$ or a path inequality constraint $\psi_e$ becomes active (i.e. equal to zero). Such a discontinuity at a time $t_d$ is expressed with a multiplier $\nu_d$ of the constraint $\psi_d$ (which is either $\psi_p$ or $\psi_e$).

$$p\left(t_d^+\right) = p\left(t_d^-\right) - \frac{\partial \psi_d}{\partial x_d}\nu_d, \quad H\left(t_d^+\right) = H\left(t_d^-\right) + \frac{\partial \psi_d}{\partial t_d}\nu_d \tag{11.45}$$

**Remark 11.1** ***(Sign Convention)*** *The original formulation of the PMP (1962) assumes a negative multiplier $\nu_0$, which leads to the maximization of the Hamiltonian. Here we choose the opposite convention ($\nu_0 \geq 0$) leading to the minimization of the Hamiltonian.*

#### 11.4.3.3 Hamiltonian Value

In some cases, the resolution is simplified by the following properties.

The total derivative of the Hamiltonian is: $\frac{dH}{dt} = \frac{\partial H}{\partial x}\dot{x} + \frac{\partial H}{\partial p}\dot{p} + \frac{\partial H}{\partial u}\dot{u} + \frac{\partial H}{\partial t}$.

By substituting $\dot{x}$ and $\dot{p}$ given by (11.42), the first two terms cancel and we have

$$\frac{dH}{dt} = \frac{\partial H}{\partial u}\dot{u} + \frac{\partial H}{\partial t}$$

If u is not bounded, the condition (11.41) gives $\frac{\partial H}{\partial u} = 0$ and we are left with

$$\frac{dH}{dt} = \frac{\partial H}{\partial t} = p^t \frac{\partial f}{\partial t}.$$

If the system is also autonomous, its evolution defined by the function f does not explicitly depend on time, so that

$$\frac{dH}{dt} = 0 \quad \Rightarrow \quad H = Cte$$

For such systems, which are very common in practice, the Hamiltonian is constant along any extremal. This property may ease the resolution of the control problem.

#### 11.4.3.4 Solution Structure

In the presence of inequality constraints, the solution may be composed of an unknown number of saturated ($\psi_e = 0$) and unsaturated ($\psi_e < 0$) legs, inducing discontinuities at their junctions. The resolution may become very complicated.

The practical approach consists of first solving without inequality constraints and observing the possible nonadmissible portions of the flight. The trajectory is then segmented according to the active constraints that are treated as equality interior constraints. This segmentation process is described in Section 11.4.5.

### 11.4.4 Numerical Methods

Optimal problem controls such as (11.39) can rarely be solved analytically, so that numerical solution methods must be used. This section presents a short overview of the main methods. For more technical descriptions, the reader can consult [R3] or [R9]. Chapter 10 in the second volume of [R9] entitled "Numerical methods in optimal control" details the main algorithms used for practical applications.

The numerical methods are split into indirect or direct methods.

#### 11.4.4.1 Indirect Methods

Indirect methods seek to solve the optimality conditions derived from the PMP.

These methods require a prior trajectory segmentation as described earlier in order to remove the inequality constraints. With this assumption, the PMP conditions yield a system of nonlinear equations forming a two-point boundary value problem (TPBVP). The unknowns are the initial costate, the constraint multipliers and the interior and final times (if free). The constraints apply at the interior and final times on the state and costate vectors and on the Hamiltonian value. The unknowns and constraints are linked as follows.

- The initial costate $p_0$ is linked to the transversality conditions applied on the final costate $p_f$.
- The final time $t_f$ is linked to the transversality condition applied on the final Hamiltonian $H_f$.
- The multipliers $\nu_f$ are linked to the final constraints $\psi_f$.
- The multipliers $\nu_d$ are linked to the interior constraints $\psi_d$.
- The times $t_d$ are linked to the Hamiltonian discontinuities $H_d^- - H_d^+$.

The nonlinear system is of the form $F(z) = 0$ by noting $z = (p_0, t_f, \nu_f, t_d, \nu_d)$ the vector of the unknowns. This system is called shooting problem, because it requires the integration of the state and costate equations (11.42) from the initial to the final time with the control derived from (11.41). The function F is called the shooting function.

The solution is generally sought by Newton's method, which is based on a first-order expansion at the current iterate $z_k$

$$F(z_{k+1}) \approx F(z_k) + \nabla F(z_k)^t (z_{k+1} - z_k) = 0$$
$$\Rightarrow \quad z_{k+1} = z_k - \nabla F(z_k)^{-t} F(z_k) \tag{11.46}$$

with iteration until finding a zero of the shooting function.

The difficulties inherent to Newton's method (calculation of the gradient of F, choice of the starting point $z_0$) are accentuated by the numerical integration of the Hamiltonian system (11.42), which is highly sensitive to the initial conditions. This sensitivity is intrinsic to the Hamiltonian system (11.42), whatever the dynamics formulation. For further analysis, the reader can consult [R9, Vol II].

Several precautions can improve the efficiency of the shooting method [R3]:

- using a numerical integration method with error control;
- fixing an error tolerance consistent with the gradient calculation method;
- scaling (or normalizing) the state and costate components;
- detecting accurately the switching times during the numerical integration.

When the shooting method fails, other techniques can be used.

- Multiple shooting consists of dividing the trajectory into segments.

The segment number k is controlled by its own vector of unknowns $z^{(k)}$ and junction conditions are imposed on the successive segments. This reduces the overall sensitivity at the expense of increasing the nonlinear system dimension.

- Homotopy (or continuation) methods consists of solving a sequence of neighboring problems $F_\alpha(z) = 0$ depending on a parameter $\alpha$, which varies from 0 to 1.

The first problem $F_0(z) = 0$ is defined so that its solution $z_0$ is simple to obtain. The solution $z_\alpha$ of the problem $F_\alpha(z) = 0$ is used as initial guess for the next problem $F_{\alpha+\delta\alpha}(z) = 0$. Starting from $z_\alpha$, the convergence can be expected, provided that the increment $\delta\alpha$ is sufficiently small. Nevertheless, various issues, either theoretical or numerical, may arise depending on the initial and final problems considered. In particular, there may be no continuous zero path between the initial and final problem. For further analysis, the reader is referred to [R9, Vol I].

#### 11.4.4.2 Direct Methods

Direct methods are preferable when the shooting problem is too difficult to solve.

The optimal control problem (11.39) is transcribed in finite dimension by discretizing the control law on N times $(t_k)_{k=1,N}$. The control law is represented by discretized values $(u_k)_{k=1,N}$ that become the unknowns of the problem. The interpolation between these values is usually piecewise constant or linear. It is also possible to discretize the state $(x_k)_{k=1,N}$ to avoid the numerical integration of the differential equations. Collocation methods use a polynomial (spline) approximation with junction conditions at the times $(t_k)_{k=1,N}$ called nodes.

The nonlinear optimization problem takes the form

$$\min_{(u_k)_{k=1,N}, t_f} \varphi(x_f, t_f) \text{ s.t. } \dot{x} = f(x, u, t) \text{ with } \begin{cases} x(t_0) = x_0 \text{ fixed} \\ \psi(x_f, t_f) = 0 \end{cases} \tag{11.47}$$

Direct methods do not involve the PMP necessary conditions and they avoid the numerical issues associated with the costate equations. Nevertheless, they may lead to large-size non-linear optimization problems. The fineness of discretization is the result of a compromise between solution quality and computation time [R3].

### 11.4.5 Trajectory Segmentation

For the launch trajectory, several interior and state constraints are active, especially during the first stage flight. This leads to the split of the trajectory into an endo-atmospheric phase and an exo-atmospheric phase.

#### 11.4.5.1 Endo-atmospheric Phase

During the endo-atmospheric phase, the launcher must fly at zero angle of attack in order to minimize the aerodynamic transverse loads and ensure the attitude control capability. This attitude constraint is managed as follows.

After a vertical ascent, the launcher makes a rapid tilting maneuver of an angle $\gamma_0$ in a direction defined by the launch azimuth $\chi_0$. The attitude is then kept constant until reaching a zero angle of attack, with the velocity aligned with the launcher axis. From that time, the thrust remains constantly directed along the relative velocity ($\vec{T} \parallel \vec{v}_r$) and the trajectory gradually curves downward under the effect of gravity (gravity turn). The atmospheric flight is thus completely determined by the two angles $\gamma_0$ and $\chi_0$. Exiting the dense atmosphere at the separation of the first stage requires a sufficiently small tilting angle $\gamma_0$ and a sufficiently long stage combustion time $t_c$ (at least two minutes). This combustion time depends on the propellant mass and the thrust level. It must be carefully chosen during the preliminary design studies (see Section 11.3.3).

The azimuth $\chi_0$ corresponding to an inclination i from a launch pad of latitude $\varphi$ (Figure 11.8) is given by

$$\sin \chi_0 = \frac{\cos i}{\cos \varphi} \tag{11.48}$$

A direct launch in the targeted orbital plane is possible only if the launch pad latitude allows it ($-i \le \varphi \le i$) and if the optimal azimuth (11.48) is within the range allowed by the safety. If not, out-of-plane thrust phases will be necessary during the flight of the upper stages, inducing higher incidence losses (see Section 11.3.1).

#### 11.4.5.2 Exo-atmospheric Phase

During the exo-atmospheric phase, the state constraints relate to the heat flux level and the visibility from ground stations. These constraints are transformed into interior equality constraints at the time of maximum flux and minimum elevation. The trajectory is thus composed of a series of unconstrained sequences connected with discontinuities expressed by the equations (11.45).

Let us detail the PMP equations for this exo-atmospheric phase.

The equations of motion (11.17) are

$$\dot{\vec{r}} = \vec{v}\,, \quad \dot{\vec{v}} = \vec{g} + \frac{T}{m}\vec{u}\,, \quad \dot{m} = -\frac{T}{v_e} \tag{11.49}$$

where $\vec{g}$ is the gravity acceleration and the unit vector $\vec{u}$ is the thrust direction.

The costate vector has components $\vec{p}_r$ (position costate), $\vec{p}_v$ (velocity costate) and $p_m$ (mass costate). The Hamiltonian of the system is

$$H = \vec{p}_r \cdot \vec{v} + \vec{p}_v \cdot \vec{g} + T\left(\frac{\vec{p}_v \cdot \vec{u}}{m} - \frac{p_m}{v_e}\right) \underset{\text{def}}{=} H_0 + TS \tag{11.50}$$

where $H_0$ is the part independent of the control $\vec{u}$.

Minimizing the Hamiltonian with respect to $\vec{u}$ (unit vector) yields

$$\min_{\vec{u}} H \quad \Rightarrow \quad \vec{u} = -\frac{\vec{p}_v}{\|\vec{p}_v\|} \tag{11.51}$$

The optimal thrust direction is opposed to the velocity costate $\vec{p}_v$ known also as the primer vector [R25]. By substituting $\vec{u}$ into (11.50), the Hamiltonian becomes

$$H = H_0 + TS \quad \text{with} \quad S \underset{\text{def}}{=} \frac{p_v}{m} + \frac{p_m}{v_e} \quad \text{and} \quad H_0 = \vec{p}_r \cdot \vec{v} + \vec{p}_v \cdot \vec{g} \tag{11.52}$$

In the case of a throttleable thrust engine ($T_{min} \le T \le T_{max}$), the minimization of the Hamiltonian (11.52) with respect to T yields

$$\begin{cases} T = T_{min} & \text{if } S > 0 \\ T = T_{max} & \text{if } S < 0 \\ T = ? & \text{if } S = 0 \end{cases} \tag{11.53}$$

The control is of the bang type with its value determined from the sign of the switching function S. A trajectory leg with $S = 0$ is called a singular arc. In such cases, the optimal control can no longer be determined from the PMP first-order conditions, and further analyses are required.

In practice, only regular solutions are considered, as they do not require a continuous modulation of the thrust level. This simplifies the trajectory optimization, the engine design and the in-flight guidance. The trajectory is composed of maximum thrust arcs ($T = T_{max}$) and coast arcs ($T = T_{min} = 0$).

The switching times are determined from the switching function sign (11.53) and the thrust direction is determined from the costate vector (11.51). This requires the integration of the costate equations together with the state equations (11.49). With the Hamiltonian (11.52), the costate equations take the form

$$\dot{\vec{p}}_r = -\frac{\partial \vec{g}}{\partial \vec{r}} \vec{p}_v \,, \quad \dot{\vec{p}}_v = -\vec{p}_r \,, \quad \dot{p}_m = \frac{T p_v}{m^2} \tag{11.54}$$

The targeted orbit is defined by the angular momentum vector $\vec{h}$ and the eccentricity vector $\vec{e}$. The transversality conditions (11.43) are expressed with the multipliers $\vec{\nu}_h, \vec{\nu}_e$ associated with the constraints on these vectors.

$$\vec{p}_r(t_f) = \frac{\partial \vec{h}}{\partial \vec{r}_f}\vec{\nu}_h + \frac{\partial \vec{e}}{\partial \vec{r}_f}\vec{\nu}_e \,, \quad \vec{p}_v(t_f) = \frac{\partial \vec{h}}{\partial \vec{v}_f}\vec{\nu}_h + \frac{\partial \vec{e}}{\partial \vec{v}_f}\vec{\nu}_e \,, \quad p_m(t_f) = \nu_0 = 1 \tag{11.55}$$

To calculate the derivatives of $\vec{h}$ and $\vec{e}$, the antisymmetric matrix $A_{\vec{a}}$ associated with the vector product by $\vec{a}$ is used: $\vec{a} \wedge \vec{b} = A_{\vec{a}}\vec{b}$. The calculations are detailed in [R1] and yield

$$\begin{cases} \dfrac{\partial \vec{h}}{\partial \vec{r}} = -A_{\vec{v}} \,, & \dfrac{\partial \vec{e}}{\partial \vec{r}} = \dfrac{\vec{r}\vec{r}^t - r^2 I}{r^3} - \dfrac{A_{\vec{v}}A_{\vec{v}}}{\mu} \\ \dfrac{\partial \vec{h}}{\partial \vec{v}} = A_{\vec{r}} \,, & \dfrac{\partial \vec{e}}{\partial \vec{v}} = \dfrac{A_{\vec{v}}A_{\vec{r}} - A_{\vec{h}}}{\mu} \end{cases} \quad \text{with } A_{\vec{a}} \underset{\text{def}}{=} \begin{pmatrix} 0 & -a_z & a_y \\ a_z & 0 & -a_x \\ -a_y & a_x & 0 \end{pmatrix} \tag{11.56}$$

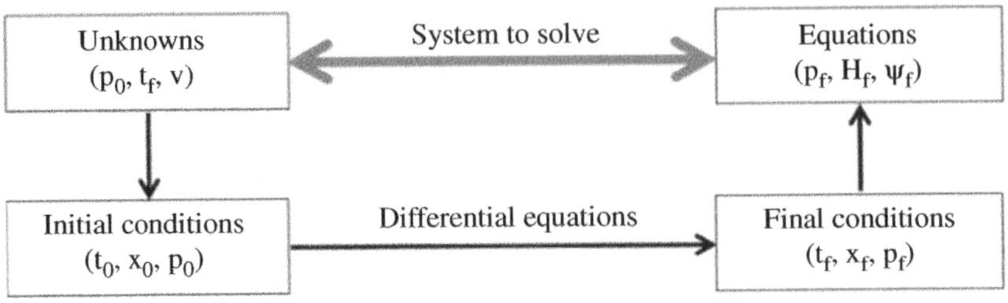


**Figure 11.9** Shooting method.

If the constraints apply only to a part of orbital parameters, and not to the full vectors $\vec{h}$ and $\vec{e}$, the corresponding transversality conditions can be calculated from the aforementioned derivatives of $\vec{h}$ and $\vec{e}$ [R1].

#### 11.4.5.3 Shooting Problem

Assuming only regular arcs, the shooting method can be implemented in a simple way. To simplify the presentation, we consider in a first step only the last stage flight and we furthermore assume that no interior constraint applies.

The state equations (11.49) and the costate equations (11.54) are integrated from the stage ignition at $t_0$ to the shutdown at $t_f$, with the optimal control defined by (11.51) and (11.53). The thrust direction is opposed to the vector $\vec{p}_v$ and the switching times (if any) are determined by the sign of S. The differential system is of dimension fourteen.

The transversality conditions (11.55) are calculated from the derivatives of the final constraints with respect to the position and the velocity. By noting n the dimension of the state (here $n = 7$), q the number of final constraints and $\nu$ the multipliers of the final constraints, the shooting problem is a system of nonlinear equations of dimension $n + 1 + q$. The unknowns are $(p_0, t_f, \nu)$ and the equations are $(p_f = \Phi_x, H_f = -\Phi_t, \psi_f = 0)$.

The shooting method is depicted in Figure 11.9.

The shooting problem is of the form $F(z) = 0$ with $z = (p_0, t_f, \nu)$ is solved by Newton's method as explained in Section 11.4.4.1.

#### 11.4.5.4 Global Trajectory Optimization

The shooting method extends easily for several stages and interior constraints by adding the corresponding costate discontinuities and multipliers.

The overall trajectory optimization concerns the atmospheric flight commanded by the initial tilting angle $\gamma_0$ and the launch azimuth $\chi_0$. The goal is either to find the maximal payload mass that can be brought to the targeted orbit, or to minimize the mass consumption for a given payload mass. The problem unknowns are thus $(m_u, \gamma_0, \chi_0, z)$ and the cost is the final mass to be maximized. The constraints on the endo-atmospheric phase (dynamic pressure) are added to the shooting problem constraints $F(z) = 0$. It is possible to reduce the global problem to variables $(m_u, \gamma_0, \chi_0)$ by an embedded resolution of the shooting problem, provided that a reliable and fast solver is available.

## 11.5 Guidance

The launch trajectory is optimized before the flight assuming nominal launcher and environment models. The actual conditions on the day of the flight may be different from the nominal conditions. The role of the on-board guidance is to correct the control law during the flight depending on the perturbations encountered. The guidance targets are the interior points associated with the nominal trajectory segmentation. The correction can be made by implicit or explicit methods.

### 11.5.1 Implicit Guidance

An implicit guidance assumes that the perturbed trajectory remains close to the nominal trajectory. The correction is formulated in terms of state $\delta x$ and control $\delta u$ deviations. For that purpose, the dynamic (11.17) is linearized in the vicinity of the nominal trajectory.

$$\begin{cases} \dot{\vec{r}} = \vec{v} \\ \dot{\vec{v}} = \vec{g} + a\vec{u} \end{cases} \Rightarrow \begin{cases} \delta\dot{\vec{r}} = \delta\vec{v} \\ \delta\dot{\vec{v}} = G\delta\vec{r} + a\delta\vec{u} \end{cases} \text{ with } G = \frac{\partial \vec{g}}{\partial \vec{r}} \tag{11.57}$$

The equation followed by the state deviation $\delta x = (\delta\vec{r}, \delta\vec{v})$ is of the form

$$\delta\dot{x} = F\delta x + G\delta u \quad \text{with } F = \begin{pmatrix} O & I \\ G & O \end{pmatrix}, \quad G = \begin{pmatrix} O \\ I \end{pmatrix} \tag{11.58}$$

Let us assume that the state deviation observed at some time $t_0$ is $\delta x_0$ and that the current trajectory segment ends at a time $t_d$. The goal is to find a control deviation history $\delta u\,(t)_{t_0 \le t \le t_d}$ that will cancel out the final deviation $\delta x(t_d)$ while keeping the deviations $\delta x(t)_{t_0 \le t \le t_d}$ as small as possible. On the other hand, the control deviations $\delta u\,(t)_{t_0 \le t \le t_d}$ must also be kept as small as possible.

This leads to define a quadratic function with weighting matrices A and B on the state and control deviations. The control variation $\delta u(t)_{t_0 \le t \le t_d}$ is sought as the solution of the following linear quadratic optimal control problem (LQ).

$$\min_{\delta u} J = \frac{1}{2}\int_{t_0}^{t_d} (\delta x^t A \delta x + \delta u^t B \delta u)dt \quad \text{s.t.} \quad \begin{cases} \delta\dot{x} = F\delta x + G\delta u \\ \delta x(t_d) = 0 \end{cases} \tag{11.59}$$

This optimal control problem is addressed by applying the PMP conditions. For a linear quadratic problem, the optimality conditions can be explicitly solved and the command is expressed as a state feedback control with a gain matrix K. The detailed calculations can be found in [R6], [R9], [R37], [R38].

$$\delta u = -K\delta x \quad \text{with } K = B^{-1}G^t(S - RQ^{-1}R^t) \tag{11.60}$$

The matrices S, R, and Q matrices are obtained by backward integration of a Riccati differential system of the form.

$$\begin{cases} \dot{S} = -SF - F^tS + SGB^{-1}G^tS + A & \text{with } S(t_d) = 0 \\ \dot{R} = (SGB^{-1}G^t - F^t)R & \text{with } R(t_d) = I \\ \dot{Q} = R^tGB^{-1}G^tR & \text{with } Q(t_d) = 0 \end{cases} \tag{11.61}$$

The gain matrix K is precomputed from the optimal trajectory and stored in the on-board computer with a suitable time step. The control deviation is directly assessed by the single matrix multiplication (11.60). Implicit guidance has the advantage of not requiring any in-flight calculations at the expense of substantial data storage. The drawback is a limited performance, both in terms of convergence domain (since the correction assumes that the actual trajectory is in the vicinity of the nominal trajectory) and of optimality (since the correction aims to follow the nominal trajectory, rather than reoptimizing the remaining part of the flight). The behavior may deteriorate for large deviations from the nominal trajectory. In such cases, an explicit guidance method is preferable.

### 11.5.2 Explicit Guidance

An explicit guidance updates the control without reference to a nominal trajectory. In order to achieve a real-time resolution, the optimization problem must be considerably simplified. This section presents several possible approaches.

#### 11.5.2.1 Linear Control

A first simplifying approach consists of assuming a constant gravity $\vec{g}$. This assumption is justified for short trajectory segments during which the launcher position will not vary too much (typically for a flight range angle of less 90 deg).

With this assumption, the costate equations (11.54) have a linear solution.

$$\begin{cases} \dot{\vec{p}}_r = \vec{0} \\ \dot{\vec{p}}_v = -\vec{p}_r \end{cases} \Rightarrow \begin{cases} \vec{p}_r = \vec{p}_{r1} \\ \vec{p}_v = \vec{p}_{v1} - (t - t_1)\vec{p}_{r1} \end{cases} \tag{11.62}$$

The thrust direction $\vec{u}$ (11.51) is defined by the unknown costates $\vec{p}_{r1}$ and $\vec{p}_{v1}$ at an arbitrary time $t_1$. By noting $m_0$ the mass at $t_0$, the dynamic equations (11.17) take the form

$$\begin{cases} \dot{\vec{r}} = \vec{v} \\ \dot{\vec{v}} = \vec{g} + \vec{a}_T \end{cases} \quad \text{with } \vec{a}_T = \frac{T}{m}\vec{u} = \frac{qv_e}{m_0 - q(t - t_0)} \frac{\vec{p}_v}{\|\vec{p}_v\|} \tag{11.63}$$

The flow rate q and the exhaust velocity $v_e$ are assumed to be constant (mean values) from the current time $t_0$ to the segment end at $t_d$. By making the time variable change defined by $q\tau = m_0 - q(t - t_0)$, the velocity variation $\Delta\vec{v}_T$ due to thrust from $t_0$ to any time t is given by the integral

$$\Delta\vec{v}_T = -v_e \int_{\tau_0}^{\tau} \frac{\vec{p}_v(\tau)}{\tau\|\vec{p}_v(\tau)\|} d\tau \quad \text{with } \vec{p}_v(\tau) = \vec{p}_{v1} + (\tau - \tau_1)\vec{p}_{r1} \tag{11.64}$$

The modulus of $\vec{p}_v$ is $\|\vec{p}_v(\tau)\| = \sqrt{(\vec{p}_{v1} - \tau_1\vec{p}_{r1})^2 + (\tau\vec{p}_{r1})^2 + 2\tau\vec{p}_{r1}\cdot(\vec{p}_{v1} - \tau_1\vec{p}_{r1})}$.

If $\vec{p}_{r1} = \vec{0}$, $\vec{p}_v$ is constant and the integral (11.64) gives $\Delta\vec{v}_T = \left(v_e \ln \frac{m_0}{m}\right)\vec{p}_{v1}$.

If $\vec{p}_{r1} \neq \vec{0}$, the arbitrary time $\tau_1$ is chosen to cancel the term in $\tau$ in $\|\vec{p}_v(\tau)\|$.

$$\tau_1 = \frac{\vec{p}_{r1}\cdot\vec{p}_{v1}}{p_{r1}^2} \Rightarrow \|\vec{p}_v(\tau)\| = \|\vec{p}_{r1}\|\sqrt{\tau^2 + c^2} \quad \text{with } c = \frac{\|\vec{p}_{v1} - \tau_1\vec{p}_{r1}\|}{\|\vec{p}_{r1}\|} \tag{11.65}$$

This choice of $\tau_1$ allows to express the variation $\Delta\vec{v}_T$ (11.64) in the form

$$\Delta\vec{v}_T = -\frac{v_e}{p_{r1}} \int_{\tau_0}^{\tau} \frac{\vec{p}_{v1} + (\tau - \tau_1)\vec{p}_{r1}}{\tau\sqrt{\tau^2 + c^2}} d\tau \tag{11.66}$$

The change of position $\Delta\vec{r}_T$ due to thrust is expressed by integrating by parts.

$$\Delta\vec{r}_T = \int_{t_0}^{t} \Delta\vec{v}_T(t)dt = t\Delta\vec{v}_T(t) - \int_{t_0}^{t} t\vec{a}_T(t)dt \tag{11.67}$$

These integrals are computed using the following primitives with $X = \tau^2 + c^2$.

$$\int \frac{d\tau}{\sqrt{X}} = \ln(\tau + \sqrt{X}), \quad \int \frac{d\tau}{\tau\sqrt{X}} = -\frac{1}{c} \ln \frac{c + \sqrt{X}}{\tau}, \quad \int \frac{\tau d\tau}{\sqrt{X}} = \sqrt{X} \tag{11.68}$$

On the other hand, the velocity variation $\Delta\vec{v}_G$ due to the gravity can be estimated by Hermite's formula, which is exact in the case of a polynomial of degree three.

$$\Delta\vec{v}_G = \int_{t_0}^{t} \vec{g}(t)dt \approx \frac{t-t_0}{2}[\vec{g}(t_0)+\vec{g}(t)] + \frac{(t-t_0)^2}{12}[\dot{\vec{g}}(t_0)-\dot{\vec{g}}(t)]$$

$$\text{with } \vec{g} = -\frac{\mu}{r^3}\vec{r} \text{ and } \dot{\vec{g}} = \frac{3\mu}{r^5}(\vec{r}.\vec{v})\vec{r} - \frac{\mu}{r^3}\vec{v} \tag{11.69}$$

Hermite's formula applied for the position variation $\Delta\vec{r}_G$ then gives

$$\Delta\vec{r}_G = \int_{t_0}^{t} \Delta\vec{v}_G(t)dt \approx \frac{t-t_0}{2}\Delta\vec{v}_G(t) + \frac{(t-t_0)^2}{12}[\vec{g}(t_0)+\vec{g}(t)] \tag{11.70}$$

by using $\Delta\vec{v}_G(t_0) = \vec{0}$ and $\Delta\dot{\vec{v}}_G = \vec{g}$.

The assessment is iterative because the gravity integrals (11.69) and (11.70) depend on the final position and velocity. The algebraic formulas (11.66)–(11.70) give the final conditions associated with the unknown vectors $\vec{p}_{r1}$ and $\vec{p}_{v1}$, which fully define the control until the current segment ends. A system of equations must then be solved to satisfy the constraints imposed at the end of the segment.

#### 11.5.2.2 Local Control on Angular Momentum and Eccentricity

A second simplifying approach is based on the angular momentum vector $\vec{h}$ and the eccentricity vector $\vec{e}$ (11.37), which define the targeted orbit.

$$\vec{h} = \vec{r} \wedge \vec{v}\ ,\quad \vec{e} = \frac{\vec{v} \wedge \vec{h}}{\mu} - \frac{\vec{r}}{r} \tag{11.71}$$

The current vectors $(\vec{h}_0, \vec{e}_0)$ at the current position and velocity $(\vec{r}_0, \vec{v}_0)$ present deviations $(\Delta\vec{h}_0, \Delta\vec{e}_0)$ from the targeted vectors $(\vec{h}_d, \vec{e}_d)$ to be reached at the end of the segment. The vectors $(\vec{h}, \vec{e})$ are constant on a Keplerian motion. Their variations are due to the thrust acceleration noted $\vec{a}_T$. By direct derivation of (11.71), we have

$$\dot{\vec{h}} = \vec{r}_0 \wedge \vec{a}_T\ ,\quad \mu\dot{\vec{e}} = \vec{a}_T \wedge \vec{h}_0 + \vec{v}_0 \wedge (\vec{r}_0 \wedge \vec{a}_T) \tag{11.72}$$

The local thrust direction noted $\vec{a}_h$ that reduces at most the deviation $\Delta\vec{h}_0$ is the one that maximizes the current scalar product $\Delta\vec{h}_0 \cdot \dot{\vec{h}}$. Similarly, the local thrust direction noted $\vec{a}_e$ that reduces at most the deviation $\Delta\vec{e}_0$ is the one that maximizes the current scalar product $\Delta\vec{e}_0 \cdot \dot{\vec{e}}$.

These local thrust directions are unit vectors calculated as follows

$$\begin{cases} \max\limits_{\vec{a}_T} \Delta\vec{h}_0 \cdot \dot{\vec{h}} = [\Delta\vec{h}_0 \wedge \vec{r}_0] \cdot \vec{a}_T \\ \qquad \Rightarrow \vec{a}_h \parallel [\Delta\vec{h}_0 \wedge \vec{r}_0] \\ \max\limits_{\vec{a}_T} \Delta\vec{e}_0 \cdot \dot{\vec{e}} = [\vec{h}_0 \wedge \Delta\vec{e}_0 + (\Delta\vec{e}_0 \wedge \vec{v}_0) \wedge \vec{r}_0] \cdot \vec{a}_T \\ \qquad \Rightarrow \vec{a}_e \parallel [\vec{h}_0 \wedge \Delta\vec{e}_0 + (\Delta\vec{e}_0 \wedge \vec{v}_0) \wedge \vec{r}_0] \end{cases} \tag{11.73}$$

with the current deviations given by $\Delta\vec{h}_0 = \vec{h}_d - \vec{h}_0$ and $\Delta\vec{e}_0 = \vec{e}_d - \vec{e}_0$.

A linear combination of the thrust directions $\vec{a}_h$ and $\vec{a}_e$ will produce a local velocity variation $\Delta\vec{v}_r$ and local variations $(\Delta\vec{h}_r, \Delta\vec{e}_r)$ calculated by (11.72).

Let us note $k_h$ and $k_e$ the coefficients of the linear combination.

$$\Delta\vec{v}_r = k_h\vec{a}_h + k_e\vec{a}_e \quad \Rightarrow \quad \begin{cases} \Delta\vec{h}_r = \vec{r}_0 \wedge \Delta\vec{v}_r \\ \mu\Delta\vec{e}_r = \Delta\vec{v}_r \wedge \vec{h}_0 + \vec{v}_0 \wedge (\vec{r}_0 \wedge \Delta\vec{v}_r) \end{cases} \tag{11.74}$$

The coefficients $k_h$ and $k_e$ are chosen to minimize the residual deviations on the vectors $\vec{h}$ and $\vec{e}$. For that purpose, these deviations are normalized and weighted by an arbitrary coefficient $\alpha \in [0; 1]$.

The cost function to minimize is then defined as:

$$\min_{k_h,k_e} F(k_h, k_e) \underset{def}{=} \alpha\left(\frac{\Delta\vec{h}_0 - \Delta\vec{h}_r}{h_c}\right)^2 + (1-\alpha)\left(\frac{\Delta\vec{e}_0 - \Delta\vec{e}_r{}^2}{1+e_c}\right)^2 \tag{11.75}$$

The optimal values $(k_h, k_e)$ are obtained analytically by deriving the quadratic function $F(k_h, k_e)$. For that purpose, we first expand (11.75)

$$F(k_h, k_e) = Ak_h^2 + Bk_e^2 + 2Ck_hk_e - 2Dk_h - 2Ek_e \tag{11.76}$$

with the coefficients calculated by

$$\begin{cases} \vec{A}_h = \vec{r}_0 \wedge \vec{a}_h, \quad \vec{B}_h = \dfrac{1}{\mu}[\vec{a}_h \wedge \vec{h}_0 + \vec{v}_0 \wedge (\vec{r}_0 \wedge \vec{a}_h)], \quad \alpha_h = \dfrac{\alpha}{h_c^2} \\ \vec{A}_e = \vec{r}_0 \wedge \vec{a}_e, \quad \vec{B}_e = \dfrac{1}{\mu}[\vec{a}_e \wedge \vec{h}_0 + \vec{v}_0 \wedge (\vec{r}_0 \wedge \vec{a}_e)], \quad \alpha_e = \dfrac{1-\alpha}{(1+e_c)^2} \end{cases} \tag{11.77}$$

$$\text{and} \begin{cases} A = \alpha_h A_h^2 + \alpha_e B_h^2, \quad D = \alpha_h \vec{A}_h \cdot \Delta\vec{h}_0 + \alpha_e \vec{B}_h \cdot \Delta\vec{e}_0 \\ B = \alpha_h A_e^2 + \alpha_e B_e^2, \quad E = \alpha_h \vec{A}_e \cdot \Delta\vec{h}_0 + \alpha_e \vec{B}_e \cdot \Delta\vec{e}_0 \\ C = \alpha_h \vec{A}_h \cdot \vec{A}_e + \alpha_e \vec{B}_h \cdot \vec{B}_e \end{cases}$$

By canceling the derivatives of $F(k_h, k_e)$ (11.76), we obtain the values of $(k_h, k_e)$.

$$k_h = \frac{BD - CE}{AB - C^2}, \quad k_e = \frac{AE - CD}{AB - C^2} \tag{11.78}$$

The formulas (11.74) give the velocity variation $\Delta\vec{v}_r$ and the associated vector variations $(\Delta\vec{h}_r, \Delta\vec{e}_r)$. The resulting orbital parameters $(a, e, i, \Omega, \omega)_r$ are assessed from the vectors $(\vec{h}_r = \vec{h}_0 + \Delta\vec{h}_r, \vec{e}_r = \vec{e}_0 + \Delta\vec{e}_r)$ and they depend on the arbitrary weight $\alpha \in [0; 1]$.

The value of $\alpha$ can then be tuned to minimize the square deviation with respect to the targeted orbital parameters $(a, e, i, \Omega, \omega)_c$, each one being affected by the desired precision. This minimization with a single variable $\alpha$ is done numerically by a classical method such as the golden ratio method.

This procedure allows an on-line calculation of the thrust direction during a propelled segment targeting specified orbital parameters.

### 11.5.3 Flight Control

Flight control requires navigation, guidance and control systems. Navigation estimates the state of the launcher (position, velocity, attitude and angular velocity) from internal and external measurements and on-board models for the gravity. The internal measurements are carried out by an inertial unit with accelerometers and gyrometers. The external measurements may come from satellites like GPS. The measurement and model information are filtered according to their respective uncertainties. The guidance updates the control law according to the state estimate and the constraints to be met at the end of each flight segment. The guidance frequency of about 1 Hz depends on the flight phase. The control system orientates the launcher in the direction defined by the guidance. The steering control depends on the rotation to be performed, the current angular velocity and the characteristics of the launcher (inertia, aerodynamics), which vary during the flight. The actual steering

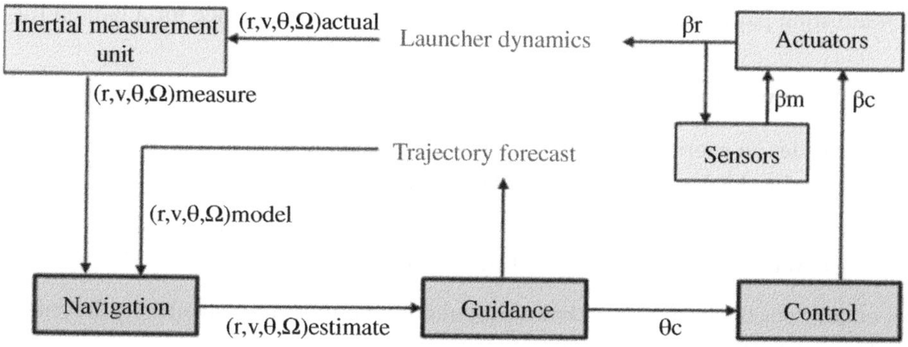


**Figure 11.10** Guidance – navigation – control system.

achieved is checked by a control loop with a frequency approximately 100 times greater than that of guidance. Figure 11.10 depicts the GNC process, with the real elements (hardware) in gray (light) and the algorithmic elements (software) in orange (dark).

The upper-stage propellant reserve allows the GNC to compensate for unfavorable perturbations. This reserve represents unused propellant on the nominal flight, which directly penalizes the launcher's performance. The post-flight exploitations progressively improve the launcher modeling and help to reduce the propellant reserves. The capability of in-flight thrust modulation can also contribute to this reduction by compensating in real time for propulsive dispersions.

## 11.6 Conclusion

### 11.6.1 The Key Points

- The endo-atmospheric flight is controlled at zero angle of attack to ensure the launcher's integrity and control.
- Staging reduces the launcher's gross mass by jettisoning progressively the useless masses of the empty stages.
- The optimization of the trajectory is carried out by a segmentation according to the constraints.
- The thrust direction undergoes discontinuities linked to the active constraints.
- The on-board guidance updates the control in real-time in order to pass through the targeted segment endpoints.

### 11.6.2 To Go Further

- BOLONKIN A. A. – *Non-Rocket Space Launch and Flight*. Elsevier (2006).

This book describes many futuristic systems that could be envisioned to launch satellites in orbit. A mathematical model is presented for each system.

- CALISE A. J., MELAMED N., LEE S. – *Design and Evaluation of a Three-Dimensional Optimal Ascent Guidance Algorithm*. Journal of guidance, Control and Dynamics, Vol 21, No. 6, Nov-Dec 1998.

A very interesting paper describing the principles of on-board guidance for a launch trajectory.

- FORTESCUE P., SWINERD G., STARK J. – *Spacecraft Systems Engineering*. Wiley (2011).

This book provides an overview of many aspects involved in the conception of a spacecraft. It addresses the technological, operational and cost aspects.

- LEEGHIM H., CHO D. H., JO S.-J., KIM D. – *Generalized Guidance Schemes for Low-Thrust Orbit Transfer*. Mathematical Problems in Engineering, Vol 2014, Article ID 407087 (2014).

This paper develops the local control law on angular momentum and eccentricity described in Section 11.5.2.2, with application cases.

- MARTY D. – *Conception des véhicules spatiaux*. Masson (1986).

This book in French provides an overview of the technological aspects involved in the conception of a launcher. The basic models associated with each discipline are clearly presented, with numerical data and an introduction to more complex simulation methods.

- SUTTON G. P., BIBLARZ O. – *Rocket Propulsion Elements*. Wiley (2001).

This book proposes a comprehensive presentation of all kinds of rocket engines (chemical, electrical and hybrid), covering physical models with their equations, technological aspects and detailed description of existing engines.

- TEWARI A. – *Advanced Control of Aircraft, Spacecraft and Rockets*. Wiley (2011).

This book proposes a very clear introduction to the optimization, guidance and control methods for launchers and spacecrafts.

- WERTZ J. R., LARSON W. J. – *Space Mission Analysis and Design*. Microcosm Press (2010).

This book proposes a comprehensive overview of all disciplines involved in the conception and operation of space systems.

# 12

# Launcher Staging

## 12.1 Introduction

A space launcher consists of propulsion stages able to reach orbital velocities of at least 8 km/s. Anaerobic propulsion is mandatory, as ambient oxygen becomes insufficient above 30 km altitude. All the fuel is carried by the launcher from the lift-off resulting in a high mass. Rocket propulsion uses solid or liquid propellants. Liquid propellants consist of an oxidizer and a reducer stored separately.

A single-stage launcher is penalized by the dry mass retained until injection into orbit. Staging consists of dividing the propellants into stages (between 2 and 4) with their own engines, tanks, structures, and equipment. The empty stages are jettisoned at the end of combustion to lighten the launcher's remaining mass. Design choices concern the number and dimensions of stages, general architecture, propellant, and structural technologies. These choices determine the development of the launcher, which takes several years, and it is very difficult to change them later. A poor initial estimate can result in an underperforming launcher or even a dead end. Margins must be sufficient to anticipate unfavorable cases, without leading to an oversized launcher.

The specifications define one or more reference missions characterized by a given payload mass to be brought into a given orbit. They may or may not impose certain technological choices (e.g. to reuse existing boosters) and authorize launcher versatility (e.g. changing the number of boosters or the last stage) for very dissimilar reference missions.

The aim is to find the minimum cost design for the specified mission(s). The global launcher optimization is a difficult problem, involving continuous (masses and thrust levels) and discrete variables (number of stages and propellant types). Multidisciplinary approaches tackle this problem with many variants, but are often complex. This chapter describes the standard methods used for preliminary design studies. The main goal is to define the propellant masses and the thrust levels.

The first section introduces the formulation of staging and trajectory problems.

The classical method presented in the second section is based on impulsive modeling that allows to decouple the design and the trajectory problems, then proceed by iterations.

The coupled method presented in the third section solves both problems simultaneously. It can handle versatile configurations and take into account uncertainties in staging assumptions.

The fourth section deals with the propellant reserves needed to guarantee a successful flight with a sufficient level of probability.

*Space Trajectories: Basic and Advanced Topics*, First Edition. Max Cerf.

## 12.2 Staging Problem

This section presents the models used for a staging study. These models, although simplified, are sufficiently representative of a preliminary design. More complex models are used in later design phases to refine the performance evaluation.

### 12.2.1 Launcher Modeling

The launcher is composed of several propulsion stages and it carries a payload of mass $m_u$ to be put into orbit. By convention, masses denoted with lowercase letters refer to a propulsive stage, whereas masses denoted with uppercase letters refer to the whole launcher.

The stage number j has a propellant mass $m_{p,j}$ and a dry mass $m_{d,j}$. The launcher masses at ignition and shutdown of stage j are noted $M_{i,j}$ and $M_{f,j}$, respectively.

The propellant mass $m_{p,j}$ is consumed during the stage flight: $M_{f,j} = M_{i,j} - m_{p,j}$, then the dry mass $m_{d,j}$ is jettisoned at the engine shutdown: $M_{i,j+1} = M_{f,j} - m_{d,j}$.

Figure 12.1 depicts a three-stage launcher. Stage j has a propellant mass $m_{p,j}$ and a dry mass $m_{d,j}$ separated at the shutdown. The launcher's total take-off mass is noted as $M_{i,1} = M_0$ and the payload mass is assimilated to the initial mass of a fictitious fourth stage: $M_{i,4} = m_u$.

The dry mass $m_d$ of each stage consists of a part dependent on the propellant mass (tanks) and a fixed part (engines, equipment, skirts, …). The ratio between the dry mass and the propellant mass of a stage is called the constructive or structural index of the stage. This index noted k depends, in the first order, on the propellant mass $m_p$ and on the engine thrust level T.

$$\frac{m_d}{m_p} = k(m_p, T) \tag{12.1}$$

The function $k(m_p, T)$ is developed for each stage by interpolation or response surface from actual launchers and/or design studies.

The propellant flow rate q and the combustion time $t_c$ of each stage are deduced from the thrust and mass by

$$T = qg_0 Isv - A_s p_a, \quad m_e = qt_c \tag{12.2}$$

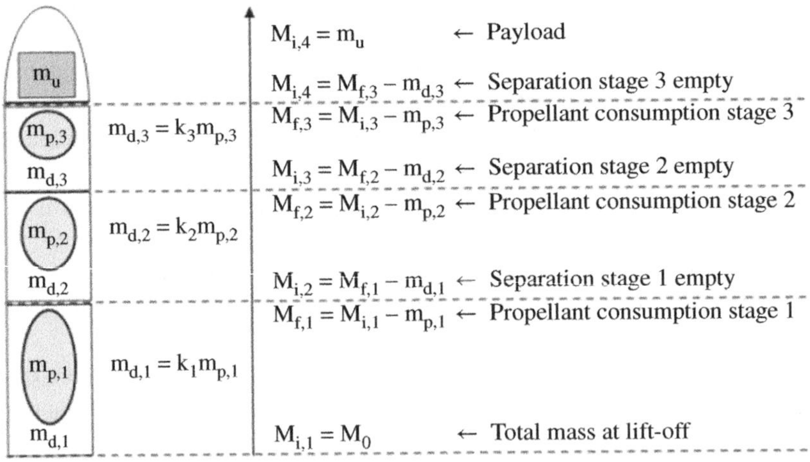


**Figure 12.1** Launcher staging.

where Isv (s) is the specific impulse in the vacuum, $A_s$ ($m^2$) is the exit section of the nozzle, $p_a$ (Pa) is the atmospheric pressure, and $g_0 = 9.806\,65\ m/s^2$.

For a preliminary design study, some simplifying assumptions are often made.

- The thrust and specific impulse of each stage are assumed to be constant.
- For the first stage operating in the atmosphere, the loss of thrust due to the nozzle section term $A_s p_a$ is taken into account approximately by considering a specific vacuum impulse reduced by a factor of 10–20%. This allows using for all stages the vacuum formula $T = qg_0 Isv$, which does not depend on the trajectory followed. This approximation is only used to initiate the staging process, before moving on to simulations that take precisely into account the effect of the atmosphere.
- The mass of the fairing is taken into account in the dry mass of the first stage.

The nongravitational acceleration must not exceed a maximum value $a_{max}$ tolerable by the payload (typically about 5 g). The acceleration increases during the stage flight (as the launcher's total mass decreases) and it reaches a maximum at the engine shutdown. The thrust $T_j$ of the engine of stage number j is thus bounded by

$$T_j \le M_{f,j} a_{max} \tag{12.3}$$

For the staging problem considered here, the number of stages n and the specific impulses $(Isv_j)_{j=1,\ldots,n}$ are fixed by the propellant choices. The staging variables are the propellant masses $(m_{p,j})_{j=1,\ldots,n}$ and the engine thrust levels $(T_j)_{j=1,\ldots,n}$.

#### 12.2.1.1 Linear or Parallel Staging

The launcher is said to be either linear if the stages operate successively, or parallel if several stages operate simultaneously. Parallel propulsion mainly concerns heavy launchers, requiring boosters for the take-off and the atmospheric phase.

The parallel case can be reduced to the linear case by summing the thrusts, the flow rates, and the nozzle sections, in order to define an equivalent engine. The equivalent specific impulse is calculated from the equivalent thrust and flow. For two thrusters in parallel, we have thus

$$\begin{cases} (T)_{equi} = T_1 + T_2 \\ (q)_{equi} = q_1 + q_2 \\ (A_s)_{equi} = A_{s1} + A_{s2} \end{cases} \Rightarrow (Isv)_{equi} = \frac{q_1 Isv_1 + q_2 Isv_2}{q_1 + q_2} \tag{12.4}$$

#### 12.2.1.2 Versatility

The staging of a launcher should be optimized for one or more reference missions.

Each mission is defined by a payload mass to be carried into a given orbit. The reference orbits are GTO (Geostationary Transfer Orbit), MEO (Medium Earth Orbit), LEO (Low Earth Orbit), or SSO (Sun-Synchronous Orbit). For missions that differ in terms of payload or orbit, it is customary to define different versions of the launcher differing by only one stage. This characterizes the versatility of the launcher (ability to perform various missions with reduced configuration changes).

The adaptation to each mission is usually done, either by changing the number of boosters or by changing the size of the last stage. The staging problem then consists of finding the least expensive versatile configuration, the versions of which allow to cover the various reference missions.

#### 12.2.1.3 Margins and Reserves

Uncertainties have to be taken into account both for the preliminary design and for the preparation of a flight.

For a preliminary design study, the uncertainties relate to the thrust and mass models of the future launcher. The discrepancy between the forecast design and the final one must be anticipated by margins, in order to avoid deadlocks during development. These margins on specific impulses and structural indexes are taken into account in the staging optimization to ensure that the final launcher will have the expected performance with a sufficient level of probability.

When preparing for a flight, the launcher (masses, thrust levels) and its environment (atmosphere) are not perfectly known. Propellant reserves must be provided to ensure that the payload can actually reach its orbit with a sufficient level of probability, usually set at 99%. The reserves mainly concern the last stage, which must compensate for the perturbations encountered during the flight. They are used only in unfavorable circumstances. In the case of a nominal flight, they represent an unused propellant mass that could have increased the launcher's performance. It is therefore important to minimize them.

### 12.2.2 Trajectory Modeling

The launcher is modeled as a material point having position $\vec{r}$, velocity $\vec{v}$, and mass m in an inertial geocentric reference frame. It is subject to the acceleration of the Earth's gravity $\vec{g}$, to the engine thrust $\vec{T}$, and to the aerodynamic drag $\vec{D}$ during atmospheric flight. The thrust is oriented along the unit vector $\vec{u}$, which is used to control the trajectory. The dynamic equations are

$$\begin{cases} \dot{\vec{r}} = \vec{v} \\ \dot{\vec{v}} = \vec{g} + \dfrac{T}{m}\vec{u} + \dfrac{\vec{D}}{m} \\ \dot{m} = -q \end{cases} \tag{12.5}$$

When the propellant is exhausted, the engine stops, the dry mass of the stage is jettisoned, and the engine of the next stage is ignited. These state and dynamic discontinuities (thrust changes) occur at each separation.

For a preliminary staging study, the simplified models described below for the forces, the flight sequences and the constraints are sufficiently representative to correctly estimate the launcher performance. More complex models are used in later design phases to refine the performance evaluation.

#### 12.2.2.1 Force Models

The Earth is spherical with a radius $R_T = 6378.137$ km. Earth's gravity is central $\vec{g}(\vec{r}) = -\frac{\mu}{r^3}\vec{r}$ with the gravitational constant $\mu = 3.986\,007{\cdot}10^{14}\ m^3/s^2$.

The drag depends on the reference surface $S_{ref}$ (usually equal to the launcher section), on the drag coefficient $C_x$, on the atmospheric density $\rho$ and on the launcher's relative velocity $v_r$ (velocity relative to the atmosphere).

It is expressed by $\vec{D} = -\frac{1}{2}\rho v_r S_{ref} C_x \vec{v}_r$.

The density is estimated by an exponential atmosphere model of the form $\rho = \rho_0 e^{-\frac{r-R_T}{H}}$, where $\rho_0 = 1.293\ kg/m^3$ is the density at the ground and $H = 7100$ m is a scale factor.

#### 12.2.2.2 Sequences and Constraints

For a preliminary design study, the trajectory is assumed to be planar, in the plane corresponding to the targeted inclination. The flight begins by a vertical take-off followed by an initial tilting maneuver of an angle $\gamma_1$, a convergence phase until reaching a null angle of attack and a zero angle of attack atmospheric flight until the separation of the first stage. The tilting angle $\gamma_1$ fully defines the flight of the first stage. To ensure the structural integrity and the controllability of the launcher, the dynamic pressure $Q = \frac{1}{2}\rho v_r^2$ shall not exceed a given threshold $Q_{max}$ of about 40–50 kPa. The separation of the first stage shall occur below a given dynamic pressure threshold $Q_{sep}$ of about 2–5 kPa. The fairing is jettisoned shortly after the first stage when the convective heat flux falls below a given threshold $\Phi_{sep}$ of about 1300–1500 W/m$^2$ tolerable by the satellite.

The flight of the upper stages is controlled by the direction of thrust. For a planar trajectory, the thrust direction $\vec{u}$ is defined by the pitch angle $\theta$ between the local horizontal and the axis of the launcher. The heat flux after the fairing jettisoning shall remain below the level $\Phi_{sep}$ tolerable by the payload. The final stage shall inject the payload into the targeted orbit defined by its altitudes of apogee $Z_A$ and perigee $Z_P$, the inclination being controlled by the selection of an appropriate launch azimuth.

#### 12.2.2.3 Optimal Trajectory

For a given launcher, the trajectory at minimum consumption for the exo-atmospheric phase is the solution to the following optimal control problem:

$$\min_{\vec{u}(t),t_f} J = -m(t_f) \text{ s.t. } \begin{cases} \dot{\vec{r}} = \vec{v} \\ \dot{\vec{v}} = \vec{g}(\vec{r}) + \dfrac{T}{m}\vec{u} \\ \dot{m} = -q \end{cases} \tag{12.6}$$

The form of the final constraints (orbital parameters) is not specified here, as they are not useful for the analysis that follows.

The Hamiltonian H of this problem is expressed with the costates $\vec{p}_r$, $\vec{p}_v$, and $p_m$ associated with position, velocity, and mass, respectively.

$$H = \vec{p}_r \cdot \dot{\vec{r}} + \vec{p}_v \cdot \dot{\vec{v}} + p_m \dot{m} = \vec{p}_r \cdot \vec{v} + \vec{p}_v \cdot \left(\vec{g} + \frac{T}{m}\vec{u}\right) - q p_m \tag{12.7}$$

Applying Pontryaguin Maximum Principle (PMP) yields the optimal thrust direction opposite to the velocity costate.

$$\min_{\vec{u}} H \quad \Rightarrow \quad \vec{u} = -\frac{\vec{p}_v}{p_v} \quad \text{with } p_v = \|\vec{p}_v\| \tag{12.8}$$

With the optimal control (12.8), the Hamiltonian (12.7) is expressed as

$$H = \vec{p}_r \cdot \vec{v} + \vec{p}_v \cdot \vec{g} - T\left(\frac{p_v}{m} + \frac{p_m}{g_0 Isv}\right) \tag{12.9}$$

The costates are obtained by integrating their differential equations (called costate equations) and they must satisfy transversality conditions associated with the final constraints noted $\psi(\vec{r}_f, \vec{v}_f, t_f)$, with multipliers noted $\nu$.

$$\begin{cases} \dot{\vec{p}}_r = -\dfrac{\partial H}{\partial \vec{r}} = -\dfrac{\partial \vec{g}}{\partial \vec{r}}\vec{p}_v, & \vec{p}_r(t_f) = \nu\dfrac{\partial \psi}{\partial \vec{r}_f} \\ \dot{\vec{p}}_v = -\dfrac{\partial H}{\partial \vec{v}} = -\vec{p}_r, & \vec{p}_v(t_f) = \nu\dfrac{\partial \psi}{\partial \vec{v}_f} \\ \dot{p}_m = -\dfrac{\partial H}{\partial m} = -\dfrac{T}{m^2}p_v, & p_m(t_f) = -\dfrac{\partial J}{\partial m_f} = 1 \end{cases} \tag{12.10}$$

The unknowns of the problem are the initial costates, the constraint multipliers, and the final time. These unknowns must satisfy a system of equations formed by the final constraints and the transversality conditions. This defines a shooting problem, which is numerically difficult to solve because of its sensitivity to the initial costates.

In addition, discontinuities of the state (mass separations) and of the dynamics (ignition of a new engine), as well as path constraints (heat flux) may induce discontinuities of the costate and thus of the thrust direction. These control discontinuities occur in the presence of velocity-dependent constraints (such as the heat flux). These constraints and their unknown multipliers must then be added to the shooting problem.

### 12.2.3 Global Problem

The objective is to design the least expensive launcher capable of performing a set of reference missions. Each mission is defined by a given payload mass to be brought into a given orbit while respecting all trajectory constraints such as dynamic pressure and heat flux. It is assumed here that the number of stages is predetermined, as well as the specific impulses of the engines (corresponding to the choice of propellants). Since these discrete variables greatly complicate the staging problem it is preferable to treat them by sweeping and optimizing separately the staging for each combination of a number of stages and propellant type.

Let us assume that the values of propellant mass $m_{p,j}$ and of thrust level $T_j$ are given for each stage. Equations (12.1) and (12.2) allow the calculation of the dry masses, the flow rates, and the combustion times. In this way, the launcher configuration is completely defined by giving, on the other hand, a reference surface and a drag coefficient (these quantities have a secondary influence on the problem).

The cost of the launcher depends mainly on the stage masses and on the thrust levels of the engines (with the associated technology). This cost function K is freely choosen. It may be defined from the prices of existing launchers and engines, or simply represent the total mass of the launcher (the most direct choice).

The global staging+trajectory problem is then formulated as a hybrid optimal control problem (with a control law and scalar variables) subject to path and final constraints.

$$\min_{\substack{m_{p,j},T_j \\ \gamma_1,\vec{u}(t)}} K(m_{p,j}, T_j) \text{ s.t.} \begin{cases} \text{Non-gravitational acceleration} \leq a_{max} \\ \text{Maximum dynamic pressure} \leq Q_{max} \\ \text{Separation dynamic pressure} \leq Q_{sep} \\ \text{Flux after fairing jettisoning} \leq \Phi_{sep} \\ \text{Final apogee, perigee} = Z_a, Z_p \end{cases} \tag{12.11}$$

The optimization variables are, on the one hand, the masses and thrust levels, which completely define the launcher configuration as explained above, and on the other hand, the initial tilting angle $\gamma_1$ and the control law $\vec{u}(t)$ starting at the second stage ignition. This control law must solve the trajectory problem (12.6), which is inherently difficult.

Solving the problem (12.11) by conventional optimization methods is almost impossible. In addition to the intrinsic difficulty of the shooting problem, the staging variables have a much stronger influence on the trajectory than the control law and make the problem very ill-conditioned. Sections 12.3 and 12.4 present two solution methods circumventing these issues.

## 12.3 Impulsive Method

The impulsive method consists of replacing in problem (12.11) the trajectory calculation with an analytical evaluation of the velocity increment. This evaluation, based on the concept of propulsive velocity increment, decouples the staging problem from the trajectory problem and it allows an iterative solution process.

### 12.3.1 Propulsive Velocity Increment

The velocity gain is obtained by integrating the equation (12.5) in $\dot{\vec{v}}$ after having projected it along the velocity. During the atmospheric flight, the drag is opposed to the thrust, and it becomes null in exo-atmospheric flight. By noting $\alpha$ the absolute angle of attack (between the thrust and the absolute velocity) and $\gamma$ the local slope (between the local horizontal and the absolute velocity), and integrating between the times $t_1$ of a stage ignition and $t_2$ of its separation gives

$$v(t_2) - v(t_1) = \int_{t_1}^{t_2} \frac{T - D}{m} \cos \alpha dt - \int_{t_1}^{t_2} g \sin \gamma dt \tag{12.12}$$

The thrust is expressed by $T = qg_0Isv = -\dot{m}v_e$ considering the equivalent specific impulse for the first stage and with the exhaust velocity defined by $v_e = g_0Isv$.

The variables D, g, $\alpha$, $\gamma$ are dependent on the trajectory followed, which does not allow an analytical integration. However, it is possible to rewrite (12.12) by isolating an integral in T/m.

$$v(t_2) - v(t_1) = \int_{t_1}^{t_2} \frac{T}{m} dt - \int_{t_1}^{t_2} \frac{T}{m}(1 - \cos \alpha) dt - \int_{t_1}^{t_2} \frac{D}{m} \cos \alpha dt - \int_{t_1}^{t_2} g \sin \gamma dt \tag{12.13}$$

The first integral, called the propulsive velocity increment, is noted ΔV. It can be calculated using $T = -\dot{m}v_e$, which gives the formula of Tsiolkovski (1905).

$$\Delta V_{prop} = v_e \ln \frac{m_1}{m_2} \tag{12.14}$$

The other integrals are called velocity losses, since they are subtracted from the propulsive ΔV. They depend on the trajectory and have the following orders of magnitude:

- drag losses (D term): 50–100 m/s depending on the launcher shape;
- gravity losses (g term): 700–1500 m/s depending on the altitude variation;
- incidence losses (T term): very variable.

Incidence losses occur when the thrust is not aligned with the absolute velocity (non-null absolute incidence angle $\alpha$). A part of the thrust is used to rotate the velocity instead of increasing it and the loss is measured by the term in $1 - \cos\alpha$. During the atmospheric flight at zero relative angle of attack, the thrust is aligned with the *relative velocity* (velocity in the rotating Earth frame) and not with the absolute velocity, resulting in incidence losses of the order of 500 m/s.

The total propulsive $\Delta V$ required for the mission includes the actual velocity increment between the take-off (at velocity $v_0$ due to the launch pad rotating with the Earth) and the injection in orbit (at velocity $v_f$ corresponding to the perigee of the targeted orbit) and the losses incurred during the ascent trajectory.

$$\Delta V_{req} = v_f - v_0 + \Delta V_{loss} \tag{12.15}$$

This required $\Delta V$ is initially estimated with a loss guess which can range from 1500 to 2500 m/s. This allows us to define a first staging of the launcher (Section 12.3.2) and assess a first ascent trajectory. The losses calculated on this trajectory are then used to update the initial guess (Section 12.3.3).

### 12.3.2 Staging Problem

The propulsive $\Delta V$ delivered by the stage number j is $\Delta V_j = v_{e,j} \frac{M_{i,j}}{M_{f,j}}$, where $M_{i,j}$ and $M_{f,j}$ are the launcher masses at the ignition and the shutdown of the stage j.

The aim of staging optimization is to find the lightest launcher that is able to bring a given payload of mass $m_u$ to the targeted orbit. The optimization variables are the propellant masses $m_{p,j}$ of the stages. They determine the mass of the launcher (with the structural indexes $k_j$) and its propulsive $\Delta V$, which must be equal to the $\Delta V$ required to reach the orbit (12.15).

The staging optimization problem is formulated as

$$\min_{m_{p,j}} M_0 \text{ s.t. } \Delta V_{tot} = \Delta V_{req} \text{ with } \begin{cases} M_0 = m_u + \sum_j (1+k_j) m_{p,j} \\ \Delta V_{tot} = \sum_j v_{e,j} \ln \frac{M_{i,j}}{M_{f,j}} \end{cases} \tag{12.16}$$

To solve this problem, we introduce new variables $x_j = \frac{M_{i,j}}{M_{f,j}}$ with the convention $M_{i,n+1} = m_u$ and we take as a cost function the ratio between the payload mass and the launcher total lift-off mass.

This mass ratio, noted J, is expressed as:

$$J = \frac{m_u}{M_0} = \frac{M_{i,n+1}}{M_{i,1}} = \prod_j \left( \frac{M_{i,j+1}}{M_{i,j}} \right) = \prod_j \left( \frac{1+k_j}{x_j} - k_j \right) \tag{12.17}$$

because $\frac{M_{i,j+1}}{M_{i,j}} = \frac{M_{i,j} - m_{e,j}(1+k_j)}{M_{i,j}} = \frac{M_{i,j} - (M_{i,j} - M_{f,j})(1+k_j)}{M_{i,j}} = \frac{1+k_j}{x_j} - k_j$.

The staging problem is thus reformulated as:

$$\max_{x_j} \prod_j \left( \frac{1+k_j}{x_j} - k_j \right) \text{ s.t. } \sum_j v_{e,j} \ln x_j = \Delta V_{req} \tag{12.18}$$

This problem has no analytical solution because the structural indexes depend on the unknown propellant masses by (12.1).

Nevertheless, by assuming *constant structural indexes* (independent of the masses $m_{p,j}$), this problem can be solved by the conditions of Karush, Kuhn, and Tucker.

The Lagrangian is defined with a multiplier $\lambda$ of the constraint.

$$L(x_j, \lambda) = \prod_j \left( \frac{1+k_j}{x_j} - k_j \right) + \lambda \left( \sum_j v_{e,j} \ln x_j - \Delta V_{req} \right) \tag{12.19}$$

The KKT conditions give a system of equations of dimension n+1.

$$\begin{cases} \dfrac{\partial L}{\partial x_j} = 0 \\ \dfrac{\partial L}{\partial \lambda} = 0 \end{cases} \Rightarrow \begin{cases} \dfrac{\partial L}{\partial x_j} = -\dfrac{1+k_j}{x_j^2} \dfrac{J}{\frac{1+k_j}{x_j} - k_j} + \lambda \dfrac{v_{e,j}}{x_j} = 0, \quad j = 1 \text{ to } n \\ \sum_j v_{e,j} \ln x_j = \Delta V_{req} \end{cases} \tag{12.20}$$

This system yields after simplifications

$$\begin{cases} v_{e,j} \left( 1 - \dfrac{k_j}{1+k_j} x_j \right) = \dfrac{J}{\lambda} = \text{Cte}, \quad j = 1 \text{ to } n \\ \sum_j v_{e,j} \ln x_j = \Delta V_{req} \end{cases} \tag{12.21}$$

The first n equations allow to express the n−1 variables $(x_1, \ldots, x_{n-1})$ as functions of the single variable $x_n$, then replace them in the last equation giving $\Delta V$.

This equation with one unknown $x_n$ is solved numerically by a Newton or dichotomy method. The variables $x_j$ are then obtained, and starting from the payload mass $M_{i,n+1} = m_u$, we get successively:

- the gross masses $M_{i,j}$ and $M_{f,j}$ by $x_j = \frac{M_{i,j}}{M_{f,j}}$ and $\frac{M_{i,j+1}}{M_{i,j}} = \frac{1+k_j}{x_j} - k_j$;
- the propellant and dry masses by $m_{p,j} = M_{i,j} - M_{f,j}$ and $m_{d,j} = k_j m_{p,j}$.

Let us illustrate this calculation with the example of the Ariane1 launcher (first flight at Christmas 1979), which was designed exactly by following this staging optimization process. The data below correspond to the actual launcher design.

The specific impulses and structural indexes are as follows (Table 12.1).

The $\Delta V$ required for a GTO mission from Kourou is $\Delta V_{req} = 11\,527$ m/s, including losses of 1725 m/s (obtained on the nominal trajectory of the launcher).

The payload mass is $m_u = 1700$ kg. The resolution gives the following results, which correspond to the staging of the actual launcher (Table 12.2).

### 12.3.3 Staging Loops

The analytical solution defined by (12.20) assumes constant structural indexes $k_j$ and it is also based on an estimate of the losses along the ascent trajectory.

**Table 12.1** Ariane1 specific impulses and structural indexes.

| | **Stage 1** | **Stage 2** | **Stage 3** |
|---|---|---|---|
| Equivalent Isv | 270 s | 298 s | 443 s |
| Structural index k | 11.01% | 15.32% | 21.54% |

**Table 12.2** Ariane1 masses.

| | Stage 1 (L145) | Stage 2 (L33) | Stage 3 (H8) |
|---|---|---|---|
| Variable $x_j$ | 3.3273 | 2.9360 | 3.2949 |
| Launcher mass $M_{i,j}$ (kg) | 208 729 | 47 344 | 11 343 |
| Mass propellant $m_{e,j}$ (kg) | 145 378 | 31 218 | 7934 |

The actual indexes depend on the propellant masses, which are unknown at the beginning of the resolution, and the losses depend on the optimal trajectory, which can only be computed when the configuration is defined. It is therefore necessary to iterate at a first level on the structural indexes by updating them according to the propellant masses found (12.1), then at a second level on the losses when the optimal trajectory has been calculated, based on the motion equations (12.5).

In addition, the thrust levels (if they are to be defined) of the stage must be "tuned" to comply with the maximum levels of acceleration and dynamic pressure (during the first stage flight and at its separation). These updates are to be done empirically depending on the results obtained.

The staging loop (Figure 12.2) consists of the following steps.

- Optimization of the masses for fixed losses $\Delta V_{loss}$ and structural indexes k
- Correction of thrust levels or masses to satisfy the acceleration constraints
- Update on the indexes k from the propellant masses and thrust levels T
- Optimization of the trajectory giving the dynamic pressure levels and losses
- Recalibration of thrust levels T, then update of the losses $\Delta V_{loss}$

This impulsive method based on embedded iterations and updates simplifies the calculations, but it has two major weaknesses.

- The updates of the indexes and of the losses are carried out by fixed point methods. This approach does not guarantee convergence toward an optimal solution, and it may even very easily diverge (depending on the features of the fixed point problem).
- The thrust levels are not taken into account by the impulsive formula and they must be set empirically. Unfortunately, the thrust has a major effect on the performance of the launcher and on the loads encountered along the flight (structural loads, controllability, and thermal fluxes).

The following section presents a coupled staging-trajectory approach that avoids these difficulties inherent to the impulsive method.

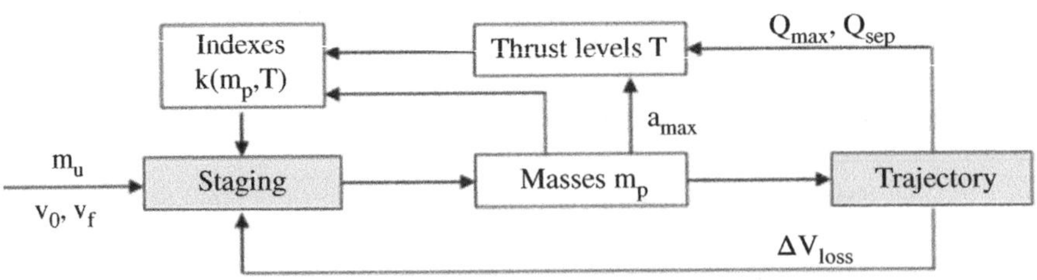


**Figure 12.2** Staging loops.

## 12.4 Coupled Method

The coupled method directly addresses the problem (12.11) by introducing a simplified control law. The simplification based on analytical costate formulas reduces the control law to two variables.

### 12.4.1 Control Law

Two major issues must be overcome to solve the global problem (12.11):

- determining the optimal control law requires solving a shooting problem, which is difficult due to its high sensitivity to initial conditions;
- the coupled staging-trajectory problem is ill-conditioned because staging variables (masses, thrust levels) have a much greater effect on the trajectory than the control.

These numerical difficulties can be circumvented by reducing the law of control by the following two simplifications.

The first simplification is to use an analytical formula for the initial costate. This formula has been obtained in Chapter 8 by considering an optimized parametric thrust profile. The initial costate given by (8.45) has the following expression:

$$\begin{cases} \vec{p}_r(t_2) = \omega_2 \begin{pmatrix} -\cos(\theta_2 - \varphi_2) \\ \sin(\theta_2 - \varphi_2) \end{pmatrix} & \text{with } \omega_2 = \sqrt{\dfrac{\mu}{r_2^3}(1 - 3\sin^2\theta_2)} \\ \vec{p}_v(t_2) = -\begin{pmatrix} \sin(\theta_2 - \varphi_2) \\ \cos(\theta_2 - \varphi_2) \end{pmatrix}, & p_m(t_2) = -\dfrac{g_0 Isv_2}{m_2} \end{cases} \tag{12.22}$$

In these equations, $t_2$ is the second-stage ignition time, $m_2$ is the launcher mass at that time, $(r_2, \varphi_2)$ are the polar coordinates in the inertial geocentric frame and $\theta_2$ is the local pitch angle.

This angle $\theta_2$ is determined by equation (8.29) as a function of the kinematic coordinates (position $r_2$, velocity $v_2$, and local slope $\gamma_2$) at $t_2$.

$$\sin(\theta_2 - \gamma_2) = \pm\frac{v_{c2}}{v_2}\frac{\sin\theta_2}{\sqrt{1 - 3\sin^2\theta_2}} \quad \text{with } v_{c2} = \sqrt{\frac{\mu}{r_2}} \tag{12.23}$$

These formulas have been established in Chapter 8 for a single planar boost with final orbit constraints. The global staging-trajectory problem considered here is more complex, as it involves several successive stages and the trajectory is subject to path and current constraints. Nevertheless, since the thrust levels of the stages are to be optimized, it can be hoped that the optimal costate will be close to the analytical solution given by (12.22). This justifies the use of the formulas (12.22), but keeping the initial pitch angle $\theta_2$ as a free optimization variable instead of calculating it by (12.23) from the kinematic coordinates at the stage ignition.

The second simplification consists of neglecting the costate discontinuities induced by path and current constraints. These discontinuities due to dynamic pressure or heat flux constraints are usually small. With this assumption, we can integrate the costate equations (12.10) in one shot from the second stage ignition until the last stage shutdown. The thrust direction $\vec{u}(t)$ given by (12.8) is thus fully defined for the whole exo-atmospheric flight by the single variable $\theta_2$.

With these simplifications, the control for the whole trajectory depends only on the two variables $\gamma_1$ for the first stage and $\theta_2$ for the upper stages. The global staging-trajectory problem (12.11) is reduced to a nonlinear optimization problem formulated as

$$\min_{\substack{m_{pj},T_j\\ \gamma_1,\theta_2}} K(m_{pj},T_j) \text{ s.t. } \begin{cases} \text{Non-gravitational acceleration} \leq a_{max} \\ \text{Maximum dynamic pressure} \leq Q_{max} \\ \text{Separation dynamic pressure} \leq Q_{sep} \\ \text{Flux after fairing jettisoning} \leq \Phi_{sep} \\ \text{Final apogee, perigee} = Z_a, Z_p \end{cases} \tag{12.24}$$

### 12.4.2 Iterations and Margins

Solving the problem (12.24) by nonlinear programming software does not raise any initialization or convergence issues. The solution simultaneously gives the launcher configuration (masses, thrust levels) and the trajectory associated with the simplified control defined by the variables $\gamma_1$ and $\theta_2$. This configuration cannot be directly considered as the optimal solution for two reasons.

- The simplified control law defined by $\gamma_1$ and $\theta_2$ is suboptimal.

The configuration obtained from (12.24) has therefore an actual performance (maximum payload mass $m_u$ that can be brought to the targeted orbit) higher than the targeted one. The targeted payload for the problem (12.24) has to be lowered so that the configuration found has exactly the desired performance.

- Margins on model uncertainties, in particular on the specific impulses and the structural indexes, should be taken into account for the preliminary staging.

The iterations to obtain a launcher configuration having the required performance with sufficient robustness to model uncertainties are carried out as follows:

- The problem (12.24) is solved for the first time by targeting the specified payload mass. We get the first configuration of the launcher.
- The actual performance of this configuration is then calculated by solving the classical launcher trajectory problem (12.6) without any simplifications.

This solution considers an optimal control law and it accounts for operational constraints (safety, visibility) that are not considered in problem (12.24). Even if considering an optimal control increases the performance, these additional constraints have an adverse effect that mitigates the performance gain.

- The performance derivatives with respect to the model uncertainties (specific impulses and structural indexes) are evaluated by finite differences, varying one model parameter at a time and solving the problem each time (12.6).
- The problem (12.24) is solved a second time by changing the payload target. The payload shift takes into account the excess performance (if any) observed with the first configuration and a desired level of performance probability with respect to the model uncertainties.

This process results in a configuration having a "guaranteed" performance (at the chosen probability level) equal to the required performance. If the design models used for the preliminary staging prove to be correct during the launcher development, the final configuration will have a performance higher than desired. However, this is rarely the case in practice, and the above procedure helps set the right level of margin for the preliminary staging, in order to limit the risks of unfeasibility during the launcher development.

### 12.4.3 Versatile Configuration

The previous approach can be easily extended to the optimization of a versatile configuration. Let us assume that the specifications define two reference missions, allowing the adaptation of one launcher element per mission. This element is usually either the number of boosters or the loading of the last stage.

#### 12.4.3.1 Problem Formulation

The problem (12.24) is formulated by considering simultaneously the two launcher stagings and the two associated trajectories. The two stagings are defined by the same variables $m_{p,j}$ for the masses and $T_j$ for the thrust levels, with the exception of the versatile stage for which the variable(s) are duplicated.

For example, the last stage propellant mass will be defined by $(m_{e,3})_1$ for the mission 1 and by $(m_{e,3})_2$ for the mission 2. The control variables are duplicated: $(\gamma_1,\theta_2)_1$ for the trajectory of mission 1 and $(\gamma_1,\theta_2)_2$ for the trajectory of mission 2.

The global problem (12.24) involves the simulation of both trajectories with their respective controls and constraints. The cost function can be defined as a linear combination of the costs of both configurations according to their respective priorities. Despite the increase in the size of the problem (12.24) with the simultaneous optimization of two trajectories, the resolution with standard nonlinear programming software does not raise any particular difficulties.

#### 12.4.3.2 Application Cases

In the following sections, the resolution procedure is illustrated on a three-stage versatile configuration.

The launcher configuration has to achieve two reference missions (payload and orbit) by varying only one configuration element. The staging options considered are either changing the number of boosters or changing the last stage propellant mass.

The structural indexes are assumed to be constant (independent of the masses and the thrust levels). Their values and the equivalent specific impulses for each stage are given in Table 12.3.

The aerodynamic reference surface is set to $S_{ref} = 10\ m^2$ and the drag coefficient is constant $C_x = 1$. The aerodynamic characteristics are assumed to be fixed for this example, but in practice, they may be in the form of a drag law ($C_x$ as a function of the Mach number) and they may also vary as a function of the launcher staging.

**Table 12.3** Equivalent specific impulses and structural indexes.

| | Stage 1 | Stage 2 | Stage 3 |
|---|---|---|---|
| Isv equivalent | 280 s | 430 s | 350 s |
| Structural index k | 10% | 15% | 20% |

For each versatile configuration, the two reference trajectories associated with the two reference missions must satisfy the following constraints:

- nongravitational acceleration lower than 5 g;
- maximum dynamic pressure lower than 50 kPa;
- first stage separation altitude higher than 50 km (this is equivalent to a dynamic pressure constraint);
- exo-atmospheric altitude higher than 120 km (this is equivalent to a heat flux constraint).

It is assumed that the fairing is jettisoned with the first stage and that its mass is taken into account in the structural index of the first stage. The exo-atmospheric altitude constraint higher than 120 km applies once the launcher has reached this altitude. Any subsequent downward arc must then remain over 120 km in order to avoid too high heat flux on the unprotected payload (since the fairing has been jettisoned). The cost function K to be minimized is defined as the mass of the heaviest configuration.

### 12.4.4 Application Case 1: Number of Boosters

This first application case considers a versatile three-stage launcher with two configurations differing by the number of boosters. Both configurations have identical second and third stages (same masses and engines).

- The four-booster configuration must bring a payload of mass 12 tons on a GTO with an apogee at 36 000 km and a perigee at 200 km.
- The two-booster configuration must bring a payload of mass 5 tons on the same GTO orbit.

The two reference missions are summarized in Table 12.4.

All boosters are identical and each booster has a propellant mass $m_{p,1}$ and a thrust level $T_1$. The design problem (12.24) is formulated with 10 unknowns.

- The six staging variables are the stage propellant masses $m_{p,1}$, $m_{p,2}$, $m_{p,3}$ and thrust levels $T_1$, $T_2$, and $T_3$, which are identical for both configurations.
- The two control variables for the trajectory of mission 1 are $\gamma_{1,1}$ and $\theta_{2,1}$.
- The two control variables for the trajectory of mission 2 are $\gamma_{1,2}$ and $\theta_{2,2}$.

For mission 1 (resp. mission 2), the payload is $m_{u,1} = 12$ tons (resp. $m_{u,2} = 5$ tons), the first stage propellant mass is $4m_{p,1}$ (resp. $2m_{p,1}$), the first stage thrust level is $4T_1$ (resp. $2T_1$), and the lift-off mass is $M_{0,1}$ (resp. $M_{0,2}$).

The launcher masses for the two configurations are thus

$$\begin{cases} M_{0,1} = 4m_{p,1}(1+k_1) + m_{p,2}(1+k_2) + m_{p,3}(1+k_3) + m_{u,1} \\ M_{0,2} = 2m_{p,1}(1+k_1) + m_{p,2}(1+k_2) + m_{p,3}(1+k_3) + m_{u,2} \end{cases} \tag{12.25}$$

**Table 12.4** Mission definition depending on the number of boosters.

| | Number of boosters | Payload mass $m_u$ (kg) | Altitude apogee $Z_A$ (km) | Altitude perigee $Z_P$ (km) |
|---|---|---|---|---|
| Mission 1 | 4 | 12 000 | 36 000 | 200 |
| Mission 2 | 2 | 5000 | 36 000 | 200 |

The cost function is associated with the heaviest configuration: $K = M_{0,1}$.

Each trajectory must satisfy four inequality constraints (acceleration, dynamic pressure, separation altitude, and exo-atmospheric altitude) and two equality constraints (final apogee and perigee altitudes). The design problem has thus eight inequality constraints and four equality constraints. It is solved by a classical gradient-based nonlinear programming method from a rough initial guess.

The three-step solution procedure described in Section 12.4.2 is applied.

#### 12.4.4.1 Step 1: First Resolution Without Margins

The problem is first solved considering the targeted payload masses of Table 12.4 with the reference design data of Table 12.3. The launcher masses obtained for the 12 and 5 tons missions are $M_{0,1} = 981\,591$ kg and $M_{0,2} = 591\,707$ kg, respectively.

The staging variables are given in Table 12.5. The gross mass, the total impulsive velocity, and the trajectory constraints for each mission are given in Table 12.6.

The altitude, acceleration, and local pitch evolutions for the two missions are plotted in Figure 12.3.

#### 12.4.4.2 Step 2: Actual Performance and Derivative Assessment

The performance and robustness of both configurations are checked as follows.

Each configuration is considered separately and its maximal performance on the targeted orbit is computed. The optimal control problem (12.6) is solved without command law simplifications, and considering the same constraints as for problem (12.24). This classical launcher problem with a fixed configuration is solved by a shooting method, with the second stage's initial costate as unknown.

Since the optimal control problem is now exactly solved, the maximal payload found will be higher than the targeted payload. Indeed, the configuration has been optimized by considering a sub-optimal control law based on the formula (12.22).

The sensitivities of the performance to the main design uncertainties are also computed in the same way. The uncertainties considered here are associated with the specific

**Table 12.5** Configurations for GTO 12t/GTO 5t.

| | Stage 1 (per booster) | Stage 2 | Stage 3 |
|---|---|---|---|
| Propellant mass | 174 038 kg | 153 131 kg | 23 102 kg |
| Thrust level | 3413 kN | 2731 kN | 187 kN |
| Burn duration | 140.0 s | 236.5 s | 425.0 s |

**Table 12.6** Performances for GTO 12t/GTO 5t.

| | Launcher gross mass (kg) | Impulsive velocity (m/s) | Separation altitude (km) | Max acceleration (m/s²) | Max dyn pressure (Pa) |
|---|---|---|---|---|---|
| Mission 1 | 981 591 | 11 595 | 75.4 | 49.0 | 23 303 |
| Mission 2 | 591 707 | 12 212 | 56.1 | 49.0 | 12 875 |

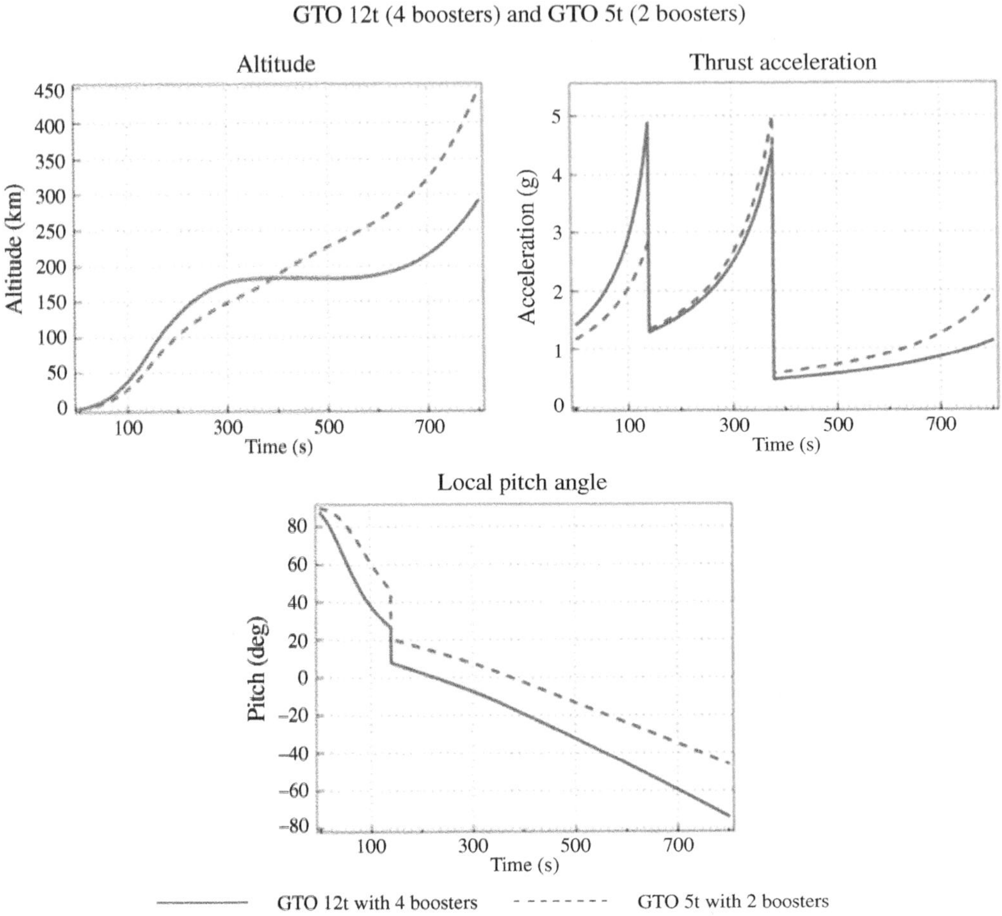


**Figure 12.3** Trajectories for GTO 12t/GTO 5t.

**Table 12.7** Performances and derivatives for GTO 12t/GTO 5t.

| Configuration | Max payload (kg) | Derivative $Isv_1$ (kg/s) | Derivative $Isv_2$ (kg/s) | Derivative $Isv_3$ (kg/s) | Derivative $k_1$ (kg/%) | Derivative $k_2$ (kg/%) | Derivative $k_3$ (kg/%) |
|---|---|---|---|---|---|---|---|
| GTO 12t 4 boosters | **13 209** | 69 | 78 | 57 | −383 | −528 | −236 |
| GTO 5t 2 boosters | **6241** | 39 | 40 | 34 | −138 | −263 | −227 |

impulse and structural index values. The maximal performance and the sensitivities of both configurations are given in Table 12.7.

The sub-optimal control has led to performance underestimates of 1209 and 1241 kg for the four-booster and two-booster configurations, respectively.

#### 12.4.4.3 Step 3: Staging Update

The actual performances and sensitivities are used to refine the configuration design. The goal is to achieve the targeted performance (payload mass) at a given probability level with respect to the design uncertainties. Assuming independent uncertainties and a Gaussian distribution of the performance with respect to these uncertainties, the worst payload at a probability level of 90% is estimated by

$$m_{u,90\%} = m_{u,nom} - 1.28\sqrt{\sum_{\alpha}\left(\frac{\partial m_u}{\partial \alpha}\sigma_\alpha\right)^2} \quad (12.26)$$

where $\alpha$ is an uncertain model parameter, $\sigma_\alpha$ is its standard deviation, $\frac{\partial m_u}{\partial \alpha}$ is the corresponding performance sensitivity (given in Table 12.7), and $m_{u,nom}$ is the nominal targeted performance (given in Table 12.4) assumed to be equal to the mean value of the probabilistic performance.

In this application case, assuming standard deviations of 1 s for the specific impulses and 1% for the structural indexes, the worst performance at a 90% probability level would be:

- $m_{u,90\%} = 13\,209 - 1.28 \times 704 = 12\,308$ kg for the four-booster configuration, corresponding to a performance excess of 308 kg;
- $m_{u,90\%} = 6241 - 1.28 \times 379 = 5755$ kg for the two-booster configuration, corresponding to a performance excess of 755 kg.

In order to come closer to the required performances of 12 000 kg and 5000 kg, with a 90% probability level, the design problem is solved again with the targeted payloads shifted respectively:

- to 12 000 – 308 = 11 692 kg for the four-booster configuration;
- to 5000 – 755 = 4245 kg for the two-booster configuration.

With these updated payload targets, the configuration masses are lowered to:

- $M_{0,1} = 870\,404$ kg (instead of 981 591 kg) for the four-booster configuration;
- $M_{0,2} = 537\,687$ kg (instead of 591 707 kg) for the two-booster configuration.

These updated configurations are compliant with the probabilistic performance requirement based on sensitivity computations. They can be used as starting points for multidisciplinary design loops based on high-fidelity models, including Monte Carlo simulations, to check the performance robustness.

### 12.4.5 Application Case 2: Upper Stage Loading

The second application case considers a three-stage configuration with a variable propellant mass of the upper stage. Both configurations have identical first and second stages (same masses and engines).

- The first configuration must bring a payload of mass 5 tons on a GTO with an apogee at 36 000 km and a perigee at 200 km.
- The second configuration must bring a payload of mass 15 tons on a LEO circular at an altitude of 400 km.

The upper stage uses the same engine for both configurations, but with a different propellant mass: $(m_{p,3})_1$ for the mission 1 and $(m_{p,3})_2$ for the mission 2.

The dry mass changes accordingly, so that the upper stages have a different design for both missions. The design data of Table 12.3 are still used.

The two reference missions are summarized in Table 12.8.

The design problem (12.24) is formulated with eleven unknowns.

- The five common staging variables are the stage propellant masses $m_{p,1}$ and $m_{p,2}$ and thrust levels $T_1$, $T_2$, and $T_3$.

**Table 12.8** Mission definition depending on the upper stage loading.

| | Upper stage propellant | Payload mass $m_u$ (kg) | Altitude apogee $Z_A$ (km) | Altitude perigee $Z_P$ (km) |
|---|---|---|---|---|
| Mission 1 | $(m_{p,3})_1$ | 5000 | 36 000 | 200 |
| Mission 2 | $(m_{p,3})_2$ | 15 000 | 400 | 400 |

- The propellant masses for missions 1 and 2 are $(m_{p,3})_1$ and $(m_{p,3})_2$, respectively.
- The two control variables for the trajectory of mission 1 are $\gamma_{1,1}$ and $\theta_{2,1}$.
- The two control variables for the trajectory of mission 2 are $\gamma_{1,2}$ and $\theta_{2,2}$.

The launcher masses for the two configurations are thus

$$\begin{cases} M_{0,1} = m_{p,1}(1+k_1) + m_{p,2}(1+k_2) + (m_{p,3})_1(1+k_3) + m_{u,1} \\ M_{0,2} = m_{p,1}(1+k_1) + m_{p,2}(1+k_2) + (m_{p,3})_2(1+k_3) + m_{u,2} \end{cases} \quad (12.27)$$

The cost function is here associated with the GTO configuration: $K = M_{0,1}$, considering that this GTO mission has a higher priority.

Each trajectory must satisfy four inequality constraints (acceleration, dynamic pressure, separation altitude, and exo-atmospheric altitude) and two equality constraints (final apogee and perigee altitudes). The design problem has thus eight inequality constraints and four equality constraints. It is solved by a classical gradient-based nonlinear programming method from a rough initial guess.

The three-step solution procedure described in Section 12.4.2 is applied.

#### 12.4.5.1 Step 1: First Resolution Without Margins

The problem is first solved considering the targeted payload masses of Table 12.8 with the reference design data of Table 12.3. The launcher masses obtained for the GTO and LEO missions are $M_{0,1} = 467\,779$ kg and $M_{0,2} = 473\,610$ kg, respectively.

The staging variables are given in Table 12.9. The gross mass, the total impulsive velocity, and the trajectory constraints for each mission are given in Table 12.10.

The altitude, acceleration, and local pitch evolutions for the two missions are plotted in Figure 12.4.

#### 12.4.5.2 Step 2: Actual Performance and Derivative Assessment

Similarly to the previous example, the performances of both configurations are assessed afterward, as well as sensitivities to uncertainties on the specific impulse and on the structural index. The maximal performance and the sensitivities of both configurations are given in Table 12.11.

**Table 12.9** Configurations for GTO 5t/LEO 15t.

| | Stage 1 | Stage 2 | Stage 3 Mission 1 | Stage 3 Mission 2 |
|---|---|---|---|---|
| Propellant mass | 297 412 kg | 95 857 kg | 21 158 kg | 17 683 kg |
| Thrust level | 8149 kN | 1991 kN | 98 kN | 98 kN |
| Burn duration | 100.2 s | 203.1 s | 741.9 s | 620.1 s |

**Table 12.10** Performances for GTO 5t/LEO 15t.

| | Launcher gross mass (kg) | Impulsive velocity (m/s) | Separation altitude (km) | Max acceleration ($m/s^2$) | Max dyn pressure (Pa) |
|---|---|---|---|---|---|
| Mission 1 | 467 779 | 11 690 | 62.1 | 47.9 | 36 255 |
| Mission 2 | 473 610 | 9496 | 61.6 | 46.3 | 35 152 |

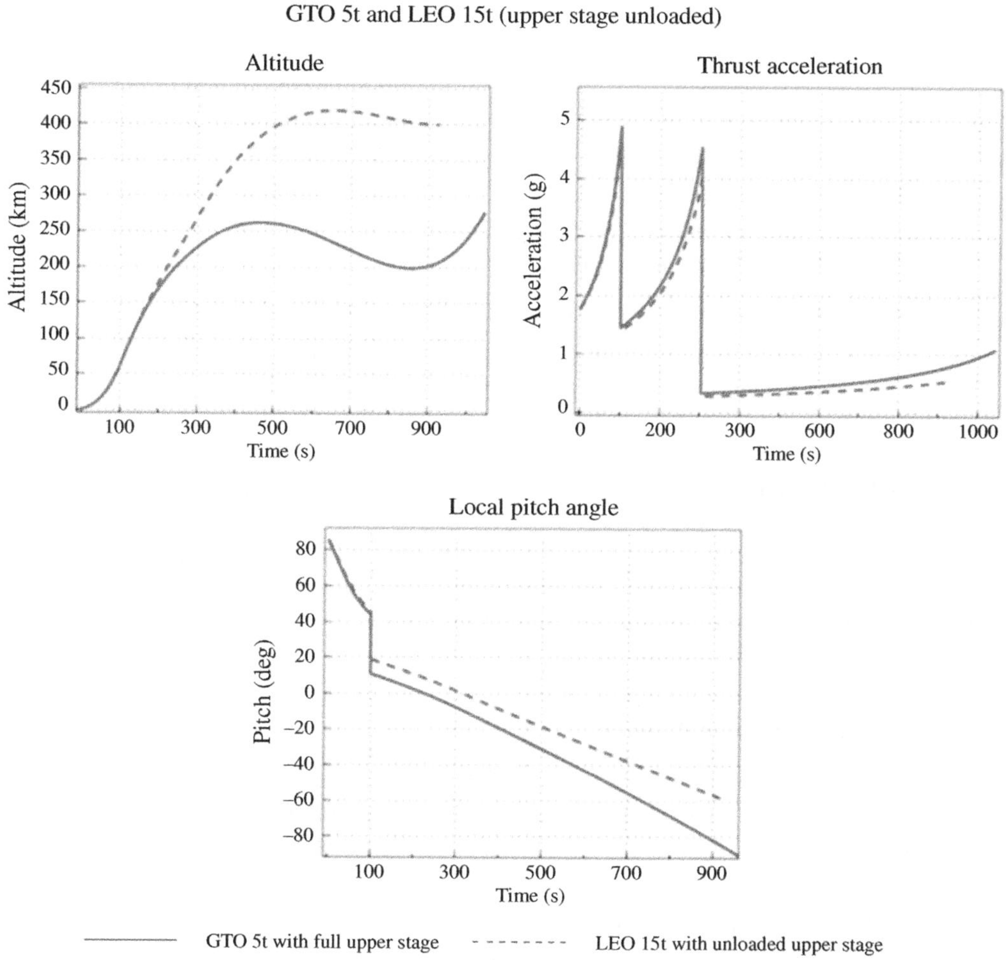


**Figure 12.4** Trajectories for GTO 5t/LEO 15t.

#### 12.4.5.3 Step 3: Staging Update

For standard deviations of 1 s on the specific impulses and 1% on the structural indexes, the worst performance at a 90% probability level would be:

- $m_{u,90\%} = 6141 - 1.28 \times 357 = 5684$ kg for the GTO configuration, corresponding to a performance excess of 684 kg;
- $m_{u,90\%} = 16\,523 - 1.28 \times 859 = 15\,423$ kg for the LEO configuration, corresponding to a performance excess of 423 kg.

**Table 12.11** Performances and derivatives for GTO 5t/LEO 15t.

| Configuration | Max payload (kg) | Derivative $Isv_1$ (kg/s) | Derivative $Isv_2$ (kg/s) | Derivative $Isv_3$ (kg/s) | Derivative $k_1$ (kg/%) | Derivative $k_2$ (kg/%) | Derivative $k_3$ (kg/%) |
|---|---|---|---|---|---|---|---|
| GTO 5t $m_{p,3} = 21\,158$ kg | **6141** | 156 | 127 | 69 | −114 | 215 | −152 |
| LEO 15t $m_{p,3} = 17\,683$ kg | **16 523** | 193 | 30 | 367 | −434 | −409 | −342 |

In order to come closer to the required performances of 12 000 and 5000 kg, with a 90% probability level, the design problem is solved again with the targeted payloads shifted respectively:

- to 5000 − 684 = 4316 kg for the GTO configuration;
- to 15 000 − 423 = 14 577 kg for the LEO configuration.

With these updated payload targets, the configuration masses are lowered to:

- $M_{0,1} = 459\,562$ kg (instead of 467 779 kg) for the GTO configuration;
- $M_{0,2} = 465\,706$ kg (instead of 473 610 kg) for the LEO configuration.

## 12.5 Propellant Reserve

The configuration of the launcher and the environment of the flight are not perfectly known on the day of launch. To ensure a sufficient probability of reaching orbit, it is necessary to have propellant reserves on board, especially for the last stage. The reserve will be consumed only in unfavorable flight conditions.

### 12.5.1 Performance and Reserve

Let us note $m_p$ as the total propellant mass of the last stage, $m_d$ as its dry mass, $m_r$ as the reserve mass, $m_u$ as the payload mass, and $v_e$ as the exhaust velocity. The stage's initial and final masses are $M_i$ and $M_f$, and the impulse delivered by the stage is

$$\Delta V_{prop} = v_e \ln \frac{M_i}{M_f} = v_e \ln \frac{m_p + m_d + m_u}{m_r + m_d + m_u} \tag{12.28}$$

The propellant mass is fixed, given by the tank size, as well as the dry mass.

Let us assume that the reserve varies by $\delta m_r$. To maintain the nominal impulse associated with the optimal trajectory, the payload must vary by $\delta m_u$, which is expressed from (12.28).

$$\delta m_u = -\delta m_r \frac{1}{1 - e^{-\frac{\Delta V}{v_e}}} = -\delta m_r \frac{M_i}{M_i - M_f} \tag{12.29}$$

This formula shows that increasing the reserve of $\delta m_r$ leads to a loss of payload mass $\delta m_u$ greater than $\delta m_r$. This is explained by the fact that the reserve takes the place of the payload (direct effect) and furthermore is no longer available for the propulsion on the nominal trajectory (indirect effect).

In practice, we have exchange coefficients of the order of

- $\delta m_u \approx -1.5\,\delta m_r$ for $\Delta V \approx v_e$;
- $\delta m_u \approx -2.5\,\delta m_r$ for $\Delta V \approx \frac{1}{2} v_e$.

The reserve must be limited to the amount strictly necessary to ensure the required success probability. The following sections present two methods of calculation. These methods are presented for the upper stage, but they can be easily transposed to the lower stages.

### 12.5.2 Iterative Method

The iterative method consists of decoupling the trajectory optimization and the reserve calculation, and then iterating between the two problems.

- The trajectory is first optimized considering the nominal configuration data (masses, thrust levels) and the nominal flight environment (atmosphere). The consumption allowed to the upper stage is set to $m_c = m_p - m_r$, where $m_p$ is the upper stage total propellant mass and $m_r$ is the mass of the reserve (arbitrary initial estimate). A nominal trajectory is obtained with a maximal payload $m_u$ associated with the reserve $m_r$.
- The payload mass $m_u$ is fixed to its maximal value and the variations of the consumed mass $m_c$ are assessed in presence of perturbations on the launcher configuration and the flight environment. These variations are used to assess the distribution of the consumed mass as a random variable. The reserve mass must then compensate for overconsumption at a given probability level.

The variations in consumed mass can be assessed either by Monte Carlo with random drawings of the perturbations, simulation of the trajectories, and analysis of the consumed mass distribution, or by derivatives of the consumed mass (by varying the parameters one by one) and quadratic summation. For a probability of nonexhaustion of 99%, the reserve is assessed by

$$m_r = 2.33\sqrt{\sum_{\alpha}\left(\frac{\partial m_c}{\partial \alpha}\sigma_\alpha\right)^2} \tag{12.30}$$

where $\alpha$ is a perturbed parameter and $\sigma_\alpha$ is its standard deviation.

This formula assumes that the perturbations are Gaussian and independent and that the mass consumed is a linear function of the perturbations. More complex formulas may be used with different assumptions.

The optimization of the nominal trajectory is then resumed with the new reserve mass. This yields a new maximal payload mass and new variations of the consumed mass. The process is iterated until stabilization.

This iterative approach simplifies the calculations, but it has two weaknesses.

- The reserve is updated by a fixed point method. This approach does not ensure the convergence toward an optimum in terms of payload mass.
- When the trajectory is sensitive to some constraints, the variations in the consumed mass may be highly nonlinear and the process may diverge.

The following section presents an embedded method that avoids these difficulties.

### 12.5.3 Embedded Method

We consider the flight of the upper stage, assuming that there is only one boost. In the case of multiple boosts, the following analysis applies in a similar way by considering separately each boost with its own targeted orbit and an associated probability level. The embedded method assesses the reserve directly within the trajectory optimization. This assessment is based on the cost derivatives.

#### 12.5.3.1 Cost Function Derivatives

The upper stage engine has an exhaust velocity $v_e$, a mass flow rate q, and a thrust level $T = qv_e$. These propulsive features may be constant or time dependent.

The engine is ignited at time $t_0$ with initial conditions $(\vec{r}_0, \vec{v}_0, m_0)$. The trajectory reaching the orbit at minimum consumption is obtained by solving the optimal control problem (12.6). The optimality conditions (12.7)–(12.10) are solved by a shooting method. This yields the costate evolution, the control law, and the consumption on the optimal trajectory.

The consumption depends mainly on the flow rate q, on the exhaust velocity $v_e$ and on the initial velocity $v_0$. All other data have a secondary influence. The propellant reserve must compensate for the uncertainties on q, $v_e$, and $v_0$. It is assessed by the formula (12.30) using the consumption derivatives with respect to these parameters. For an optimal control problem like (12.6), the first variation of the cost function with respect to a model parameter α is given by (see [R9], [R21])

$$\delta J = \int_{t_0}^{t_f} \frac{\partial H}{\partial \alpha} \delta\alpha \, dt \tag{12.31}$$

This formula is obtained by analyzing the first-order optimality conditions. The integral is to be calculated along the optimal trajectory. The Hamiltonian obtained in (12.9) is rewritten so that its dependency on the flow rate and on the exhaust velocity appears explicitly.

$$H = \vec{p}_r \cdot \vec{v} + \vec{p}_v \cdot \vec{g} - q\left(\frac{v_e p_v}{m} + p_m\right) \tag{12.32}$$

The following formulas obtained by applying (12.31) with q and $v_e$ as model parameters give the final mass derivatives with respect to q and $v_e$.

$$\begin{cases} \dfrac{\partial m_f}{\partial q} = \displaystyle\int_{t_0}^{t_f} \frac{\partial H}{\partial q} dt = -\int_{t_0}^{t_f} \left(\frac{v_e p_v}{m} + p_m\right) dt \\ \dfrac{\partial m_f}{\partial v_e} = \displaystyle\int_{t_0}^{t_f} \frac{\partial H}{\partial v_e} dt = -\int_{t_0}^{t_f} \frac{q p_v}{m} dt \end{cases} \tag{12.33}$$

On the other hand, the final mass derivative with respect to the initial velocity $v_0$ is given by the initial velocity costate, since the costate represents at any time the cost function derivative with respect to the current state (see [R6], [R9], [R21]).

$$\frac{\partial m_f}{\partial v_e} = p_v(t_0) \tag{12.34}$$

The formulas (12.33) and (12.34) depend only on the state and costate history along the optimal trajectory. They can thus be calculated immediately after the optimal control problem has been solved by a shooting method. This provides an "on-line" evaluation of the derivatives during the optimization of the trajectory.

#### 12.5.3.2 Reserve Assessment

The uncertainties on q, $v_e$, and $v_0$ have different origins. In first approximation:

- the flow rate depends on the combustion chamber feeding system;
- the exhaust velocity depends on the chamber pressure and the nozzle shape;
- the initial velocity depends on the dispersions during the previous stage flight.

Due to their different causes, these uncertainties can reasonably be considered as Gaussian and independent. Noting $\delta q$, $\delta v_e$, and $\delta v_0$ their respective standard deviations, the final mass standard deviation is estimated by quadratic summation.

$$\delta m_f = \sqrt{\left(\frac{\partial m_f}{\partial q}\delta q\right)^2 + \left(\frac{\partial m_f}{\partial v_e}\delta v_e\right)^2 + (p_v(t_0)\delta v_0)^2} \tag{12.35}$$

and the propellant reserve for a probability of nonexhaustion of 99% is

$$m_r = 2.33\,\delta m_f \tag{12.36}$$

The guaranteed performance (payload mass injected into the targeted orbit with a probability level of 99%) is obtained by subtracting from the final mass the reserve $m_r$ calculated "on-line" by the aforementioned formulas, with the exchange coefficient given by the formula (12.29).

This guaranteed payload mass is taken as a cost function (instead of the actual payload mass) for the optimization of the launch trajectory. It results in a more robust launch trajectory because the sensitivities to uncertainties are taken into account via the penalty due to the reserve. The subsequent update of the reserve mass (calculated a-posteriori with all perturbation contributors) is then quite small compared to the iterative method presented in Section 12.5.2.

### 12.5.4 Application Case

A three-stage launcher described in Table 12.12 is considered. The gross lift-off mass is 500 000 kg without the payload. The launch takes place from the equator toward a circular polar orbit at 800 km. The first stage follows a zero angle of attack flight after an initial optimized tilting maneuver. The second stage is controlled by optimized linear laws on the pitch and yaw angles. The upper stage is reignitable and it follows the optimal control law defined by (12.8) derived from the shooting problem solution. The orbit is reached after a coast arc and a circularization maneuver.

**Table 12.12** Three-stage launcher.

| | Stage 1 | Stage 2 | Stage 3 |
|---|---|---|---|
| Propellant mass (kg) | 288 000 | 120 000 | 32 000 |
| Dry mass (kg) | 32 000 | 20 000 | 8000 |
| Specific impulse (s) | 280 | 430 | 350 |
| Flow rate (kg/s) | 2400 | 300 | 40 |
| Duration (s) | 120 | 400 | 800 |

#### 12.5.4.1 First Method: Nominal Performance and A-posteriori Reserve Assessment

The trajectory is first optimized without accounting for any upper-stage propellant reserve. The total mass injected on the targeted orbit is 3578 kg. The altitude and the pitch angle evolutions along the optimal trajectory are plotted as bold lines in Figures 12.5 and 12.6.

The performance derivatives along this optimal trajectory are obtained from the costate history using the formulas (12.33) and (12.34). The upper stage reserve is then assessed by

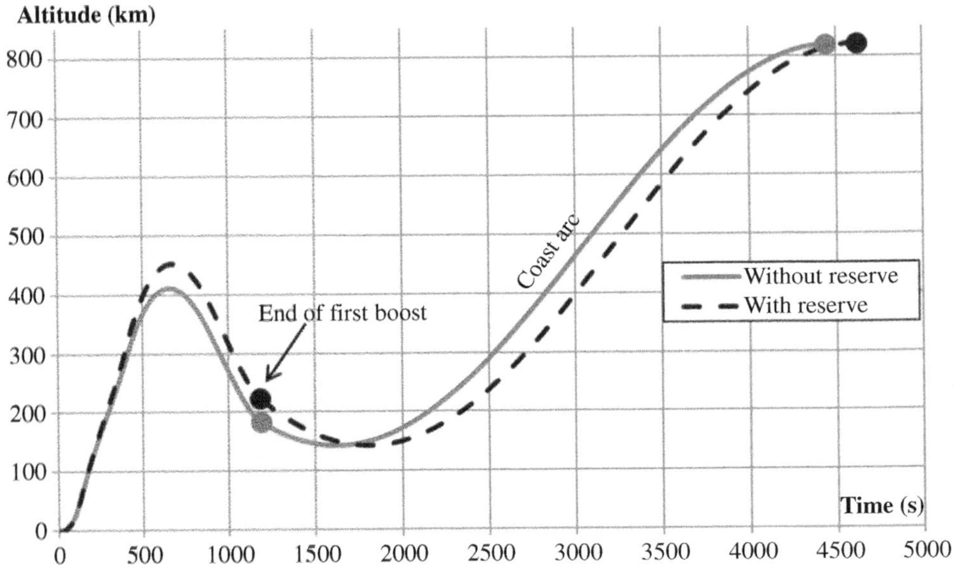


**Figure 12.5** Altitude along the optimal trajectory.

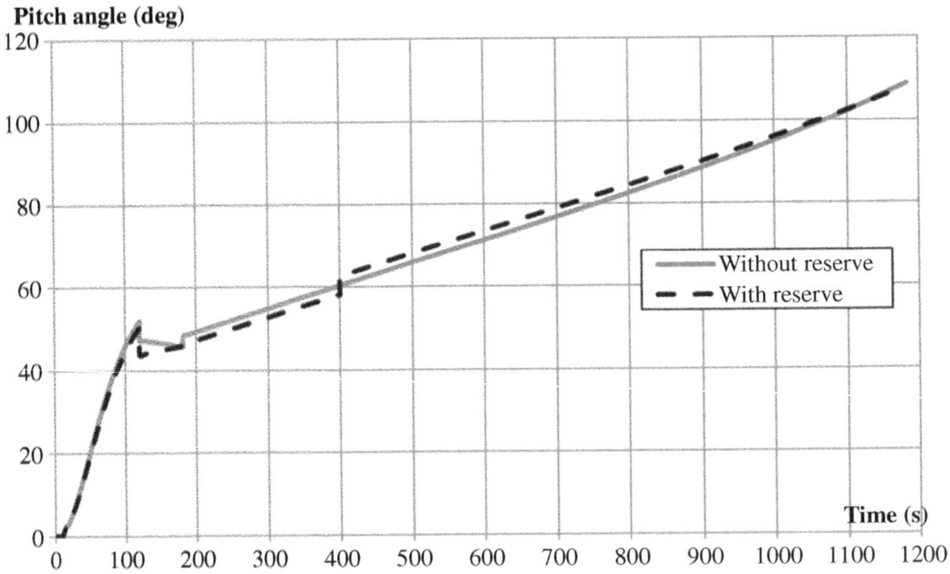


**Figure 12.6** Pitch angle along the optimal trajectory.

**Table 12.13** Derivatives, reserve, and guaranteed performance.

| | Optimization without reserve | Optimization with reserve |
|---|---|---|
| Derivative wrt flow rate (kg/(kg/s)) | 256.6 | 218.8 |
| Derivative wrt specific impulse (kg/s) | 63.6 | 60.5 |
| Derivative wrt initial velocity (kg/(m/s)) | 7.9 | 7.0 |
| Total injected mass (kg) | 3578 | 3543 |
| Reserve mass (kg) | 651 | 565 |
| Guaranteed payload mass (kg) | **2927** | **2978** |

the formulas (12.35) and (12.36) with the following uncertainties at 2.33σ:

$$\delta q_m = 2\ \text{kg/s}, \delta Isv = 1\ \text{s}, \delta v_0 = 50\ \text{m/s}$$

This a-posteriori assessment yields a reserve mass of 651 kg and a guaranteed payload mass of 2927 kg on the targeted orbit. The results are given in the first column of Table 12.13 (entitled "optimization without reserve").

#### 12.5.4.2 Second Method: Trajectory Optimization with Embedded Reserve Assessment

The trajectory is now optimized using the embedded reserve assessment. The cost function for this optimization is directly the guaranteed payload mass. The optimal trajectory shape changes and it is plotted as dotted lines in Figures 12.5 and 12.6.

This new trajectory has a lower total mass injected on orbit (3543 kg), but also lower derivatives. The reserve mass is reduced to 565 kg resulting finally in a gain of 51 kg on the guaranteed payload. The results are given in the second column of Table 12.13 (entitled "optimization with reserve").

This example shows that the optimization "with reserve" allows a significant gain in the guaranteed payload. This assessment accounts for the main uncertainties that influence the upper stage consumption, namely the flow rate, the specific impulse, and the initial velocity. The uncertainty on the initial velocity depends on the previous stage flight. An approximate value can be derived from impulsive formulas accounting for the previous stage uncertainties.

The reserve assessed by formulas (12.33)–(12.36) is a preliminary estimate. It does not require time-consuming extra computations since it is based on the costate history along the trajectory. This approach is applicable to design loops or to performance calculations on a range of missions. For a real flight preparation, a reference flight is first defined by trajectory optimization. Monte Carlo simulations are then run considering the detailed flight dynamics, the vehicle and environment uncertainties, and the on-board control. The actual reserve is defined from the Monte Carlo results as presented in Section 12.5.2. This actual reserve differs from the analytical estimate, but the offset is usually small, as formulas (12.33)–(12.36) already incorporate the main last stage perturbations.

Embedding the reserve calculation within the trajectory optimization allows a gain in the guaranteed payload. The same approach applies to the preparation of an actual flight with a fixed payload mass. By taking the estimated reserve (12.36) as a cost function to be minimized, we enhance the robustness of the trajectory with respect to the perturbations (by reducing the consumption derivatives). This favors the on-board guidance system responsible for correcting in-flight perturbations. Since these perturbations have less effect, the required corrections are smaller, and the overconsumption needed by the guidance decreases.

## 12.6 Conclusion

### 12.6.1 The Key Points

- A simplified near-optimal control law reduces the coupled staging + trajectory problem to a nonlinear programming problem with a dozen variables at most.
- Several configurations and missions are optimized simultaneously in a single optimization problem. A versatile launcher configuration can be optimized for several missions with margins adapted to the level of a preliminary design.
- The resulting configuration is updated using an optimal control law to achieve the targeted performance with margins accounting for model uncertainties.
- The propellant reserve is assessed directly within the trajectory optimization in order to guarantee the performance at a given probability level.
- These methods can be used to quickly define an optimized staging and trajectory with some robustness against uncertainties. This reduces the number of iterations and limits the risk of subsequent unfeasibility.

### 12.6.2 To Go Further

- CIVEK-COSKUN E. – *Space Launcher Design with Thrust Profile and Trajectory Optimization*. LAP LAMBERT Academic Publishing (2018).

This book describes some procedures and optimization methods applied to the coupled staging + trajectory optimization of a launcher.

- HAYNES T. – *Multidisciplinary Design Optimization*. NY Research Press (2017).

This book presents an interesting panorama of multidisciplinary optimization methods applied to various engineering problems.

### 12.6.3 Authorizations

- Examples in Section 12.4 are reprinted by permission of the American Institute of Aeronautics and Astronautics, Inc., from "Versatile Launcher Preliminary Design" published in Journal of Spacecrafts and Rockets Volume 56, Number 2, March 2019. Copyright c AIAA.
- Examples in Section 12.5 are reprinted by permission of the American Institute of Aeronautics and Astronautics, Inc., from "Launcher Performance with Reserves" published in Journal of Spacecrafts and Rockets Volume 55, Number 6, November 2018. Copyright c AIAA.

# 13

# Flat Earth Solutions

## 13.1 Introduction

Space trajectory problems can rarely be solved analytically. Assuming a flat Earth with constant gravity, with no atmosphere or with a simplified atmosphere model greatly simplifies the equations of motion, which in some cases allows explicit solutions to be found. These solutions are useful for quickly giving orders of magnitude of performance in preliminary design phases or for initializing a numerical resolution with more realistic models. This chapter presents the solution of four classical problems of launch trajectories in flat Earth, followed by a method for passing the solution from the flat Earth to the round Earth model.

The first problem deals with the maximum-range ballistic launch at constant acceleration. The flat Earth solution with no atmosphere shows that the optimal thrust direction is constant. The optimal pitch angle is obtained by solving an equation with one unknown, which depends only on the acceleration level.

The second problem deals with the injection into orbit at constant acceleration. The flat Earth solution with no atmosphere shows that the thrust direction follows a linear tangent law. The problem is reduced to an equation with one unknown. The solution makes it possible to evaluate the required propulsive impulse depending on the acceleration level and the targeted altitude.

The third problem deals with the maximum-apogee vertical launch assuming variable thrust. This is Goddard's problem, the solution of which may involve a nonmaximum variable thrust arc in the presence of aerodynamic braking. This singular arc is determined analytically as a function of the rocket drag features.

The fourth problem deals with the injection into orbit assuming variable thrust. The optimal solution in a flat Earth with no atmosphere consists of a single arc with maximum thrust. The optimal control problem is reduced to a system with two unknowns that can easily be solved by a shooting method.

The fifth section deals with the transition from a flat Earth and constant gravity to a round Earth and variable gravity. The transformation is carried out by a change of coordinates and a continuation method, allowing the gradual introduction of the curvature of the Earth and variable gravity.

## 13.2 Ballistic Launch at Constant Acceleration

The first problem of this chapter concerns a ballistic launch on a flat Earth with no atmosphere. The aim is to maximize the range of a constant acceleration rocket launched

*Space Trajectories: Basic and Advanced Topics*, First Edition. Max Cerf.

from the ground. Although simplified, this model can be used as a first approximation, considering the average acceleration of the rocket during its flight.

### 13.2.1 Modeling

The trajectory takes place in the vertical plane (Oxz), as shown in Figure 13.1.

The Earth is assumed to be flat, with no atmosphere; the gravity is constant and directed downward along the vertical.

The rocket is a material point of position $\vec{r}\,(x, z)$ and velocity $\vec{v}\,(v_x, v_z)$. It takes off from the ground at the time $t_0 = 0$ and stops thrusting at the fixed time $t_f$.

The thrust acceleration is constant of modulus a. The direction of thrust is oriented by the pitch angle θ (angle with the horizontal Ox), which is a function of time. After the thrust has stopped, the trajectory continues on a ballistic arc under the influence of gravity until the impact on the horizontal axis Ox.

During the propelled phase, the state vector $(x, z, v_x, v_z)$ evolves according to the differential equations

$$\begin{cases} \dot{x}(t) = v_x(t) \\ \dot{z}(t) = v_z(t) \\ \dot{v}_x(t) = a\cos\theta(t) \\ \dot{v}_z(t) = a\sin\theta(t) - g \end{cases} \quad \text{with} \quad \begin{cases} x(t_0) = 0 \\ z(t_0) = 0 \\ v_x(t_0) = 0 \\ v_z(t_0) = 0 \end{cases} \tag{13.1}$$

The aim is to determine the control law $\theta(t)_{t_0 \le t \le t_f}$ that maximizes the range R at impact.

The range depends on the position and velocity conditions reached at the end of propulsion. These conditions are noted $(x_f, z_f, v_{xf}, v_{zf})$. With a constant gravity, the range is easily calculated yielding the expression

$$R = x_f + \frac{v_{xf}}{g}\left[v_{zf} + \sqrt{v_{zf}^2 + 2gz_f}\right] \tag{13.2}$$

The optimal control problem is formulated as:

$$\min_{\theta(t)} \; -R(x_f, z_f, v_{xf}, v_{zf}) \;\; \text{s.t.} \; \begin{cases} \dot{x}(t) = v_x(t) \\ \dot{z}(t) = v_z(t) \\ \dot{v}_x(t) = a\cos\theta(t) \\ \dot{v}_z(t) = a\sin\theta(t) - g \end{cases}, \quad \begin{cases} x(t_0) = 0 \\ z(t_0) = 0 \\ v_x(t_0) = 0 \\ v_z(t_0) = 0 \end{cases} \tag{13.3}$$

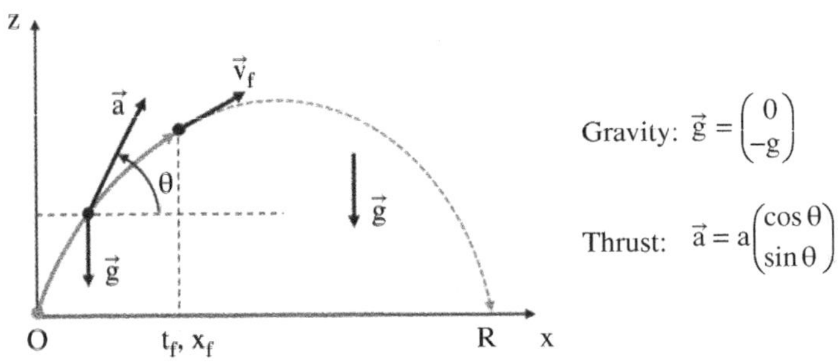


**Figure 13.1** Maximum range at constant acceleration.

### 13.2.2 Optimal Solution

To simplify the notation, the time dependence is no longer explicitly noted in the calculations that follow, except where it is useful.

The problem (13.3) is solved by applying Pontryaguin Maximum Principle (PMP).

For this, the Hamiltonian H is formed by introducing the costate vector of components ($p_x$, $p_z$, $p_{vx}$, $p_{vz}$).

$$H = p_x v_x + p_z v_z + p_{vx} a \cos\theta + p_{vz}(a \sin\theta - g) \tag{13.4}$$

The optimal control θ is given by the Hamiltonian minimization.

$$\min_{\theta} H \to \tan\theta = \frac{p_{vz}}{p_{vx}} \tag{13.5}$$

To determine the control θ(t), we integrate the costate equations.

$$\begin{cases} \dot{p}_x = -H_x = 0 & \to p_x(t) = A \\ \dot{p}_z = -H_z = 0 & \to p_z(t) = B \\ \dot{p}_{vx} = -H_{vx} = -p_x & \to p_{vx}(t) = C + A(t_f - t) \\ \dot{p}_{vz} = -H_{vz} = -p_z & \to p_{vz}(t) = D + B(t_f - t) \end{cases} \tag{13.6}$$

The four integration constants (A, B, C, D) are determined from the transversality conditions on the final costate.

$$\begin{cases} p_x(t_f) = -\dfrac{\partial R}{\partial x_f} & \to A = -1 \\ p_z(t_f) = -\dfrac{\partial R}{\partial z_f} & \to B = -\dfrac{v_{xf}}{\sqrt{v_{zf}^2 + 2gz_f}} \\ p_{vx}(t_f) = -\dfrac{\partial R}{\partial v_{xf}} & \to C = -\dfrac{v_{zf} + \sqrt{v_{zf}^2 + 2gz_f}}{g} \\ p_{vz}(t_f) = -\dfrac{\partial R}{\partial v_{zf}} & \to D = -\dfrac{v_{xf}}{g}\left(1 + \dfrac{v_{zf}}{\sqrt{v_{zf}^2 + 2gz_f}}\right) \end{cases} \tag{13.7}$$

Substituting into (13.5) the costates ($p_{vx}$, $p_{vz}$) given by (13.6) with the constants given by (13.7), and observing that $D = -BC$, we obtain

$$\tan\theta(t) = \frac{p_{vz}(t)}{p_{vx}(t)} = \frac{D + B(t_f - t)}{C + A(t_f - t)} = -B\frac{C - (t_f - t)}{C - (t_f - t)} = -B = \frac{v_{xf}}{\sqrt{v_{zf}^2 + 2gz_f}} \tag{13.8}$$

The optimal pitch angle θ is therefore constant and the maximum range is achieved by thrusting in a fixed direction. It can be observed that the calculations (13.4)–(13.8) and the resulting conclusion on the constant thrust direction are still valid if the acceleration law a(t) is time varying instead of constant.

To find the optimal value of θ, we integrate the motion equations (13.1), knowing that θ is constant. By assuming that the acceleration is constant, these equations can be integrated analytically.

$$\begin{cases} \dot{x} = v_x \\ \dot{z} = v_z \\ \dot{v}_x = a \cos\theta \\ \dot{v}_z = a \sin\theta - g \end{cases} \to \begin{cases} x_f = (a \cos\theta)\, t_f^2/2 \\ z_f = (a \sin\theta - g)\, t_f^2/2 \\ v_{xf} = (a \cos\theta)\, t_f \\ v_{zf} = (a \sin\theta - g)\, t_f \end{cases} \tag{13.9}$$

Substituting ($z_f$, $v_{xf}$, $v_{zf}$) into equation (13.8), we get an equation in θ.

$$\tan\theta\sqrt{(a\sin\theta-g)^2+g(a\sin\theta-g)}=a\cos\theta \tag{13.10}$$

which can be put into the form

$$\sin^3\theta-2n\sin^2\theta+n=0 \quad \text{with } n=\frac{a}{g} \tag{13.11}$$

The variable n represents the rocket acceleration expressed in g.

### 13.2.3 Application Case

Let us look for the solutions of equation (13.11) depending on n. The angle θ varies between 0 deg (horizontal thrust) and 90 deg (vertical thrust). Figure 13.2 shows the function $f(\theta)=\sin^3\theta-2n\sin^2\theta+n$ plotted for four different values of n. The optimal angle is the one at which the function cancels.

Depending on the acceleration level, the following results are obtained.

- For $n<1$, equation (13.11) has no solution. Indeed, the take-off is impossible with a thrust lower than the weight ($a<g$).
- For $n=1$ , the optimal angle is 90 deg. The thrust must be strictly vertical in order to take off ($a=g$). The resulting range is null.
- For $n>1$, equation (13.11) has a single solution with a thrust angle comprised between 45 and 90 deg.
- For $n\to\infty$, the optimal angle is 45 deg. This solution corresponds to an initial velocity impulse given at the ground. This result can be retrieved directly from equation (13.2) by putting $z_f=0$. The maximum range is then obtained for $v_{xf}=v_{zf}$, which corresponds to an angle of 45 deg.

Once the value of θ is known, equations (13.9) yield the position and velocity at injection and equation (13.2) yields the maximum range achieved.

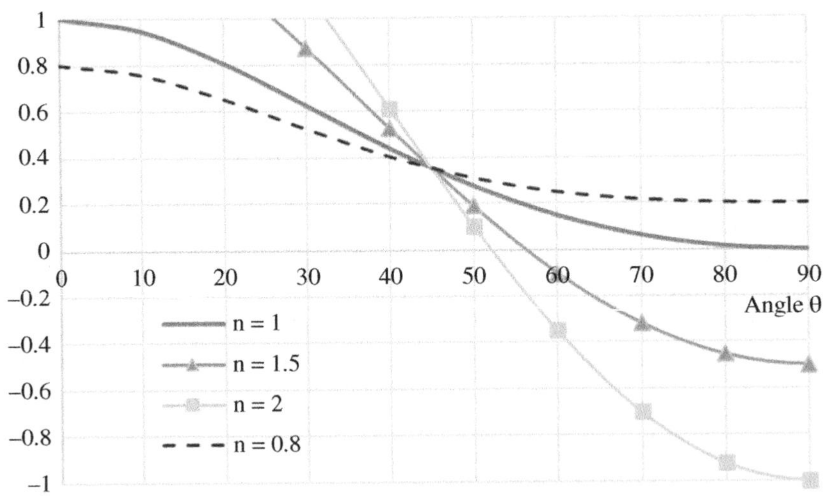


**Figure 13.2** Solution as a function of the acceleration level.

## 13.3 Injection into Orbit at Constant Acceleration

The second problem of this chapter concerns an injection into orbit on a flat Earth with no atmosphere. The orbit is defined by an altitude and a horizontal velocity. The aim is to reach the orbit in minimum time with a rocket at constant acceleration. This problem is presented in [R6].

### 13.3.1 Modeling

The modeling is similar to that of the previous problem. The trajectory takes place in the vertical plane (Oxz), as shown in Figure 13.3. The Earth is assumed to be flat, with no atmosphere; the constant gravity is directed downward along the vertical.

The rocket is a material point of position $\vec{r}\,(x, z)$ and velocity $\vec{v}\,(v_x, v_z)$. It takes off from the ground at the time $t_0 = 0$ and stops thrusting at a free time $t_f$.

The thrust acceleration is constant of modulus a. The direction of thrust is oriented by the pitch angle $\theta$ (angle with the horizontal Ox). The unit vector along the thrust direction is noted $\vec{u}$.

The state vector $(\vec{r}, \vec{v})$ evolves according to the differential equations

$$\begin{cases} \dot{\vec{r}} = \vec{v} \\ \dot{\vec{v}} = \vec{g} + a\vec{u} \end{cases}, \quad \begin{cases} \vec{r}(t_0) = \vec{0} \\ \vec{v}(t_0) = \vec{0} \end{cases} \tag{13.12}$$

The aim is to find the control law $\theta(t)_{t_0 \le t \le t_f}$ to reach the targeted altitude $z_c$ and horizontal velocity $v_c$ by minimizing the final time $t_f$. The final abscissa is free.

### 13.3.2 Optimal Solution

The optimal control problem is formulated as:

$$\min_{\vec{u}, t_f} J = \int_{t_0}^{t_f} 1\,dt \quad \text{with} \quad \begin{cases} \dot{\vec{r}} = \vec{v} \\ \dot{\vec{v}} = \vec{g} + a\vec{u} \end{cases} \quad \text{and} \quad \begin{cases} \vec{r}(t_0) = \vec{0} \\ \vec{v}(t_0) = \vec{0} \end{cases}, \quad \begin{cases} z(t_f) = z_c \\ v_x(t_f) = v_c \\ v_z(t_f) = 0 \end{cases} \tag{13.13}$$

The problem (13.13) is solved by applying the PMP.

For this, the Hamiltonian H is formed by introducing the costate vector of components $(\vec{p}_r, \vec{p}_v)$.

$$H = 1 + \vec{p}_r \cdot \vec{v} + \vec{p}_v \cdot (\vec{g} + a\vec{u}) \tag{13.14}$$

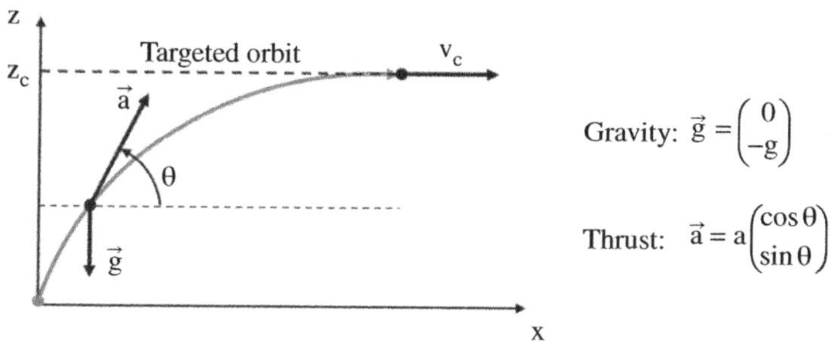


**Figure 13.3** Injection in orbit at constant acceleration.

The optimal control $\vec{u}$ is given by the Hamiltonian minimization.

$$\min_{\vec{u}} H \quad \rightarrow \quad \vec{u} = -\frac{\vec{p}_v}{\|\vec{p}_v\|} \text{ (unit vector)} \tag{13.15}$$

To determine $\vec{u}(t)$, we integrate the costate equations.

$$\begin{cases} \dot{\vec{p}}_r = -H_{\vec{r}} = \vec{0} & \rightarrow \quad \vec{p}_r = \vec{B} \\ \dot{\vec{p}}_v = -H_{\vec{v}} = -\vec{p}_r & \rightarrow \quad \vec{p}_v = \vec{B}(t_f - t) + \vec{A} \end{cases} \text{ with constants } \vec{A}, \vec{B} \tag{13.16}$$

Since the final abscissa $x(t_f)$ is free, we have the transversality condition

$$p_x(t_f) = 0 \quad \Rightarrow \quad B_x = 0 \tag{13.17}$$

#### 13.3.2.1 Control and Trajectory

By substituting the costates (13.16) into the expression of $\vec{u}$ (13.15), we can assess the pitch angle $\theta$.

$$\vec{u} = -\frac{\vec{p}_v}{\|\vec{p}_v\|} = -\frac{\vec{A} + \vec{B}(t_f - t)}{\|\vec{A} + \vec{B}(t_f - t)\|} \quad \rightarrow \quad \tan\theta = \frac{u_z}{u_x} = \frac{A_z + B_z(t_f - t)}{A_x} \tag{13.18}$$

This control follows a **linear tangent law** of the form: $\tan\theta = \tan\theta_0 - ct$.

By making the change of variable from the time t to the pitch angle $\theta$, the differential equations of motion (13.12) are rewritten as:

$$\tan\theta = \tan\theta_0 - ct \quad \Rightarrow \quad dt = -\frac{d\theta}{c\cos^2\theta} \quad \rightarrow \quad \begin{cases} \dfrac{d\vec{v}}{dt} = -c\cos^2\theta\dfrac{d\vec{v}}{d\theta} = \vec{g} + a\vec{u} \\ \dfrac{d\vec{r}}{dt} = -c\cos^2\theta\dfrac{d\vec{r}}{d\theta} = \vec{v} \end{cases} \tag{13.19}$$

The components of velocity are first integrated into variable $\theta$. The result is

$$\begin{cases} \dfrac{dv_x}{d\theta} = -\dfrac{a}{c\cos\theta} \\ \dfrac{dv_z}{d\theta} = -\dfrac{a\sin\theta}{c\cos^2\theta} - g\dfrac{dt}{d\theta} \end{cases} \rightarrow \begin{cases} v_x = \dfrac{a}{c}\ln\left(\tan\theta_0 + \dfrac{1}{\cos\theta_0}\right) - \dfrac{a}{c}\ln\left(\tan\theta + \dfrac{1}{\cos\theta}\right) \\ v_z = \dfrac{a}{c}\left(\dfrac{1}{\cos\theta_0} - \dfrac{1}{\cos\theta}\right) - gt \end{cases} \tag{13.20}$$

The components of position are then integrated into variable $\theta$. The calculations are more tedious and the result is

$$\begin{cases} x = \dfrac{a}{c^2}\left[\dfrac{1}{\cos\theta_0} - \dfrac{1}{\cos\theta} + \tan\theta\ln\left(\tan\theta + \dfrac{1}{\cos\theta}\right) - \tan\theta\ln\left(\tan\theta_0 + \dfrac{1}{\cos\theta_0}\right)\right] \\ z = \dfrac{a}{2c^2}\left[\dfrac{\tan\theta_0}{\cos\theta_0} + \dfrac{\tan\theta}{\cos\theta} - \dfrac{2\tan\theta}{\cos\theta_0} + \ln\left(\tan\theta + \dfrac{1}{\cos\theta}\right) - \ln\left(\tan\theta_0 + \dfrac{1}{\cos\theta_0}\right)\right] - \dfrac{gt^2}{2} \end{cases} \tag{13.21}$$

#### 13.3.2.2 Final Conditions

The final conditions concern the components $(v_x, v_z, z)$, whereas the control at the final time is given by $\tan\theta_f = \tan\theta_0 - ct_f$. This defines a nonlinear system of four equations whose unknowns are $(\theta_0, \theta_f, t_f, c)$.

$$\begin{cases} \tan\theta_f = \tan\theta_0 - ct_f \\ v_x(t_f) = \dfrac{a}{c}\ln\left(\tan\theta_0 + \dfrac{1}{\cos\theta_0}\right) - \dfrac{a}{c}\ln\left(\tan\theta_f + \dfrac{1}{\cos\theta_f}\right) = v_c \\ v_z(t_f) = \dfrac{a}{c}\left(\dfrac{1}{\cos\theta_0} - \dfrac{1}{\cos\theta_f}\right) - gt_f = 0 \\ z(t_f) = \dfrac{a}{2c^2}\left[\dfrac{\tan\theta_0}{\cos\theta_0} + \dfrac{\tan\theta_f}{\cos\theta_f} - \dfrac{2\tan\theta_f}{\cos\theta_0} + \ln\left(\tan\theta_f + \dfrac{1}{\cos\theta_f}\right) - \ln\left(\tan\theta_0 + \dfrac{1}{\cos\theta_0}\right)\right] \\ \quad - \dfrac{gt_f^2}{2} = z_c \end{cases} \tag{13.22}$$

#### 13.3.2.3 Reduction of the System

The unknowns $(c, t_f)$ can be eliminated with the first two equations.

There remains a system in $(\theta_0, \theta_f)$.

$$\begin{cases} \dfrac{a}{g}\left(\dfrac{1}{\cos\theta_0} - \dfrac{1}{\cos\theta_f}\right) = \tan\theta_0 - \tan\theta_f \\ \dfrac{\tan\theta_0}{\cos\theta_f} - \dfrac{\tan\theta_f}{\cos\theta_0} - \ln\left[\dfrac{\tan\theta_0 + \dfrac{1}{\cos\theta_0}}{\tan\theta_f + \dfrac{1}{\cos\theta_f}}\right] - \dfrac{2az_c}{v_c^2}\ln^2\left[\dfrac{\tan\theta_0 + \dfrac{1}{\cos\theta_0}}{\tan\theta_f + \dfrac{1}{\cos\theta_f}}\right] = 0 \end{cases} \tag{13.23}$$

The first equation gives $\theta_0$ as a function of $\theta_f$ (equation of degree 2 in $\cos\theta_0$).

There remains an equation in $\theta_f$ that must be solved numerically. From the value of $\theta_f$, we obtain $\theta_0$ and then $t_f$ by

$$t_f \ln\left[\dfrac{\tan\theta_0 + \dfrac{1}{\cos\theta_0}}{\tan\theta_f + \dfrac{1}{\cos\theta_f}}\right] = \frac{v_c}{a}(\tan\theta_0 - \tan\theta_f) \tag{13.24}$$

The propulsive impulse delivered during the trajectory is assessed by $\Delta V = at_f$.

The velocity losses (by gravity and incidence) are the difference between this propulsive $\Delta V$ and the actual final velocity $v_c$.

### 13.3.3 Application Case

The targeted final conditions are $z_c = 200$ km and $v_c = 7784\ \mathrm{m\cdot s^{-1}}$. This value corresponds to the velocity in a circular orbit at 200 km. For this application case, the constant gravity is set to $g = 9\ \mathrm{m\cdot s^{-2}}$.

| Acceleration | a = 2 g | a = 3 g | a = 5 g |
|---|---|---|---|
| $\theta_0$ (deg) | 45.64 | 43.90 | 49.24 |
| $\theta_f$ (deg) | 11.46 | −9.20 | −31.75 |
| $t_f$ (s) | 509.5 | 319.1 | 195.4 |
| $\Delta V$ (m/s) | 9171 | 8616 | 8792 |

The above table gives the solutions for accelerations of 2, 3 and 5 g.

A higher acceleration reduces the final time and the total impulse.

The corresponding trajectories are plotted in Figure 13.4. The arrows indicate the thrust direction on the trajectory at 3 g. It is observed that the thrust direction corresponds to an angle of attack (angle between the velocity and the thrust direction) positive at the beginning of the trajectory and a negative at the end.

The next step is to determine the acceleration level that minimizes the propulsive impulse $\Delta V = at_f$. The problem is solved for accelerations ranging from 1.5 to 6 g and for orbits at altitudes of 200, 300, and 400 km, respectively.

The velocity in a circular orbit at altitude $z_c$ is given by $v_c = \sqrt{g(R_T + z_c)}$ taking for the Earth radius $R_T = 6\ 378$ km and for the average gravity $g = 9\ m \cdot s^{-2}$.

The results plotted in Figure 13.5 show that the optimal acceleration depends on the target altitude and that the impulse becomes quite sensitive to the acceleration level when it is below 2 g.

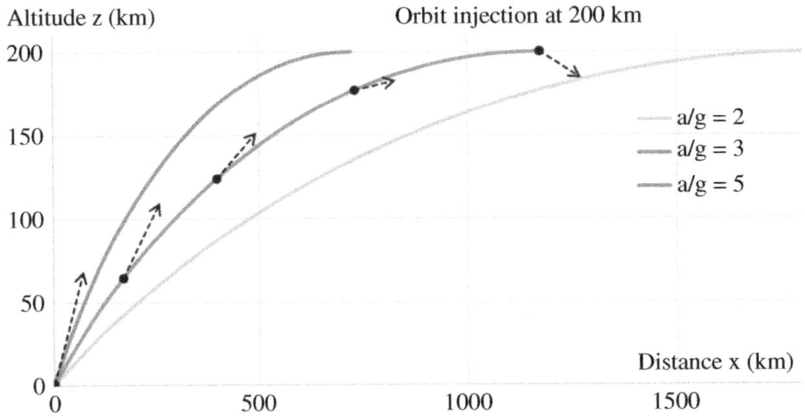


**Figure 13.4** Trajectories for orbit injection at 200 km altitude.

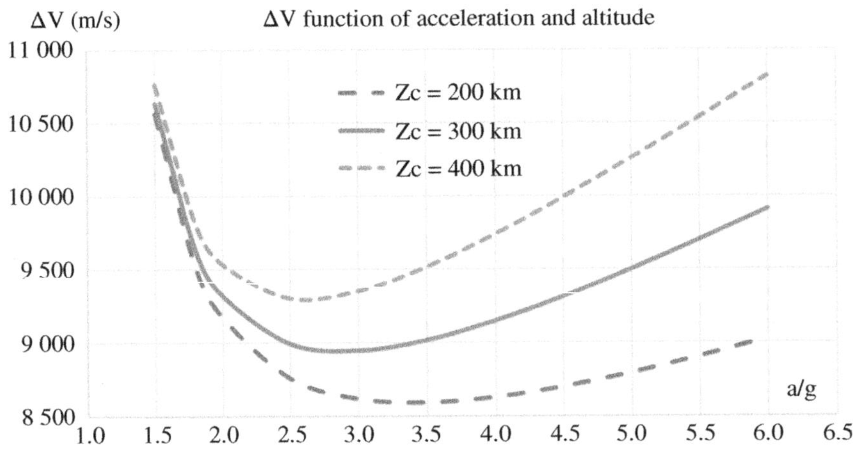


**Figure 13.5** Total impulse as a function of acceleration.

## 13.4 Vertical Launch at Variable Thrust

The American engineer and physicist R. H. Goddard is the inventor of the liquid rocket engine. After a successful first flight in 1926, Goddard tried to adjust the engine thrust to reach the highest possible altitude with a given rocket mass. This third section concerns Goddard's problem, which consists of finding the optimal thrust law accounting for atmospheric braking. This problem is presented in [R6].

### 13.4.1 Modeling

The rocket is represented by a material point of altitude h, velocity v, and mass m. It takes off from the ground ($h_0 = 0$, $v_0 = 0$) in the vertical direction. The initial total mass $m_0$ and the mass of propellant $m_p$ are given. The forces acting on the rocket are the weight W, the thrust T, and the drag D. All forces are vertical as depicted in the following figure and they are modeled as follows.

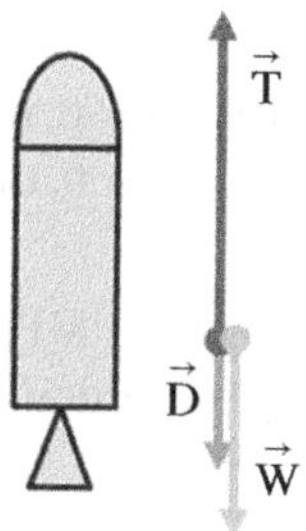


#### 13.4.1.1 Weight

The weight modulus is $W = mg$. The acceleration of gravity g is assumed to be constant.

#### 13.4.1.2 Drag

The drag modulus is $D = \frac{1}{2}\rho v^2 S C_D$.

- S is the reference surface (section of the rocket).
- $C_D$ is the drag coefficient depending on the Mach number.
- $\rho$ is the atmospheric density depending on the altitude.

In order to derive an analytical solution to Goddard' problem, we assume that the density follows an exponential model $\rho = \rho_0 e^{-\frac{h}{L}}$, where $\rho_0$ is the density at ground and L is the atmosphere scale factor. We assume also that the drag coefficient is constant (representing the average drag coefficient along the actual trajectory).

#### 13.4.1.3 Thrust

The thrust level T(t) can vary freely during the flight, remaining between 0 and an upper bound $T_{max}$.

The mass decreases according to $m(t) = m_0 - qt$, where $q = \frac{T}{v_e}$ is the propellant flow rate and $v_e$ is the exhaust velocity of the burnt gases, assumed constant.

When all propellants are consumed (at time $t_c = \frac{m_p}{q}$), the thrust is canceled. The trajectory continues under the influence of weight and drag until the apogee.

The state vector (h, v, m) evolves according to the differential equations

$$\begin{cases} \dot{h} = v \\ \dot{v} = \dfrac{T - D - W}{m} \\ \dot{m} = -\dfrac{T}{v_e} \end{cases} \tag{13.25}$$

### 13.4.2 Optimal Solution

The aim is to find the thrust law T(t) and the final time $t_f$ to maximize the altitude reached by rocket. The optimal control problem is formulated as:

$$\min_{T,t_f} J = -h(t_f)$$
$$\text{s.t.} \begin{cases} \dot{h} = v \\ \dot{v} = (T - D - W)/m \\ \dot{m} = -T/v_e \end{cases} \text{ and } \begin{cases} h(t_0) = 0, m(t_0) = m_0 \\ v(t_0) = 0, m(t_f) = m_0 - m_p \end{cases} \tag{13.26}$$

The problem (13.26) is solved by applying the PMP.

For this, the Hamiltonian H is formed by introducing the costate vector of components ($p_h$, $p_v$, $p_m$).

$$H = p_h v + p_v \frac{T - D - W}{m} - p_m \frac{T}{v_e} \tag{13.27}$$

By defining the function $\Phi = v_e p_v - m p_m$, the Hamiltonian takes the form

$$H = p_h v - p_v \left(\frac{D}{m} + g\right) + \frac{\Phi}{m v_e} T \tag{13.28}$$

The application of the PMP results in

$$\min_{0 \le T \le T_{max}} H \Rightarrow \begin{cases} T = 0 & \text{if } \Phi > 0 \\ T = T_{max} & \text{if } \Phi < 0 \end{cases} \text{ and } T = ? \text{ if } \Phi = 0 \tag{13.29}$$

The optimal thrust depends on the switching function $\Phi = v_e p_v - m p_m$.

A singular arc may occur if the function $\Phi$ cancels over a time interval.

#### 13.4.2.1 Search for Singular Arcs

Let us look for conditions on the costate ($p_h$, $p_v$, $p_m$) to get a singular arc.

If $\Phi$ cancels over an interval, its successive derivatives must be zero. On the other hand, the Hamiltonian is constant (because the system is autonomous) and null (because the final time is free).

We have thus the following system noted (S):

$$\begin{cases} H = 0 & \rightarrow \quad p_h v + p_v \dfrac{T - D - W}{m} - p_m \dfrac{T}{v_e} = 0 \\ \Phi = 0 & \rightarrow \quad v_e p_v - m p_m = 0 \\ \dot{\Phi} = 0 & \rightarrow \quad v_e \dot{p}_v - \dot{m} p_m - m \dot{p}_m = 0 \end{cases} \tag{13.30}$$

The costate equations are, by noting $D_h$ and $D_v$ the partial derivatives of the drag with respect to altitude and velocity (with the assumption of constant $C_D$)

$$\begin{cases} \dot{p}_h = -H_h = \dfrac{p_v D_h}{m} \\ \dot{p}_v = -H_v = -p_h + \dfrac{p_v D_v}{m} \\ \dot{p}_m = -H_m = p_v \dfrac{T-D}{m^2} \end{cases} \quad \text{with} \quad \begin{cases} D = \dfrac{1}{2}\rho v^2 S C_D = \dfrac{1}{2}\rho_0 e^{-\frac{h}{L}} v^2 S C_D \\ D_h = \dfrac{\partial D}{\partial h} = \dfrac{1}{2} v^2 S C_D \dfrac{\partial \rho}{\partial h} = -\dfrac{D}{L} \\ D_v = \dfrac{\partial D}{\partial v} = \rho v S C_D = \dfrac{2D}{v} \end{cases} \tag{13.31}$$

Substituting $\dot{m}$, $\dot{p}_v$, $\dot{p}_m$ into the expression of $\dot{\Phi}$, and using $\Phi = v_e p_v - m p_m = 0$ to replace $p_m$, we have

$$\begin{aligned} \dot{\Phi} &= v_e \dot{p}_v - \dot{m} p_m - m \dot{p}_m \\ &= v_e \left(-p_h + \frac{p_v D_v}{m}\right) + \frac{T}{v_e} p_m - p_v \frac{T-D}{m} = \frac{p_v}{m}\left(1 + 2\frac{v_e}{v}\right) D - v_e p_h \end{aligned} \tag{13.32}$$

The system (S) is thus reduced to

$$\begin{cases} p_h v - p_v \dfrac{D+W}{m} = 0 \\ p_v v_e - p_m m = 0 \\ p_h m - p_v \left(\dfrac{2}{v} + \dfrac{1}{v_e}\right) D = 0 \end{cases} \tag{13.33}$$

This system admits the null solution, but according to the PMP, the costate $(p_h, p_v, p_m)$ cannot be identically null. For another non-null solution to exist, the matrix of the linear system (S) must be singular and its determinant must be null.

$$\begin{vmatrix} v & -\dfrac{D+W}{m} & 0 \\ 0 & v_e & -m \\ m & -\left(\dfrac{2}{v} + \dfrac{1}{v_e}\right) D & 0 \end{vmatrix} = mW - mD\left(1 + \frac{v}{v_e}\right) = 0 \tag{13.34}$$

A singular arc is possible only if the function $\Psi \underset{\text{def}}{=} W - D\left(1 + \frac{v}{v_e}\right)$ cancels over a time interval. The singular surface is the set of states $(h, v, m)$ for which $\Psi = 0$.

Let us study the evolution of the function $\Psi$ along the flight. The velocity is zero at the beginning (take-off) and at the end of the flight (apogee).

The drag is therefore zero at these endpoints, so that $\begin{cases} \Psi(t_0) = m_0 g > 0 \\ \Psi(t_f) = m_f g > 0 \end{cases}$.

At the beginning of the flight, the weight W decreases (mass loss), whereas the velocity and drag increase. The function $\Psi$ is decreasing and it may or may not cancel.

Depending on the evolution of $\Phi$ and $\Psi$, we get three possible control structures:

- either maximum thrust, then ballistic arc;
- or maximum thrust, then singular arc, then ballistic arc;
- or maximum thrust, then singular arc, then maximum thrust, then ballistic arc.

#### 13.4.2.2 Expression of the Thrust on a Singular Arc

Let us look for the expression of the thrust on a singular arc. To do this, we derive the switching function $\Phi$ until the control T appears.

The first derivative has already been computed in (13.32):

$$\dot{\Phi} = \frac{p_v}{m}\left(1 + 2\frac{v_e}{v}\right)D - v_e p_h$$

This derivative is zero on the singular arc.

$$\dot{\Phi} = 0 \quad \Rightarrow \quad p_v G - m p_h = 0 \quad \text{with } G \underset{\text{def}}{=} \left(\frac{2}{v} + \frac{1}{v_e}\right) D \tag{13.35}$$

By differentiating a second time, we have

$$\dot{p}_v G + p_v \dot{G} - \dot{m} p_h - m \dot{p}_h = 0 \quad \text{with } \dot{G} = G_h \dot{h} + G_v \dot{v} \tag{13.36}$$

Substituting then $(\dot{h}, \dot{v}, \dot{m}, \dot{p}_h, \dot{p}_v)$ from (13.25) and (13.31), we have

$$T\left(G_v + \frac{G}{v_e}\right) = \frac{D}{v_e} G + (D + W) G_v - m v G_h + m D_h \tag{13.37}$$

Then by substituting G with its expression (13.35), we obtain after calculations the following equation satisfied by the thrust on a singular arc.

$$T(1 + 4\alpha + 2\alpha^2) = D(1 + 2\alpha) + \left(2D + 2W + \frac{mv^2}{H}\right)\alpha(1 + \alpha) \quad \text{with } \alpha \underset{\text{def}}{=} \frac{v_e}{v} \tag{13.38}$$

This equation defines the thrust as a function of the state along the singular arc. This function is noted $T = T_{sing}(h, v, m)$.

The trajectory necessarily begins with a regular arc at maximum thrust $T_{max}$, because $\Psi(t_0) > 0$. A singular arc is possible if the functions $\Phi$ (switching) and $\Psi$ (singular surface) cancel simultaneously. The thrust then becomes variable defined by the formula $T = T_{sing}(h, v, m)$ associated with equation (13.38) and the trajectory follows the singular surface. The singular arc is followed as long as $T_{sing} \le T_{max}$ or until the possible exhaustion of the propellant.

The application case presented in the next section shows a solution structure of the form: $T_{max} - T_{sing} - T_{max} - 0$.

### 13.4.3 Application Case

The numerical data are summarized in the following table (rounded values).

| $m_0$ | $m_p$ | $T_{max}$ | $v_e$ | $SC_D$ | $\rho_0$ | H | g |
|---|---|---|---|---|---|---|---|
| 100 kg | 90 kg | 2000 N | 2000 m/s | 0.1 m$^2$ | 1.2 kg/m$^3$ | 5000 m | 10 m/s$^2$ |

The solution of the form $T_{max} - T_{sing} - T_{max} - 0$ is plotted in Figure 13.6.

The structure of the solution depends on the available propellant mass.

- If $m_p > 81.3$ kg, the solution is of the form: $T_{max} - T_{sing} - T_{max} - 0$.
- If 16.1 kg $< m_p <$ 81.3 kg, the solution is of the form: $T_{max} - T_{sing} - 0$.
- If $m_p < 16.1$ kg, the solution is of the form: $T_{max} - 0$.

In the case $m_p$ = 90 kg shown in Figure 13.6, a control $T_{max} - 0$ would result in a burn time of 90 s (instead of 125 s) and the trajectory would reach a maximum altitude of 236.8 km (instead of 300.6 km).

The solution $T_{max} - 0$ in this case is not as good as $T_{max} - T_{sing} - T_{max} - 0$.

These two solutions are compared in Figure 13.7.

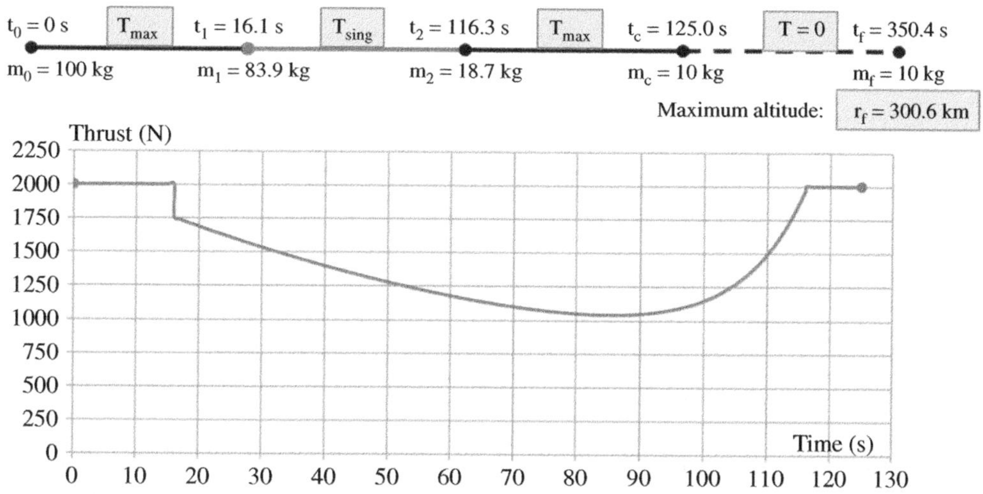


**Figure 13.6** Thrust law along a singular arc.

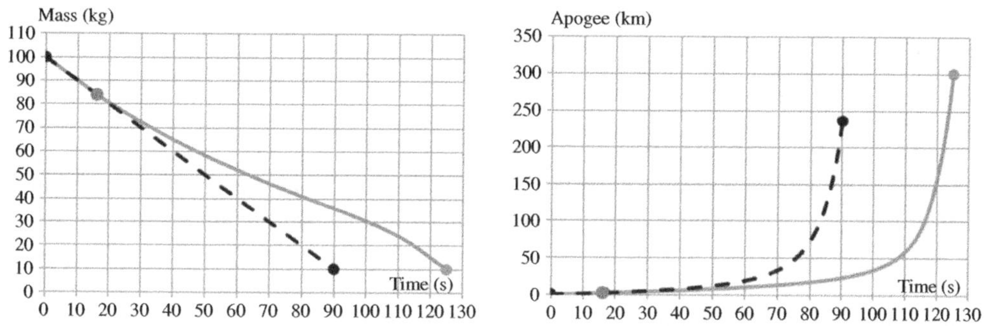


**Figure 13.7** Comparison of singular solution and maximum thrust solution.

The data in this example have been deliberately chosen in order to obtain a significant singular arc. In practical cases, there is usually no singular arc, or only a short one. It is then operationally simpler to maintain a constant thrust level, even if this is not strictly optimal.

## 13.5 Injection into Orbit at Variable Thrust

The fourth problem of this chapter concerns an injection into orbit on a flat Earth with no atmosphere. The orbit is defined by an altitude and a horizontal velocity. As with the second problem, the aim is to reach the orbit in minimum time, but here with the possibility of varying the engine thrust.

### 13.5.1 Modeling

The modeling is similar to that of the second problem. The trajectory takes place in the vertical plane (Hxz), as shown in Figure 13.8. The Earth is assumed to be flat, with no atmosphere, and the gravity is constant downward.

The rocket is modeled as a material point of position $\vec{r}\,(d, h)$, velocity $\vec{v}\,(v_d, v_h)$ and mass m. It starts from given initial conditions $(\vec{r}_0, \vec{v}_0, m_0)$ at time $t_0 = 0$. The targeted orbit is defined by the altitude $h_c$ and the horizontal velocity $v_c$.

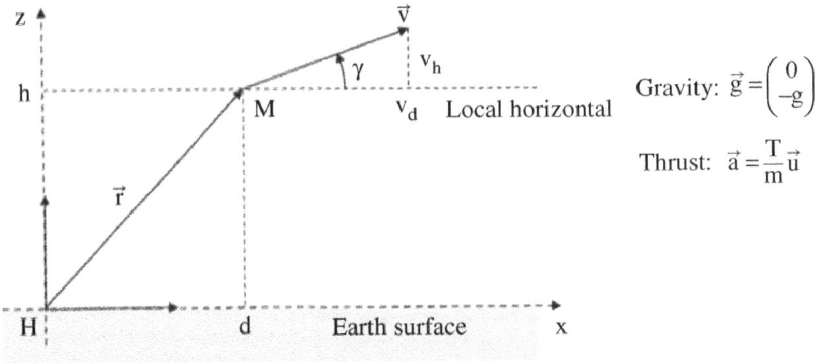


**Figure 13.8** Injection into orbit with variable thrust.

The control is the thrust vector $\vec{T}(t)$, whose modulus is comprised between 0 and an upper bound $T_{max}$. The thrust direction is oriented by the unit vector $\vec{u}(t)$. The thrust is related to the mass flow rate q by $T = qv_e$, where $v_e$ is the exhaust velocity of the burnt gases.

The state vector $(\vec{r}, \vec{v}, m)$ evolves according to the differential equations

$$\begin{cases} \dot{\vec{r}} = \vec{v} \ , \ \dot{\vec{v}} = \vec{g} + \dfrac{T}{m}\vec{u} \\ \dot{m} = -\dfrac{T}{v_e} \end{cases} , \quad \begin{cases} \vec{r}(t_0) = \vec{r}_0 \\ \vec{v}(t_0) = \vec{v}_0 \\ m(t_0) = m_0 \end{cases} \tag{13.39}$$

The aim is to determine the control law $\vec{T}(t)_{t_0 \le t \le t_f}$ and the final time $t_f$ to reach the targeted altitude $h_c$ and horizontal velocity $v_c$, while maximizing the final mass $m(t_f)$. The final abscissa $d(t_f)$ is free.

**Remark 13.1** *For the transition to the round Earth model (presented in Section 13.6), the notations adopted here for the axes are "d" for the horizontal component (like downrange) and "h" for the vertical component (like height). The frame origin is noted H in order to keep the notation O for the center of the Earth.*

#### Assumption on the Initial Conditions

We assume that the initial conditions $(\vec{r}_0, \vec{v}_0)$ do not allow to reach the targeted altitude $h_c$ by a purely ballistic arc. The initial osculating apogee satisfies then the condition: $h_{max} = h_0 + \frac{v_{h0}^2}{2g} < h_c$. This assumption is consistent with practical applications and it simplifies the analysis of the optimal solution.

### 13.5.2 Optimal Solution

The optimal control problem is formulated as:

$$\min_{T(t),\vec{u}(t),t_f} J = -m(t_f)$$

$$\text{s.t.} \begin{cases} \dot{\vec{r}}(t) = \vec{v}(t) \ , \ \dot{\vec{v}}(t) = \vec{g} + \dfrac{T(t)}{m}\vec{u}(t) \\ \dot{m}(t) = -\dfrac{T(t)}{v_e} \end{cases} , \quad \begin{cases} \vec{r}(t_f) = \begin{pmatrix} \text{free} \\ h_c \end{pmatrix} \\ \vec{v}(t_f) = \begin{pmatrix} v_c \\ 0 \end{pmatrix} \end{cases} \tag{13.40}$$

Subsequently, the time dependence is omitted to simplify the equations.

The problem (13.40) is solved by applying the PMP.

For this, the Hamiltonian H is formed by introducing the costate vector of components $(\vec{p}_r, \vec{p}_v, p_m)$.

$$H = \vec{p}_r \cdot \vec{v} + \vec{p}_v \cdot \left(\vec{g} + \frac{T}{m}\vec{u}\right) - p_m \frac{T}{v_e} \tag{13.41}$$

According to the PMP, the optimal control $(T, \vec{u})$ minimizes the Hamiltonian.

The minimization of H with respect to the thrust direction $\vec{u}$ yields

$$\min_{\vec{u}} H = \min_{\vec{u}} \left(\vec{p}_r \cdot \vec{v} + \vec{p}_v \cdot \vec{g} - p_m \frac{T}{v_e} + \frac{T}{m}\vec{p}_v \cdot \vec{u}\right) \quad \rightarrow \quad \vec{u} = -\frac{\vec{p}_v}{\|\vec{p}_v\|} \tag{13.42}$$

By substituting this expression of $\vec{u}$ into (13.41), the Hamiltonian takes the form

$$H = \vec{p}_r \cdot \vec{v} + \vec{p}_v \cdot \vec{g} - T\Phi \tag{13.43}$$

where the function $\Phi$ is defined by

$$\Phi = \frac{p_v}{m} + \frac{p_m}{v_e} \text{ with the notation } p_v = \|\vec{p}_v\| \tag{13.44}$$

The minimization of H (13.43) with respect to the thrust modulus T yields

$$\min_{T} H = \min_{T} (\vec{p}_r \cdot \vec{v} + \vec{p}_v \cdot \vec{g} - T\Phi) \quad \rightarrow \quad \begin{cases} T = 0 & \text{if } \Phi < 0 \\ T = T_{max} & \text{if } \Phi > 0 \end{cases} \tag{13.45}$$

The structure of the solution (thrust and ballistic arcs) depends on the variations of the function $\Phi$ (switching function).

To study this, we integrate the costate equations associated with the position and the velocity.

$$\begin{cases} \dot{\vec{p}}_r = -H_{\vec{r}} = \vec{0} & \rightarrow \quad \vec{p}_r(t) = Cte \\ \dot{\vec{p}}_v = -H_{\vec{v}} = -\vec{p}_r & \rightarrow \quad \vec{p}_v(t) = \vec{p}_v(t_f) + \vec{p}_r(t_f - t) \end{cases} \tag{13.46}$$

Since the final abscissa $d(t_f)$ is free, we have the transversality condition of on the associated costate: $p_d(t_f) = 0$.

The costate vectors $\vec{p}_r(t)$ and $\vec{p}_v(t)$ are thus of the form

$$\vec{p}_r(t) = \begin{pmatrix} 0 \\ p_h \end{pmatrix}, \quad \vec{p}_v(t) = \begin{pmatrix} p_{vd} \\ p_{vhf} + p_h(t_f - t) \end{pmatrix} \tag{13.47}$$

where $(p_h, p_{vd}, p_{vhf})$ are constants to be determined.

On the other hand, the mass costate $p_m(t)$ satisfy the equations

$$\dot{p}_m = -H_m = -\frac{Tp_v}{m^2} \quad \text{with } p_m(t_f) = \frac{\partial J}{\partial m(t_f)} = -1 \tag{13.48}$$

where J is the cost function of the control problem (13.40).

The problem (13.40) is autonomous and with free final time. The Hamiltonian is therefore constant and null.

Expanding the Hamiltonian with the costates (13.47) yields an additional equation

$$H = \vec{p}_r \cdot \vec{v} + \vec{p}_v \cdot \vec{g} - T\Phi = p_h v_h - g p_{vh} - T\Phi = 0, \quad \forall t \tag{13.49}$$

The costate $\vec{p}_v$ cannot cancel over an interval. Otherwise, we would have: $\vec{p}_r = \dot{\vec{p}}_v = \vec{0}$ and with (13.49) and (13.44): $T\Phi = T\frac{p_m}{v_e} = 0$. The two possible solutions would then be:

- either $T = 0$ (no thrust), which cannot allow to reach the targeted altitude (this is excluded by the assumption on the initial conditions at the end of Section 13.5.1);
- or $p_m = 0$, and the costate $(\vec{p}_r, \vec{p}_v, p_m)$ would cancel over an interval, which is discarded by the PMP.

### 13.5.3 Switching Function

The derivative of the switching function $\Phi$ is calculated using the costate (13.46) and (13.48), and knowing that $\vec{p}_v$ cannot be zero (as established just earlier).

The derivative of $\|\vec{p}_v\|$ is obtained by: $p_v^2 = \vec{p}_v \cdot \vec{p}_v \Rightarrow \dot{p}_v p_v = \dot{\vec{p}}_v \cdot \vec{p}_v = -\vec{p}_r \cdot \vec{p}_v$.

Substituting the derivatives of $(p_v, p_m, m)$, we obtain after simplifications the derivative of $\Phi$.

$$\Phi = \frac{p_v}{m} + \frac{p_m}{v_e} \quad \Rightarrow \quad \dot{\Phi} = -\frac{\vec{p}_r \cdot \vec{p}_v}{m p_v} \tag{13.50}$$

The sign of $\dot{\Phi}$ is that of $\Phi_1 = -\vec{p}_r \cdot \vec{p}_v$, whose derivative is: $\dot{\Phi}_1 = p_r^2 = p_h^2 \geq 0$.

We have to distinguish two cases, depending on whether $p_h$ is zero or not.

- First case

In the case $p_h = 0$, $\Phi$ is constant and cannot be null. Otherwise, equation (13.49) would give $p_{vh} = 0$ and from (13.42), the vertical thrust component $u_h$ would be constantly zero. This is not compliant with the assumption (see Section 13.5.1) that the initial osculating apogee $h_{max}$ is lower than the targeted altitude $h_c$.

Since $\Phi$ cannot be null, there is no singular arc and the only possible solution in the case $p_h = 0$ is a single arc with maximum thrust: $T(t) = T_{max}$.

- Second case

In the case $p_h \neq 0$, the derivative $\dot{\Phi}$ is increasing (because $\dot{\Phi}_1 > 0$) and the variation of the switching function is as follows.

| $t$ | | $t_m$ | |
|---|---|---|---|
| $\dot{\Phi}(t)$ | $< 0$ | $0$ | $> 0$ |
| $\Phi(t)$ | ↘ | $\Phi(t_m)$ | ↗ |

The structure of the solution depends on the sign of the minimum $\Phi(t_m)$.

- If $\Phi(t_m) > 0$, the solution is a single arc with maximum thrust: $T(t) = T_{max}$.
- If $\Phi(t_m) < 0$, the solution is "at most" of the form: $T_{max} - 0 - T_{max}$.

It is shown below that it is in fact not possible to have a second thrust arc. The only solutions are thus of the form $T_{max} - 0$ with a ballistic arc of possibly zero duration.

**Study of the Solution of the Form: $T_{max} - 0 - T_{max}$ when $p_h \neq 0$**

Let us assume that the function $\Phi$ admits a negative minimum at time $t_m$, which results in $T(t_m) = 0$.

Using (13.50), we have:

$$\dot{\Phi}(t_m) = 0 \;\Rightarrow\; \vec{p}_r(t_m)\cdot\vec{p}_v(t_m) = p_h p_{vh}(t_m) = 0 \;\Rightarrow\; p_{vh}(t_m) = 0$$

Using (13.49) we have:

$$H(t_m) = 0 \;\Rightarrow\; p_h v_h(t_m) - T(t_m)\Phi(t_m) = 0 \quad\Rightarrow\; v_h(t_m) = 0$$

The costate $\vec{p}_v$ (13.47) is expressed with $t_m$ as follows: $\vec{p}_v(t) = \begin{pmatrix} p_{vd} \\ p_h(t_m - t) \end{pmatrix}$.

Let us now see if the targeted conditions of altitude and velocity can be reached at a time later than $t_m$. For this purpose, we study the variations of the vertical velocity $v_h$ and of the osculating apogee $h_a$.

- Vertical velocity $v_h$

  The derivative of vertical velocity is $\dot{v}_h = -g + \frac{T}{m}u_h$ with $u_h = -\frac{p_{vh}}{p_v} = \frac{p_h(t-t_m)}{p_v}$ from (13.42). To express its second derivative $\ddot{v}_h$, we compute $\dot{p}_v$, and then calculate $\dot{u}_h$.

- Calculation of $\dot{p}_v$

$$p_v^2 = \vec{p}_v\cdot\vec{p}_v \;\Rightarrow\; \dot{p}_v p_v = \dot{\vec{p}}_v\cdot\vec{p}_v = -\vec{p}_r\cdot\vec{p}_v = -p_h p_{vh} \;\Rightarrow\; \dot{p}_v = -p_h\frac{p_{vh}}{p_v} = p_h u_h$$

- Calculation of $\dot{u}_h$

$$u_h = \frac{p_h(t-t_m)}{p_v} \;\Rightarrow\; \dot{u}_h = \frac{p_h p_v - \dot{p}_v p_h(t-t_m)}{p_v^2} \;\Rightarrow\; \dot{u}_h = \frac{p_h}{p_v}\left(1-u_h^2\right)$$

- Calculation of $\ddot{v}_h$

$$\ddot{v}_h = T\frac{\dot{u}_h m - u_h\dot{m}}{m^2} \quad\text{with } \dot{m} = -\frac{T}{v_e} \;\Rightarrow\; \ddot{v}_h = \frac{Tp_h}{mp_v}\left[\left(1-u_h^2\right) + \frac{T}{mv_e}(t-t_m)\right]$$

On the second thrust arc ($t > t_m$, $T = T_{max}$), $\ddot{v}_h$ has the same sign as $p_h$.
Suppose by contradiction that $p_h < 0$. Then $\ddot{v}_h < 0$ and $\dot{v}_h$ decreases after $t_m$.

· Since $\dot{v}_h(t_m) = -g < 0$, $\dot{v}_h$ remains negative and $v_h$ decreases after $t_m$.
· Since $v_h(t_m) = 0$ (shown above), $v_h$ is negative after $t_m$.

Therefore, the final condition $v_h(t_f) = 0$ cannot be reached after $t_m$.
The assumption $p_h < 0$ has led to a contradiction. In order to reach the final condition $v_h(t_f) = 0$ at some time $t_f > t_m$, we necessarily have $p_h > 0$.

- Osculating apogee $h_a$

The osculating apogee $h_a$ is the maximum ballistic altitude reached from conditions $(h, v_h)$.

$$h_a = h + \frac{v_h^2}{2g} \;\Rightarrow\; \dot{h}_a = \dot{h} + \frac{v_h\dot{v}_h}{g} = \frac{T}{mg}v_h u_h \quad\text{with } u_h = \frac{p_h(t-t_m)}{p_v}$$

Let us draw up the table of variation of $u_h$ and $v_h$ in the case $p_h > 0$.

The switching times are noted $t_1$ and $t_2$. Since the second derivative $\ddot{v}_h$ is positive after $t_m$, the function $\dot{v}_h = -g + \frac{T}{m}u_h$ (which is negative in $t_2$) increases after $t_m$ and may eventually cancel at a time $t_3 > t_2$.

| t | | $t_1$ | | $t_m$ | | $t_2$ | | $t_3$ | | $t_4$ | |
|---|---|---|---|---|---|---|---|---|---|---|---|
| T(t) | $T_{max}$ | | 0 | | 0 | | $T_{max}$ | | $T_{max}$ | | $T_{max}$ |
| $u_h$ (t) | − | | − | 0 | + | | + | | + | | + |
| $\dot{v}_h$ (t) | − | | − | | − | | − | 0 | + | | + |
| $v_h$ (t) | + | | + | 0 | − | | − | | − | 0 | − |

We observe that the product $u_h v_h$ is constantly negative, so that $h_a$ is always a decreasing function (its derivative calculated above is negative). This is not compliant with the assumption (see Section 13.5.1) that the initial osculating apogee $h_a(t_0)$ is lower than the targeted altitude $h_c$.

Therefore, the two final conditions $v_h(t_f) = 0$ and $h(t_f) = h_c$ cannot be reached simultaneously at a time $t_f > t_m$.

The optimal solution therefore consists of **only one thrust arc**.

### 13.5.4 Numerical Solution

The previous analysis has shown that the optimal solution is of the form $T_{max} - 0$.

To fully determine the solution, it is necessary to find the components of the initial costate $(\vec{p}_r, \vec{p}_v, p_m)$ at $t_0$, the switching time $t_1$ and the final time $t_f$.

#### 13.5.4.1 Reduced Solution

The two-point boundary value problem has six unknowns, which are the costate components (13.47), the switching time and the final time: $p_h, p_{vd}, p_{vh}(t_0), p_m(t_0), t_1, t_f$.

The six equations to be satisfied are:

- the three final conditions : $h(t_f) = h_c$, $v_d(t_f) = v_c$, $v_h(t_f) = 0$
- the two transversality conditions: $H(t_f) = 0$, $p_m(t_f) = -1$
- the switching condition : $\Phi(t_1) = 0$

We can reduce the problem to the two unknowns $p_{vd}, p_{vh}(t_0)$ as follows:

- The final time $t_f$ is calculated explicitly from (13.49) and (13.47).

$$H(t_f) = 0 \text{ with } \begin{cases} T(t_f) = 0 \\ v_h(t_f) = 0 \end{cases} \Rightarrow p_{vh}(t_f) = 0 \Rightarrow t_f = t_0 + \frac{p_{vh}(t_0)}{p_h} \tag{13.51}$$

- The switching time $t_1$ is detected during the numerical integration.

$$H(t_1) = 0 \text{ with } \Phi(t_1) = 0 \Rightarrow p_h v_h(t_1) + g p_h \cdot (t_1 - t_0) = g p_{vh}(t_0) \tag{13.52}$$

- Since the costate is defined to within one multiplicative factor, we can choose to normalize it by $p_h = 1$, which eliminates the unknown $p_h$.

The time $t_1$ is implicitly determined by (13.52), which does not depend on the switching function. The switching function, and therefore the mass costate, are thus no longer needed to solve the problem. The mass costate initial value $p_m(t_0)$ will be assessed a posteriori to satisfy the transversality condition $p_m(t_f) = -1$.

With the times $t_f$ and $t_1$ determined by (13.51) and (13.52), the final condition $v_h(t_f) = 0$ is automatically satisfied. Indeed, we have on the ballistic arc going from $t_1$ to $t_f$:

$$\begin{aligned} v_h(t_f) &= v_h(t_1) - (t_f - t_1)g \\ &= [v_h(t_1) + (t_1 - t_0)g] - (t_f - t_0)g = \frac{p_{vh}(t_0)g}{p_h} - (t_f - t_0)g = 0 \end{aligned}$$

The problem has been reduced to two unknowns $p_{vd}$, $p_{vh}(t_0)$ and two equations: $\begin{cases} h(t_f) = h_c \\ v_d(t_f) = v_c \end{cases}$.

This two-point boundary value problem (called shooting problem) is solved by the following procedure.

#### 13.5.4.2 Numerical Procedure

For a set of values $p_{vd}$, $p_{vh}(t_0)$, we numerically integrate the differential system

$$\begin{cases} \dot{h}(t) = v_h(t) \\ \dot{v}_d(t) = \dfrac{T_{max}}{m(t)} \dfrac{p_{vd}}{p_v(t)} \\ \dot{v}_h(t) = \dfrac{T_{max}}{m(t)} \dfrac{p_{vh}(t_0) - (t - t_0)}{p_v(t)} - g \end{cases} \quad \text{with} \quad \begin{cases} m(t) = m(t_0) - \dfrac{T_{max}}{v_e}(t - t_0) \\ p_v(t) = \sqrt{p_{vd}^2 + [p_{vh}(t_0) - (t - t_0)]^2} \end{cases} \tag{13.53}$$

from $t_0$ up to a time $t_1$ such that: $v_h(t_1) + (t_1 - t_0)g = g p_{vh}(t_0)$

The final time is then calculated by: $t_f = t_0 + p_{vh}(t_0)$ and the final state is

$$\begin{cases} h(t_f) = h(t_1) + v_h(t_1)\,(t_f - t_1) - \frac{1}{2}g(t_f - t_1)^2 \\ v_d(t_f) = v_d(t_1) \end{cases} \tag{13.54}$$

The aim is to find $p_{vd}$, $p_{vh}(t_0)$ satisfying the equations: $\begin{cases} h(t_f) = h_c \\ v_d(t_f) = v_c \end{cases}$.

This system is solved using Newton's method. In practice, the convergence is very fast without the need for a specific initial guess.

**Remark 13.2** *The signs of* $p_{vd}$, $p_{vh}(t_0)$ *are known:*

- $p_{vd} < 0$ *to have* $u_d > 0$ *and increase the horizontal velocity* $u_d$
- $p_{vh}(t_0) > 0$ *to have* $t_f > t_0$

**Remark 13.3** *The differential equations (13.53) could be analytically integrated. But since the time* $t_1$ *is not known a priori, it would still be necessary to advance by time steps until the switching is detected. As a result, this approach does not provide any gain in accuracy or computation time.*

#### 13.5.4.3 Full Solution

The solution of the reduced problem gives the values of $p_{vd}$, $p_{vh}(t_0)$ associated with the particular normalization $p_h = 1$ of the costate. The initial mass costate $p_m(t_0)$ is then obtained from the Hamiltonian and must be zero. Writing this condition at the initial time $t_0$, we obtain

$$H(t_0) = 0 \Rightarrow p_m(t_0) = \frac{v_e}{T_{max}}\left[v_h(t_0) - p_{vh}(t_0)g + \frac{T_{max}}{m(t_0)}\sqrt{p_{vd}^2 + p_{vh}(t_0)^2}\right] \tag{13.55}$$

The initial costate $(\vec{p}_r, \vec{p}_v, p_m)$ is then completely known and associated with the normalization chosen $p_h = 1$.

The solution for the usual normalization $p_m(t_f) = -1$ is retrieved as follows.

- The switching function is zero at $t_1$, which allows us to calculate $p_m(t_1)$.

$$\Phi(t_1) = 0 \Rightarrow p_m(t_1) = \frac{p_v(t_1)v_e}{m(t_1)} \tag{13.56}$$

- The mass costate is constant over the ballistic arc from $t_1$ to $t_f$ (13.48).

$$p_m(t_f) = p_m(t_1) = v_e \frac{\sqrt{p_{vd}^2 + [p_{vh}(t_0) - (t_1 - t_0)]^2}}{m(t_0) - \frac{T_{max}}{v_e}(t_1 - t_0)} \tag{13.57}$$

It is then sufficient to divide all the components of the initial costate $(\vec{p}_r, \vec{p}_v, p_m)$ by the value $-p_m(t_f)$ given by (13.57) to return to the usual normalization. The final time given by (13.51) is unaffected by this renormalization.

The above procedure allows us to easily solve the problem regardless of the targeted orbit and the vehicle features (mass, thrust and exhaust velocity).

## 13.6 Transition to the Round Earth

The previous section has shown how to find a quasi-analytical solution to the launch into orbit problem under the assumption of flat Earth and constant gravity. This section presents a method for passing from the flat Earth solution to the round Earth solution corresponding to realistic application cases.

### 13.6.1 Change of Coordinates

The equations of motion in the round Earth model are expressed with the polar coordinates $(r, \varphi, v, \gamma)$ shown in Figure 13.9. The position is determined by the radius r and the polar angle $\varphi$, and the velocity is determined by its modulus v and its local slope $\gamma$, also called flight path angle.

The thrust is directed by the unit vector $\vec{u}$ of components $(u_v, u_\gamma)$ along the velocity and the normal. The polar coordinate equations have the following form:

$$\begin{cases} \dot{r} = v \sin\gamma, \quad \dot{\varphi} = \frac{v}{r}\cos\gamma \\ \dot{v} = -g(r)\sin\gamma + \frac{Tu_v}{m} \quad \text{with} \quad g(r) = \frac{\mu}{r^2} \\ \dot{\gamma} = \left(\frac{v}{r} - \frac{g(r)}{v}\right)\cos\gamma + \frac{Tu_\gamma}{mv} \end{cases} \tag{13.58}$$

The choice of polar coordinates makes it possible to establish a geometrical link between the flat Earth model and the round Earth model.

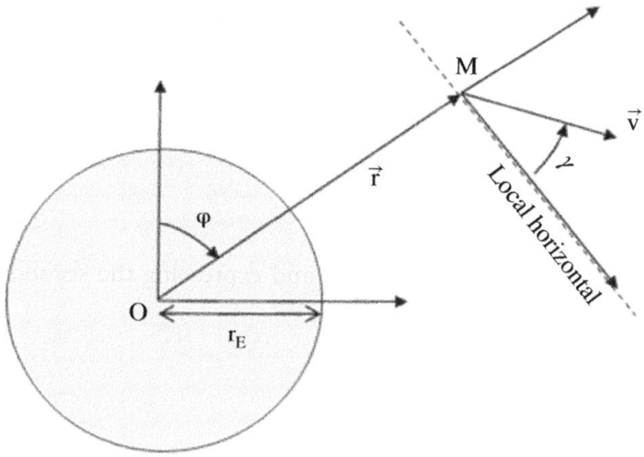


**Figure 13.9** Polar coordinates.

Indeed, these coordinates naturally transform into local horizontal and vertical coordinates $(d, h, v_d, v_h)$ used in the Flat Earth model. More specifically:

- the horizontal coordinate d (downrange) measures the curvilinear abscissa along the Earth's surface;
- the vertical coordinate h (height) measures the altitude relative to the Earth's surface;
- the velocity and the thrust are projected along the local horizontal and vertical directions.

The transition from $(r, \varphi, v, \gamma)$ to $(d, h, v_d, v_h)$ is depicted in Figure 13.10 and the relations between these coordinates are

$$\begin{cases} d = r\varphi \\ h = r - r_E \\ v_d = v\cos\gamma \\ v_h = v\sin\gamma \end{cases} \Leftrightarrow \begin{cases} r = r_E + h \\ \varphi = \dfrac{d}{r_E + h} \\ v = \sqrt{v_d^2 + v_h^2} \\ \gamma = \text{Arctan}\,\dfrac{v_h}{v_d} \end{cases} \quad \text{and} \quad \begin{pmatrix} u_d \\ u_h \end{pmatrix} = \begin{pmatrix} \cos\gamma & -\sin\gamma \\ \sin\gamma & \cos\gamma \end{pmatrix} \begin{pmatrix} u_v \\ u_\gamma \end{pmatrix} \tag{13.59}$$

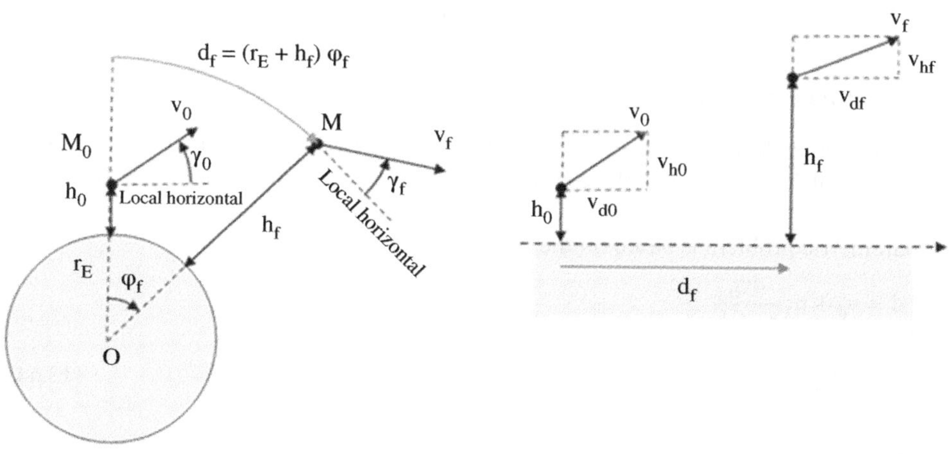


**Figure 13.10** Transition to local coordinates.

To write the differential equations of motion in the new coordinate system, we first derive equations (13.59).

$$\begin{cases} \dot{d} = \dot{r}\,\varphi + r\,\dot{\varphi} \\ \dot{h} = \dot{r} \\ \dot{v}_d = \dot{v}\cos\gamma - v\,\dot{\gamma}\sin\gamma \\ \dot{v}_h = \dot{v}\sin\gamma + v\,\dot{\gamma}\cos\gamma \end{cases} \tag{13.60}$$

By substituting the derivatives of (r, φ, v, γ) given by (13.58), and expressing the second members in function of $(d, h, v_d, v_h)$ using (13.59), we obtain

$$\begin{cases} \dot{d} = v_d + \dfrac{d v_h}{r_E + h}, \quad \dot{h} = v_h \\ \dot{v}_d = \dfrac{T}{m} u_d - \dfrac{v_d v_h}{r_E + h} \\ \dot{v}_h = \dfrac{T}{m} u_h + \dfrac{v_d^2}{r_E + h} - g(r_E + h) \end{cases} \tag{13.61}$$

These equations are equivalent to (13.58). They represent the round Earth dynamics passed in the flat Earth coordinates.

Let us compare these equations with those in flat Earth derived from (13.39), which are

$$\begin{cases} \dot{d} = v_d\,, \quad \dot{v}_d = \dfrac{T}{m} u_d \\ \dot{h} = v_h\,, \quad \dot{v}_h = \dfrac{T}{m} u_h - g \end{cases} \tag{13.62}$$

We observe two differences:

- the gravity in (13.61) depends on altitude instead of being constant in (13.62);
- coupling terms appear in (13.61) on the horizontal and vertical components.

The coupling terms related to the curvature of the Earth allow the existence of orbits (periodic trajectories without thrust) in the round Earth model. A circular orbit at altitude $h_0$ is obtained with the following initial conditions:

$$\begin{cases} h(t_0) = h_0 = r_0 - r_E \\ v_h(t_0) = 0 \\ v_d(t_0) = r_0 g(r_0) \end{cases} \Rightarrow \begin{cases} \dot{h}(t_0) = 0 \\ \dot{v}_h(t_0) = 0 \\ \dot{v}_d(t_0) = 0 \end{cases} \Rightarrow \begin{cases} h(t) = h(t_0) \\ v_h(t) = 0 \\ v_d(t) = v_d(t_0) \end{cases}, \forall t \tag{13.63}$$

### 13.6.2 Continuation Method

The continuation method consists of formulating a parametric problem, which can represent either the flat Earth model (when the coefficients are 0) or the round Earth model (when the coefficients are 1).

This parametric problem is based on the following dynamic equations:

$$\begin{cases} \dot{d} = v_d + \alpha_2 \dfrac{d v_h}{r_E + h} \\ \dot{h} = v_h \\ \dot{v}_d = \dfrac{T}{m} u_d - \alpha_2 \dfrac{v_d v_h}{r_E + h} \\ \dot{v}_h = \dfrac{T}{m} u_h + \alpha_2 \dfrac{v_d^2}{r_E + h} - g(r_E + \alpha_1 h) \end{cases} \tag{13.64}$$

- For $(\alpha_1 = 0, \alpha_2 = 0)$, we retrieve the flat Earth equations (13.62).
- For $(\alpha_1 = 1, \alpha_2 = 1)$, we retrieve the round Earth equations (13.61).

The coefficient $\alpha_1$ allows to pass from a constant gravity to a gravity as a function of the altitude h. The coefficient $\alpha_2$ allows to introduce the coupling terms in $(v_d, v_h)$ of the round Earth model.

The optimal control problem associated with the dynamic equations (13.64) is

$$\min_{\vec{u},t_f} J = -m(t_f) \quad \text{s.t.} \quad \begin{cases} \text{equations (13.64)} \\ \dot{m} = -\dfrac{T}{v_e} \end{cases}, \quad \begin{cases} d(t_f) \ \text{free} \\ h(t_f) = h_c \\ v_d(t_f) = v_c \\ v_h(t_f) = 0 \end{cases} \tag{13.65}$$

Let us apply the Pontryaguin Maximum Principe to this optimal control problem.

The Hamiltonian is defined with the costate $(p_d, p_h, p_{vd}, p_{vh}, p_m)$.

$$\begin{aligned} H &= p_d\dot{d} + p_h\dot{h} + p_{vd}\dot{v}_d + p_{vh}\dot{v}_h + p_m\dot{m} \\ &= p_d\left(v_d + \alpha_2\frac{dv_h}{r_E + h}\right) + p_h v_h + p_{vd}\left(\frac{T}{m}u_d - \alpha_2\frac{v_d v_h}{r_E + h}\right) \\ &\quad + p_{vh}\left(\frac{T}{m}u_h + \alpha_2\frac{v_d^2}{r_E + h} - g(r_E + \alpha_1 h)\right) - p_m\frac{T}{v_e} \end{aligned} \tag{13.66}$$

The costate equations are

$$\begin{cases} \dot{p}_d = -p_d\dfrac{\alpha_2 v_h}{r_E + h} \\ \dot{p}_h = p_d\dfrac{\alpha_2 d v_h}{(r_E + h)^2} - p_{vd}\dfrac{\alpha_2 v_d v_h}{(r_E + h)^2} + p_{vh}\left[\dfrac{\alpha_2 {v_d}^2}{(r_E + h)^2} + \alpha_1\dfrac{\partial g}{\partial r}(r_E + \alpha_1 h)\right] \\ \dot{p}_{vd} = -p_d + p_{vd}\dfrac{\alpha_2 v_h}{r_E + h} - p_{vh}\dfrac{2\alpha_2 v_d}{r_E + h} \\ \dot{p}_{vh} = -p_d\dfrac{\alpha_2 d}{r_E + h} - p_h + p_{vd}\dfrac{\alpha_2 v_d}{r_E + h} \\ \dot{p}_m = \dfrac{T}{m^2}(p_{vd}u_d + p_{vh}u_h) \end{cases} \tag{13.67}$$

with the central gravity $g(r) = \dfrac{\mu}{r^2} \Rightarrow \dfrac{\partial g}{\partial r} = -\dfrac{2\mu}{r^3} = -\dfrac{2g(r)}{r}$.

The transversality conditions for the free final time problem (13.65) are

$$\begin{cases} p_d(t_f) = 0 \\ p_m(t_f) = 1 \\ H(t_f) = 0 \end{cases} \tag{13.68}$$

since only the final downrange is free and all other state components are fixed.

The shooting problem derived from the PMP conditions is of dimension six.

The six unknowns are the initial costate components $(p_d, p_h, p_{vd}, p_{vh}, p_m)(t_0)$ and the final time $t_f$. The six equations are:

- the three final conditions : $h(t_f) = h_c$, $v_d(t_f) = v_c$, $v_h(t_f) = 0$;
- the three transversality conditions: $p_d(t_f) = 0$, $p_m(t_f) = 1$, $H(t_f) = 0$.

The continuation method consists of starting from the flat Earth problem corresponding to $(\alpha_1 = 0, \alpha_2 = 0)$. The solution to this problem is easily obtained by the method described in Section 13.5.4.

The coefficient $\alpha_1$ is first progressively increased by small steps from 0 to 1. For each new value of $\alpha_1$, the shooting problem is solved using Newton's method initialized with the previous solution. During this continuation, the final time is free, since the flat Earth problem has a finite solution with free final time.

When the solution for $\alpha_1 = 1$ is reached, the coefficient $\alpha_2$ is in turn progressively increased by small steps from 0 to 1. During this continuation, the final time is fixed to the value $t_{flat}$ found at the end of the first continuation. Indeed, the minimum consumption transfer in round Earth has no finite solution with free final time, since increasing indefinitely the final time would allow the convergence to an impulsive solution with one impulse per revolution. The shooting problem during the second continuation is therefore of dimension five.

This second continuation yields the solution of the actual control problem in round Earth and variable gravity ($\alpha_1 = 1, \alpha_2 = 1$) with the final time set to $t_{flat}$.

In order to improve this solution (i.e. increase the final mass), a third continuation is added, which consists in progressively increasing the final time. The shooting problem is still of dimension five. During this continuation, the final mass increases. This provides a series of solutions for varying final times, and helps to make the compromise between time and mass.

For all problems, the thrust level is defined by the switching function sign (13.45). The initial solution in flat Earth has always the control form $T_{max} - 0$. This form evolves during the continuation process and the final solution in round Earth has generally the form $T_{max} - 0 - T_{max}$. The main interest of this process is to solve the problem in round Earth without having to look for an initial costate guess. This initialization is indeed a hard issue associated with the shooting method.

### 13.6.3 Application Case

The continuation procedure is illustrated on a transfer of an upper stage toward a circular orbit at an altitude of 800 km.

The initial apogee and perigee altitudes are 204 and −4194 km respectively, and the initial anomaly is 178 deg. This corresponds to an altitude of 200 km, a velocity of 5500 m/s and a flight path angle of 2 deg.

The initial gross mass is 40 000 kg, the engine thrust is 180 kN, and the specific impulse is 450 s, corresponding to an exhaust velocity of 4413 m/s.

The continuation procedure is applied with the following stages:

- solution of the flat Earth constant gravity problem;
- costate renormalization to have $p_m(t_f) = 1$;
- continuation on the variable gravity ($\alpha_1 = 0 \rightarrow 1$);
- continuation on the round Earth dynamics ($\alpha_2 = 0 \rightarrow 1$);
- continuation on the final time with a parameter ($\alpha_3 = 0 \rightarrow 1$) to pass from $t_{flat}$ to a prescribed value $t_{max}$.

Table 13.1 presents the numerical results after each stage.

The total number of simulations is about 2300 and the computation takes 10 s on a standard laptop:

- 41 simulations for the initialization stage;
- 953 simulations for the gravity continuation stage;
- 741 simulations for the dynamics continuation stage;
- 593 simulations for the final time continuation stage.

**Table 13.1** Solutions during the continuation stages.

| | Final time (s) | Final mass (kg) | Costate $p_h(t_0)$ | Costate $p_{vd}(t_0)$ | Costate $p_{vh}(t_0)$ | Costate $p_m(t_0)$ |
|---|---|---|---|---|---|---|
| Initialization | 1433.3 | 1676.2 | 1.000 | 96.349 | 1433.314 | NA |
| Renormalization | 1433.3 | 1676.2 | 0.756 | 72.808 | 1083.108 | −0.137 |
| Gravity | 1483.2 | 1504.8 | 3.850 | 69.874 | 2198.207 | −0.236 |
| Dynamics | 1483.2 | 18 915.4 | 12.211 | 6821.530 | 5225.629 | 0.310 |
| Final time | 3183.2 | 21294.1 | 6.500 | 6098.6 | 1912.1 | 0.475 |

Figure 13.11 plots the evolution of the costate components during the continuation process. The abscissa is the "global" continuation parameter defined as the sum $\alpha = \alpha_1 + \alpha_2 + \alpha_3$, which varies from 0 to 3.

Figure 13.12 plots the evolution of the boost durations during the continuation process. We observe that the initial control structure $T_{max} - 0$ changes several times during the second continuation ending with the form $T_{max} - 0 - T_{max}$. More precisely, this structure is

- $T_{max} - 0$ for $0 < \alpha_2 < 0.41$
- $T_{max} - 0 - T_{max}$ for $0.41 < \alpha_2 < 0.78$
- $T_{max} - 0$ for $0.78 < \alpha_2 < 0.86$
- $T_{max} - 0 - T_{max}$ for $0.86 < \alpha_2 < 1$

When the optimal thrust strategy changes, the zero path of the shooting function makes corners. It can be observed numerically, by reducing the homotopy steps, that the path remains continuous, but no longer differentiable. This phenomenon is due to the occurrence in the continuation process of a switching time (i.e. a zero of the switching function) coinciding with the final time. This occurs three times for $\alpha_2$ taking the values 0.41, 0.78, and 0.86.

A large mass increase (from 1504 to 18 915 kg) is achieved during the dynamics continuation stage (see Table 13.1). This could be expected because increasing the altitude is easier

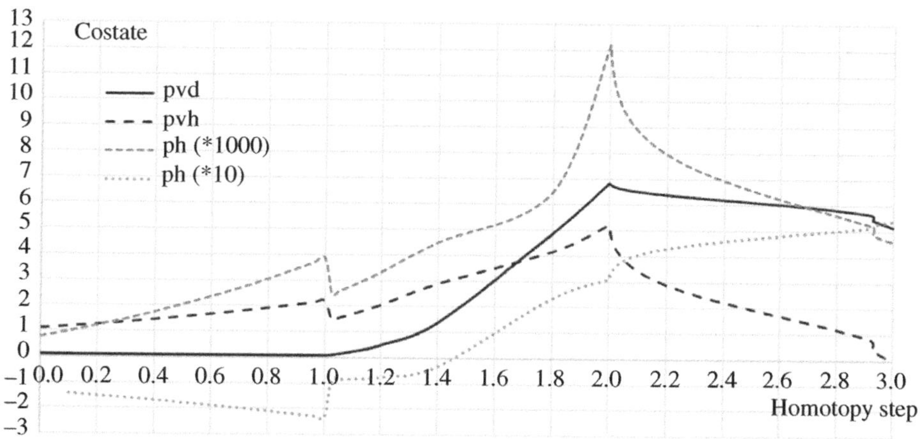


**Figure 13.11** Costate components during the continuation.

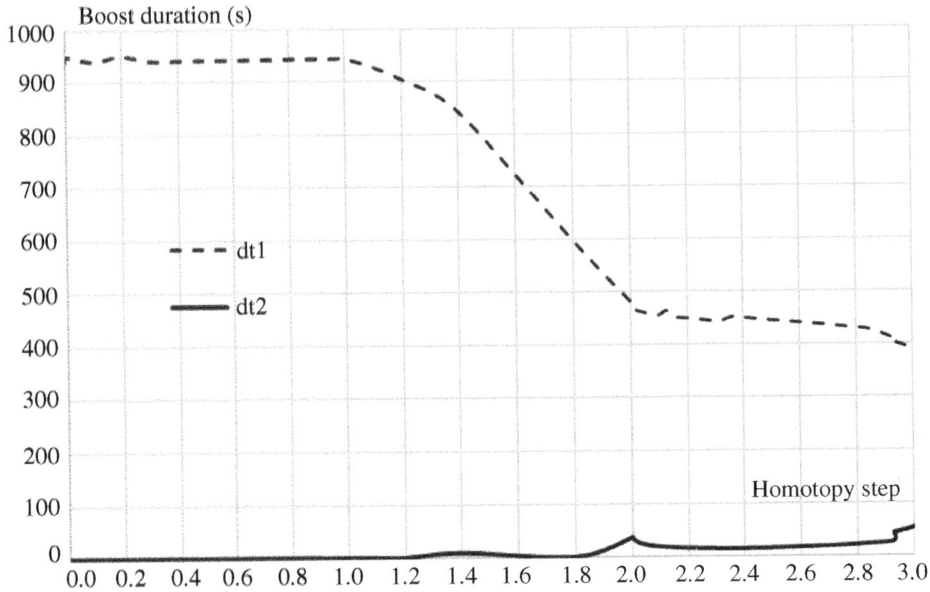


**Figure 13.12** Boost durations during the continuation.

in the round Earth model than in the flat Earth model. Indeed, the Earth's surface curvature makes the altitude naturally increase on a free straight-line motion (i.e. without external forces) as depicted in Figure 13.13.

This effect of Earth surface curvature is linked to the centrifugal term $\frac{v}{r}\cos\gamma$, in the flight path angle equation (13.58).

Consistently, the mass gain is mostly due to the reduction of the first boost duration. This can be observed in Figure 13.12 when the continuation parameter passes from 1 to 2, corresponding to the dynamics continuation. Introducing the Earth's curvature helps pass "naturally" from the initial altitude of 200 km to the targeted altitude of 800 km.

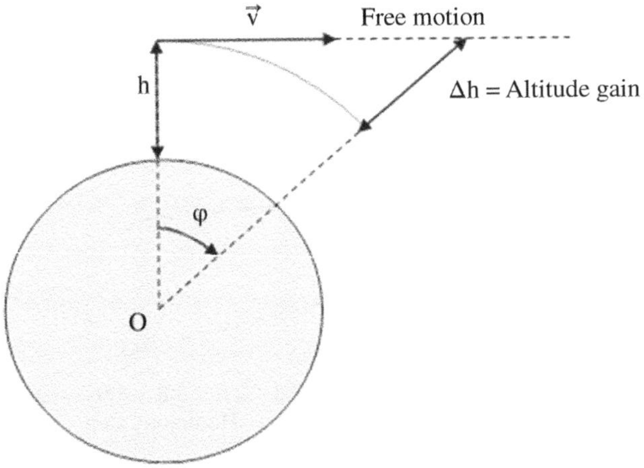


**Figure 13.13** Natural altitude increase in the round Earth model.

As shown in Table 13.1, increasing the final time yields a significant performance improvement (of more than 2300 kg in this case). The solution selected (costate and time) can be used as a starting point for a refined optimization with more complex models and additional path or final constraints.

## 13.7 Conclusion

### 13.7.1 The Key Points

- The maximum range in flat Earth is obtained with a constant thrust direction.
- The influence of the acceleration level can be estimated from the minimum-time injection into orbit on flat Earth.
- The thrust profile to reach the maximum altitude in vertical launch (Goddard problem) may present singular arcs (nonmaximum thrust) in the presence of aerodynamic braking.
- The optimal thrust law for injection into orbit in flat Earth is a maximum thrust arc followed by a ballistic arc.
- The flat Earth solution can be transformed progressively into the round Earth solution by a two-parameter continuation procedure, introducing variable gravity and Earth curvature.

### 13.7.2 To Go Further

- BRYSON A. E., HO Y.-C. – *Applied Optimal Control*. Hemisphere Publishing Corporation (1975).

This book is a reference for engineers having to deal with optimal control applications. The methods are very clearly explained and numerous application cases are covered in detail. The three first problems of this chapter are retrieved from this book.

- CERF M., TRELAT E., HABERKORN T. – *Continuation from a Flat to a Round Earth Model in the Coplanar Orbit Transfer Problem*. Optimal Control Application and Methods, 36 pages, DOI: 10.1002/oca (2011).

A paper presenting the continuation method to pass from flat Earth solutions to round Earth solutions.

- GODDARD R. H. – *A Method for Reaching Extreme Altitudes*. Smithsonian Int. Misc. Collections 71, (1919).

Goddard's original paper.

# 14

# Interplanetary Trajectory

## 14.1 Introduction

The exploration of the solar system began in the 1960s in parallel with the race to the Moon. The first interplanetary probes (Mariners for the United States and Venera for the USSR) were designed to carry out simple fly-bys of Venus and Mars, which are the closest planets to the Earth. The next major steps were the first landing on Venus by Venera 7 in 1970, the first orbital injection around Mars by Mariner 9 in 1971, and the first landing on Mars by Viking 1 in 1976.

Unlike Venus and Mars, which can be reached by direct launches, planets further away than Venus and Mars require more complex interplanetary routes using the concept of fly-bys and gravity assist to increase progressively the spacecraft orbit within the solar system. The Pioneer 10 and 11 and Voyager 1 and 2 probes launched between 1972 and 1977 made such successful flights to the outer planets. These probes continued their mission long after the planned date and they have now left the solar system. Among further noticeable missions, we can mention Galileo (1995, exploration of Jupiter and its satellites), Cassini–Huygens (2004, exploration of Saturn and landing on Titan), Mars exploration rovers, and Rosetta (2014, landing on the comet Churyumov–Gerasimenko).

During an interplanetary mission, the vehicle is subject to the attraction of all the bodies of the Solar System. This N-body problem has no analytical solution and the calculation of the trajectory requires the use of numerical integration methods. These methods do not provide physical understanding and are not appropriate for the study of the mission scenario.

The concept of sphere of influence allows us to decompose the interplanetary path as a succession of two-body problems. This modeling highlights the phenomenon of gravity assist, which modifies the spacecraft trajectory in the heliocentric frame without propelled maneuvers. The goal is then to define an interplanetary route using such fly-bys in order to minimize propellant consumption.

The first section introduces the trajectory modeling using the patched conic method, which divides the trajectory into heliocentric and planetocentric phases.

The second section details the calculations associated with escape orbits in terms of infinite velocities and fly-by maneuvers.

The third section presents the main aspects of a mission scenario, including the launch phase, the optimization of the interplanetary path, and the targeting correction maneuvers (TCMs).

*Space Trajectories: Basic and Advanced Topics*, First Edition. Max Cerf.

## 14.2 Trajectory Modeling

The study of interplanetary missions is based on the concept of sphere of influence to split trajectory into a succession of two-body problems. This section presents the patched conics approach based on the concept of sphere of influence.

### 14.2.1 Patched Conics

A vehicle moving through the solar system is subject to the attraction of numerous celestial bodies, the most influential of which are the Sun and the planets. Such a problem with several attracting bodies is not tractable for preliminary studies. We saw in Chapter 3 that even the three-body problem raises difficult questions.

As the planets are much smaller than the Sun, and quite far apart, one simplifying approach is to divide space into regions where only one attracting body acts. Within each region, the motion reduces to a two-body problem (the attracting body and the vehicle), whose analytical solution is a Keplerian conic.

The method of patched conics consists in linking such conics to form the global interplanetary path as depicted in Figure 14.1. The main body is the Sun in S. Three planets in $C_1$, $C_2$ and $C_3$ are overflown at times $t_1$, $t_2$ and $t_3$, respectively.

The region of influence of each planet is represented by the dotted circle. Inside this circle, the vehicle follows a Keplerian conic relative to the planet (shown as blue dotted lines). This conic is necessarily a hyperbola since the vehicle-free trajectory continues far from the planet. Between the regions of influence of the planets, the main attracting body is the Sun. These legs form the majority of the interplanetary path (the figure is not to scale) and they consist of heliocentric elliptical conics (shown as green solid lines). The planetocentric and heliocentric conics connect at the boundaries of the region of influence of the planets.

Building a mission scenario consists of selecting which planets to fly over and the associated times. Throughout the trajectory, the vehicle may fly over the same planet several times or make several revolutions around the Sun before entering the sphere of influence of a planet. This involves solving successive Lambert's problems for the heliocentric arcs joining

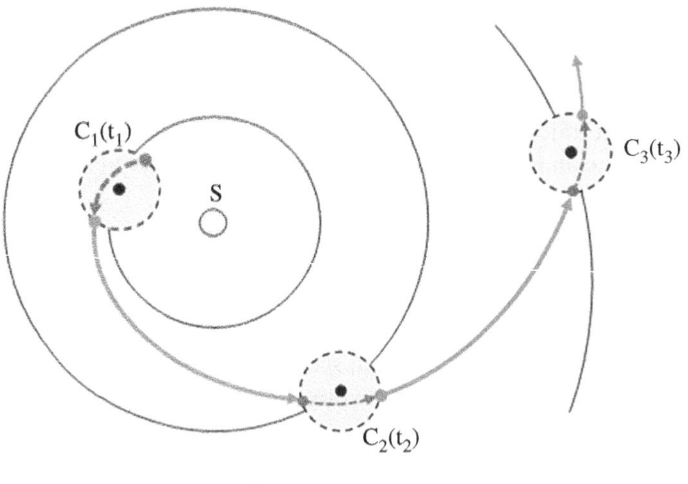


**Figure 14.1** Patched conics.

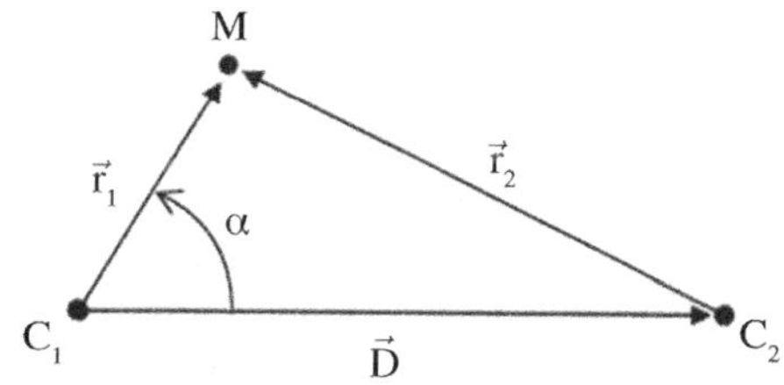


**Figure 14.2** Gravitational interaction of three bodies.

the planets at given times and making frame changes from heliocentric to planetocentric (and conversely) when entering or leaving a planet region of influence. These calculations and the definition of the region of influence are addressed in the following sections.

### 14.2.2 Sphere of Influence

A spacecraft in M is subject to the attraction of two celestial bodies in $C_1$ and $C_2$. The three bodies, assumed to be point-like, are in gravitational interaction.

The masses of $C_1$, $C_2$ and M are noted $m_1$, $m_2$ and m, whereas their positions in an inertial reference frame are noted $\vec{R}_1$, $\vec{R}_2$ and $\vec{R}$, respectively.

The relative positions are noted $\vec{r}_1 = \vec{R} - \vec{R}_1$, $\vec{r}_2 = \vec{R} - \vec{R}_2$, and $\vec{D} = \vec{R}_2 - \vec{R}_1$.

These vectors are depicted in Figure 14.2.

The equations of motion for $C_1$, $C_2$ and M are given by Newton's second law.

$$\begin{cases} \ddot{\vec{R}}_1 = Gm\dfrac{\vec{r}_1}{r_1^3} + Gm_2\dfrac{\vec{D}}{D^3} & (C_1 \text{ attracted by M and } C_2) \\ \ddot{\vec{R}}_2 = Gm\dfrac{\vec{r}_2}{r_2^3} - Gm_1\dfrac{\vec{D}}{D^3} & (C_2 \text{ attracted by M and } C_1) \\ \ddot{\vec{R}} = -Gm_1\dfrac{\vec{r}_1}{r_1^3} - Gm_2\dfrac{\vec{r}_2}{r_2^3} & (\text{M attracted by } C_1 \text{ and } C_2) \end{cases} \tag{14.1}$$

where $G = 6.674\,08 \cdot 10^{-11}\ m^3\ kg^{-1}\ s^{-2}$ is the universal gravitational constant.

The equations of motion of M with respect to $C_1$ (respectively $C_2$) are obtained by subtracting the first (respectively second) equation from the third.

$$\begin{cases} \ddot{\vec{r}}_1 = -G(m_1 + m)\dfrac{\vec{r}_1}{r_1^3} - Gm_2\left(\dfrac{\vec{r}_2}{r_2^3} + \dfrac{\vec{D}}{D^3}\right) \underset{\text{noted}}{=} \vec{a}_1 + \vec{a}_{1p} \\ \ddot{\vec{r}}_2 = -G(m_2 + m)\dfrac{\vec{r}_2}{r_2^3} - Gm_1\left(\dfrac{\vec{r}_1}{r_1^3} - \dfrac{\vec{D}}{D^3}\right) \underset{\text{noted}}{=} \vec{a}_2 + \vec{a}_{2p} \end{cases} \tag{14.2}$$

The first equation shows that M acceleration relative to $C_1$ is the sum of two terms:

- the direct attraction of $C_1$ (main term $\vec{a}_1$);
- the *difference in attraction* exerted by $C_2$ on M and $C_1$ (perturbation $\vec{a}_{1p}$).

If M is close to $C_1$, this perturbation term is negligible and the equation becomes that of a two-body problem. M then follows a Keplerian motion around $C_1$. The motion of M relative to $C_2$ can be interpreted in a similar way, as the sum of a main term $\vec{a}_2$ and a perturbation term $\vec{a}_{2p}$.

The **gravisphere** of $C_1$ with respect to $C_2$ is the region of space in which $C_1$ can be considered the principal body, $C_2$ being the perturbing body. One possible definition is to compare the ratios of the perturbation term to the principal term.

With this approach, the body $C_1$ is considered as the main body if $\frac{a_{1p}}{a_1} < \frac{a_{2p}}{a_2}$; otherwise, the body $C_2$ is considered as the main body.

The limit of the gravisphere is then the set of points M satisfying

$$\frac{a_{1p}}{a_1} = \frac{a_{2p}}{a_2} \Rightarrow \frac{m_2}{m_1 + m} r_1^2 \left\| \frac{\vec{r}_2}{r_2^3} + \frac{\vec{D}}{D^3} \right\| = \frac{m_1}{m_2 + m} r_2^2 \left\| \frac{\vec{r}_1}{r_1^3} - \frac{\vec{D}}{D^3} \right\| \tag{14.3}$$

The shape of this surface is complex in the general case.

In practical applications, point M is a vehicle of negligible mass, body $C_1$ is a planet (or the Moon) and body $C_2$ is the Sun (or the Earth).

When the mass of $C_1$ is small compared with the mass of $C_2$, which is the case of a planet with respect to the Sun, the gravisphere of $C_1$ is of small dimension compared with the distance $C_1C_2$, allowing the approximation $r_1 \ll D$.

The norms in (14.3) are expanded to order 1 in $s = \frac{r_1}{D}$ with the angle $\alpha = (\vec{D}, \vec{r}_1)$ depicted in Figure 14.1.

- For the term $\|\vec{r}_2\|$, we have:

$$\|\vec{r}_2\| = \|\vec{r}_1 - \vec{D}\| = \sqrt{r_1^2 + D^2 - 2r_1 D \cos\alpha} \approx D(1 - s\cos\alpha)$$

- For the term $\left\| \frac{\vec{r}_2}{r_2^3} + \frac{\vec{D}}{D^3} \right\|$, we have:

$$\left\| \frac{\vec{r}_2}{r_2^3} + \frac{\vec{D}}{D^3} \right\| = \frac{1}{D^2} \left\| \frac{D^3}{r_2^3}\frac{\vec{r}_1}{D} + \left(1 - \frac{D^3}{r_2^3}\right)\frac{\vec{D}}{D} \right\| \text{ with } \frac{D^3}{r_2^3} \approx 1 + 3s\cos\alpha$$

$$= \frac{1}{D^2}\sqrt{(1 + 3s\cos\alpha)^2 s^2 + (3s\cos\alpha)^2 - 6s^2\cos^2\alpha\,(1 + 3s\cos\alpha)}$$

$$\approx \frac{s}{D^2}\sqrt{1 + 3\cos^2\alpha} \quad \text{by keeping only the terms in } s^2$$

- For the term $\left\| \frac{\vec{r}_1}{r_1^3} - \frac{\vec{D}}{D^3} \right\|$, we have:

$$\left\| \frac{\vec{r}_1}{r_1^3} - \frac{\vec{D}}{D^3} \right\| = \frac{1}{r_1^2} \left\| \frac{\vec{r}_1}{r_1} - \frac{r_1^2}{D^2}\frac{\vec{D}}{D} \right\| = \frac{1}{r_1^2}\sqrt{1 + s^4 - 2s^2\cos\alpha} \approx \frac{1}{r_1^2}$$

Substituting these expansions into (14.3), we obtain the gravisphere equation in the case $m_1 \ll m_2$. This equation defines the radius $r_1$ as a function of the angle $\alpha$.

$$\left(\frac{r_1}{D}\right)^5 \sqrt{1 + 3\cos^2\alpha} = \left(\frac{m_1}{m_2}\right)^2 \tag{14.4}$$

The radius $r_1$ depends slightly on the angle $\alpha$, with a maximum difference of 15% between $\alpha = 0$ deg and $\alpha = 90$ deg. The surface is thus close to a sphere of radius

$$r_1 = D\left(\frac{m_1}{m_2}\right)^{\frac{2}{5}} \tag{14.5}$$

This sphere is called the **sphere of influence** of the lighter body $C_1$ relative to the heavier body $C_2$. The radii of the sphere of influence for the Moon/Earth and the planet/Sun systems are given in the following table (in millions of km).

**Table 14.1** Spheres of influence of the planets.

| System | Moon Earth | Mercury Sun | Venus Sun | Earth Sun | Mars Sun | Jupiter Sun | Saturn Sun | Uranus Sun | Neptune Sun |
|---|---|---|---|---|---|---|---|---|---|
| R (Mkm) | 0.0661 | 0.112 | 0.616 | 0.925 | 0.577 | 48.2 | 54.8 | 51.7 | 86.6 |

This definition of the sphere of influence based on the ratio of perturbative accelerations is due to Tisserand (1845–1896). Other definitions of the gravisphere are possible ([R12], [R29], [R39]). In all cases, these are approximations intended for preliminary scenario studies.

### 14.2.3 Heliocentric Phase

The spheres of influence of the planets are very small compared to their distances to the Sun (Table 14.1). When studying a preliminary scenario, their dimensions are neglected and the planets are considered as points, whose positions are given by ephemerides. The interplanetary path is defined by the series of planets to fly over and the associated times.

Flying from a planet $C_1$ at time $t_1$ to a planet $C_2$ at time $t_2$ is Lambert's problem, consisting of finding a heliocentric Keplerian conic passing through the positions of the planets at fixed times. Lambert's problem has no analytical solution, but it can be reduced to a single equation with one unknown, as shown in Section 6.4.3. This equation must be solved using a numerical method.

#### 14.2.3.1 Transfer Types

Depending on the planets' positions and the time of flight, four types of transfers can be obtained as depicted in Figure 14.3. The planet orbits are dotted and the transfer ellipse is solid.

- Outward transfers are the fastest. They start from $C_1$ and can be either short (with an arrival in $C_2$ from below) or long (with an arrival in $C'_2$ from above).
- Inward transfers are the longest. They start from $C'_1$ and can be either short (with an arrival in $C_2$ from below) or long (with an arrival in $C'_2$ from above).

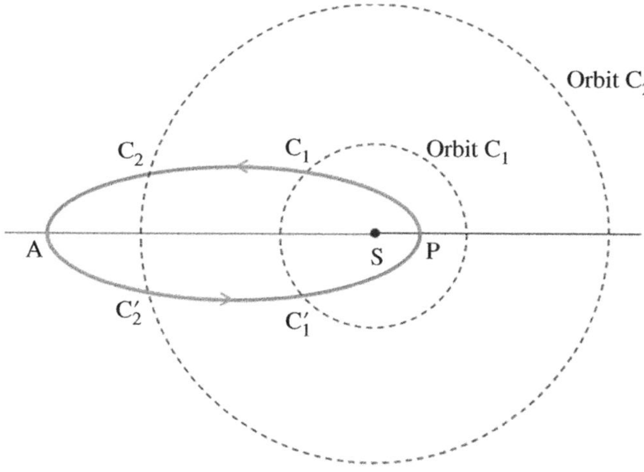


**Figure 14.3** Transfer ellipse.

Solving Lambert's problem from $C_1(t_1)$ to $C_2(t_2)$ under the attraction of the Sun S yields the transfer ellipse and the associated velocities at departure and arrival. The vehicle must perform a first propelled maneuver in $C_1$ to inject itself into the transfer ellipse. This maneuver depends on the vehicle arrival velocity in $C_1$.

A second propelled maneuver must also be performed in $C_2$, which depends on the required velocity after the passage in $C_2$. Planet $C_2$ may indeed be the final target of the interplanetary path, or a planet flown over along the trajectory.

#### 14.2.3.2 Inclination Effect

Let us consider a direct trajectory with no fly-by, such as an Earth–Mars mission.

The vehicle velocity at departure (before the transfer) is that of planet $C_1$ and its velocity at arrival (after the transfer) is that of planet $C_2$. Moreover, the planet orbits are almost circular and coplanar in the ecliptic plane. In this situation, the optimal transfer from $C_1$ to $C_2$ would be the Hohmann transfer (Section 6.2.3.1) and the transfer ellipse has its perihelion P in $C_1$ and its aphelion A in $C_2$.

However, this conclusion does not hold in the case of inclined planet orbits.

Figure 14.4 depicts a transfer from a planet $C_1$ to a planet $C_2$. Their orbital planes make an angle $i_2$ and they cross along the node line SN. Let us take as reference plane the orbit of $C_1$ and as reference axis the node line SN.

- At departure, the vehicle is on planet $C_1$ with a right ascension $\alpha_1$ and a null declination. The transfer orbit inclination is $i_1$ and the transfer angle is $\nu_1$.
- At arrival, the vehicle is on planet $C_2$ with a right ascension $\alpha_2$ and a declination $\delta$. The planet $C_2$ is at the angular distance $\nu_2$ from the node N.

From spherical trigonometry, we have in the triangle ($NHC_2$)

$$\begin{cases} \cos\delta = \cos\nu_2 \cos\alpha_2 + \sin\nu_2 \sin\alpha_2 \cos i_2 \\ \sin\delta = \sin\nu_2 \sin i_2 \\ \cos\nu_2 = \cos\delta \cos\alpha_2 \end{cases} \tag{14.6}$$

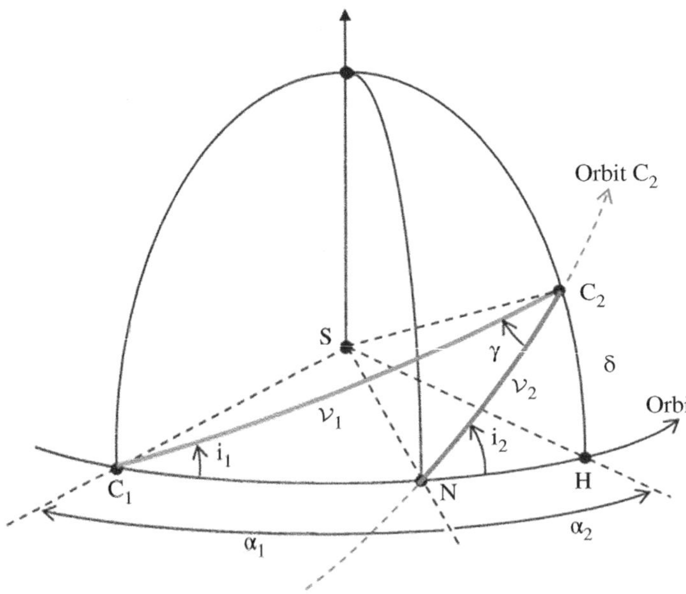


**Figure 14.4** Inclination of the transfer plane.

Substituting $\sin\nu_2$ and $\cos\nu_2$ into the first equation yields

$$\tan\delta = \tan i_2 \ \sin\alpha_2 \tag{14.7}$$

We have similarly in the triangle ($C_1HC_2$)

$$\begin{cases} \cos\delta = \cos\nu_1 \cos(\alpha_2 - \alpha_1) + \sin\nu_1 \sin(\alpha_2 - \alpha_1)\cos i_1 \\ \sin\delta = \sin\nu_1 \ \sin i_1 \\ \cos\nu_1 = \cos\delta \ \cos(\alpha_2 - \alpha_1) \end{cases} \tag{14.8}$$

Substituting $\sin\nu_1$ and $\cos\nu_1$ into the first equation yields

$$\tan\delta = \tan i_1 \ \sin(\alpha_2 - \alpha_1) \tag{14.9}$$

From (14.7) and (14.9), we obtain a relation giving the transfer orbit inclination.

$$\tan i_1 = \tan i_2 \frac{\sin\alpha_2}{\sin(\alpha_2 - \alpha_1)} \tag{14.10}$$

This relation shows that the transfer orbit inclination $i_1$ can take any value even if the planet mutual inclination $i_2$ is very small. In the worst case, it is 90 deg when the transfer departure and arrival are opposite with respect to the Sun.

$$\alpha_2 = \alpha_1 + 180 \text{ deg} \quad \Rightarrow \quad i_1 = 90 \text{ deg} \ \text{ if } i_2 \neq 0 \text{ deg} \tag{14.11}$$

This situation corresponding to an "inclined" Hohmann transfer is therefore very costly, since it induces large plane changes at departure and arrival. In this situation, the optimal configurations correspond to transfer angles shifted on either side of the Hohmann configuration ($\theta < 180$ deg or $\theta > 180$ deg), making a compromise between the optimal transfer angle ($\theta = 180$ deg) and the induced plane change.

This optimal transfer angle can be estimated by considering a semi-tangential strategy as in Section 7.2.3. For that purpose, we assume that planets $C_1$ and $C_2$ are on circular orbits of radii $r_1$ and $r_2$ respectively. The velocities before and after the departure impulse from $C_1$ are $\vec{w}_1$ and $\vec{v}_1$, respectively. For a semi-tangential transfer, both vectors are normal to the radius in $C_1$ and the velocity $v_1$ is given by (7.13) with the eccentricity given by (7.23). We have thus

$$v_1 = w_1 \sqrt{\frac{1 - \cos\nu_1}{\frac{1}{\rho} - \cos\nu_1}} \quad \text{with } \rho = \frac{r_2}{r_1} \tag{14.12}$$

The departure impulse to pass from $\vec{w}_1$ to $\vec{v}_1$, which form an angle $i_1$, is then

$$\Delta v^2 = (\vec{v}_1 - \vec{w}_1)^2 = v_1^2 + w_1^2 - 2v_1 w_1 \ \cos i_1 \tag{14.13}$$

For a given planet configuration (Figure 14.4), the angles ($\alpha_1, \alpha_2, i_2$) are given.

The angle $i_1$ between the velocities is assessed by (14.10), whereas the transfer angle $\nu_1$ is assessed from formulas (14.8) and (14.9), which yield

$$\tan\nu_1 = \frac{\tan(\alpha_2 - \alpha_1)}{\cos i_1} \tag{14.14}$$

Substituting into (14.13) $i_1$ (14.10), $\nu_1$ (14.14), and $v_1$ (14.12), we obtain an expression of the departure impulse as a function of ($\alpha_1, \alpha_2, i_2$) and of the radius ratio $\rho$. This impulse has a minimum that depends on these four variables.

Figure 14.5 plots the ratio $\frac{\Delta v}{w_1}$ as a function of the longitude difference $\alpha_2 - \alpha_1$ for radius ratios $\rho$ corresponding to Earth–Mars, Earth–Venus, and Earth–Jupiter transfers. These curves are plotted for an Earth's initial longitude $\alpha_1 = -90$ deg.

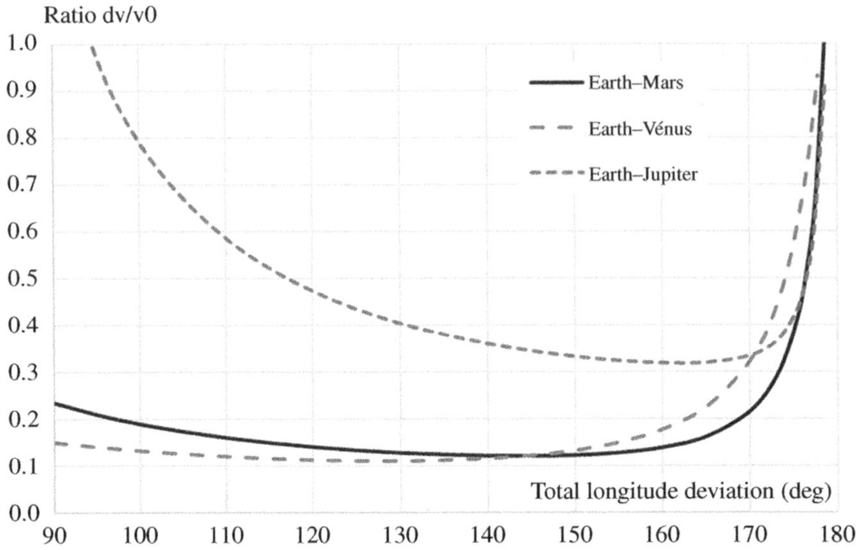


**Figure 14.5** Departure impulse for planetary transfer.

For a launch toward Mars, the minimum is around 145 deg with an inclination of 1.85 deg. The required impulse increases very strongly around 180 deg, which corresponds to a quasi-Hohmann configuration. This simplified calculation shows the influence of the planet configuration and the existence of favorable launch opportunities on either side of the planet's opposition. These favorable opportunities are repeated with the synodic period of the planets.

#### 14.2.3.3 Synodic Period

For planets in circular orbits of mean motions $n_1$ and $n_2$, respectively, the anomaly difference evolution is given by

$$\begin{cases} \theta_1(t) = \theta_1(t_0) + n_1(t - t_0) \\ \theta_2(t) = \theta_2(t_0) + n_2(t - t_0) \end{cases} \Rightarrow \Delta\theta(t) = \Delta\theta(t_0) + (n_1 - n_2)\Delta t \tag{14.15}$$

After one synodic period $T_{syn}$, the same configuration is repeated, so that the synodic period is given by

$$\Delta\theta(t_0 + T_{syn}) = \Delta\theta(t_0) + 2\pi \Rightarrow \frac{1}{T_{syn}} = \frac{1}{T_1} - \frac{1}{T_2} \tag{14.16}$$

In the case of an Earth–Mars transfer, the favorable opportunities are repeated with the synodic period of the two planets, which is of 2.13 years.

### 14.2.4 Planetocentric Phase

Flying over a planet produces a change in velocity with respect to the Sun. This effect, called **gravity assist** (or swing-by), is analyzed as follows.

The vehicle in heliocentric orbit has an arrival velocity $\vec{v}_{a/S}$ relative to the Sun when it enters the sphere of influence of planet C. The planet and its sphere of influence are moving around the Sun with the drive velocity $\vec{v}_C$. Entering the planet reference frame induces a composition of velocities, and the vehicle arrival velocity relative to the planet is the difference $\vec{v}_{a/C} = \vec{v}_{a/S} - \vec{v}_C$ (Figure 14.6).

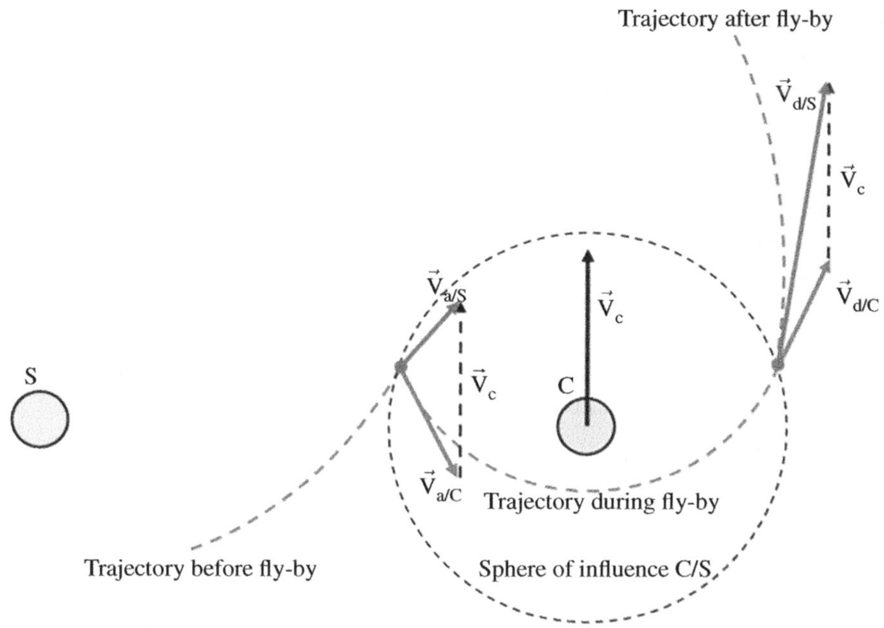


**Figure 14.6** Gravity assist.

The vehicle arrives from outside the sphere of influence, which means that it has enough energy to escape the attraction of the planet C. Its trajectory relative to C is a hyperbola and the entry into the sphere of influence takes place along the arrival asymptote, with a velocity assimilated to the infinite velocity $\vec{v}_{\infty a} = \vec{v}_{a/C}$. Similarly, the velocity at the exit of the sphere of influence is assimilated to the infinite velocity along the departure asymptote of the hyperbola $\vec{v}_{\infty d} = \vec{v}_{d/C}$.

By conservation of energy along the Keplerian motion, the velocities at entry and exit have the same modulus since both points are located on the boundary of the sphere of influence, therefore at the same distance from the planet C. Between entry and exit, the velocity has simply changed direction by an angle depending on the pericenter radius. The calculation of this angle is detailed in Section 14.3.3.

At the exit of the sphere of influence, the vehicle comes back in the inertial heliocentric frame, inducing the converse composition of velocities. The departure velocity $\vec{v}_{d/C}$ relative to the planet is added to the drive velocity $\vec{v}_C$ and the vehicle departure velocity relative to the Sum is the sum $\vec{v}_{d/S} = \vec{v}_{d/C} + \vec{v}_C$.

Figure 14.6 illustrates (with deliberate exaggeration) this gravity-assisted effect. The heliocentric velocity variation during the fly-by depends on the angle between the planet's velocity and the infinite velocities at arrival and departure. If the planet is flown "from behind" as in Figure 14.6, the infinite velocity is deflected along to the planet's velocity and the result is an increase in velocity relative to the Sun. Conversely, flying the planet "from the front" decreases the heliocentric velocity.

This variation of the velocity vehicle with respect to the Sun is not free. In fact, the vehicle and the planet are mutually attracted during the fly-by. The vehicle–planet system retains its total momentum, and the gain in vehicle velocity is offset by a loss in planet velocity, imperceptible due to their mass ratio.

## 14.3 Escape Conditions

### 14.3.1 Keplerian Motion

When vehicle M comes inside the sphere of influence of a celestial body C, the attraction of other bodies can be neglected to a first approximation. We find ourselves in a two-body problem involving only bodies C and M, and the solution is a Keplerian motion (the two bodies are assumed to be point-like).

The trajectory of vehicle M relative to body C is a conic characterized by the semi-major axis a, the eccentricity e, the parameter p, the angular momentum h, and energy w per unit mass.

These constants are linked to the position in polar coordinates (r, θ) and to the velocity V by the following relations established in Section 1.2.

$$\begin{cases} r = \dfrac{p}{1 + e\cos\theta}, \quad h = r^2\dot{\theta}, \quad h^2 = \mu p = \mu a(1 - e^2) \\ w = \dfrac{v^2}{2} - \dfrac{\mu}{r} = -\dfrac{\mu}{2a} \end{cases} \tag{14.17}$$

where $\mu = Gm_c$ is the gravitational constant of body C of mass $m_c$.

The radius at pericenter is $r_p = a(1 - e)$ and the true anomaly θ is the angle between the pericenter position P and the current position M.

The type of conic depends on the energy w.

- If the kinetic energy is lower than the potential energy ($w < 0$), the vehicle remains captive to the attracting body. The conic is an ellipse with semi-major axis $a > 0$ and eccentricity $e < 1$. The radius at the apocenter is $r_a = a(1 + e)$.
- If the kinetic energy is greater than the potential energy ($w > 0$), the vehicle escapes the body attraction and goes away to infinity. The conic is a hyperbola with semi-major axis $a < 0$ and eccentricity $e > 1$. Formulas (14.17) for $r \to \infty$ give the anomaly $\theta_\infty$ of the asymptote (>90 deg), the velocity at infinity $v_\infty$ and the expression of h as a function of $\theta_\infty$ and $v_\infty$.

$$\cos\theta_\infty = -\frac{1}{e}, \quad v_\infty^2 = -\frac{\mu}{a}, \quad h = -\frac{\mu}{v_\infty}\tan\theta_\infty \tag{14.18}$$

The conic is positioned in the inertial reference frame (CXYZ) linked to the attracting body (Figure 14.7a). The origin C is the body center of mass, the axis CZ is the pole axis and the plane CXY is the equatorial plane, with the X axis pointing to a fixed direction in space (the vernal point for the Earth).

The orbital plane containing the conic is deduced from the equatorial plane CXY by a rotation of axis Z and angle Ω (RAAN, right ascension of the ascending node), then a rotation of axis X′ and angle i (inclination). The axis X′, called the line of nodes, is the intersection of the orbital plane and the equator.

The ascending node N, opposite the descending node N′, is the point at which the trajectory crosses the equator from south to north. The angular momentum $\vec{h}$ is normal to the orbital plane. The direction of the pericenter P is obtained from the axis X′ by a rotation of axis $\vec{h}$ and angle ω (argument of the pericenter).

Any point M on the orbit is located by its latitude argument $\nu = \omega + \theta$, which represents the angular distance from the ascending node.

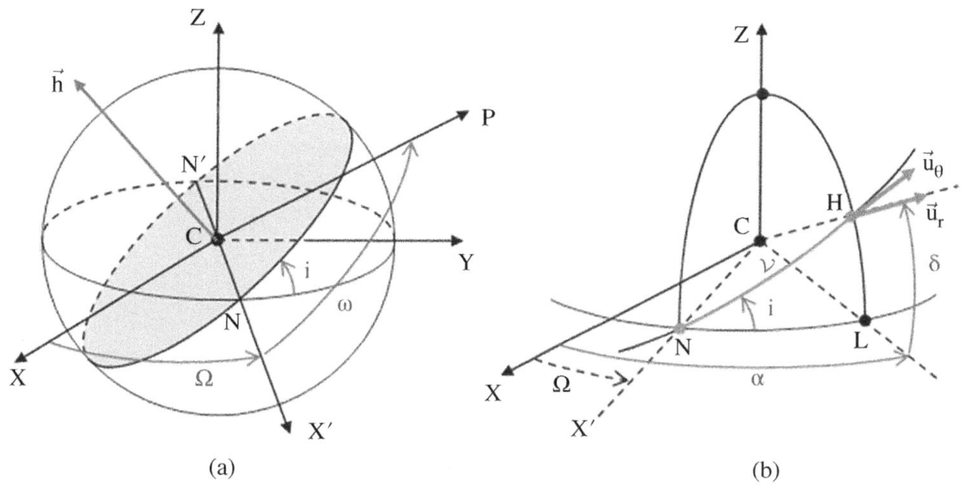


**Figure 14.7** Orbital parameters (a), right ascension, and declination (b).

The components of the radial $\vec{u}_r$ and orthoradial $\vec{u}_\theta$ unit vectors in M are derived from the formulas (1.114) and (1.116) with $\nu = \omega + \theta$.

$$\vec{u}_r = \begin{pmatrix} \cos\Omega\,\cos\nu - \sin\Omega\,\sin\nu\,\cos i \\ \sin\Omega\,\cos\nu + \cos\Omega\,\sin\nu\,\cos i \\ \sin\nu\,\sin i \end{pmatrix}, \quad \vec{u}_\theta = \begin{pmatrix} -\cos\Omega\,\sin\nu - \sin\Omega\,\cos\nu\,\cos i \\ -\sin\Omega\,\sin\nu + \cos\Omega\,\cos\nu\,\cos i \\ \cos\nu\,\sin i \end{pmatrix} \tag{14.19}$$

The vector $\vec{u}_r$ can also be defined in the frame (CXYZ) by the right ascension α along the equator and the declination δ along the meridian. For that purpose, let us consider the spherical triangle (NLH) depicted in Figure 14.7b, where H is the projection of M on the unit sphere and L is the point at the Equator on M meridian.

From spherical trigonometry relations in the triangle (NLH), we have

$$\vec{u}_r = \begin{pmatrix} \cos\alpha\,\cos\delta \\ \sin\alpha\,\cos\delta \\ \sin\delta \end{pmatrix} \quad \text{with} \quad \begin{cases} \sin i\,\sin\nu = \sin\delta \\ \cos i\,\tan\nu = \tan(\alpha - \Omega) \end{cases} \tag{14.20}$$

### 14.3.2 Hyperbolic Orbit

The hyperbola asymptote is defined by its anomaly $\theta_\infty$ (angle with the pericenter) and its distance $r_b$ from the attracting center, as depicted in Figure 14.8a.

The point B (as in “Body”) is the projection of C on the asymptote. The plane normal to the asymptote and passing through the center C of the attracting body is called the **B-plane** (shaded in Figure 14.8a).

For a vehicle arriving from infinity with the velocity $\vec{v}_{\infty a}$, the position of the target point B completely determines the hyperbolic trajectory around the body C.

The angular momentum $\vec{h} = \vec{r}_b \wedge \vec{v}_{\infty a}$ defines indeed the orbital plane (shown in Figure 14.8b), whereas its modulus $h = r_b v_{\infty a}$ allows to assess $(\theta_\infty, e, a)$ from formulas (14.18).

Let us detail these calculations.

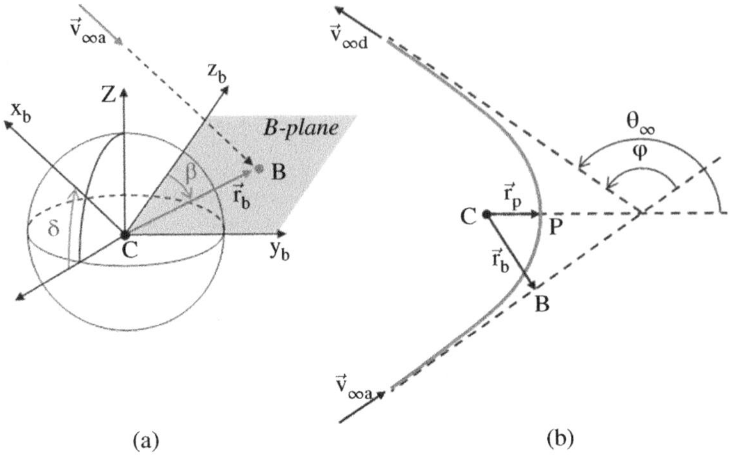


**Figure 14.8** (a) Arrival asymptote. (b) Transfer plane.

We first define the arrival reference frame $(C, \vec{x}_b, \vec{y}_b, \vec{z}_b)$ depicted in Figure 14.8a:

- the axis $\vec{x}_b$ is opposite to the arrival infinite velocity $\vec{v}_{\infty a}$, whose direction is defined by the right ascension α and the declination δ;
- the axis $\vec{y}_b$ is in the equatorial plane, directed along $\vec{x}_b \wedge \vec{z}$;
- the axis $\vec{z}_b$ completes the trihedron.

The components of these vectors are

$$\vec{x}_b = \begin{pmatrix} \cos\alpha\ \cos\delta \\ \sin\alpha\ \cos\delta \\ \sin\delta \end{pmatrix}, \quad \vec{y}_b = \begin{pmatrix} -\sin\alpha \\ \cos\alpha \\ 0 \end{pmatrix}, \quad \vec{z}_b = \begin{pmatrix} -\cos\alpha\ \sin\delta \\ -\sin\alpha\ \sin\delta \\ \cos\delta \end{pmatrix} \tag{14.21}$$

The position of B noted $\vec{r}_b = \overrightarrow{CB}$ makes an angle β with $\vec{z}_b$. These vectors both lie in the B-plane $(\vec{y}_b, \vec{z}_b)$. The angular momentum $\vec{h} = \vec{r}_b \wedge \vec{v}_{\infty a}$ is assessed by

$$\begin{cases} \vec{r}_b = r_b(\vec{z}_b\ \cos\beta - \vec{y}_b\ \sin\beta) \\ \vec{v}_{\infty a} = v_{\infty a}\ \vec{x}_b \end{cases} \Rightarrow \vec{h} = r_b v_{\infty a}(-\vec{y}_b\ \cos\beta + \vec{z}_b\ \sin\beta) \tag{14.22}$$

The inclination of the orbital plane is the angle $(\vec{h}, \vec{z})$.
It is assessed from $\vec{h} \cdot \vec{z} = h\ \cos i$, which using (14.22) gives

$$\cos i = \sin\beta\ \cos\delta \tag{14.23}$$

The argument of the pericenter ω and the right ascension of the ascending node Ω are then determined by applying formulas (14.20) to the vector $\vec{v}_{\infty a}$ considered as the position at infinity.

As this vector is directed toward the attracting body, the angles $(\theta_\infty, \alpha, \delta)$ must be taken with a minus sign $(\varepsilon = -1)$ in these formulas.

$$\begin{cases} \sin i\ \sin(\omega + \varepsilon\theta_\infty) = \varepsilon\ \sin\delta \\ \cos i\ \tan(\omega + \varepsilon\theta_\infty) = \varepsilon\ \tan(\alpha + \Omega) \end{cases} \tag{14.24}$$

The formulas (14.24) yield ω and Ω from the known values of $(\theta_\infty, \alpha, \delta, i)$.

The coordinates $(\alpha_p, \delta_p)$ of the pericenter are then assessed from the same formulas (14.20) with $\theta_p = 0$, corresponding to $\nu_p = \omega$.

$$\begin{cases} \sin i \ \sin\omega = \sin\delta_p \\ \cos i \ \tan\omega = \tan(\alpha_p - \Omega) \end{cases} \tag{14.25}$$

By conservation of energy, the arrival $\vec{v}_{\infty a}$ and departure $\vec{v}_{\infty d}$ velocities have the same modulus. They make an angle $\varphi = 2\theta_\infty - \pi$, as can be seen in Figure 14.8b.

The deviation angle $\varphi$ and the pericenter radius $r_p = a(1-e)$ can be expressed as a function of $r_b$ and $v_\infty$. By using (14.18) to substitute a and e, we obtain

$$\tan\frac{\varphi}{2} = \frac{\mu}{r_b v_\infty^2}, \quad r_p = \frac{\mu}{v_\infty^2}\left(\sqrt{1 + \frac{r_b^2 v_\infty^4}{\mu^2}} - 1\right) \tag{14.26}$$

These formulas help position the target point B at the right distance with respect to the body.

Indeed, the passage distance $r_p$ to the body must be sufficiently large to take account of navigation inaccuracies and to avoid atmospheric effects, if any. This minimum bound $r_m$ limits the deviation angle $\varphi$. The maximal achievable deviation $\varphi_m$ associated with $r_m$ is obtained by combining both formulas (14.26), which yields

$$\sin\frac{\varphi_m}{2} = \left(1 + \frac{r_m v_\infty^2}{\mu}\right)^{-1} \tag{14.27}$$

### 14.3.3 Escape Velocity

Let us assume that the vehicle is at a position $\vec{r}$ and that we wish to place it in a hyperbolic orbit with a specified infinite velocity $\vec{v}_\infty$. This section describes the calculation of the required escape velocity $\vec{v}$ associated with $\vec{r}$ and $\vec{v}_\infty$.

The infinite velocities at arrival and departure are noted $\vec{v}_{\infty a}$ and $\vec{v}_{\infty d}$, respectively. These velocities are along the hyperbola asymptotes, whose unit vectors are noted $\vec{u}_{\infty a}$ (anomaly $-\theta_\infty$) and $\vec{u}_{\infty d}$ (anomaly $+\theta_\infty$). The angles between the current position $\vec{r}$ and the asymptotes are noted $\psi_a$ and $\psi_d$ (see Figure 14.9).

In order to obtain formulas applicable on arrival and departure, we retrieve the convention already used in equation (14.24)

$$\begin{cases} \varepsilon = -1 \ \text{ for the arrival asymptote} \\ \varepsilon = +1 \ \text{ for the departure asymptote} \end{cases} \tag{14.28}$$

which gives the relations

$$\begin{cases} \vec{v}_{\infty a} = -v_\infty \, \vec{u}_{\infty a} \\ \vec{v}_{\infty d} = +v_\infty \, \vec{u}_{\infty d} \end{cases} \quad \rightarrow \quad \vec{v}_\infty = \varepsilon \, \vec{u}_\infty$$

$$\begin{cases} \psi_a = (\vec{u}_{\infty a}, \vec{u}_r) = \theta_\infty + \theta \\ \psi_d = (\vec{u}_r, \vec{u}_{\infty d}) = \theta_\infty - \theta \end{cases} \quad \rightarrow \quad \theta = \varepsilon(\theta_\infty - \psi)$$

$$\begin{cases} \vec{u}_{\infty a} = \cos\psi_a \, \vec{u}_r - \sin\psi_a \, \vec{u}_\theta \\ \vec{u}_{\infty d} = \cos\psi_d \, \vec{u}_r + \sin\psi_d \, \vec{u}_\theta \end{cases} \quad \rightarrow \quad \vec{u}_\infty = \cos\psi \, \vec{u}_r + \varepsilon \, \sin\psi \, \vec{u}_\theta \tag{14.29}$$

where $\vec{u}_r$ and $\vec{u}_\theta$ are the radial and orthoradial unit vectors given by (14.19) and $\theta$ is the true anomaly at the position $\vec{r}$.

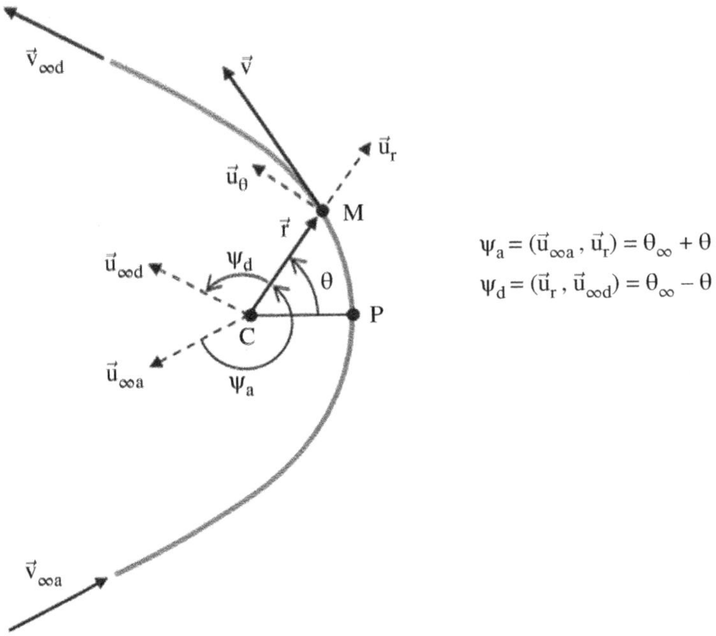


**Figure 14.9** Escape velocity.

The velocity $\vec{v}$ is the derivative of the position $\vec{r} = r\vec{u}_r$.

$$\vec{v} = \dot{\vec{r}} = \dot{r}\vec{u}_r + r\dot{\vec{u}}_r = \dot{r}\vec{u}_r + r\dot{\theta}\vec{u}_\theta \tag{14.30}$$

The derivatives $\dot{r}$ and $\dot{\theta}$ are expressed from the Keplerian formulas (14.17).

$$\begin{cases} \dot{\theta} = \dfrac{h}{r^2} \quad \text{with } p = \dfrac{h^2}{\mu} \\ \dot{r} = \dfrac{pe\dot{\theta}\sin\theta}{(1+e\cos\theta)^2} = \dfrac{r^2 e\dot{\theta}\sin\theta}{p} = \dfrac{he\sin\theta}{p} = \dfrac{\mu e\sin\theta}{h} \end{cases} \tag{14.31}$$

Substituting into (14.30), we get the expression of the velocity at the position $\vec{r}$.

$$\vec{v} = \frac{\mu e}{h}\sin\theta\,\vec{u}_r + \frac{h}{r}\vec{u}_\theta \tag{14.32}$$

The goal is to express $\vec{v}$ directly in terms of $\vec{r}$ and $\vec{v}_\infty$ without using the orbital parameters. The calculation takes place in two steps.

The first step is to express $\theta_\infty$ in terms of $(r, v_\infty)$ and the transfer angle $\psi$.

From the equation of the conic $r(1 + e\cos\theta) = p$, and using the formulas (14.29) and (14.18) to substitute $(\theta, p, e)$

$$\theta = \varepsilon(\theta_\infty - \psi), \quad p = \frac{\mu}{v_\infty^2}\tan^2\theta_\infty, \quad e = -\frac{1}{\cos\theta_\infty} \tag{14.33}$$

we get the equation satisfied by $\theta_\infty$ when $(r, \psi, v_\infty)$ are given.

$$\frac{\mu}{rv_\infty^2}\tan^2\theta_\infty + \sin\psi\,\tan\theta_\infty + \cos\psi - 1 = 0 \tag{14.34}$$

This equation of the second degree in $\tan\theta_\infty$ admits two roots of opposite signs.

$$\tan\theta_\infty = -\frac{rv_\infty^2\sin\psi}{2\mu}(1+s) \quad \text{with } s = \pm\sqrt{1 + \frac{4\mu}{rv_\infty^2(1+\cos\psi)}} \tag{14.35}$$

The negative solution ($\tan\theta_\infty < 0$) must be chosen because of (14.18).

The asymptote anomaly is indeed always greater than 90 deg.

The sign of s in the formula (14.35) is that of $\sin\psi$. The transfer angle $\psi$ is lower than 180 deg when the vehicle is close to the attracting body (see Figure 14.9), so that $\sin\psi > 0$ in most practical cases.

The second step of the calculation is to remove the orbital parameters from the velocity expression (14.32).

By substituting ($\theta$, p, e) from (14.33), $\vec{u}_\theta = \frac{\varepsilon}{\sin\psi}(\vec{u}_\infty - \cos\psi\,\vec{u}_r)$ from (14.29), and using equation (14.34) to substitute the term into $\tan\theta_\infty^2$, formula (14.32) gives

$$\vec{v} = -\varepsilon v_\infty \vec{u}_r - \frac{\varepsilon\mu}{r v_\infty}\frac{\tan\theta_\infty}{\sin\psi}(\vec{u}_r + \vec{u}_\infty) \tag{14.36}$$

Finally, with $\tan\theta_\infty$ given by (14.33), we obtain the expression of the velocity $\vec{v}$ required at position $\vec{r}$ to attain the desired infinite velocity $\vec{v}_\infty$.

$$\vec{v} = \varepsilon\frac{v_\infty}{2}[(s-1)\vec{u}_r + (s+1)\vec{u}_\infty] \quad \text{with } s = \sqrt{1 + \frac{4\mu}{r v_\infty^2(1+\cos\psi)}} \tag{14.37}$$

This formula is valid for both arrival ($\varepsilon = -1$) and departure ($\varepsilon = +1$), with the transfer angle $\psi$ counted positively in the direction of motion (Figure 14.9).

If the angle $\psi$ is greater than 180 deg, the sign of s shall be taken negative.

In practice, the maneuvers always take place in the vicinity of the pericenter so that this angle is lower than 180 deg.

### 14.3.4 Fly-by Maneuver

When flying over a planet, the arrival and departure velocities are obtained from Lambert's transfers (one before arrival and one after departure) solved in the heliocentric frame. The infinite velocities $\vec{v}_{\infty a}$ at entry and $\vec{v}_{\infty d}$ at exit of the planet sphere of influence are therefore given. Passing naturally from $\vec{v}_{\infty a}$ to $\vec{v}_{\infty d}$ is possible only if the infinite velocities have the same modulus and if their angle is lower than the maximal possible deviation assessed by (14.27). In most cases, a complementary propelled maneuver is necessary to achieve the targeted exit velocity. The goal is then to orientate the arrival hyperbola in order to minimize the required corrective maneuver.

#### 14.3.4.1 Problem Formulation

At any position $\vec{r}$ on the flyover path, we can assess by (14.37) the velocities $\vec{v}_a$ and $\vec{v}_d$ associated with the arrival $\vec{v}_{\infty a}$ and departure $\vec{v}_{\infty d}$ conditions (with $\varepsilon = -1$ for arrival and $\varepsilon = +1$ for departure).

The impulsive maneuver $\Delta\vec{v}$ in position $\vec{r} = r\vec{u}_r$ to pass from $\vec{v}_{\infty a}$ to $\vec{v}_{\infty d}$ is the difference $\Delta\vec{v} = \vec{v}_d - \vec{v}_a$. By noting $s_a$ and $s_d$ the variables in (14.37) associated with the arrival and departure branches, and with $\vec{v}_\infty = \varepsilon\vec{u}_\infty$, the impulse is

$$\Delta\vec{v} = \frac{1}{2}\vec{u}_r[(s_d-1)v_{\infty d} + (s_a-1)v_{\infty a}] + \frac{1}{2}[(s_d+1)\vec{v}_{\infty d} - (s_a+1)\vec{v}_{\infty a}] \tag{14.38}$$

The objective is to find the position $\vec{r} = r\vec{u}_r$ minimizing the modulus: $\min_{\vec{u}_r, r} \|\Delta\vec{v}\|$.

For given vectors $\vec{v}_{\infty a}$ and $\vec{v}_{\infty d}$, the impulse modulus is minimized by placing the vector $\vec{u}_r$ in the plane defined by ($\vec{v}_{\infty a}, \vec{v}_{\infty d}$). The fly-by trajectory then takes place in the plane ($C, \vec{v}_{\infty a}, \vec{v}_{\infty d}$) called the fly-by plane.

The vector $\vec{u}_r$ is positioned in this plane by the angle $\psi_a = (\vec{u}_{\infty a}, \vec{u}_r)$ depicted in Figure 14.9. The angle $\psi_d = (\vec{u}_r, \vec{u}_{\infty d})$ is derived from this. The minimization problem thus reduces to two variables $(\psi_a, r)$ defining the maneuver location.

On the other hand, the flyover trajectory associated with the maneuver location must pass at a sufficient distance from the planet. This implies two altitude constraints, one on the arrival branch before the maneuver and the other on the departure branch after the maneuver. These constraints complicate the problem, so that no analytical solution can be found. This motivates the following simplification.

#### 14.3.4.2 Simplified Solution

A quasi-analytical solution is determined by setting the pericenter radius of the two branches to the minimum acceptable value: $r_{pa} = r_{pd} = r_m$. This value depends on the navigation inaccuracies and the planet's atmosphere, if any. By this way, the minimum distance to the planet is directly respected. Since the infinite velocities and the radius at pericenter are fixed for the arrival and departure hyperboles, the anomalies $\theta_{\infty a}$ and $\theta_{\infty d}$ of the asymptotes can be assessed by (14.18).

$$\cos\theta_\infty = -\frac{1}{e} \quad \text{with } e = 1 + \frac{r_p v_\infty^2}{\mu} \tag{14.39}$$

The angle $\Delta\theta$ between the major axes of the arrival and departure hyperboles (Figure 14.8) is then

$$\Delta\theta = \varphi + \pi - (\theta_{\infty a} + \theta_{\infty d}) \tag{14.40}$$

where $\varphi$ is the known angle between the arrival $\vec{v}_{\infty a}$ and departure $\vec{v}_{\infty d}$ velocities, as depicted in Figure 14.10.

The point of maneuver M lies at the intersection of the arrival and departure hyperboles. Its polar coordinates are $(r, \theta_a)$ with respect to the arrival hyperbola major axis $CP_a$ and $(r, \theta_d)$ with respect to the departure hyperbola major axis $CP_d$.

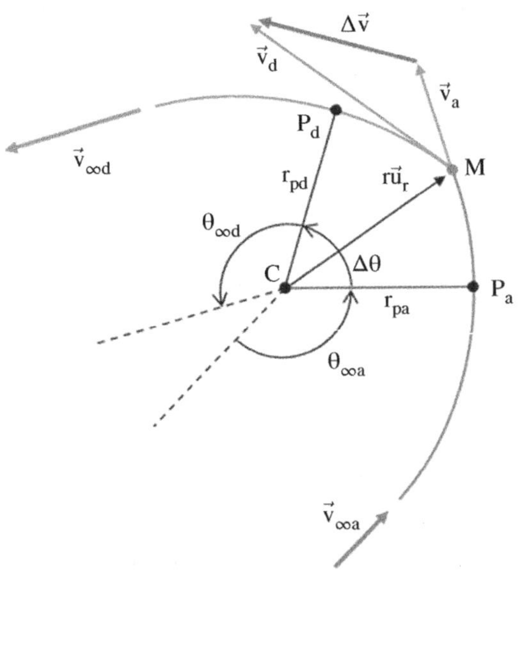


**Figure 14.10** Fly-by maneuver.

These polar coordinates satisfy the respective conic equations

$$r = \frac{p_a}{1 + e_a \cos\theta_a} = \frac{p_d}{1 + e_d \cos\theta_d} \quad \text{with } \theta_d = \theta_a - \Delta\theta \tag{14.41}$$

By substituting: $p = a(1 - e^2) = r_p(1 + e)$ and $e = -\frac{1}{\cos\theta_\infty}$, this equation becomes

$$\frac{\cos\theta_{\infty a} - 1}{\cos\theta_{\infty a} - \cos\theta_a} = \frac{\cos\theta_{\infty d} - 1}{\cos\theta_{\infty d} - \cos\theta_d} \quad \text{with } \theta_d = \theta_a - \Delta\theta \tag{14.42}$$

Since the angles $\theta_{\infty a}$, $\theta_{\infty d}$, $\Delta\theta$ are known from (14.39) and (14.40), the resolution of this equation yields the anomaly $\theta_a$ of the maneuver position on the arrival branch. The impulse $\Delta\vec{v}$ is then assessed by formula (14.38).

In the special case where the arrival and departure velocities have the same modulus, the anomalies $\theta_{\infty a}$ and $\theta_{\infty d}$ evaluated by (14.39) are equal, and the solution is $\theta_a = -\theta_d = \frac{\Delta\theta}{2}$. Both branches are symmetrical with respect to the bisector of the asymptotes.

The solution obtained from (14.42) is not strictly optimal, as it imposes the passage distance on both branches. However, it provides good estimates of fly-by maneuvers during the preliminary search for a mission scenario.

## 14.4 Mission Scenario

The method of patched conics highlights the phenomenon of gravity assist that provides "free" velocity gains in heliocentric reference. Propelled maneuvers are required when the gravity assist effect is not sufficient to reach the targeted infinite velocity. This section presents the concepts used for the preliminary optimization of an interplanetary mission scenario.

### 14.4.1 Launch Strategy

An interplanetary mission can start either by a direct launch on an escape orbit or by a passage through a parking orbit.

#### 14.4.1.1 Direct Launch

The direct launch strategy consists of injecting the vehicle directly into the escape orbit defined by the departure infinity velocity. The vector $\vec{v}_\infty$ is defined in the geocentric frame by the modulus $v_\infty$, the right ascension $\alpha$ and the declination $\delta$. These coordinates are associated with a particular day of launch and the positions of the planets in the solar system. If the launch cannot take place on the scheduled day, the vector $\vec{v}_\infty$ must be updated to take account of changes in the configuration of the planets. The launch time depends on the targeted right ascension of the infinite velocity. The optimal launch time is associated with a planar (as far as possible) trajectory until injection in order to maximize the launcher's performance. Any launch time delay requires out-of-plane corrections during the flight that induce overconsumption of propellants.

The direct launch strategy is optimal in terms of launcher performance, but it requires the launch to take place precisely on the scheduled day and time.

#### 14.4.1.2 Parking Orbit

Passing through a parking orbit is an alternative strategy to gain flexibility. The vehicle is launched a few days before the interplanetary launch into a low orbit. The waiting phase on this parking orbit makes it possible to ensure the proper functioning of the vehicle and to prepare the escape maneuver independently of the launcher flight. The goal is to determine the parking orbit and the maneuver location in order to minimize the required escape impulse. This optimization problem must be solved numerically in the general case.

In practice, the parking orbit is a low circular orbit of given radius r and the departure maneuver is in the horizontal plane, which imposes the perigee of the departure hyperbola at $r_p = r$. These choices simplify the baseline scenario and make it more robust to uncertainties. With these assumptions the optimization problem can be solved analytically as follows.

The departure impulse is the difference between the targeted escape velocity $\vec{v}_1$ and the initial velocity $\vec{v}_0$ on the parking orbit. Since the orbit is circular with a fixed radius r, the modules of these velocities are

$$v_0^2 = \frac{\mu}{r}, \quad v_1^2 = v_\infty^2 + 2\frac{\mu}{r} \tag{14.43}$$

and the impulse modulus is

$$\Delta v^2 = (\vec{v}_1 - \vec{v}_0)^2 = v_1^2 + v_0^2 - 2\vec{v}_0 \cdot \vec{v}_1 = v_\infty^2 + 3v_0^2 - 2\vec{v}_0 \cdot \vec{v}_1 \tag{14.44}$$

The velocity in the circular orbit is orthoradial: $\vec{v}_0 = v_0\vec{u}_\theta$. Since the maneuver takes place at the perigee of the hyperbola, the angle of transfer $\psi = (\vec{u}_r, \vec{u}_\infty)$ is equal to $\theta_\infty$ (see Figure 14.9), which is calculated by (14.39).

On the other hand, the required escape velocity is assessed by (14.37).

$$\vec{v}_1 = \frac{v_\infty}{2}[(s-1)\vec{u}_r + (s+1)\vec{u}_\infty] \quad \text{with } s = \frac{1-\cos\theta_\infty}{1+\cos\theta_\infty} \tag{14.45}$$

By substituting $\vec{v}_0$ and $\vec{v}_1$ into (14.44), the modulus of the impulse is expressed as

$$\Delta v^2 = v_\infty^2 + 3v_0^2 - 2\frac{v_0 v_\infty}{1+\cos\theta_\infty}\vec{u}_\theta \cdot \vec{u}_\infty \tag{14.46}$$

The plane of the parking orbit is defined in the geocentric frame by the right ascension of ascending node $\Omega$ and the inclination i. The maneuver takes place at the argument of latitude $\nu$ and the direction of the departure asymptote is defined by the right ascension $\alpha$ and the declination $\delta$ (see Figure 14.7).

The scalar product $\vec{u}_\theta \cdot \vec{u}_\infty$ is expressed with the components of the vectors given by (14.19) and (14.20). To simplify the calculations, the frame is rotated by the angle $\alpha$ around Z, so that the asymptote lies in the plane (XZ). This is equivalent to setting $\alpha = 0$ and considering $\Omega$ as the angle between the meridian of the asymptote and the node (angle LN in Figure 14.7).

$$\vec{u}_\theta \cdot \vec{u}_\infty = -(\cos\Omega \sin\nu + \sin\Omega \cos\nu \cos i)\cos\delta + \cos\nu \sin i \sin\delta \tag{14.47}$$

The values of $(v_0, v_\infty, \theta_\infty, \delta)$ are given. Minimizing the impulse $\Delta v$ (14.46) is equivalent to maximizing $\vec{u}_\theta \cdot \vec{u}_\infty$, therefore minimizing the angle $(\vec{u}_\theta, \vec{u}_\infty)$.

The optimization variables are $(\Omega, i, \nu)$. Let us first determine $(\Omega, \nu)$, then i.

The derivatives of $\vec{u}_\theta \cdot \vec{u}_\infty$ with respect to $(\Omega, \nu)$ must be zero.

$$\begin{cases} (\sin\Omega \sin\nu - \cos\Omega \cos\nu \cos i)\cos\delta = 0 \\ (-\cos\Omega \cos\nu + \sin\Omega \sin\nu \cos i)\cos\delta - \sin\nu \sin i \sin\delta = 0 \end{cases} \tag{14.48}$$

The matrix of second derivatives must be positive, which determines the correct quadrant of the angles $(\Omega, \nu)$. Solving the system (14.48) gives rise to different cases to be distinguished.

This discussion is not detailed here. It results in the solution below, that gives the angles $(\Omega, \nu)$ in all cases, with the convention: $\tan(\pm 90 \text{ deg}) = \pm\infty$.

$$\tan\nu = \frac{\cos i}{\tan\Omega} \quad \text{and} \quad \begin{cases} \sin\Omega = -\dfrac{\tan\delta}{\tan i} & \text{if } |\delta| \le |i| \quad \rightarrow \quad \vec{u}_\theta \cdot \vec{u}_\infty = 1 \\ \cos\Omega = 0 & \text{if } |\delta| > |i| \quad \rightarrow \quad \vec{u}_\theta \cdot \vec{u}_\infty = \cos(\delta - i) \end{cases} \tag{14.49}$$

The inclination of the parking orbit may be restricted between bounds depending on the launch site location and the feasible launch azimuths.

If the inclination is greater than the targeted declination, then it is possible to cancel the angle $(\vec{u}_\theta, \vec{u}_\infty)$. Otherwise, this angle cannot be lower than $|\delta - i|$, which results in an additional impulse in (14.44).

The consequence is that the inclination of the parking orbit should be, if possible, greater than the targeted declination. The parking orbit node is to be positioned relative to the meridian of the asymptote using formula (14.49). The targeted value at launch shall furthermore take into account the nodal precession (2.103) between the time of injection into a parking orbit and the time of interplanetary departure.

#### 14.4.1.3 Launch Opportunities

The trajectories of the planets around the Sun are almost circular and very close to the plane of the ecliptic. In the perfect case of coplanar circular orbits, the Hohmann transfer would be the most economical. However, even low inclinations make the Hohmann transfer very expensive, as the transfer plane becomes orthogonal to the initial velocity. The optimal transfer angle results from a compromise between the initial velocity modulus and the angle with the initial orbit. This point has been illustrated in Section 14.2.3.2 with a simplified model.

This inclination effect explains the shape of the contours of the departure velocity for transfers between planets. Figure 14.11 plots the contours of the departure velocity for an Earth–Mars transfer, with departure dates ranging from August 2026 to January 2027 and arrival dates ranging from Mars 2027 to April 2028. Two minima can be observed on either side from a diagonal corresponding to quasi-Hohmann transfers, with the planets in opposite positions. The required departure velocity increases drastically when approaching this diagonal.

These contours help in choosing transfer opportunities. The geometric configuration of the planets $C_1$ and $C_2$ with respect to the Sun is repeated with their synodic period, inducing of translation of the level lines with time.

### 14.4.2 Tisserand Criterion

Fly-bys allow to reduce the propellant consumption, especially for missions to planets further away than Venus or Mars. The search for an "optimal" path poses a combinatorial problem involving the selection of planets and the associated times of fly-by. A useful method to select the most interesting possibilities is based on the Tisserand criterion.

Let us consider a fly-by over a planet in a circular orbit around the Sun. The motion of the spacecraft M subjected to the attraction of the planet P and of the Sun S is a three-body problem studied in Chapter 3. For such a circular restricted three-body problem (CR3BP), the Jacobi integral is constant.

This integral is expressed in the rotating frame RC (Cross frame) associated with the planet's motion around the Sun. This frame shown in Figure 3.1 has for origin the center of

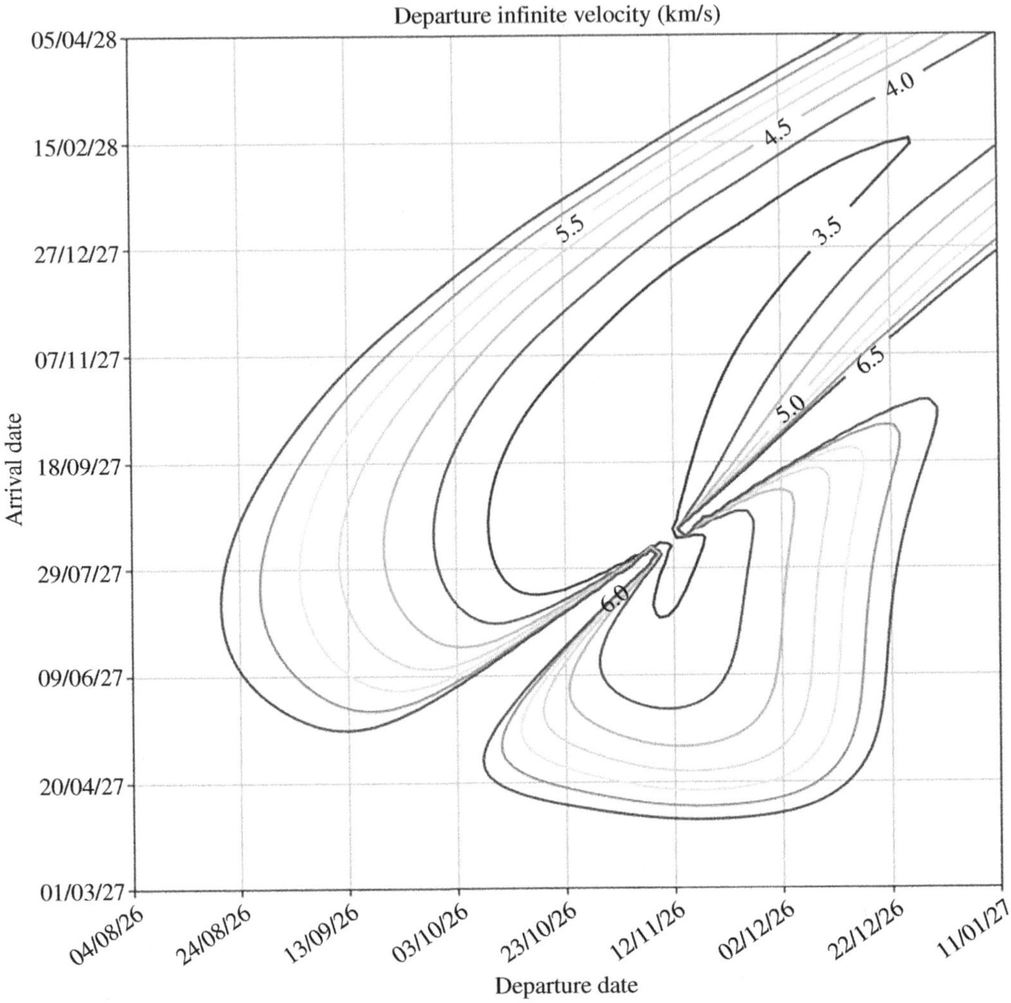


**Figure 14.11** Contours of departure velocity for an Earth–Mars transfer.

mass C of the Sun-planet system. The axis x is along the Sun–planet vector and the axis z is the axis of rotation of the planet around the Sun.

Retrieving formula (3.13), the Jacobi integral is given by

$$2U - v_C^2 = K = Cte \tag{14.50}$$

In this formula, K is Jacobi constant, $\vec{v}_C$ is the vehicle velocity in Cross frame, and U is the potential defined by

$$U = \frac{\mu_S}{r_S} + \frac{\mu_P}{r_P} + \frac{1}{2}(\vec{\Omega}_C \wedge \vec{r}_C)^2 \tag{14.51}$$

where $\mu_S$ and $\mu_P$ are the gravitational constants of the Sun and the planet, $r_S = SM$ is the distance Sun-vehicle, $r_P = PM$ is the distance planet-vehicle, $\vec{r}_C = \overrightarrow{CM}$ is the vehicle position in Cross frame, and $\vec{\Omega}_C$ is the angular velocity of Cross frame.

Tisserand criterion is obtained by passing the Jacobi integral (14.50) from the rotating Cross frame to the inertial heliocentric frame. The vehicle velocity at position $\vec{r}_S$ in the heliocentric frame is noted $\vec{v}_S$. It is the sum of the velocity $\vec{v}_C$ in Cross frame and the drive velocity $\vec{\Omega}_C \wedge \vec{r}_S$ at that position.

$$\vec{v}_S = \vec{v}_C + \vec{\Omega}_C \wedge \vec{r}_S \tag{14.52}$$

Since all planets are of negligible mass with respect to the Sun, we can consider that the center of mass C of the Sun-planet system is the Sun S ($\vec{r}_C \approx \vec{r}_S$). Furthermore, the planet attraction in the potential (14.51) can be neglected before and after the fly-by.

Substituting (14.51) and (14.52) into (14.50) with these approximations, we have

$$2\frac{\mu_S}{r_S} + (\vec{\Omega}_C \wedge \vec{r}_S)^2 - (\vec{v}_S - \vec{\Omega}_C \wedge \vec{r}_S)^2 = K$$

$$\Rightarrow \quad 2\frac{\mu_S}{r_S} - v_S^2 + 2\vec{\Omega}_C \cdot (\vec{r}_S \wedge \vec{v}_S) = K \tag{14.53}$$

The planet's motion is circular of radius d in the ecliptic plane, so that

$$\vec{\Omega}_C = \Omega_C \begin{pmatrix} 0 \\ 0 \\ 1 \end{pmatrix} \quad \text{with } \Omega_C = \sqrt{\frac{\mu_S}{d^3}} \tag{14.54}$$

On the other hand, the orbital parameters of the vehicle in the heliocentric frame are (a, e, i, $\Psi$, $\omega$), where the notation $\Psi$ is used for the right ascension of the ascending node in order not to confuse it with Cross-frame angular velocity.

The angular momentum components in the heliocentric frame are given by (2.22).

$$\vec{h} = \vec{r}_S \wedge \vec{v}_S = h \begin{pmatrix} \sin i \ \sin \Psi \\ -\sin i \ \cos \Psi \\ \cos i \end{pmatrix} \quad \text{with } h = \sqrt{\mu_S a(1 - e^2)} \tag{14.55}$$

Substituting $\vec{\Omega}_C$ (14.54) and $\vec{h}$ (14.55) into (14.53), we obtain

$$2\frac{\mu_S}{r_S} - v_S^2 + 2\sqrt{\frac{\mu_S}{d^3}}\sqrt{\mu_S a(1 - e^2)} \cos i = K \tag{14.56}$$

which using $v_S^2 = \mu_S \left(\frac{2}{r_S} - \frac{1}{a}\right)$ yields Tisserand formula

$$\frac{1}{a} + 2\sqrt{\frac{a(1 - e^2)}{d^3}} \cos i = \frac{K}{\mu_S} \tag{14.57}$$

The first member is a constant (prime integral) along the vehicle motion. It relates the orbital parameters $(a, e, i)_1$ before and $(a, e, i)_2$ after the fly-by. Tisserand applied it to the trajectory of comets deviated by planets, in order to determine whether the object observed before and after the fly-by was indeed the same.

Tisserand formula can be used to elaborate a preliminary interplanetary scenario using multiple fly-bys. For that purpose, we express the semi-major axis a and the eccentricity e as functions of the aphelion $r_A$ and the perihelion $r_P$.

$$a = \frac{r_A + r_P}{2}, \quad e = \frac{r_A - r_P}{r_A + r_P} \quad \Rightarrow \quad p = a(1 - e^2) = \frac{2r_A r_P}{r_A + r_P} \tag{14.58}$$

By defining scaled distances with respect to the distance d of the planet to the Sun

$$\rho_A \underset{\text{def}}{=} \frac{r_A}{d}, \quad \rho_P \underset{\text{def}}{=} \frac{r_P}{d} \tag{14.59}$$

and substituting into (14.57), we obtain an equivalent form of the Tisserand formula

$$\frac{1}{\rho_A + \rho_P} + \sqrt{\frac{2\rho_A\rho_P}{\rho_A + \rho_P}} \cos i = \frac{Kd}{2\mu_S} \underset{\text{noted}}{=} K' \tag{14.60}$$

This formula relates the aphelion and perihelion radii before and after a fly-by. Assuming the trajectory remains in the ecliptic plane ($i = 0$), we can plot contour lines of the first member of (14.60) for different levels of the constant $K'$.

The fly-by is possible only if the perihelion scaled radius $\rho_P$ is lower than 1 and the aphelion scaled radius $\rho_A$ is greater than 1. The limit cases are

- for $\rho_P = \rho_A = 1$ (circular orbit) $\rightarrow$ $K' = 1.5$;
- for $\rho_P = 1$, $\rho_A \rightarrow \infty$ (heliocentric escape) $\rightarrow$ $K' = \sqrt{2}$;
- for $\rho_P = 0$, $\rho_A = 1$ (rectilinear ellipse) $\rightarrow$ $K' = 1$.

Figure 14.12 plots contours of the Tisserand criterion for $K'$ ranging from 0.5 to 1.5.

Each level line in Figure 14.12 shows the possible variation in aphelion and perihelion by repeated fly-bys over a planet, starting from a given initial orbit. A single fly-by generates a step on the level line. This step depends on the arrival conditions. It is assessed as follows.

Let us consider an arrival orbit of aphelion $r_A$ and perihelion $r_P$ on a planet at a distance d from the Sun. The semi-major axis a, eccentricity e and parameter p of the arrival orbit are determined by (14.58) and the true anomaly $\theta$ is obtained from the conic equation.

$$r = \frac{p}{1 + e\cos\theta} = d \Rightarrow \cos\theta = \frac{1}{e}\left(\frac{p}{d} - 1\right) \tag{14.61}$$

This defines the position $\vec{r} = \vec{d}$, where the fly-by takes place.

The velocity $\vec{v}_a$ of the spacecraft on arrival is assessed by formula (1.37).

$$\vec{v}_a = \sqrt{\frac{\mu_S}{p}}\vec{u}_h \wedge (\vec{e} + \vec{u}_r) \text{ with } \vec{u}_r = \frac{\vec{d}}{d}, \quad \vec{e} = e(\cos\theta\,\vec{u}_r - \sin\theta\,\vec{u}_\theta) \tag{14.62}$$

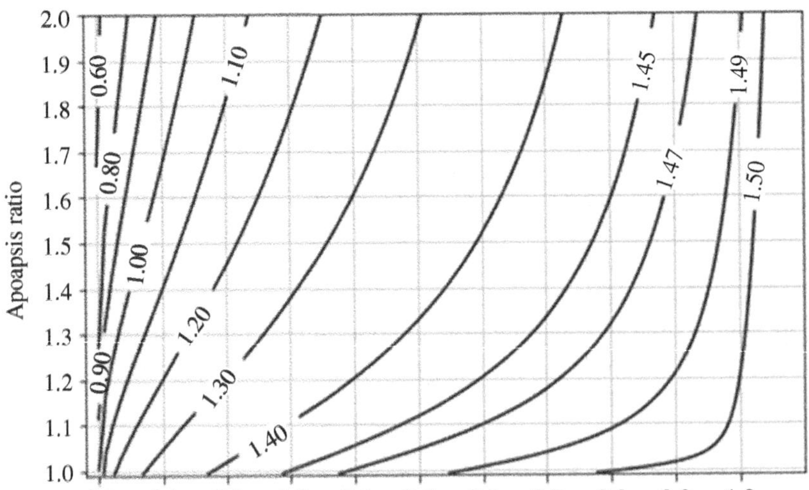


**Figure 14.12** Levels line of the Tisserand criterion.

where $\vec{e}$ is the eccentricity vector of the arrival orbit, $\vec{u}_r$ and $\vec{u}_\theta$ are the radial and orthoradial unit vectors associated with the arrival position and $\vec{u}_h$ is the unit vector normal to the orbital plane.

The velocity $\vec{v}_p$ of the planet in circular orbit of radius d around the Sun is

$$\vec{v}_p = \sqrt{\frac{\mu_S}{d}}\vec{u}_\theta \tag{14.63}$$

and the arrival infinite velocity $\vec{v}_{\infty a}$ is the difference

$$\vec{v}_{\infty a} = \vec{v}_a - \vec{v}_p = \sqrt{\frac{\mu_S}{p}} e \sin\theta\, \vec{u}_r + \left(\sqrt{\frac{\mu_S}{p}}(1 + e \cos\theta) - \sqrt{\frac{\mu_S}{d}}\right)\vec{u}_\theta \tag{14.64}$$

The departure infinite velocity $\vec{v}_{\infty d}$ has the same modulus as $\vec{v}_{\infty a}$.

The fly-by effect is to rotate the infinite velocity by an angle $\varphi$ (see Figure 14.8), which depends on the pericenter radius by formula (14.26). The maximum deviation is obtained for the minimum pericenter radius, which is bounded by the planet radius augmented by the atmosphere height (if any) and margins for navigation inaccuracies.

Assuming a planar fly-by (in the plane of the arrival orbit), the velocity variation is obtained by applying a rotation matrix of angle $\frac{\varphi}{2}$ to $\vec{v}_{\infty a}$ (see Figure 14.8).

$$\Delta\vec{v} = \vec{v}_{\infty d} - \vec{v}_{\infty a} = 2\sin\frac{\varphi}{2}\begin{pmatrix} \cos\frac{\varphi}{2} & \sin\frac{\varphi}{2} & 0 \\ -\sin\frac{\varphi}{2} & \cos\frac{\varphi}{2} & 0 \\ 0 & 0 & 1 \end{pmatrix}\vec{v}_{\infty a} \tag{14.65}$$

Coming back to the heliocentric frame, the departure velocity $\vec{v}_d$ undergoes the same variation.

$$\vec{v}_d = \vec{v}_a + \Delta\vec{v} \tag{14.66}$$

The angular momentum and eccentricity vectors are then calculated for this new velocity $\vec{v}_d$ at the fly-by position $\vec{d}$, from which the new aphelion and perihelion are deduced. These updated apses define the new point on the level line.

This method helps to define interplanetary paths by choosing the most interesting fly-by sequences. However, it does not take planetary ephemeris into account, which necessitates a simulation-based method, as presented in the next section.

### 14.4.3 Scenario Optimization

The scenario of an interplanetary mission is defined by the departure conditions, the sequence of bodies flown over, the times of fly-by, and the arrival conditions. The departure strategy can be a direct launch on an escape orbit or the injection on a low Earth parking orbit before an escape maneuver. The arrival strategy can be a direct descent or an orbital insertion around the targeted body.

During preliminary analyses, the departure time is swept in a large range in order to prepare alternative scenarios in case of a mission delay. The arrival time shall take into account the maximum duration allocated to the mission, which is related to the interplanetary spacecraft design. For missions to Mars with a descent and landing, the arrival must also take place outside of the stormy seasons.

Between the departure at $t_0$ and the arrival at $t_f$, the vehicle may fly over a sequence of n bodies selected from a list of N candidates. The same body can be flown over several times. The numbers of the bodies flown over are noted $(k_1, k_2, \ldots, k_n)$ with $1 \le k_j \le N$, and the fly-bys take place on successive times $(t_0<)\ t_1 < t_2 < \ldots < t_n\ (<t_f)$.

For a given combination of bodies and times $\{n, (k_j)_{j=1,n}, (t_j)_{j=1,n}\}$, the scenario is fully defined. The transfers between successive bodies are Lambert's problems, whose solution determines the infinite velocities at the arrival and departure of each body, and the complementary fly-by impulses $(\Delta v_j)_{j=1,n}$ to be performed. These impulses are assessed by formula (14.38).

The goal is to choose the bodies and times to minimize the sum of the impulses, including the departure $\Delta v_0$ and the arrival $\Delta v_f$ impulses.

$$\min_{n,k_j,t_j} \sum_{j=1}^{n} \Delta v_j + \Delta v_0 + \Delta v_f \tag{14.67}$$

Lambert's transfers can be direct or with several revolutions. These possibilities are taken into account in the resolution, either by adding to the optimization variables the number of revolutions for each transfer, or by scanning a selection of "reasonable" options and keeping the best one for each evaluation of the cost function (14.67).

The optimization problem (14.67) mixes integer and real variables. There are no theoretical optimality conditions for such problems and only approximate solutions can be hoped. This problem is most often addressed by metaheuristics (such as evolutionary algorithms, particle swarms, simulated annealing…) to explore the space of solutions and identify the most interesting scenarios [R13].

In some cases, trajectories returning naturally to Earth without propelled maneuvers (free return) may be desired. These solutions are to be preferred for a manned mission to the Moon or Mars, as they guarantee the return of the crew in the event of a failure after departure into the escape orbit [R22]. A constraint of free return can be taken into account in formulation (14.67) by adding to the path the return arc to Earth and penalizing the associated fly-by impulse.

During a fly-by, the vehicle must be able to communicate with the Earth in order to control the maneuver and correct it if necessary. This communication is not feasible if the planet being flown over is in opposition to the Earth with respect to the Sun. It is therefore necessary to introduce a constraint on the Sun–Earth–planet angle that must be sufficiently large. Optional additional maneuvers, called DSM (Deep Space Maneuver), may also be considered during heliocentric arcs. These maneuvers can reduce the total cost, especially if the duration of the mission is bounded. These additional constraints and maneuvers complicate the optimization problem [R13].

The choice of a path among the many possible solutions is the result of a trade-off between the propellant consumption required by the maneuvers and the duration of the mission. For example, the path of Cassini/Huygens to Saturn/Titan took seven years from 1997 to 2004 with four fly-bys (Venus–Venus–Earth–Jupiter) and the path of Rosetta to the comet Churyumov-Gerasimenko took ten years from 2004 to 2014 with four fly-bys (Earth–Mars–Earth–Earth).

### 14.4.4 Correction Maneuvers

When a mission scenario has been defined, the transfer between the successive planets on the path is assessed by solving Lambert's problems in the heliocentric frame. The impulsive maneuvers are then refined by simulating the trajectory with accurate models of ephemeris and forces. This process results in a reference nominal trajectory.

The actual trajectory of the vehicle may deviate from the nominal trajectory. These deviations are due to maneuver inaccuracies and modeling uncertainties (ephemeris, attraction of solar system bodies, solar radiation pressure …).

Correction maneuvers, called TCM, are required to accurately control the arrival conditions over each planet and obtain the expected gravity assist effect. These maneuvers are calculated by assuming small deviations and by linearizing the equations of motion in the vicinity of the nominal trajectory.

#### 14.4.4.1 Maneuver Calculation

The position and velocity on the nominal path are noted $(\vec{r}_{nom}, \vec{v}_{nom})$ and the corresponding deviations are noted $(\delta\vec{r}, \delta\vec{v})$.

The acceleration along a heliocentric arc is mainly due to the Sun's gravity and it depends on the position. This acceleration noted $\vec{g}(\vec{r})$ is expanded to first order as:

$$\vec{g}(\vec{r}_{nom} + \delta\vec{r}) \approx \vec{g}(\vec{r}_{nom}) + G\delta\vec{r} \quad \text{with } G = \frac{\partial\vec{g}}{\partial\vec{r}} \tag{14.68}$$

where G is the gravity gradient matrix.

For a central gravity $\vec{g} = -\frac{\mu}{r^3}\vec{r}$, this matrix given by (2.86) is $G = \frac{\mu}{r^5}(3\vec{r}\vec{r}^t - r^2 I)$.

The differential equations of the position-velocity deviations are obtained by expanding the equations of motion to first order with the matrix G calculated on the nominal trajectory. These variational equations are

$$\begin{cases} \dot{\vec{r}} = \vec{v} \\ \dot{\vec{v}} = \vec{g} \end{cases} \Rightarrow \begin{cases} \delta\dot{\vec{r}} = \delta\vec{v} \\ \delta\dot{\vec{v}} = G\delta\vec{r} \end{cases} \Rightarrow \begin{pmatrix} \delta\dot{\vec{r}} \\ \delta\dot{\vec{v}} \end{pmatrix} = \begin{pmatrix} 0 & I \\ G & 0 \end{pmatrix} \begin{pmatrix} \delta\vec{r} \\ \delta\vec{v} \end{pmatrix} \underset{noted}{=} F \begin{pmatrix} \delta\vec{r} \\ \delta\vec{v} \end{pmatrix} \tag{14.69}$$

The solution of this linear differential system of matrix F(t) is expressed with its transition matrix $\Phi(t, t_0)$. The deviations $(\delta\vec{r}, \delta\vec{v})$ at time t depend on the initial deviations $(\delta\vec{r}_0, \delta\vec{v}_0)$ at time $t_0$ by

$$\begin{pmatrix} \delta\vec{r} \\ \delta\vec{v} \end{pmatrix} = \Phi(t, t_0) \begin{pmatrix} \delta\vec{r}_0 \\ \delta\vec{v}_0 \end{pmatrix} \tag{14.70}$$

The transition matrix satisfies the differential equation

$$\dot{\Phi}(t, t_0) = F(t)\Phi(t, t_0) \quad \text{with } \Phi(t_0, t_0) = I, \quad F(t) = \begin{pmatrix} 0 & I \\ G[\vec{r}_{nom}(t)] & 0 \end{pmatrix} \tag{14.71}$$

In the case of complex dynamics, this equation must be numerically integrated to obtain the transition matrix over time.

In the case of a two-body problem, the transition matrix can be calculated analytically by deriving the Lagrange coefficients presented in Section 1.3.4.

$$\begin{cases} \vec{r} = f_r\,\vec{r}_0 + g_r\,\vec{v}_0 \\ \vec{v} = f_v\,\vec{r}_0 + g_v\,\vec{v}_0 \end{cases} \tag{14.72}$$

The Lagrange coefficients are noted here $(f_r, g_r, f_v, g_v)$ in order not to be confused with the F and G matrices in (14.69). These coefficients are analytic functions of the position, the velocity and the orbital parameters. By deriving them with respect to $(\vec{r}_0, \vec{v}_0)$, the components of the transition matrix are obtained. This tedious calculation is detailed in [R1] and not reproduced here.

The resulting transition matrix can be partitioned with respect to position and velocity as:

$$\Phi(t,t_0)=\begin{pmatrix}\dfrac{\partial\vec{r}}{\partial\vec{r}_0} & \dfrac{\partial\vec{r}}{\partial\vec{v}_0}\\ \dfrac{\partial\vec{v}}{\partial\vec{r}_0} & \dfrac{\partial\vec{v}}{\partial\vec{v}_0}\end{pmatrix}\underset{\text{noted}}{=}\begin{pmatrix}\Phi_{rr} & \Phi_{rv}\\ \Phi_{vr} & \Phi_{vv}\end{pmatrix} \tag{14.73}$$

The transition matrix has several properties that make it easier to use.

- Composition : $\Phi(t_2,t_1)\Phi(t_1,t_0)=\Phi(t_2,t_0)$
- Inversion : $\Phi^{-1}(t_2,t_1)=\Phi(t_1,t_2)$
- Symplectic : $\Phi^t J\Phi = J$ with $J=\begin{pmatrix}0 & I\\ -I & 0\end{pmatrix}$

The two first properties are obvious. The third one is demonstrated by deriving $\Phi^t(t,t_0)\,J\,\Phi(t,t_0)$ and using (14.71).

$$\frac{d}{dt}(\Phi^t J\Phi)=\Phi^t F^t J\Phi+\Phi^t JF\Phi=0 \text{ since } F^t J=\begin{pmatrix}-G^t & 0\\ 0 & I\end{pmatrix}=-JF \tag{14.74}$$

Since the derivative of $\Phi^t(t,t_0)\,J\,\Phi(t,t_0)$ is zero, this matrix is constant equal to its initial value: $\Phi^t(t,t_0)\,J\,\Phi(t,t_0)=\Phi^t(t_0,t_0)\,J\,\Phi(t_0,t_0)=J$.

This property makes it possible to calculate the inverse by $\Phi^{-1} = -J\Phi^t J$, which gives considering the form (14.73)

$$\Phi^{-1}(t,t_0)=\Phi(t_0,t)=\begin{pmatrix}\Phi^t_{vv} & -\Phi^t_{rv}\\ -\Phi^t_{vr} & \Phi^t_{rr}\end{pmatrix} \tag{14.75}$$

The two main uses of the transition matrix are the following:

- Predictions of deviations resulting from initial errors (for navigation)

  The covariance matrix of the error is defined by

$$P=E\left[\begin{pmatrix}\delta\vec{r}\\ \delta\vec{v}\end{pmatrix}\begin{pmatrix}\delta\vec{r}\\ \delta\vec{v}\end{pmatrix}^t\right] \tag{14.76}$$

  With the deviations (14.53), we get the expression of P(t) as a function of $P(t_0)$

$$P(t)=\Phi(t,t_0)P(t_0)\Phi(t,t_0)^t \tag{14.77}$$

- Calculation of correction maneuvers to cancel the final error (for guidance)

  Deviations $(\delta\vec{r},\delta\vec{v})$ at some time t will induce a final position deviation given by $\delta\vec{r}_f=\Phi_{rr}(t_f,t)\delta\vec{r}+\Phi_{rv}(t_f,t)\delta\vec{v}$.

  To cancel this final deviation, the impulse maneuver to be performed at time t is

$$\delta\vec{v}_c=-\Phi_{rv}^{-1}(t_f,t)\Phi_{rr}(t_f,t)\delta\vec{r}-\delta\vec{v} \tag{14.78}$$

  The arrival time $t_f$ is similarly controlled. This time of arrival on a planet can be optimized to reduce the fly-by impulse required. It can also help control the precise time of passage at the planet's pericenter in order to satisfy lightening constraints. Such constraints may be imposed, either for a braking maneuver to insert the spacecraft into a stable orbit around the planet, or for a direct descent and landing on the planet's ground.

#### 14.4.4.2 Times of Maneuvers

The times of the corrective maneuvers should be chosen to minimize the propellant consumption.

A simple model consists of assuming that the vehicle can command a thrust acceleration $a_T$ comprised between 0 and $a_m$, directed along the unit vector $\vec{u}$. Starting from initial deviations $(\delta\vec{r}_0, \delta\vec{v}_0)$ at time $t_0$, the goal is to cancel the final deviation at time $t_f$ by minimizing the acceleration integral, which is equivalent to the total impulse delivered.

$$\min_{a,\vec{u}} \int_{t_0}^{t_f} a dt \quad \text{s.t.} \quad \begin{cases} \delta\dot{\vec{r}} = \delta\vec{v}, & \delta\vec{r}(t_f) = \vec{0} \\ \delta\dot{\vec{v}} = G\delta\vec{r} + a\vec{u}, & \delta\vec{v}(t_f) = \vec{0} \end{cases} \tag{14.79}$$

This optimal control problem is solved by applying Pontryaguin Maximum Principle (see [R9], [R38]). The costate vector has components $(\vec{p}_r, \vec{p}_v)$ associated with the state vector $(\delta\vec{r}, \delta\vec{v})$ and follows the costate equations.

$$\begin{cases} \dot{\vec{p}}_r = -G\vec{p}_v \\ \dot{\vec{p}}_v = -\vec{p}_r \end{cases} \tag{14.80}$$

The Hamiltonian of the problem is

$$H = a + \vec{p}_r \cdot \delta\vec{v} + \vec{p}_v \cdot (G\delta\vec{r} + a\vec{u}) \tag{14.81}$$

The minimization of H with respect to the unit vector $\vec{u}$ leads to

$$\min_{\vec{u}} H \Rightarrow \vec{u} = -\frac{\vec{p}_v}{p_v} \Rightarrow H = a(1 - p_v) + \vec{p}_r \cdot \delta\vec{v} + \vec{p}_v \cdot G\delta\vec{r} \tag{14.82}$$

The minimization of H with respect to the modulus $a_T$ leads to

$$\min_{a_T} H \Rightarrow a_T = \begin{cases} 0 & \text{if } p_v < 1 \\ a_m & \text{if } p_v > 1 \end{cases} \tag{14.83}$$

The maneuvers are thus determined by the modulus of the velocity costate.

For a trajectory leg significantly shorter than the orbital period, the gravitational field $g(\vec{r})$ may be considered as constant and its gradient is zero: $G = 0$.

The costate equations (14.80) can then be integrated to give

$$\begin{cases} \vec{p}_r = \vec{p}_{r0} \\ \vec{p}_v = \vec{p}_{v0} - (t - t_0)\vec{p}_{r0} \end{cases} \tag{14.84}$$

The evolution of $p_v$ is then analyzed by deriving $p_v^2 = \vec{p}_v \cdot \vec{p}_v$.

$$\dot{p}_v p_v = -\vec{p}_r \cdot \vec{p}_v = (t - t_0)p_{r0}^2 - \vec{p}_{r0} \cdot \vec{p}_{v0} \tag{14.85}$$

If the derivative $\dot{p}_v$ is canceled at a time $t_1$, it is necessarily negative before $t_1$ and positive after $t_1$. The costate modulus $p_v$ may be higher than 1 only at the beginning and/or at the end of the trajectory. Corrective maneuvers are thus optimal at the beginning and/or at the end of the trajectory [R2]. In practice, it is necessary to ensure that the vehicle does not deviate too much from the nominal trajectory and corrective maneuvers shall be carried out regularly along the path, as soon as the deviations exceed acceptable tolerances.

## 14.5 Conclusion

### 14.5.1 The Key Points

- The sphere of influence of a planet is the region where it is the main attracting body, which allows us to consider a two-body problem.

- The gravity assist is the free velocity change observed with respect to the Sun when flying over a planet.
- The patched conic method decomposes the interplanetary path as a succession of heliocentric elliptical and planetocentric hyperbolic conics.
- The heliocentric phases are solved as Lambert's problem between planets, whereas the planetocentric phases are solved by imposing the fly-by maneuver at the pericenter.
- The optimization of the mission scenario mixes integer variables, which are the planets to fly over, and continuous variables, which are the fly-by times.

### 14.5.2 To Go Further

- CONWAY Bruce A. – *Spacecraft Trajectory Optimization*. Cambridge University Press (2010).

Several chapters of this book are devoted to interplanetary trajectories. Chapters 7 and 8 presents metaheuristics applied to multiple gravity assist optimization. Chapters 9 and 10 deal with low-thrust transfers applied to lunar missions.

- CHOBOTOV Vladimir – *Orbital Mechanics (Third edition)*. AIAA (2002).

Chapter 12 introduces interplanetary and lunar trajectories, with examples of real missions.

- KEMBLE Stephen – *Interplanetary Mission Analysis and Design*. Springer, Praxis Publishing (2006).

This comprehensive book covers all aspects of interplanetary trajectories, from the basics of dynamics to advanced topics such as transfers in the three-body problems or optimization using low-thrust propulsion. Numerous examples are presented for all kinds of missions in the solar system.

- LABUNSKY Alexei V., PAPKOV Oleg V., SUKHANOV Konstantin G. *Multiple Gravity Assist Interplanetary Trajectories*. Gordon and Breach Science Publishers (1998).

This book presents some very interesting methods for developing preliminary scenarios for complex interplanetary missions. These methods are mostly based on graphical analysis, with very few calculations.

- ROY Archie Edmiston – *Orbital Motion*. Adam Hilger (1988).

Chapter 12 introduces interplanetary and lunar trajectories.

# 15

# Atmospheric Reentry

## 15.1 Introduction

Some space missions include a reentry phase. This is particularly the case for manned flights (return of astronauts from the space station), for exploration probes landing on the surface of planets with an atmosphere (Mars and Titan) or for the recovery of reusable launcher stages. Depending on the case, the trajectory may end at sea or on land, vertically or horizontally.

The reentry goal is to dissipate the initial kinetic energy, of the order of 25–30 MJ/kg in orbit, in order to land on the ground at near zero velocity. The use of propulsion means would lead to excessive propellant masses, incompatible with the size of an orbiting spacecraft. The most economical solution is to use aerodynamic braking and to dissipate the energy in the form of heat absorbed by the thermal protection. The aerodynamic capabilities of the vehicle make it possible to control the rate of descent in order to regulate the thermal and mechanical loads due to the atmosphere pressure. The trajectory must be compatible with the vehicle's thermal and mechanical design, while arriving at the landing site with a low residual velocity. These constraints are closely linked to the conditions at the reentry beginning, which must be specifically adjusted.

The first section establishes the general motion equations in spherical coordinates in a rotating Earth frame, which is best suited to the analysis of reentry performances.

The second section concerns the deorbitation maneuver and the orbital arc prior to reentry, with their analytical solutions.

The third section derives simplified analytical models that allow to estimate the thermal and mechanical load levels, as well as the accessible ground area depending on the vehicle's aerodynamical characteristics. These models are useful in preliminary design phases and during the preparation of mission scenarios.

## 15.2 Motion Equations

### 15.2.1 Reentry Missions and Vehicles

The goal of an atmospheric reentry is to recover a spacecraft safely on the ground. This may involve the return of a crew from a space station (Figure 15.1), the reuse of a launcher stage (Figure 15.2), or the retrieval of an interplanetary probe.

The gliding capabilities of a reentry vehicle depends mainly on its **lift-to-drag ratio** noted L/D (aerodynamic lift divided by drag). A null L/D corresponds to a ballistic object with no

*Space Trajectories: Basic and Advanced Topics*, First Edition. Max Cerf.

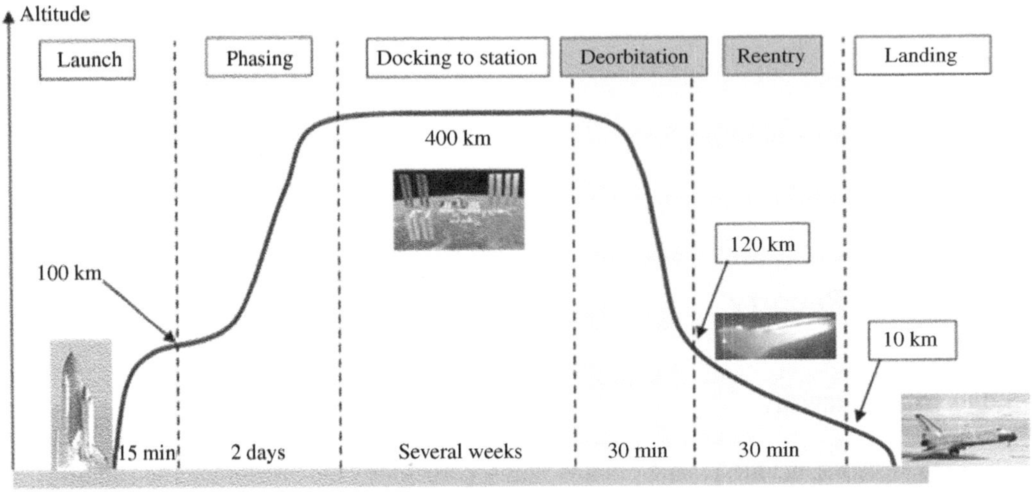


**Figure 15.1** Inhabited mission toward a space station credit-Jiamin Zhu et.al., (2017)/Springer Nature/CC BY 4.0.

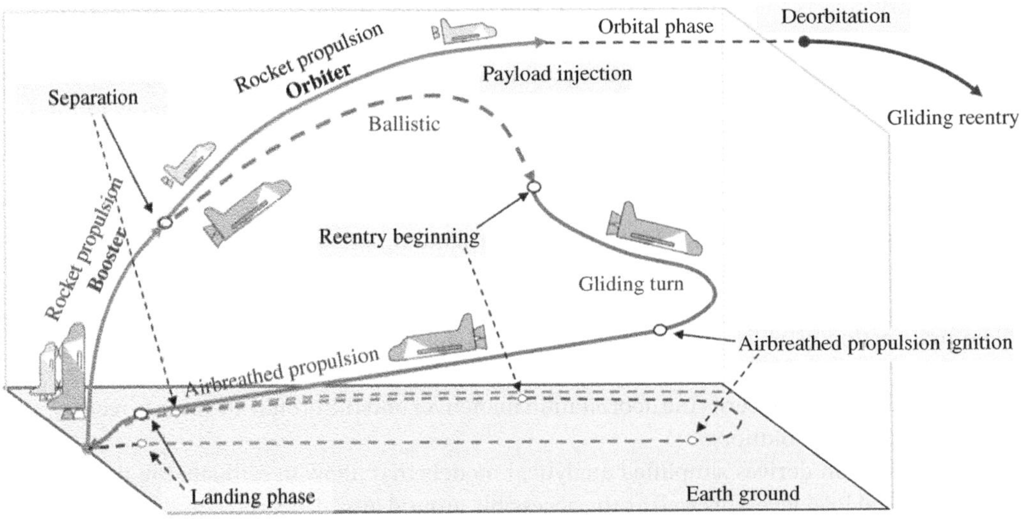


**Figure 15.2** Recovery of reusable launcher stages.

maneuvering capability. The best gliders may have an L/D up to 70, to be compared to 20 for an Airbus A380. Figure 15.3 shows the three main shapes of reentry vehicles with their associated L/D ratio.

Capsule (L/D = 0.3) Lifting body (L/D = 0.7) Space shuttle (L/D = 1.5)

In order to land safely, the total kinetic energy at the atmosphere entry must be dissipated as heat. The initial velocity is about 5 km/s for a suborbital trajectory, 8 km/s for a return from a low Earth orbit, and up to 12 km/s for an interplanetary mission return. The entry slope must be carefully controlled in order to both ensure a sufficient atmospheric braking while limiting heat and mechanical loads. The position of the entry point is chosen according to the targeted return site and the flight distance that can be achieved by the gliding vehicle.

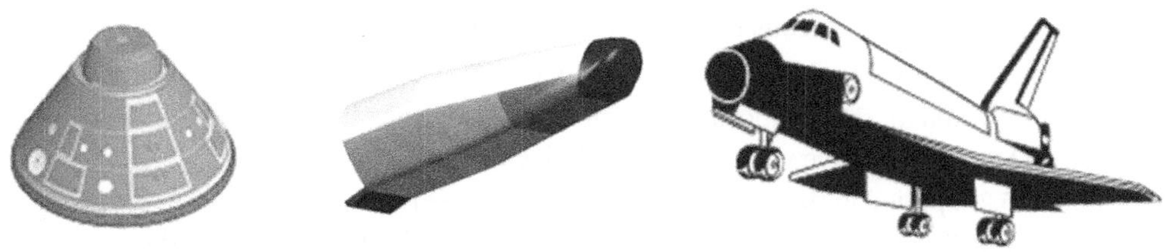

Capsule (L/D = 0.3) Lifting body (L/D = 0.7) Space shuttle (L/D = 1.5)

**Figure 15.3** Lift-to-drag ratio of space vehicles.

During an atmospheric reentry, the physical quantities of interest depend on the vehicle's kinematics relative to the Earth and the atmosphere. This leads us to write the motion equations in spherical coordinates in the rotating geocentric frame.

## 15.2.2 Rotating Geocentric Frame

### 15.2.2.1 Galilean (or Inertial) Frame

The equations of motion are given by Newton's second law in a Galilean frame (also called inertial or absolute frame). For an atmospheric reentry, the reference inertial frame, noted RA, has its origin O at the Earth's center, its axis $\vec{z}_A$ along the polar axis directed northward and the axes $\vec{x}_A$ and $\vec{y}_A$ in the equatorial plane. The plane $(\vec{x}_A, \vec{z}_A)$ is the Greenwich meridian at the reentry initial time. The position $\vec{r}_A$ and velocity $\vec{v}_A$ in the inertial frame RA are said "absolute."

Since no propulsion is used, the vehicle mass m is assumed to be constant during the reentry. The only forces acting on the vehicle are the weight $m\vec{g}$ and the aerodynamic force $\vec{F}_a$. The equations of motion in RA are

$$\left.\frac{d\vec{r}_A}{dt}\right|_{RA} = \vec{v}_A \; , \quad \left.\frac{d\vec{v}_A}{dt}\right|_{RA} = \vec{g} + \frac{\vec{F}_a}{m} \tag{15.1}$$

where the notation $\left.\frac{d\vec{u}}{dt}\right|_{R1}$ represents the derivative of the vector $\vec{u}$ with respect to the frame $R_1$.

These two vector equations are called the **kinematic equation** (relation between position derivative and velocity) and the **dynamic equation** (relation between velocity derivative and forces), respectively.

### 15.2.2.2 Rotating Frame

The analysis of an atmospheric reentry is eased by using a rotating frame linked to the Earth, noted RT. This frame RT rotates with respect to RA with the angular rate $\vec{\Omega}_T$ around the pole axis $\vec{z}_A$ (Figure 15.4).

The position $\vec{r}_R$ and velocity $\vec{v}_R$ in RT are referred to as "relative." Since the frames RA and RT have the same origin O, the absolute and relative position vectors coincide and they will be noted as $\vec{r} = \vec{r}_A = \vec{r}_R$.

The equations of motion in RT are obtained by applying to equations (15.1) the frame change formula

$$\left.\frac{d\vec{u}}{dt}\right|_{R1} = \left.\frac{d\vec{u}}{dt}\right|_{R2} + \vec{\Omega}_{R2/R1} \wedge \vec{u} \tag{15.2}$$

where $\vec{\Omega}_{R2/R1}$ is the instantaneous rotation rate of $R_2$ relative to $R_1$.

Formula (15.2) applied to the position vector gives: $\left.\frac{d\vec{r}}{dt}\right|_{RA} = \left.\frac{d\vec{r}}{dt}\right|_{RT} + \vec{\Omega}_T \wedge \vec{r}$.

This is the formula for the composition of velocities, which can be written as: $\vec{v}_A = \vec{v}_R + \vec{v}_E$, where $\vec{v}_E = \vec{\Omega}_T \wedge \vec{r}$ is the drive velocity of RT relative to RA.

Formula (15.2) applied to the velocity gives: $\left.\frac{d\vec{v}_A}{dt}\right|_{RA} = \left.\frac{d\vec{v}_A}{dt}\right|_{RT} + \vec{\Omega}_T \wedge \vec{v}_A$.

By substituting $\vec{v}_A = \vec{v}_R + \vec{v}_E$ into the second term, we obtain the formula for the composition of accelerations, which can be written as: $\vec{a}_A = \vec{a}_R + \vec{a}_E + \vec{a}_C$, where:

- $\vec{a}_A = \left.\frac{d\vec{v}_A}{dt}\right|_{RA}$ and $\vec{a}_R = \left.\frac{d\vec{v}_R}{dt}\right|_{RT}$ are the absolute and relative acceleration;
- $\vec{a}_E = \dot{\vec{\Omega}}_T \wedge \vec{r} + \vec{\Omega}_T \wedge (\vec{\Omega}_T \wedge \vec{r})$ is the **drive** acceleration;
- $\vec{a}_C = 2\vec{\Omega}_T \wedge \vec{v}_R$ is the **Coriolis** acceleration, depending on the velocity in RT.

The angular rotation rate of the Earth is constant ($\Omega_T = 7.2921 \cdot 10^{-5}$ rad/s).

The drive acceleration $\vec{a}_E$ generates an upward deviation (centrifugal effect).

The Coriolis acceleration $\vec{a}_C$ generates a deviation to the right in the northern hemisphere and to the left in the southern hemisphere.

The motion equations (15.1) passed in the rotating frame RT are finally

$$\left.\frac{d\vec{r}}{dt}\right|_{RT} = \vec{v}_R, \quad \left.\frac{d\vec{v}_R}{dt}\right|_{RT} = \vec{g} + \frac{\vec{F}_a}{m} - \vec{a}_E - \vec{a}_C \tag{15.3}$$

These differential equations can be projected into Cartesian coordinates in RT and integrated by a numerical method.

### 15.2.3 Spherical Coordinates

The choice of spherical coordinates for position and velocity allows a physical analysis of motion and the development of simplified analytical models associated with the different phases of reentry. These spherical coordinates involve the local geographic frame RG and the kinematic frame RC depicted in Figure 15.4.

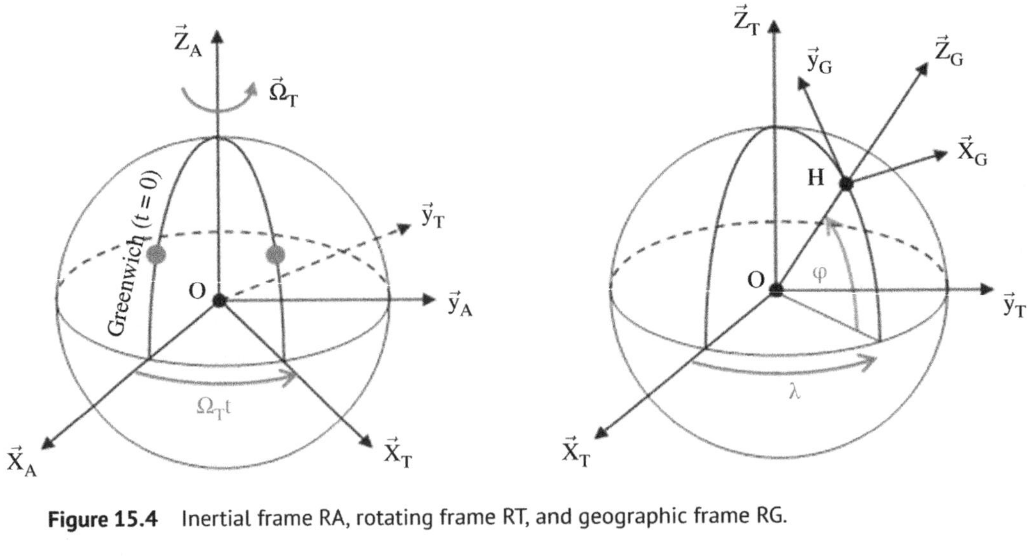


**Figure 15.4** Inertial frame RA, rotating frame RT, and geographic frame RG.

#### 15.2.3.1 Geographic Frame

The local geographic frame RG is based on the current vehicle position $\vec{r}$. The axis $\vec{z}_G$ is vertical along $\vec{r}$, whereas the axes $\vec{x}_G$ to the east and $\vec{y}_G$ to the north form the local horizontal plane.

The spherical coordinates of position are the radius r (distance to the Earth center), the longitude λ (angle between Greenwich meridian and the vehicle meridian), and the geocentric latitude φ (angle between the equator and the vehicle position).

The transition from RT to RG is achieved by three successive rotations of angle $\frac{\pi}{2}+\lambda$ around $\vec{z}_T$, then angle $-\varphi$ around $\vec{x}_G$, which yields

$$\begin{cases} \vec{x}_T = -\sin\lambda\,\vec{x}_G - \cos\lambda\,\sin\varphi\,\vec{y}_G + \cos\lambda\,\cos\varphi\,\vec{z}_G \\ \vec{y}_T = \cos\lambda\,\vec{x}_G - \sin\lambda\,\sin\varphi\,\vec{y}_G + \sin\lambda\,\cos\varphi\,\vec{z}_G \\ \vec{z}_T = \cos\varphi\,\vec{y}_G + \sin\varphi\,\vec{z}_G \end{cases} \tag{15.4}$$

#### 15.2.3.2 Kinematic Frame

The kinematic frame RC (Figure 15.5) is based on the relative velocity $\vec{v}_R$. The axis $\vec{x}_C$ is along $\vec{v}_R$, the axis $\vec{z}_C$ is upward in the local vertical plane $(\vec{z}_G, \vec{v}_R)$, the axis $\vec{y}_C$ is to the left of the velocity in the local horizontal plane. The spherical coordinates of velocity are the modulus $v_R$ of relative velocity, the local slope γ (angle between the horizontal plane and the velocity, also called flight path angle), and the local azimuth χ (angle between the north $\vec{y}_G$ and the projection $\vec{x}_H$ of the velocity on the horizontal plane). By convention, the azimuth is positive eastward.

The transition from RG to RC is achieved by three successive rotations of angle $\frac{\pi}{2}-\chi$ around $\vec{z}_G$ , then angle $-\gamma$ around $\vec{y}_C$, which yields

$$\begin{cases} \vec{x}_G = \sin\chi\,\vec{x}_H - \cos\chi\,\vec{y}_C \\ \vec{y}_G = \cos\chi\,\vec{x}_H + \sin\chi\,\vec{y}_C \\ \vec{z}_G = \sin\gamma\,\vec{x}_C + \cos\gamma\,\vec{z}_C \end{cases} \tag{15.5}$$

$$\text{with} \begin{cases} \vec{x}_H = \cos\gamma\,\vec{x}_C - \sin\gamma\,\vec{z}_C \\ \vec{x}_C = \cos\gamma\sin\chi\,\vec{x}_G + \cos\gamma\cos\chi\,\vec{y}_G + \sin\gamma\,\vec{z}_G \end{cases}$$

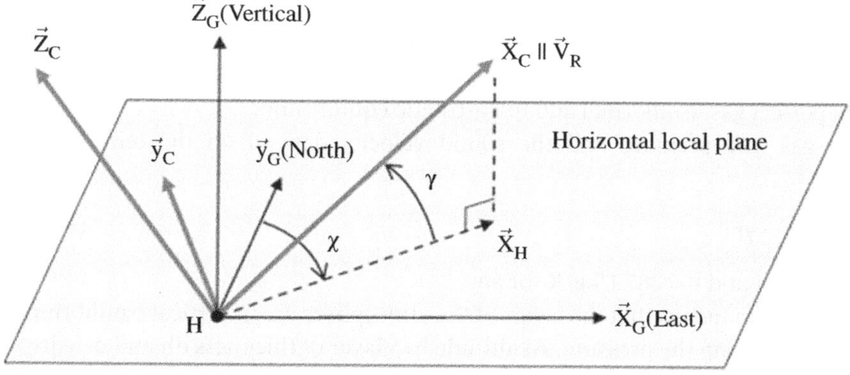


**Figure 15.5** Kinematic frame RC.

### 15.2.4 Forces Applied

This section details the models used for the calculation of the gravity, the aerodynamic force and the atmosphere parameters.

#### 15.2.4.1 Gravity

The gravity acceleration $\vec{g}$ derives from the gravitational potential $U(r, \varphi, \lambda)$. The components of $\vec{g}$ in RG are obtained by considering a variation of the spherical coordinates $(dr, d\varphi, d\lambda)$ producing an elementary move $d\vec{r}$.

$$\begin{cases} \vec{g} = g_r \vec{x}_G + g_\varphi \vec{y}_G + g_\lambda \vec{z}_G \\ d\vec{r} = r\cos\varphi\, d\lambda\, \vec{x}_G + r\, d\varphi\, \vec{y}_G + dr\, \vec{z}_G \end{cases} \tag{15.6}$$

The variation of the potential is: $dU(r, \phi, \lambda) = \frac{\partial U}{\partial r}dr + \frac{\partial U}{\partial \phi}d\phi + \frac{\partial U}{\partial \lambda}d\lambda \underset{\text{def}}{=} -\vec{g}\cdot d\vec{r}$.

By identifying the terms in $(dr, d\varphi, d\lambda)$, we obtain the components $(g_r, g_\varphi, g_\lambda)$ as functions of the partial derivatives of U.

$$\vec{g} = -\frac{\partial U}{\partial r}\vec{z}_G - \frac{1}{r}\frac{\partial U}{\partial \varphi}\vec{y}_G - \frac{1}{r\cos\varphi}\frac{\partial U}{\partial \lambda}\vec{x}_G \tag{15.7}$$

For an atmospheric reentry, the Earth can be approximated by an ellipsoid of revolution flattened at the poles and of equatorial radius $R_T = 6\,378\,137$ m.

Its gravitational potential expressed in spherical coordinates is (2.98)

$$U = -\frac{\mu_T}{r}\left[1 - \frac{1}{2}J_2\left(\frac{R_T}{r}\right)^2(3\sin^2\varphi - 1)\right] \tag{15.8}$$

where $\mu_T = 3.986005 \cdot 10^{14}$ m$^3$/s$^2$ is the Earth's gravitational constant and $J_2 = 0.001\,082\,63$ is the first zonal coefficient.

The perturbation in $J_2$ due to Earth flattening is about a thousand times smaller than the central term and it is directed toward the equatorial bulge.

By applying (15.7) with the potential (15.8), we obtain the acceleration of gravity components in RG.

$$\vec{g} = -\frac{\mu_T}{r^2}\vec{z}_G + 3J_2\frac{\mu_T}{r^2}\left(\frac{R_T}{r}\right)^2\left[\frac{1}{2}(3\sin^2\varphi - 1)\,\vec{z}_G - \sin\varphi\cos\varphi\vec{y}_G\right] \tag{15.9}$$

#### 15.2.4.2 Atmosphere

The aerodynamic force depends on the parameters of the atmosphere: density $\rho$, static pressure $P_s$, temperature T and sound velocity c. The simplest model assumes that the atmosphere is a perfect gas, isothermal and in barostatic equilibrium.

For a perfect gas, the pressure and the sound velocity depend on the temperature according to

$$P_s = \rho r T, \quad c^2 = \gamma r T \tag{15.10}$$

with constants: $\gamma = 1.4$ and $r = 287$ J/kg/K for air.

Barostatic equilibrium means that the layers of the atmosphere are in vertical equilibrium, with their weight balancing the pressure. At altitude h, a layer of thickness dh and density $\rho$ is subject to an upward pressure $P_s(h)$ on its lower face and a downward pressure $P_s(h + dh)$ on its upper face.

The vertical equilibrium yields

$$P_s(h) - P_s(h + dh) - \rho g dh = 0 \quad \Rightarrow \quad dP_s = -\rho g dh \tag{15.11}$$

By substituting $P_s$ given by (15.10) and assuming constant gravity and constant temperature in the atmosphere, the resulting equation in $\rho$ can be integrated.

We obtain the **exponential atmosphere model**

$$\rho(h) = \rho_0 e^{-\frac{h}{H}} \quad \text{with} \quad H = \frac{rT}{g} \tag{15.12}$$

where $\rho_0 = 1.225\,\text{kg/m}^3$ is the ground density and $H = 8400\,\text{m}$ is the height scale at ground. These constants can be adjusted for a nonisothermal Earth atmosphere by taking $\rho_0 = 1.752\,\text{kg/m}^3$ and $H = 6700\,\text{m}$.

The exponential model (15.12) gives an approximation of the atmospheric parameters. More accurate evaluations are based on tabulated models derived from theoretical equations with corrections from measurements. The atmospheric density depends on many parameters: geodetic altitude, geographical location, date (season), time (day/night), solar activity, magnetic field… A precise estimation is difficult because many physical phenomena are poorly known, especially at high altitudes. More and more comprehensive atmospheric models have been developed since the 1960s, like USSA76, JACCHIA77, CIRA90, GRAM99, and NRLMSIS00.

#### 15.2.4.3 Aerodynamics

The aerodynamic force $\vec{F}_a$ depends on the vehicle's shape, velocity and attitude relative to the atmosphere. The aerodynamic velocity noted $\vec{w}_R$ (relative to the atmosphere) is the sum of the relative velocity $\vec{v}_R$ (relative to the Earth) and the wind $\vec{w}$. The vehicle frame RE (Figure 15.6) is defined by the longitudinal or roll axis $\vec{x}_E$, the normal or yaw axis $\vec{z}_E$, and the lateral or pitch axis $\vec{y}_E$.

Three assumptions are made to simplify the expression of the aerodynamic force and the trajectory calculation, while maintaining a good representativeness:

- the pitch plane $(\vec{x}_E, \vec{z}_E)$ is a plane of aerodynamic symmetry of the vehicle;
- the wind velocity $\vec{w}$ is null, so that: $\vec{w}_R = \vec{v}_R$;
- the velocity $\vec{v}_R$ is in the plane $(\vec{x}_E, \vec{z}_E)$, assuming a null sideslip angle $\beta$.

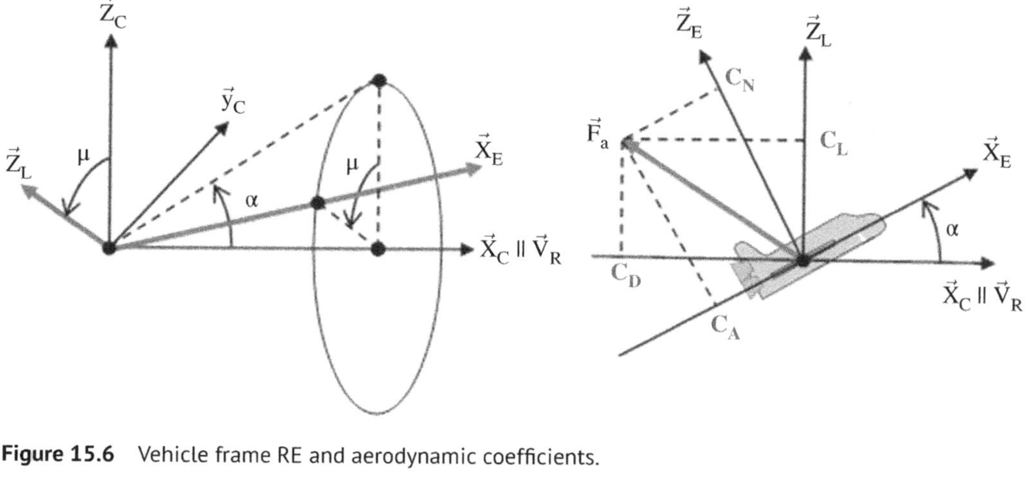


**Figure 15.6** Vehicle frame RE and aerodynamic coefficients.

Under these assumptions the aerodynamic resultant is in the pitch plane, without lateral component along $\vec{y}_E$.

The transition from the RC to the RE frame is achieved by the bank rotation of angle μ around $\vec{x}_C$ and the incidence (or angle of attack) rotation of angle α around $-\vec{y}_E$. In the general case, a third sideslip rotation β around $\vec{z}_E$ must be added. We assume in the sequel that the sideslip angle is zero.

The bank rotation defines the plane $(\vec{x}_C, \vec{z}_L)$, which contains the aerodynamic force $\vec{F}_a$. The components of $\vec{F}_a$ along the axes $-\vec{x}_C$ and $\vec{z}_L = \cos\mu\,\vec{z}_C - \sin\mu\,\vec{y}_C$ are the **drag** D and the **lift** L, respectively.

$$\vec{F}_a = -D\vec{x}_C + L\vec{z}_L \quad \text{with} \quad \begin{cases} D = P_d S C_D \\ L = P_d S C_L \end{cases}, \quad P_d = \frac{1}{2}\rho v_R^2 \tag{15.13}$$

The **dynamic pressure** $P_d$ (in Pa) depends on the atmospheric density and on the velocity relative to the atmosphere. It measures the magnitude of aerodynamic forces. The aerodynamic reference surface S is a wet surface representative of the vehicle shape (it can be freely chosen).

The drag $C_D$ and lift $C_L$ coefficients are dimensionless. They depend in first approximation on the shape of the vehicle (in particular on the deflection δ of the aerodynamic control surfaces if the vehicle is aerodynamically controlled), on its velocity via the **Mach number** $M = \frac{v_R}{c}$ (where c is the sound velocity at the current altitude) and on its angle of attack α in the external flow. The aerodynamic coefficients are derived from theoretical models tested and calibrated in wind tunnels and they are presented in the form of tables $C_D(M, \alpha, \delta)$ and $C_L(M, \alpha, \delta)$.

The projection of $\vec{F}_a$ on vehicle axes $(\vec{x}_E, \vec{z}_E)$ is obtained by a rotation of the angle of attack α (Figure 15.6). The coefficients along $-\vec{x}_E$ and $\vec{z}_E$ are noted $C_A$ (axial) and $C_N$ (normal) respectively.

$$\begin{cases} C_A = C_D \cos\alpha - C_L \sin\alpha \\ C_N = C_D \sin\alpha + C_L \cos\alpha \end{cases} \tag{15.14}$$

The lift-to-drag ratio f and the ballistic coefficient b of the vehicle are defined by

$$f = \frac{C_L}{C_D}, \quad b = \frac{SC_D}{m} \tag{15.15}$$

The lift-to-drag ratio (without dimension) measures the vehicle's capability to glide, whereas the ballistic coefficient (in $m^2/kg$) measures the capability to brake.

The aerodynamics effects consist of a force $\vec{F}_a$ and a moment $\vec{M}_a$ relative to the center of gravity of the vehicle. Under the assumptions of symmetry and zero sideslip, the torque is reduced to its pitch component: $\vec{M}_a = P_d S C_m \vec{y}_E$, where $C_m(M, \alpha, \delta)$ is the pitch torque coefficient. The cancellation of this moment gives the so-called trim condition: $C_m(M, \alpha, \delta) = 0$. This equation implicitly defines the trimmed incidence $\alpha_e$ as a function of the Mach number M and the deflection δ. The number of solutions $\alpha_e(M, \delta)$ to this equation depends on the values of M and δ. The rotational equilibrium is stable or unstable depending on the sign of the derivative $\frac{\partial C_m}{\partial \alpha}$.

These characteristics depend on the vehicle design (shape and center of gravity). The drag and lift coefficients at trimmed angle of attack are called the **trimmed coefficients** and they depend only on the Mach number and the deflection angle.

The trim condition is achieved throughout the reentry. If the vehicle is not aerodynamically controlled, the trimmed angle of attack occurs naturally and it is not controllable. If the vehicle has deflection controls, their value δ changes the aerodynamic shape and allows an indirect control of the trimmed angle of attack. For further details, the reader can consult [R30] and [R40].

### 15.2.5 Differential Equations of Motion

The differential equations of motion in spherical coordinates are obtained from the vector equations (15.3). The kinematic equation is projected into RG and the dynamic equation is projected into RC as presented in the following text. In the sequel, the modulus of the relative velocity $\vec{v}_R$ is noted $\vec{v}$.

#### 15.2.5.1 Kinematic Equation

The kinematic equation expresses the derivative in RT of the position $\vec{r} = r\vec{z}_G$ as a function of the relative velocity $\vec{v}_R = v\vec{x}_C$. The derivative is passed in RG by the formula (15.2) with the frame angular velocity: $\vec{\Omega}_{RG/RT} = \dot{\lambda}\vec{z}_T - \dot{\varphi}\vec{x}_G$.

$$\left.\frac{d\vec{r}}{dt}\right|_{RT} = \left.\frac{d\vec{r}}{dt}\right|_{RG} + \vec{\Omega}_{RG/RT} \wedge \vec{r} = \dot{r}\,\vec{z}_G + (\dot{\lambda}\,\vec{z}_T - \dot{\varphi}\,\vec{x}_G) \wedge (r\,\vec{z}_G)$$
$$= \dot{r}\,\vec{z}_G + r\dot{\lambda}\cos\varphi\,\vec{x}_G + r\dot{\varphi}\,\vec{y}_G \tag{15.16}$$

By identifying to the components of $\vec{v}_R = v\vec{x}_C$ projected in RG, with $\vec{x}_C$ given by (15.5), we obtain the differential equations of position in spherical coordinates.

$$\begin{cases} \dot{r} = v\sin\gamma & (\vec{z}_G) \\ r\,\dot{\varphi} = v\cos\gamma\cos\chi & (\vec{y}_G) \\ r\,\dot{\lambda}\,\cos\varphi = v\cos\gamma\sin\chi & (\vec{x}_G) \end{cases} \tag{15.17}$$

#### 15.2.5.2 Dynamic Equation

The dynamic equation expresses the derivative in RT of the velocity $\vec{v}_R = v\vec{x}_C$ as a function of the forces applied and the drive and Coriolis accelerations.

The derivative is passed in RC by the formula (15.2) with the frame angular velocity given by: $\vec{\Omega}_{RC/RT} = \vec{\Omega}_{RC/RG} + \vec{\Omega}_{RG/RT} = -\dot{\chi}\vec{z}_G - \dot{\gamma}\vec{y}_C + \dot{\lambda}\vec{z}_T - \dot{\varphi}\vec{x}_G$.

$$\left.\frac{d\vec{v}_R}{dt}\right|_{RT} = \left.\frac{d\vec{v}_R}{dt}\right|_{RC} + \vec{\Omega}_{RC/RT} \wedge \vec{v}_R$$
$$= \dot{v}\,\vec{x}_C + (-\dot{\chi}\vec{z}_G - \dot{\gamma}\vec{y}_C + \dot{\lambda}\vec{z}_T - \dot{\varphi}\vec{x}_G) \wedge (v\,\vec{x}_C) \tag{15.18}$$

Using the projections (15.5), and substituting $\dot{\varphi}$ and $\dot{\lambda}$ given by (15.17), we obtain the components of the relative acceleration in RC.

$$\left.\frac{d\vec{v}_R}{dt}\right|_{RT} = \dot{v}\,\vec{x}_C + \left(\frac{v^2}{r}\tan\varphi\cos^2\gamma\sin\chi - v\dot{\chi}\cos\gamma\right)\vec{y}_C + \left(v\dot{\gamma} - \frac{v^2}{r}\cos\gamma\right)\vec{z}_C \tag{15.19}$$

The drive and Coriolis accelerations have the expression

$$\begin{cases} \vec{a}_E = \vec{\Omega}_T \wedge (\vec{\Omega}_T \wedge \vec{r}) = \Omega_T^2 r\,\vec{z}_T \wedge (\vec{z}_T \wedge \vec{z}_G) \\ \vec{a}_C = 2\vec{\Omega}_T \wedge \vec{v}_R = 2\Omega_T v\,\vec{z}_T \wedge \vec{x}_C \end{cases} \tag{15.20}$$

These accelerations (15.20), the gravity (15.9), and the aerodynamic force (15.13) are projected into RC using (15.4) and (15.5).

By substituting these vectors in (15.3) with the relative acceleration given by (15.19), we obtain the differential equations of velocity in spherical coordinates.

$$\left\{\begin{aligned}
\dot{v} &= -\frac{1}{2}\rho v^2\frac{SC_x}{m} - \frac{\mu_T}{r^2}\sin\gamma \\
&\quad -\frac{3}{2}J_2\frac{\mu_T}{r^2}\left(\frac{R_e}{r}\right)^2[\sin\gamma(1-3\sin^2 f) + 2\cos\gamma\cos\chi\ \sin f\ \cos f] \\
&\quad -\Omega_T^2\, r\cos f(\sin f\ \cos\gamma\ \cos\chi - \cos f\ \sin\gamma) \qquad (\vec{x}_C) \\
v\dot{\gamma} &= \frac{1}{2}\rho v^2\frac{SC_z}{m}\cos\mu + \left(\frac{v^2}{r} - \frac{\mu_T}{r^2}\right)\cos\gamma \\
&\quad -\frac{3}{2}J_2\frac{\mu_T}{r^2}\left(\frac{R_e}{r}\right)^2[\cos\gamma\,(1-3\sin^2 f) - 2\ \sin\gamma\ \cos\chi\ \sin f\cos f] \\
&\quad +\Omega_T^2\, r\ \cos f\,(\cos f\ \cos\gamma + \sin f\ \sin\gamma\ \cos\chi) \\
&\quad +2\,\Omega_T v\ \cos f\ \sin\chi \qquad (\vec{z}_C) \\
v\dot{\chi}\cos\gamma &= \frac{1}{2}\rho v^2\frac{SC_z}{m}\sin\mu + \frac{v^2}{r}\tan f\cos^2\gamma\ \sin\chi \\
&\quad +3J_2\frac{\mu_T}{r^2}\left(\frac{R_e}{r}\right)^2\sin f\ \cos f\ \sin\chi \\
&\quad +\Omega_T^2\, r\ \sin f\ \cos f\ \sin\chi \\
&\quad +2\Omega_T v\,(\sin f\ \cos\gamma - \cos f\ \sin\gamma\ \cos\chi) \qquad (\vec{y}_C)
\end{aligned}\right. \tag{15.21}$$

These equations present singularities at the poles ($\varphi = \pm 90$ deg), for zero velocity ($v = 0$) and for vertical velocity ($\gamma = \pm 90$ deg). These conditions do not occur in practice, except in very special cases. In such cases, the use of Cartesian coordinates avoids these singularities.

These complex differential equations require a numerical integration method. During preliminary design phases, simplifications can be made based on the order of magnitudes of the terms in the second member:

- The terms associated with drag $C_D$, lift $C_L$, and central gravity $\frac{\mu_T}{r}$ are the most important, of order 10 m/s$^2$.
- The terms associated with Coriolis acceleration $\Omega_T v$ are of order 0.5 m/s$^2$.
- The terms associated with centrifugal acceleration $\Omega_T^2 r$ and gravity first zonal perturbation $J_2$ are of order 0.01 m/s$^2$.

The simplified models used during preliminary design phases are presented in Section 15.4.

### 15.2.6 Trajectory Optimization

The trajectory follows the differential equations (15.17) and (15.21) from initial conditions at the entry into the atmosphere. It can be controlled through the bank angle $\mu$ and the angle of attack $\alpha$. The bank angle orientates the aerodynamic force with respect to the vertical plane without changing its modulus. The angle of attack acts directly on the drag and lift coefficients $C_D(M,\alpha)$ and $C_L(M,\alpha)$. The flight occurs at the trimmed angle of attack that can be modulated within a certain range if the vehicle has some deflection capabilities.

The reentry aim is to recover the vehicle safely at the landing site. The thermal and mechanical loads shall remain within limits compatible with the vehicle sizing.

The precise calculation of these loads requires complex simulations. For the trajectory optimization, these limits are translated in levels of dynamic pressure $P_d$ (Pa), heat flux $\Phi_a$ ($W/m^2$), temperature $T_a$ (K), total heat Q (J), axial $g_A$ and normal $g_N$ accelerations ($m/s^2$). The quantities $P_d$, $\Phi_a$, $g_A$ and $g_N$ are defined by

$$P_d = \frac{1}{2}\rho v^2, \quad \Phi_a = c_q\sqrt{\frac{\rho}{R_a}}v^3, \quad g_A = P_d\frac{SC_A}{m}, \quad g_N = P_d\frac{SC_N}{m} \tag{15.22}$$

The convective heat flux $\Phi_a$ is assessed at the stagnation point located at the vehicle nose, whose radius of curvature is $R_a$. A standard value of the constant $c_q$ is $1.705 \cdot 10^{-4}\,J\,s^2\,kg^{-0.5}\,m^{-3}$. More specific formulations can be used depending on vehicle geometry and flight conditions. The temperature at the stagnation point (radiative equilibrium hypothesis) and the total heat absorbed during the reentry are estimated by

$$\Phi_a = \varepsilon\sigma T_a^4, \quad Q = \int \Phi_a\,dt \tag{15.23}$$

where $\sigma = 5.66\,825 \cdot 10^{-8}\,W\,m^{-2}\,K^{-4}$ is the constant of Stefan and $\varepsilon$ is the wall emissivity coefficient.

The acceptable levels of flux and temperature depend on the thermal protection material. The total heat determines the required thickness of thermal protection. To avoid heating and cooling cycles and multiple flux peaks, it is preferable that the trajectory be constantly downward without rebounds (also called skip). The velocity slope must therefore remain always negative. Skips increase the distance traveled, but make the trajectory more sensitive to atmosphere and aerodynamics uncertainties. These uncertainties are high in the upper atmosphere, which is not well known, and they must be taken into account to ensure a successful reentry.

The landing may be either vertical with parachutes, or horizontal on a runway. The landing phase begins at an altitude of a few kilometers. In order to achieve a safe landing, specific velocity conditions must be met depending on the remaining distance to the targeted landing site. These conditions represent final constraints for the hypersonic reentry phase (Mach number greater than 5).

Optimization consists either in assessing the accessible landing area or in maximizing some margins with respect to loads (15.22). The optimization of the trajectory is thus formulated as an optimal control problem. The numerical resolution is generally difficult due to the following features:

- the equations of motion depend on tabulated models of the atmosphere and of the vehicle aerodynamics;
- the final constraints on position and velocity are very stringent;
- the state constraints (dynamic pressure, flux and slope) may be active or not on the successive trajectory legs.

Direct optimization methods are preferable for this type of problem where the precision of the optimum is not critical. They offer more flexibility in taking state constraints into account. Good quality solutions can be obtained with an appropriate choice of discretization.

## 15.3 Orbital Phase

The orbital phase precedes the atmospheric reentry. The limit of the atmosphere is set at a conventional altitude of 100 km below which the aerodynamic effects become noticeable.

The position and velocity at the atmosphere entry drive the loads during the atmospheric flight and the accessible landing area. This section presents the calculation of the deorbitation maneuver and resulting ballistic range. The calculations are carried out in the inertial frame RA and they consider the vehicle's absolute velocity.

### 15.3.1 Deorbitation

The purpose of deorbiting is to place the vehicle on a trajectory entering the Earth's atmosphere. The vehicle is initially in a circular orbit of radius $r_0$. This assumption corresponds to most applications, especially for a return from a space station. The absolute velocity in the circular orbit is noted $\vec{v}_C$.

The deorbitation is achieved by an impulsive maneuver $\Delta\vec{v}$ making an angle $\alpha$ with the velocity $\vec{v}_C$. The velocity after maneuver is $\vec{v}_0 = \vec{v}_C + \Delta\vec{v}$ and its local slope is $\gamma_0$. The trajectory from the conditions $(\vec{r}_0, \vec{v}_0)$ is a Keplerian coast arc until the entry into the atmosphere. The position, velocity, and slope at the reentry are $\vec{r}_e$, $\vec{v}_e$ and $\gamma_e$, respectively (Figure 15.7).

The loads during the atmospheric phase are highly dependent on the slope $\gamma_e$, as will be analyzed in Section 15.4.2. The entry slope must be set to allow an atmospheric flight compliant with the vehicle sizing. The deorbitation optimization consists of finding the impulse $\Delta\vec{v}$ of minimum modulus achieving the targeted slope value $\gamma_e$ at the atmosphere entry.

The modulus $\Delta v$ is a function of $v_0$ and $\gamma_0$.

$$\Delta v^2 = \|\vec{v}_0 - \vec{v}_c\|^2 = v_0^2 + v_c^2 - 2v_0 v_c \cos\gamma_0 \tag{15.24}$$

The conservation of the angular momentum and energy along the Keplerian arc makes it possible to express $v_0$ and $\gamma_0$ as a function of $v_e$. By introducing the variable $s \underset{\text{def}}{=} \frac{r_0}{r_e} > 1$ and with the circular velocity given by $v_c^2 = \frac{\mu_T}{r_0}$, we have

$$\begin{cases} \dfrac{v_0^2}{2} - \dfrac{\mu_T}{r_0} = \dfrac{v_e^2}{2} - \dfrac{\mu_T}{r_e} & \Rightarrow \quad v_0^2 = v_e^2 + 2v_c^2(1-s) \\ r_0\, v_0 \cos\gamma_0 = r_e\, v_e \cos\gamma_e & \Rightarrow \quad \cos\gamma_0 = \dfrac{v_e}{s v_0} \cos\gamma_e \end{cases} \tag{15.25}$$

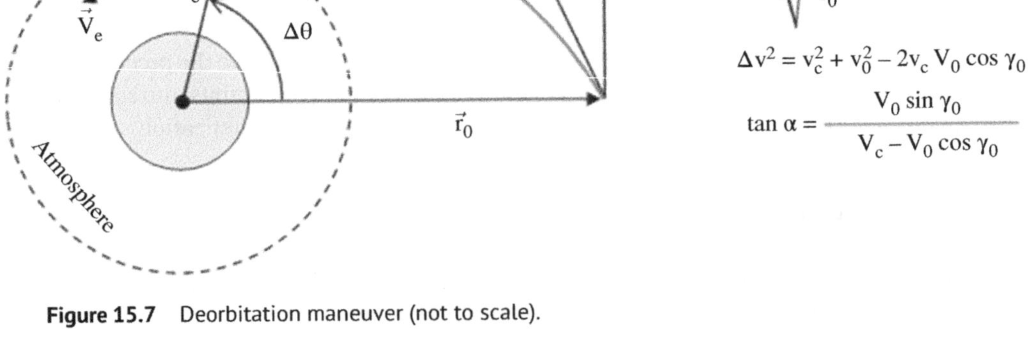


**Figure 15.7** Deorbitation maneuver (not to scale).

By substituting $v_0$ and $\gamma_0$ into (15.24), the modulus $\Delta v$ is expressed as:

$$\Delta v^2 = v_e^2 - \frac{2}{s} v_c v_e \cos\gamma_e + v_c^2(3 - 2s) \tag{15.26}$$

We look for the value of $v_e$ minimizing $\Delta v^2$ for given values of $r_0$, $r_e$, and $\gamma_e$. The range of variation of $v_e$ is bounded, since the relations (15.25) require

$$\begin{cases} v_0^2 \geq 0 & \Rightarrow \; v_e^2 \geq 2v_c^2(s-1) \\ \cos^2\gamma_0 \geq 1 & \Rightarrow \; v_e^2 \geq 2v_c^2(s-1)\dfrac{s^2}{s^2 - \cos^2\gamma_e} \end{cases} \tag{15.27}$$

The second condition is more stringent because $s > 1$.

The optimization problem is finally formulated as:

$$\begin{aligned} \min_{v_e} \; F(v_e) &\underset{\text{def}}{=} v_e^2 - \frac{2}{s} v_c v_e \cos\gamma_e + v_c^2(3 - 2s) \\ \text{s.t. } G(v_e) &\underset{\text{def}}{=} v_e^2(\cos^2\gamma_e - s^2) + 2v_c^2 s^2(s-1) \leq 0 \end{aligned} \tag{15.28}$$

The Lagrangian of this problem is: $L(v_e, \lambda) = F(v_e) + \lambda G(v_e)$, where $\lambda$ is the multiplier associated with the inequality constraint $G(v_e) \leq 0$.

The first-order necessary conditions of optimality (KKT conditions) are

$$\frac{\partial L}{\partial v_e} = \frac{\partial F}{\partial v_e} + \lambda \frac{\partial G}{\partial v_e} = 0 \quad \text{with} \quad \begin{cases} G(v_e) \leq 0 \;, \; \lambda \geq 0 \\ \lambda G(v_e) = 0 \end{cases} \tag{15.29}$$

The complementarity condition $\lambda G(v_e) = 0$ leads to two possibilities, depending on whether the inequality constraint is active or not.

- Inactive constraint: $G(v_e) < 0$ and $\lambda = 0$

  Equation (15.29) reduces to $\frac{\partial L}{\partial v_e} = \frac{\partial F}{\partial v_e} = 0$, yielding the solution $v_e = v_c \frac{\cos\gamma_e}{s}$.
  By substituting this value of $v_e$ into the constraint $G(v_e) < 0$, we get the condition

$$D \underset{\text{def}}{=} \cos^2\gamma_e(s^2 - \cos^2\gamma_e) - 2s^4(s-1) > 0 \tag{15.30}$$

This condition depends only on the given values of $r_0$, $r_e$, and $\gamma_e$. If it is satisfied (with D positive), the solution with an inactive constraint is optimal. The velocity and slope after the maneuver are given by (15.25). The maneuver angle $\alpha$ with the velocity $\vec{v}_C$ is given by the formula in Figure 15.7 and the modulus $\Delta v$ (15.26) is

$$\Delta v = v_C \sqrt{3 - 2s - \frac{\cos^2\gamma_e}{s^2}} \tag{15.31}$$

On the other hand, if the condition (15.30) is not satisfied (with D negative), the constraint is active, and the corresponding optimal solution is derived hereafter.

- Active constraint: $G(v_e) = 0$ and $\lambda \geq 0$

  The constraint $G(v_e) = 0$ yields the solution $v_e = s v_c \sqrt{\frac{2(s-1)}{s^2 - \cos^2\gamma_e}}$.
  This solution corresponds to the limit case of (15.27): $\cos\gamma_0 = 1 \Rightarrow \gamma_0 = 0$.
  The maneuver is then opposed to the velocity (retro-boost with $\alpha$ = 180 deg) and the modulus $\Delta v$ (15.26) is expressed as

$$\Delta v = v_c \left[ 1 - \cos\gamma_e \sqrt{\frac{2(s-1)}{s^2 - \cos^2\gamma_e}} \right] \tag{15.32}$$

Synthesis

The optimal solution depends on the sign of $D = \cos^2\gamma_e(s^2 - \cos^2\gamma_e) - 2s^4(s-1)$, which cancels when

$$2\cos^2\gamma_e = s^2(1 \pm \sqrt{9-8s}) \tag{15.33}$$

Figure 15.8 shows the curve $D(s, \gamma_e) = 0$ bounding the areas $D < 0$ and $D > 0$. When the ratio s exceeds 9/8 (horizontal dotted line) or when the slope $\gamma_e$ is small, D is negative and the optimal solution is a retro-boost. This solution is the most frequent in practical applications, since the targeted slope usually ranges from −1 to −7 deg, as will be seen in Section 15.4.2.

### 15.3.2 Ballistic Range

The angular range $\Delta\theta$ flown during the orbital arc is the difference between the initial anomaly $\theta_0$ and the anomaly at reentry $\theta_e = \theta_0 + \Delta\theta$ (Figure 15.7). We wish to express this range $\Delta\theta$ in terms of $r_0$, $v_0$, $\gamma_0$ and $r_e$. The Keplerian arc of eccentricity e, of angular momentum h and of parameter p has for equation: $r = \frac{p}{1+e\cos\theta}$ with $h = r_0 v_0 \cos\gamma_0$ and $p = \frac{h^2}{\mu_T}$.

The equation of the orbital arc in $\theta_e = \theta_0 + \Delta\theta$ gives by expanding the cosine

$$r_e(1 + e\cos\theta_0\cos\Delta\theta - e\sin\theta_0\sin\Delta\theta) = p \tag{15.34}$$

In order to express the anomaly $\theta_0$ as a function of $r_0$, $v_0$ and $\gamma_0$, the radial and orthoradial components of the velocity are used. From

$$\begin{cases} v_r = v\sin\gamma = \dot{r} \\ v_\theta = v\cos\gamma = r\dot{\theta} \end{cases} \quad \text{with} \quad h = r^2\dot{\theta} \quad \text{and} \quad r = \frac{p}{1+e\cos\theta} \Rightarrow \dot{r} = \frac{h}{p}e\sin\theta \tag{15.35}$$

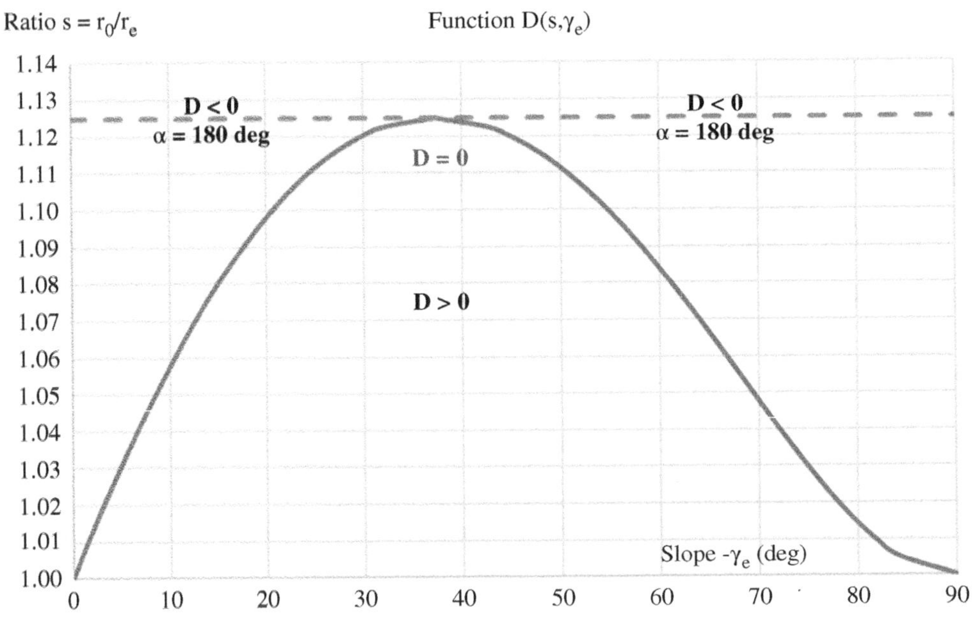


**Figure 15.8** Optimal deorbitation maneuver.

we have

$$\begin{cases} v\sin\gamma = \dfrac{h}{p} e\sin\theta \\ v\cos\gamma = \dfrac{h}{p}(1+e\cos\theta) \end{cases} \Rightarrow \begin{cases} e\sin\theta = \dfrac{p}{h} v\sin\gamma \\ e\cos\theta = \dfrac{p}{h} v\cos\gamma - 1 \end{cases} \tag{15.36}$$

By substituting these expressions (15.36) of $\sin\theta_0$ and $\cos\theta_0$ into (15.34) with $h = r_0 v_0 \cos\gamma_0$ and $h^2 = \mu_T p$, we obtain

$$\frac{\cos(\gamma_0+\Delta\theta)}{\cos\gamma_0} + \frac{\mu_T(1-\cos\Delta\theta)}{r_0 v_0^2 \cos^2\gamma_0} - \frac{r_0}{r_e} = 0 \tag{15.37}$$

This relation yields the initial velocity $v_0$ required for a range $\Delta\theta$.

$$v_0^2 = \frac{\mu_T r_e (1-\cos\Delta\theta)}{(r_0\cos\gamma_0)[r_0\cos\gamma_0 - r_e\cos(\gamma_0+\Delta\theta)]} \tag{15.38}$$

The slope $\gamma_{0m}$ minimizing this required velocity (minimum-energy trajectory) is

$$\tan(2\gamma_{0m}) = \frac{r_e\sin\Delta\theta}{r_0 - r_e\cos\Delta\theta} \Rightarrow \gamma_{0m} = \frac{\pi-\Delta\theta}{4} \text{ if } r_0 = r_e \tag{15.39}$$

The equation (15.37) also allows to assess the derivatives of the range with respect to the initial conditions $r_0$, $v_0$ and $\gamma_0$. The calculations in the general case are somewhat complex. Nevertheless, simpler formulas can be obtained with the assumption that $r_0 \approx r_e$, which corresponds to most applications.

We obtain after calculation

$$\begin{cases} \dfrac{\partial\Delta\theta}{\partial v_0} = \dfrac{2(1-\cos\Delta\theta)}{v_0\left[\sin\Delta\theta - \dfrac{r_0 v_0^2}{\mu_T}\cos\gamma_0\sin(\gamma_0+\Delta\theta)\right]} \\ \dfrac{\partial\Delta\theta}{\partial\gamma_0} = 2\left[\dfrac{\sin(2\gamma_0+\Delta\theta)}{\sin 2\gamma_0} - 1\right], \quad \dfrac{\partial\Delta\theta}{\partial r_0} = \dfrac{1}{r_0}\left[\sin\theta + \dfrac{2-\cos\theta}{\tan\gamma_0}\right] \end{cases} \tag{15.40}$$

The formulas (15.40) are useful to estimate the effect of initial errors on the range.

**Remark 15.1** *The analytical formulas in this section assume a non-rotating spherical Earth. For greater accuracy, the Earth flattening ($J_2$ term) must be considered and the targeted reentry conditions must be expressed in the rotating Earth frame. The reentry slope to be controlled is indeed that of the relative velocity in the rotating Earth frame. Since the transition in the rotating frame depends on the position, an iterative solution method is required to find the optimal deorbitation maneuver.*

### 15.3.3 Return Maneuver

The previous section dealt with the deorbitation propelled maneuver from a circular orbit. The recovery of a launcher's first stage after its separation presents a different situation. The altitude and velocity conditions correspond to a suborbital trajectory and a fallout. To recover the stage safely, a return (or toss-back) maneuver is required after separation to control the ballistic landing point.

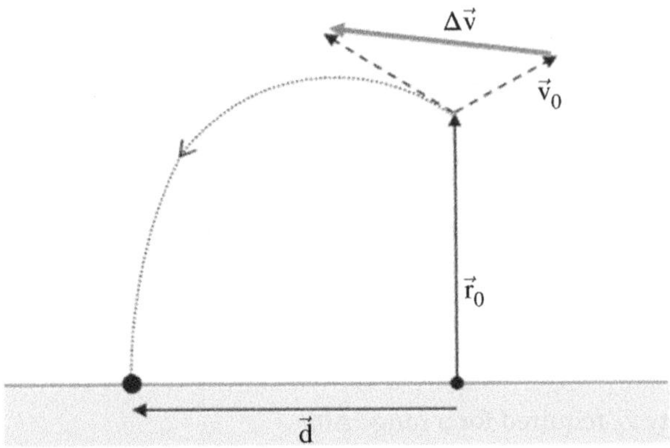


**Figure 15.9** Optimal return maneuver.

The altitude is generally low, comprised between 50 and 100 km in most cases, and the downrange until the targeted landing site does not exceed a few hundred kilometers. In such conditions, a flat Earth constant gravity model can be used to estimate the return impulse.

Let us note $\vec{r}_0$ and $\vec{v}_0$ the position and velocity of the stage at separation before maneuver, $\vec{d}$ the targeted landing position, $\Delta\vec{v}$ the impulse applied, and $t_f$ the landing time, as depicted in Figure 15.9.

With a flat Earth constant gravity model, the final position $\vec{r}_f$ at $t_f$ is given by

$$\vec{r}_f = \vec{r}_0 + (\vec{v}_0 + \Delta\vec{v})t_f + \frac{1}{2}\vec{g}t_f^2 = \vec{d} \tag{15.41}$$

The goal is to find the minimum-norm impulse $\Delta\vec{v}$ allowing to reach the targeted landing site located in $\vec{d}$. The optimization problem is formulated as:

$$\min_{\Delta\vec{v},t_f} \|\Delta\vec{v}\| \quad \text{s.t.} \quad \vec{r}_f = \vec{d} \tag{15.42}$$

where the unknowns are the impulse vector and the final time.

To solve this problem, we express from (15.41) the impulse as a function the final time

$$\Delta\vec{v} = \frac{\vec{d} - \vec{r}_0 - \vec{v}_0 t_f - \frac{1}{2}\vec{g}t_f^2}{t_f} = \frac{\vec{d} - \vec{r}_0}{t_f} - \vec{v}_0 - \frac{1}{2}\vec{g}t_f \tag{15.43}$$

The minimum impulse is obtained for

$$\frac{d}{dt_f}\|\Delta\vec{v}\|^2 = 2\Delta\vec{v} \cdot \frac{d\Delta\vec{v}}{dt_f} = 0 \tag{15.44}$$

which yields

$$\left(\frac{\vec{d} - \vec{r}_0}{t_f} - \vec{v}_0 - \frac{1}{2}\vec{g}t_f\right) \cdot \left(\frac{\vec{d} - \vec{r}_0}{t_f^2} + \frac{1}{2}\vec{g}\right) = 0 \tag{15.45}$$

The landing site is on the ground, so that $\vec{d} \cdot \vec{r}_0 = 0$ and $\vec{g} \cdot \vec{r}_0 = 0$.

Expanding the brackets in (15.45), we obtain an equation of degree four in $t_f$.

$$g^2 t_f^4 + 2\vec{v}_0 \cdot \vec{g}t_f^3 + 4\vec{v}_0 \cdot (\vec{d} - \vec{r}_0)t_f - 4(\vec{d} - \vec{r}_0)^2 = 0 \tag{15.46}$$

This quartic equation with a missing quadratic term is solved by Ferrari's method already used in Section 6.3.4. This method is explained after equation (6.71).

Once the value of $t_f$ is known, the impulse is determined by (15.43).

**Remark 15.2** *Let us consider the fictitious case of an initial position at ground ($\vec{r}_0 = \vec{0}$) and a null initial velocity ($\vec{v}_0 = \vec{0}$). Equation (15.46) yields $t_f = \sqrt{\frac{2d}{g}}$ and the optimal impulse is given by $\Delta\vec{v} = \sqrt{\frac{g}{2d}}\vec{d} - \sqrt{\frac{2d}{g}}\vec{g}$, which makes an angle of 45 deg with the horizontal. We recover the classical result associated with a ballistic range in flat Earth constant gravity model.*

The return impulse places the stage on a ballistic trajectory targeting the landing site. Additional braking impulses are necessary at the atmospheric entry to limit the loads and at the landing to cancel the velocity. The assessment of these impulses is addressed in Section 15.4.5.

## 15.4 Atmospheric Phase

The calculation of a reentry trajectory requires numerical optimization methods (to find the control law in angles $\alpha$, $\mu$) and numerical integration methods (to propagate the differential equations of motion). These methods do not provide a physical understanding of the phenomena. When designing a vehicle or preparing a mission, simple analytical models are needed to quickly assess the performances and define a preliminary reentry scenario. This section presents the most useful simplified models, that can also be found in [R30] and [R40].

### 15.4.1 Simplified Equations

The following assumptions are used to simplify the equations of motion (15.21).

- The Earth is spherical and not rotating, so that the drive, Coriolis and $J_2$ accelerations are null. They are indeed negligible compared to aerodynamic accelerations, as indicated at the end of Section 15.2.5.
- The slope remains small throughout the reentry, so that: $\sin\gamma \approx 0, \cos\gamma \approx 1$. High slopes would indeed generate excessive thermal and mechanical loads.
- The aerodynamic coefficients (b, f) are constant throughout the trajectory. This assumption proves realistic for most reentry vehicles during the hypersonic (Mach 25 to 5) and supersonic (Mach 5 to 2) phases.
- The density $\rho$ follows the exponential model (15.12), which is sufficiently representative for preliminary assessments.

These assumptions reduce the complex equations (15.21) to the following system in variables v, $\gamma$ and $\rho$:

$$\begin{cases} \dot{v} = -\dfrac{1}{2}\rho v^2 b \\[2ex] v\dot{\gamma} = \dfrac{1}{2}\rho v^2 bf\cos\mu + \dfrac{v^2 - v_c^2}{r} \\[2ex] \dot{\rho} = -\dfrac{\dot{r}}{H}\rho = -\dfrac{v\sin\gamma}{H}\rho \end{cases} \tag{15.47}$$

where $v_c = \sqrt{\frac{\mu_T}{r}}$ denotes the circular velocity at the radius r.

### 15.4.2 Reentry Corridor

The reentry corridor bounds the initial conditions ($r_e$, $v_e$, $\gamma_e$) in order to ensure the atmospheric capture while limiting the thermal and mechanical loads.

#### 15.4.2.1 Atmospheric Braking

The conditions ($r_e$, $v_e$, $\gamma_e$) at reentry come from a Keplerian conic of angular momentum h, energy w, semi-major axis a, eccentricity e and radius at perigee $r_p$.

These reentry conditions are calculated using the following relations, which are valid for ellipses (a > 0, e < 1) or hyperbolas (a < 0, e > 1).

$$\begin{cases} h = r_e v_e \cos\gamma_e = \sqrt{\mu_T a(1-e^2)} \\ w = \dfrac{v_e^2}{2} - \dfrac{\mu_T}{r_e} = -\dfrac{\mu_T}{2a} \\ r_p = a(1-e) \end{cases} \tag{15.48}$$

Figure 15.10 depicts the trajectory leg from the entry point E to the perigee P.

The velocity $v_e$ is directly defined by the mission, ranging from 8 km/s for a low Earth orbit return to 12 km/s for an interplanetary return, and it cannot be changed (excepted by propulsive means). On the other hand, the slope $\gamma_e$ can be chosen and it allows to control the perigee altitude:

- if the perigee altitude is too high, the braking is not sufficient and the vehicle will escape from the atmosphere;
- if the perigee altitude is too low, the vehicle will enter the dense layers of the atmosphere at high velocity and it will undergo severe loads.

As a first approximation, the atmospheric braking reduces the velocity without changing the altitude of the Keplerian perigee given by (15.48). The velocity variation can be estimated by dividing $\dot{v}$ by $\dot{\rho}$ in (15.47), and considering that the slope remains constant ($\gamma \approx \gamma_e$) over the arc EP. This yields

$$\frac{dv}{d\rho} = \frac{Hb}{2\sin\gamma} v \quad \Rightarrow \quad v = v_e e^{\frac{Hb}{2\sin\gamma_e}(\rho-\rho_e)} \tag{15.49}$$

The formulas (15.48) and (15.49) allow us to assess the perigee conditions ($r_p$, $v_p$) as functions of the entry velocity $v_e$ and slope $\gamma_e$ while accounting for the atmospheric braking. They are used in the two following sections to bound the admissible entry conditions.

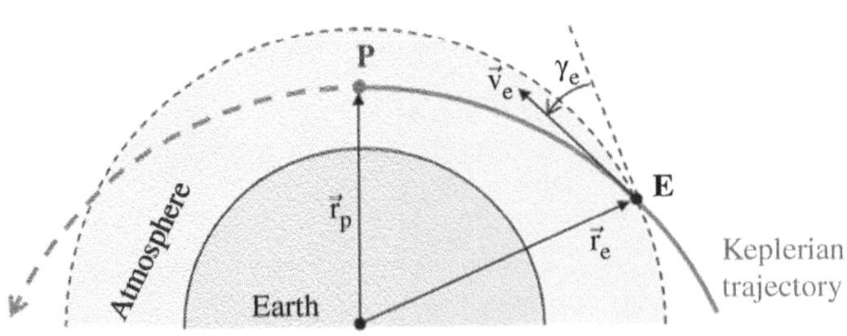


**Figure 15.10** Capture by the atmosphere.

#### 15.4.2.2 Upper Bound

For the trajectory to continue downward after the passage in P (where $\gamma_p = 0$), the derivative $\dot{\gamma}$ in (15.47) must be negative. The capture is enforced by applying a downward lift (with a bank angle $\mu = 180$ deg) that maximizes the downward acceleration. With this control, the capture condition in P is

$$r_p \rho_p b f \geq 2\left(1 - \frac{v_c^2}{v_p^2}\right) \quad \text{with} \quad v_c = \sqrt{\frac{\mu_T}{r_p}} \tag{15.50}$$

The conditions $(r_p, v_p)$ depend on $(v_e, \gamma_e)$ through equations (15.48) for $r_p$ and (15.49) for $v_p$.

Since the entry velocity $v_e$ is given (associated with the return mission studied), the inequality (15.50) yields the maximum slope $\gamma_e$ ensuring the capture. The corresponding perigee is then calculated by (15.48).

Figure 15.11 shows the maximum perigee altitude as a function of the entry velocity for a ballistic coefficient $b = 0.01\ m^2/kg$ and for lift-to-drag ratios ($f = L/D$) ranging from 0 to 1.5 (solid lines). At zero L/D, the condition (15.50) shows as expected that the velocity in P must be less than the circular velocity. In the case of the sub-orbital reentry of a launcher stage (generally the last-but-one stage) with typical characteristics $b = 0.005\ m^2/kg$ and $f = 0$, it is found that the perigee must be lower than 70 km to ensure the natural fallout.

#### 15.4.2.3 Lower Bound

The heat fluxes and the accelerations increase as the vehicle descends into the dense layers of the atmosphere. As a first approximation, their maximum is reached at the perigee P, where the velocity reaches a maximum.

If the vehicle can withstand a maximum heat flux $\Phi_m$ and a maximum drag acceleration $a_m$, the kinematic conditions $(r_p, v_p)$ in P must satisfy

$$c_q \sqrt{\rho_p} v_p^3 \leq \Phi_m, \quad \frac{1}{2}\rho_p v_p^2 b \leq a_m \tag{15.51}$$

for a reference nose radius set to $R_a = 1$ m in (15.22).

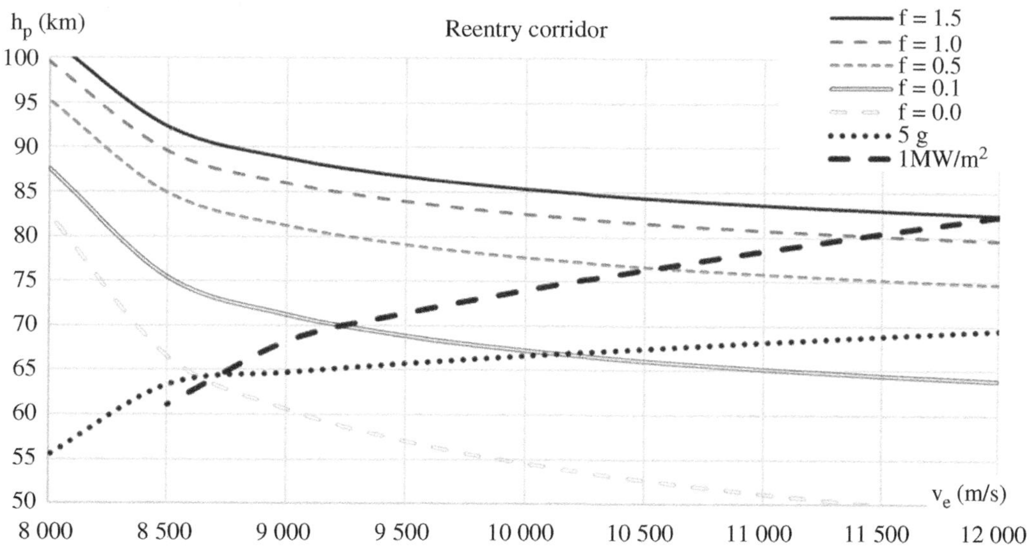


**Figure 15.11** Reentry corridor ($b = 0.01\ m^2/kg$).

The conditions $(r_p, v_p)$ depend on $(v_e, \gamma_e)$ through equations (15.48) for $r_p$ and (15.49) for $v_p$. Since the entry velocity $v_e$ is given (associated with the return mission studied), the inequalities (15.51) yield the minimum slope $\gamma_e$ compliant with the admissible load levels $\Phi_m$ and $a_m$. The corresponding perigee is calculated by (15.48) and it does not depend on the lift-to-drag ratio.

Figure 15.11 shows in dotted lines the minimum perigee altitude as a function of the entry velocity for typical limit levels $\Phi_m = 800\ \text{kW/m}^2$ and $a_m = 5\ \text{g}$. The vehicle ballistic coefficient is still $b = 0.01\ \text{m}^2/\text{kg}$.

The reentry corridor is bounded by the lower limit related to loads (the two increasing curves) and by the upper limit related to capture (the five decreasing curves depending on f).

From the entry velocity (in abscissa) and the perigee altitude (in ordinate), we can assess by (15.48) and (15.49) the admissible entry slope $\gamma_e$, which ranges from −1 to −7 deg.

This corridor narrows at high entry velocities, which require higher lift capabilities (higher L/D values). This is especially the case for interplanetary return missions.

The curves in Figure 15.11 assume that the initial perigee altitude (15.48) is not changed during the reentry beginning. The lift capability of the vehicle allows in fact to change this altitude, therefore enlarging the reentry corridor. The first skip model takes this effect into account for a reentry from a low-altitude orbit.

#### 15.4.2.4 First Skip Model

The first skip model assumes that the lift term is predominant in the equation of the slope $\dot{\gamma}$ (15.47). This is true for low-orbit reentries with a velocity close to the orbital value ($v \approx v_c$). The evolution of the slope is obtained by dividing $\dot{\gamma}$ by $\dot{\rho}$ in (15.47) and assuming a constant bank angle.

$$\frac{d\gamma}{d\rho} \sin\gamma = -\frac{1}{2} H\, b\, f \cos\mu \quad \Rightarrow \quad \cos\gamma - \cos\gamma_e = \frac{1}{2} H\, b\, f \cos\mu\, (\rho - \rho_e) \tag{15.52}$$

If the lift is upward ($\cos\mu > 0$), the skip condition ($\gamma = 0$) occurs at the altitude

$$h_{skip} = H \ln\frac{\rho_0}{\rho_{skip}} \quad \text{with} \quad \rho_{skip} = \rho_e + \frac{2(1 - \cos\gamma_e)}{Hbf \cos\mu} \tag{15.53}$$

The calculation shows that this altitude is higher than the initial Keplerian perigee.

The maximum heat flux and dynamic pressure values are obtained by deriving their expressions (15.22) and substituting $\dot{v}$ and $\dot{\rho}$ given by (15.47).

$$\begin{cases} \dot{\Phi}_a = 0 \ \Rightarrow \ \sin\gamma = -3Hb\rho \\ \dot{P}_d = 0 \ \Rightarrow \ \sin\gamma = -Hb\rho \end{cases} \tag{15.54}$$

Both maxima occur before the skip since $\gamma < 0$, and the flux peak is encountered before the dynamic pressure peak. By combining (15.52) and (15.54) and assuming a small slope, we obtain the values of the maxima.

$$\Phi_{max} = \frac{c_q |\gamma_e| v_e^3}{\sqrt{R_a Hbf \cos\mu}} e^{\frac{3\gamma_e}{f \cos\mu}} \ , \quad P_{max} = \frac{\gamma_e^2 v_e^2}{Hbf \cos\mu} e^{\frac{2\gamma_e}{f \cos\mu}} \tag{15.55}$$

### 15.4.3 Flight Corridor

The flight corridor defines the domain in altitude–velocity accessible to the vehicle according to its aerodynamic capabilities and to the constraints to be met. The descent rate can be controlled through the bank angle in order to keep admissible load levels and remain inside the flight corridor.

#### 15.4.3.1 Controllability and Constraints

The descent is controllable if the available lift is sufficient to cancel $\dot{\gamma}$ (15.47) in order to follow a horizontal flight. By orienting the lift upward, this balance condition determines the minimum required velocity to fly at a given altitude (or conversely the maximum altitude that can be flown at a given velocity).

$$\frac{1}{2}\rho v^2 bf + \frac{v^2 - v_c^2}{r} \geq 0 \tag{15.56}$$

On the other hand, the flux and acceleration limit the velocity at a given altitude.

$$c_q\sqrt{\rho}v^3 \leq \Phi_m \ , \ \frac{1}{2}\rho v^2 b \leq a_m \tag{15.57}$$

Figure 15.12 shows the flight corridor with the Mach number (equivalent to the velocity) in abscissa and the altitude in ordinate. The upper bounds (3 curves) are plotted for lift-to-drag ratio of 0.1, 0.5, and 1 with $b = 0.01\ m^2/kg$. The lower bounds (3 curves) are plotted for flux levels of 500 and 800 $kW/m^2$ assuming $R_a = 1$ m, and for a drag acceleration of 5 g.

Depending on the constraint levels, the lift-to-drag ratio must be sufficient for the corridor to exist throughout the descent.

#### 15.4.3.2 Bank Angle Control

The reentry conditions determine the flux peak (15.55), which is encountered before the skip. Once the skip condition is reached, the available lift is sufficient to control the descent rate, since a null slope has been established. The control strategy is split into two phases.

The first phase aims to dissipate energy by keeping a constant heat flux, usually set at the level of the initial peak. The bank angle control for this iso-flux phase is determined by deriving $\Phi_a$ twice to make the bank angle $\mu$ appear and then imposing these derivatives to zero.

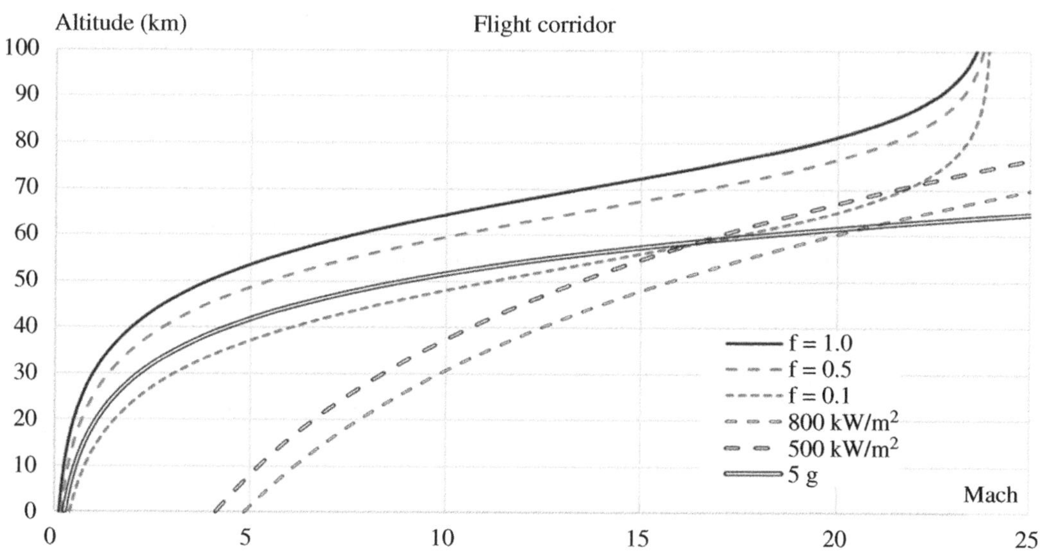


**Figure 15.12** Flight corridor ($b = 0.01\ m^2/kg$).

Using the derivatives of $\dot{v}$ and $\dot{\rho}$ given by (15.47), we obtain a feedback control with the bank angle depending on the current kinematic conditions and the vehicle's aerodynamic characteristics.

$$\begin{cases} \dot{\Phi}_a = 0 \;\Rightarrow\; \dfrac{\dot{\rho}}{2\sqrt{\rho}} v^3 + 3\rho\dot{v}v^2 = 0 \;\Rightarrow\; \sin\gamma = -3Hb\rho \\ \ddot{\Phi}_a = 0 \;\Rightarrow \dot{\gamma}\cos\gamma = -3Hb\dot{\rho} \qquad \Rightarrow\; f\cos\mu = 6\tan\gamma + \dfrac{6H}{r\tan\gamma}\left(1 - \dfrac{v_c^2}{v^2}\right) \end{cases} \tag{15.58}$$

When the velocity decreases, the control (15.58) will finally saturate ($\cos\mu > 1$), indicating that the flux is no longer controllable, which ends the first phase.

The second phase aims then to descend at constant drag acceleration, with an adjusted level depending on the remaining distance to fly until the landing site. The feedback bank control for this iso-drag phase is derived in a similar way to the flux.

$$\begin{cases} \dot{a}_D = 0 \;\Rightarrow\; \dfrac{1}{2}\dot{\rho}v^2 + \rho\dot{v}v = 0 \;\Rightarrow\; \sin\gamma = -Hb\rho \\ \ddot{a}_D = 0 \;\Rightarrow\; \dot{\gamma}\cos\gamma = -Hb\dot{\rho} \;\Rightarrow\; f\cos\mu = 2\tan\gamma + \dfrac{2H}{r\tan\gamma}\left(1 - \dfrac{v_c^2}{v^2}\right) \end{cases} \tag{15.59}$$

This descent strategy has been developed for the Space Shuttle (1981–2011) and it is used by most of the reentry vehicles. The feedback control form as a function of the current state ($r$, $v$, $\gamma$) allows the construction of a robust reentry scenario accounting for the atmospheric and aerodynamic uncertainties, which can be large at high altitudes or for complex geometries.

### 15.4.4 Accessible Area

Preparing a reentry scenario requires knowledge of the ground area accessible from the entry into the atmosphere. The downrange X and crossrange Y measure the distances traveled along the initial azimuth and in the transverse direction. The crossrange is conventionally positive to the right (as is the bank angle).

#### 15.4.4.1 Simplified Equations

In a non-rotating spherical Earth, the trajectory does not depend on the initial plane and the reentry can be assumed equatorial ($\varphi_e = 0$ deg, $\chi_e = 90$ deg). The downrange and crossrange are then given by $X = R_T\lambda$ (along the equator) and $Y = -R_T\varphi$ (along the meridian), respectively.

Their analytical assessment is based on the following simplifying assumptions:

- the variation in the radius vector is negligible during the reentry ($r \approx R_T$);
- the slope remains small ($\gamma \approx 0$), as well as the azimuth deviation noted $\psi \underset{\text{def}}{=} \chi - \chi_e = \chi - 90 \approx 0$.

Equations (15.17) and (15.21) are then reduced to

$$\begin{cases} \dot{X} = v \\ \dot{Y} = v\psi \end{cases}, \quad \begin{cases} \dot{v} = -a_D \\ v\dot{\psi} = a_D f \sin\mu \end{cases} \quad \text{with} \quad a_D = \frac{1}{2}\rho v^2 b \tag{15.60}$$

These equations are passed in variable v instead of t by dividing the derivatives $(\dot{X}, \dot{Y}, \dot{\psi})$ by $\dot{v}$.

$$\begin{cases} \dfrac{dX}{dv} = -\dfrac{v}{a_D} \\ \dfrac{dY}{dv} = -\dfrac{v}{a_D}\psi \end{cases} \quad \text{and} \quad \frac{d\psi}{dv} = -\frac{f \sin\mu}{v} \tag{15.61}$$

For a constant bank control, the azimuth is integrated as a function of the velocity.

$$\psi = f \sin\mu \, \ln\frac{v_e}{v} \tag{15.62}$$

#### 15.4.4.2 Steady Descent Model

The analytical integration of the downrange X and the crossrange Y is made with the assumption of a steady descent trajectory: the forces balance in the vertical plane and cancel the slope derivative $\dot{\gamma}$ in (15.47). With the approximation $r = R_T$ and noting $v_c$ the circular orbital velocity at the ground level, equation (15.47) gives the drag acceleration $a_D$ as a function of velocity.

$$a_D = \frac{v_c^2 - v^2}{R_T f \cos\mu} \quad \text{with} \quad v_c^2 = \frac{\mu_T}{R_T} \tag{15.63}$$

Substituting $a_D$ (15.63) and $\psi$ (15.62) into equations (15.61) gives

$$\begin{cases} \dfrac{dX}{dv} = R_T f \cos\mu \dfrac{v}{v^2 - v_c^2} \\ \dfrac{dY}{dv} = \dfrac{1}{2} R_T f^2 \sin 2\mu \dfrac{v}{v^2 - v_c^2} \ln\dfrac{v_e}{v} \end{cases} \tag{15.64}$$

These differential equations are integrated as follows.

The downrange integrates analytically yielding

$$X = \frac{1}{2} R_T f \cos\mu \left[\ln\left(v^2 - v_c^2\right)\right]_{v_i}^{v_f} \tag{15.65}$$

Let us determine the velocity bounds for assessing the total range.

- The steady descent begins as soon as the skip condition is met and it ends at almost zero velocity before the landing phase: $v_f \approx 0$.
- The initial velocity $v_i = v_{skip}$ is calculated at the skip altitude from the condition $\dot{\gamma} = 0$ in (15.47). The skip altitude, usually in the order of 70 km, is assessed from the density $\rho_{skip}$ by (15.53). Substituting into (15.47), we obtain

$$\frac{1}{2}\rho_{skip} v_{skip}^2 b = \frac{v_c^2 - v_{skip}^2}{R_T f \cos\mu} \quad \Rightarrow \quad v_{skip}^2 = v_c^2 \left(1 + \frac{1}{2} R_T \rho_{skip} b f \cos\mu\right)^{-1} \tag{15.66}$$

The maximal downrange is obtained for a zero bank angle and its value is

$$X_{max} = \frac{1}{2} R_T f \ln\left(1 + \frac{2}{R_T \rho_{skip} b f}\right) \tag{15.67}$$

On the other hand, the crossrange is not analytically integrable. Nevertheless, it is possible to assess the integral $I = \int_{v_i}^{v_f} \frac{v}{v^2 - v_c^2} \ln\frac{v}{v_e} dv$, provided that the bounds are set to $v_i = v_e = v_c$ (orbital circular velocity) and $v_f = 0$.

By the variable change $v = xv_c$, the integral becomes: $I = \int_0^1 \frac{x \ln x}{1-x^2} dx$.

The series expansion: $\frac{1}{1-x^2} = \sum_{k=0}^{\infty} x^{2k}$ allows then an integration by parts.

$$I = \sum_{k=0}^{\infty} \int_0^1 x^{2k+1} \ln x \, dx = \sum_{k=0}^{\infty} \left( \left[ \frac{x^{2k+2}}{2k+2} \ln x \right]_0^1 - \int_0^1 \frac{x^{2k+1}}{2k+2} dx \right) = \sum_{k=0}^{\infty} \left( - \left[ \frac{x^{2k+2}}{(2k+2)^2} \right]_0^1 \right)$$

We get: $I = -\frac{1}{4} \sum_{k=0}^{\infty} \frac{1}{(k+1)^2} = -\frac{1}{4} \frac{\pi^2}{6}$, which yields for the total crossrange (15.64)

$$Y = \frac{\pi^2}{48} R_T f^2 \sin 2\mu \tag{15.68}$$

The maximal crossrange is obtained for a bank angle of 45 deg and its value is

$$Y_{max} \approx 1300 \, f^2 \tag{15.69}$$

#### 15.4.4.3 Optimal Bank Angle Law

The steady descent model developed here above assumes a constant bank control and small azimuth variations. A more accurate assessment of the maximal crossrange requires solving an optimal control problem with variable bank control. This problem is formulated as before for an equatorial reentry.

The optimal control problem is formulated by choosing as state variables the current crossrange $Y = -R_T \varphi$ and its derivative noted $v_Y = \dot{Y}$. Neglecting the variation of the radius ($r \approx R_T$) and noting the azimuth deviation $\psi \underset{def}{=} \chi - 90$, equations (15.17) and (15.21) give for $v_Y$ and $\dot{v}_Y$

$$v_Y = v \cos\gamma \sin\psi \;\Rightarrow\; \dot{v}_Y = \dot{v} \cos\gamma \sin\psi - v\dot{\gamma} \sin\gamma \sin\psi + v\dot{\psi} \cos\gamma \cos\psi$$

$$\text{with} \begin{cases} \dot{v} = -a_D \\ v\dot{\gamma} = a_D f \cos\mu + a_G \cos\gamma \, , \quad a_D = \frac{1}{2}\rho v^2 b \, , \; a_G = \frac{v^2 - v_c^2}{r} \\ v\dot{\psi} \cos\gamma = a_D f \sin\mu \end{cases} \tag{15.70}$$

The optimal control problem is formulated as:

$$\max_{\mu} Y_f \text{ s.t. } \begin{cases} \dot{Y} = v_Y \\ \dot{v}_Y = a_0 + a_D f(\sin\mu \cos\psi - \cos\mu \sin\gamma \sin\psi) \end{cases} \tag{15.71}$$

where $a_0$ groups the terms that are not dependent on the control $\mu$.

The Hamiltonian of the problem is

$$H = p_Y v_Y + p_V [a_0 + a_D f(\sin\mu \cos\psi - \cos\mu \sin\gamma \sin\psi)] \tag{15.72}$$

with the costates $p_Y$ and $p_V$ associated with $Y$ and $V_Y$, respectively.

The Hamiltonian maximization with respect to $\mu$ gives

$$\max_{\mu} H \;\Rightarrow\; \tan\mu = \frac{-1}{\sin\gamma \tan\psi} \tag{15.73}$$

This feedback control exhibits the expected behavior during the gliding turn:

- when the slope is negative, the bank angle decreases from 90 to 0 deg (upward lift) when the azimuth deviation increases from 0 to 90 deg;
- when the slope is positive, the bank angle increases from 90 to 180 deg (downward lift) when the azimuth deviation increases from 0 to 90 deg;
- when the turn is achieved ($\psi = 90$ deg), the lift remains vertical ($\mu = 0$ deg).

In the presence of constraints, this control can be combined with the iso-flux or iso-drag laws developed in Section 15.4.3.

### 15.4.5 Steep Reentry

In the case of a first-stage recovery, the atmospheric entry slope is generally high and the atmospheric trajectory is nearly vertical. The stage has been placed on a ballistic retour trajectory by a return maneuver assessed in Section 15.3.3. In order to recover the stage safely at landing, a second braking maneuver is necessary to avoid excessive loads. The loads along the ballistic trajectory can be estimated from the simplified dynamic equations of a vertical trajectory with drag.

$$\begin{cases} \dot{h} = -v \\ \dot{v} = g - a_D \end{cases} \tag{15.74}$$

The drag acceleration $a_D$ is expressed with an exponential atmosphere model

$$a_D = Qb \quad \text{with} \quad Q = \frac{1}{2}\rho v^2, \quad \rho = \rho_0 e^{-\frac{h}{H}}, \quad b = \frac{SC_D}{m} \tag{15.75}$$

The maximum drag acceleration is obtained when

$$\dot{Q} = 0 \quad \Rightarrow \quad \frac{1}{2}\dot{\rho}v^2 + \rho\dot{v}v = \rho v\left(\frac{v^2}{2H} + g - a_D\right) = 0 \tag{15.76}$$

The solution of this equation is not analytical because it requires integrating the motion equations (15.74) and then solving for the time (or altitude). However, by assuming that the gravity is negligible with respect to the drag, which is the case for a steep reentry, we can integrate the velocity equation from (15.74) by passing in variable $\rho$.

$$\begin{cases} \dot{\rho} = \dfrac{-\dot{h}}{H}\rho_0 e^{-\frac{h}{H}} = \dfrac{1}{H}\rho v \\ \dot{v} = -a_D = -\dfrac{1}{2}\rho v^2 b \end{cases} \Rightarrow \quad \frac{dv}{d\rho} = -\frac{H}{2}bv \quad \Rightarrow \quad v = v_0 e^{-\frac{1}{2}Hb\rho} \tag{15.77}$$

where the initial velocity $v_0$ corresponds to the atmospheric entry with $\rho = 0$.

From (15.76), the maximum dynamic pressure is obtained for

$$\frac{v^2}{2H} = a_D \quad \Rightarrow \quad \rho_m = \frac{1}{Hb} \quad \Rightarrow \quad v_m = \frac{v_0}{\sqrt{e}} \tag{15.78}$$

and using the velocity expression (15.77), the maximum dynamic pressure is

$$Q_m = \frac{v_0^2}{2bHe} \quad \text{with} \quad e \approx 2.71828 \tag{15.79}$$

This formula can be used to estimate the required braking impulse for a specified dynamic pressure level $Q_t$ that must not be exceeded. By assuming that the impulse is delivered at the location of the maximum, we obtain

$$Q_t = \frac{1}{2}\rho_m(v_m - \Delta v)^2 \quad \Rightarrow \quad \Delta v_{pdyn} = \sqrt{2Hb}\,(\sqrt{Q_m} - \sqrt{Q_t}) \tag{15.80}$$

The landing velocity can also be estimated as follows. When the stage falls under the action of gravity and drag, the drag increases and a limit velocity will be attained (if the drag is sufficient) corresponding to $a_D = g$, which yields

$$v_{lim} = \sqrt{\frac{2g}{\rho b}} \tag{15.81}$$

The limit velocity at ground must be canceled by a landing impulse, which can be estimated by

$$\Delta v_{land} = \sqrt{\frac{2g}{\rho_0 b}} \tag{15.82}$$

This impulse is about one hundred meters per second, depending on the stage drag. A more precise estimation can be done, taking into account the boost duration $t_c = \frac{\Delta v_{land}}{a_T}$ (depending on the available thrust acceleration $a_T$) and the gravity losses $gt_c$ (which increase the velocity during the boost).

## 15.5 Conclusion

### 15.5.1 The Key Points

- The analysis of reentry trajectories is eased by writing the motion equations in spherical coordinates in the rotating geocentric frame.
- The optimal deorbitation maneuver consists of a retro-impulse targeting a specified entry slope.
- The thermal and mechanical loads depend directly on the entry slope, which must be very small (−7 deg at most).
- The reentry occurs at trimmed angle of attack, which can be controlled if the vehicle has deflection capabilities.
- The gliding reentry is controlled by the bank angle, following first an iso-flux leg to limit the heat load, then an iso-drag leg to achieve the required downrange until the landing site.

### 15.5.2 To Go Further

- REGAN Frank J., ANANDAKRISHNAN Satya M. – *Dynamics of Atmospheric Re-Entry*. AIAA (1993).

This book details all useful models for the calculations of reentry trajectories, the study of attitude motion and of accuracy analysis. Listings of source codes are also provided with numerical application cases.

- VINH Nguyen X., BUSEMANN Adolf, CULP Robert D. – *Hypersonic and Planetary Entry Flight Mechanics*. The University of Michigan Press (1980).

This book presents the modeling of hypersonic reentry, with a focus on aerothermal effects.

## Short Bibliography

**R1** *An Introduction to the Mathematics and Methods of Astrodynamics.* BATTIN Richard, AIAA (1999).

**R2** *Essais sur le mouvement des corps cosmiques.* BELETSKI Vladimir, Editions Mir (1986).

**R3** *Practical Methods for Optimal Control and Estimation Using Nonlinear Programming.* BETTS John T., SIAM (2010).

**R4** *The Accuracy of Geodetic Latitude and Height Equations.* BOWRING B.R. *Survey Review* 28(218), 202–206 (1985).

**R5** *Handbook of Mathematics (Fourth edition).* BRONSHTEIN I.N., SEMENDYAYEV K.A., MUSIOL G., MUEHLIG H., Springer (2004).

**R6** *Applied Optimal Control.* BRYSON Arthur E., Ho Yu-Chi, Hemisphere Publishing Corporation (1975).

**R7** *Satellites : de Kepler au GPS.* CAPDEROU Michel, Springer (2012).

**R8** *Spaceflight Dynamics.* CARROU Jean-Pierre, Cépadues CNES (1995).

**R9** *Optimization Techniques I and II.* CERF Max, EDP Sciences (2023).

**R10** *Applied Orbit Perturbation and Maintenance.* CHAO Chia-Chun George, Aerospace Press (2005).

**R11** *Spacecraft Collision Probability.* CHAN F. Kenneth, The Aerospace Press (2008).

**R12** *Orbital Mechanics (Third edition).* CHOBOTOV Vladimir, AIAA (2002).

**R13** *Spacecraft Trajectory Optimization.* CONWAY Bruce A., Cambridge University Press (2010).

**R14** *Orbital Mechanics for Engineering Students.* CURTIS Howard, Elsevier (2005).

**R15** *Conception générale des systèmes spatiaux – Conception des fusées porteuses.* DURET François, FROUARD Jean-Pierre, ENSAE (1980).

**R16** *Methods of Orbit Determination.* ESCOBAL Pedro Ramon, Wiley (1965).

**R17** *The Control and Use of Libration Point Satellites.* FARQUHAR Robert W., Goddard Space Flight Center, NASA TR R-346 (1970).

**R18** *Libration Point Orbits: A Survey from the Dynamical Point of View.* GOMEZ Gerard, MASDEMONT Josep J., MONDELO José Maria, International Conference on Libration Point Orbits and Applications, Girona, Spain (2002).

**R19** *An Algebraic Method to Compute the Critical Points of the Distance Function Between Two Keplerian Orbits.* GRONCHI G.F., *Celestial Mechanics and Dynamical Astronomy* 93, 295–329 (2005).

**R20** *Orbital Mechanics and Astrodynamics.* HINTZ Gerald R., Springer (2015).

**R21** *Optimal Control Theory for Applications.* HULL David G., Springer (2003).

*Space Trajectories: Basic and Advanced Topics*, First Edition. Max Cerf.

**R22** *Interplanetary Mission Analysis and Design.* KEMBLE Stephen, Springer, Praxis Publishing (2006).

**R23** *Dynamical Systems, the Three-Body Problem and Space Mission Design.* KOON Wang Sang, LO Martin W., MARSDEN Jerrold E., ROSS Shane D., Marsden Books (2008).

**R24** *Modern Astrometry (Second edition).* KOVALEVSKY Jean, Springer (2002).

**R25** *Optimal Trajectories for Space Navigation.* LAWDEN Derek F., Butterworths (1963).

**R26** *Conception des véhicules spatiaux.* MARTY Daniel, Masson (1986).

**R27** *Solar System Dynamics.* MURRAY Carl D., DERMOTT Stanley F., Cambridge University Press (1999).

**R28** *Astronomie Fondamentale.* PASCOLI Gianni, Dunod (2000).

**R29** *Essential Spaceflight Dynamics and Magnetospherics.* RAUSCHENBAKH Boris V., OVCHINNIKOV Michael Yu, McKENNA-LAWLOR Susan, Kluwer Academic Publishers (2003).

**R30** *Dynamics of Atmospheric Re-Entry.* REGAN Frank J., ANANDAKRISHNAN Satya M., AIAA (1993).

**R31** *Analytic Construction of Periodic Orbits About the Collinear Points.* RICHARDSON David L., *Celestial Mechanics* 22, 241–253 (1980).

**R32** *Orbital Motion.* ROY Archie Edmiston, Adam Hilger (1988).

**R33** *Optimal State Estimation.* SIMON Dan, Wiley (2006).

**R34** *Rocket Propulsion Elements.* SUTTON George P., BIBLARZ Oscar, Wiley (2001).

**R35** *Theory of Orbits: The Restricted Problem of Three Bodies.* SZEBEHELY Victor G., Academic Press (1967).

**R36** *Statistical Orbit Determination.* TAPLEY Byron D., SCHUTZ Bob E., BORN George H., Adam Hilger Elsevier (2004).

**R37** *Advanced Control of Aircraft, Spacecraft and Rockets.* TESHARI Ashish, Wiley (2011).

**R38** *Contrôle Optimal – Théorie et Applications.* TRELAT Emmanuel, Vuibert (2005).

**R39** *Fundamentals of Astrodynamics and Applications.* VALLADO David, Microcosm Press, Springer (2007).

**R40** *Hypersonic and Planetary Entry Flight Mechanics.* VINH Nguyen X., BUSEMANN Adolf, CULP Robert D., The University of Michigan Press (1980).

**R41** *Orbital and Celestial Mechanics.* VINTI John P., AIAA (1998)

**R42** *Trajectoires Spatiales.* ZARROUATI Olivier, Cépadues Editions (1987).

**R43** *Theory of Orbits: The Restricted Problem of Three Bodies.* SZEBEHELY Victor G., Academic Press (1967).

## Web links

**R44** NORAD Two-Line Element Sets. https://celestrak.com/NORAD/elements/

**R45** SOCRATES Conjunction Reports. https://celestrak.com/SOCRATES/

# Index


**a**

aberration 102
angular momentum 5, 35, 37, 106, 313, 321
anomaly
  eccentric 9, 35, 171
  generalized 19
  hyperbolic 9, 172
  mean 13, 15, 35
  parabolic 10, 173
  true 7, 25
apoapsis (apocenter, apogee) 8, 154
argument
  of latitude 27
  of pericenter 24
ascending node 23, 262
  change 153, 189
  precession 55
  right ascension 24, 34
atmospheric capture 429
azimuth 100, 315, 320, 415, 432

**b**

ballistic coefficient 64, 418
Barker equation 15, 173
B plane 393

**c**

conic, degenerate 5
constant
  gravitational 4, 34
  Jacobi 70, 88
control
  feedback 208, 210, 225, 246, 434
  linear tangent 360
  proportional navigation 211
  velocity to gain 212
coordinates
  apparent 102
  geodetic 28, 97
  of Lagrange points 71
Coriolis acceleration 414
costate 217, 242, 249, 254, 264, 317, 333, 357, 360, 369, 409
  analytical 225, 256, 339
crossrange 432

**d**

declination 27
deflection 307, 418
deorbitation 289, 422
downrange 375, 432
drag 303, 363, 418
  acceleration 64
drive acceleration 414
dynamic pressure 303, 314, 333, 418, 421, 430

**e**

eccentricity vector 6, 34, 105, 313, 321
eclipse 252
ecliptic 26, 251
elevation 100
ellipsoid
  equiprobability 126
  uncertainty 125
energy 4
exponential atmosphere 64, 363, 417

**f**

fallout 315
first zonal term 54, 263, 276, 416

fly-by 393
formula
  Bowring 130
  Clairaut 58
  MacCullagh 54
  Sherman-Morrison-Woodbury 115
  Tsiolkovski 230, 240, 308, 335
frame
  Cross 68
  Galilean (inertial) 26
  geographic 415
  J2000 26, 96
  kinematic 415
  local geodetic 97
  local orbital 36, 199
  perifocal 16, 23, 122
  Richardson 80

*g*

Gauss equations 41, 111, 123, 241
geoid 28, 97
gnomonic projection 103
gravisphere 386
gravity
  acceleration with first zonal term 58
  assist 92
  deviation 102
  gradient 51, 124, 222
  potential with third body 59
guidance
  explicit 324
  implicit 323

*h*

heat flux 314, 333, 421, 430
Hill-Clohessy-Wiltshire equations 200
Hill region 70

*i*

impulsive transfer 151

*k*

Kalman filter 117
Kepler
  equation (ellipse) 13, 171
  equation (generalized) 20
  equation (hyperbola) 15, 172
  first law 7
  second law 6
  third law 12

*l*

Lagrange
  brackets 42
  coefficients 16, 17, 22, 99, 109, 132, 181
  equations 41
  interpolation 108
  points 71
Lambert theorem 171
law
  marginal 127
  normal (Gaussian) 126, 138, 144
Legendre polynomials 52, 59, 80
lift 303, 418
lift-to-drag ratio 411, 418
longitude
  mean 25
  true 25

*m*

matrix
  combined covariance 136
  covariance 115, 117
  gain 114
  inertia 306
  monodromy 88
  symplectic 87, 408
  transition 86, 132, 407
mean motion 12
measurement
  equation 112
  expected 113
  reduction 101
meridian 27
method
  averaging 46, 249
  continuation 232, 378
  Cowell 45
  direct 319
  Encke 45
  Ferrari 167
  Gauss 109

Gibbs 105
Halley 183
Herrick 106
indirect 318
Laguerre 13, 183
Laplace 107
least squares 112
Poincaré 82
Richardson 80
shooting 83, 322
minimum orbit intersection distance (MOID) 129
mixture ratio 311

*n*
nutation 26, 96

*o*
obliquity 26
orbit
geostationary 29, 63, 155, 258
Halo 83
Lissajous 78
Lyapunov 83
manifold 88
osculating 37, 98, 123
sun-synchronous 30, 56, 277
orbital
classical parameters 23, 34, 98, 122, 151
equinoctial parameters 25, 43
local frame 36
period 56

*p*
parallax 102
periapsis (pericenter, perigee) 8, 154, 392
Pontryaguin Maximum Principle 217, 242, 253, 264, 317, 333, 357, 359, 364, 369, 409
potential
centrifugal 69
gravitational 49, 69, 416
precession 26, 55, 96, 153, 188, 263, 266, 276
primer vector 217
problem
linear quadratic 113, 209, 323
path 279
shooting 232, 243, 255, 265, 318
staging 309, 336

*r*
rectilinear
ellipse 9
hyperbola 10
parabola 10
refraction 100
reserve 311, 348
restricted
four-body problem 91
three-body problem 68
two-body problem 4
Riccati equation 209, 323
Rice integral 140
right ascension 27

*s*
sidereal time 27
simulated annealing 285
singular arc 265, 321, 364
skip 430
sphere of influence 74, 386
spherical harmonics 51
steady descent 433
structural index 309, 330
switching function 264, 321, 364, 369
synodic
frame 68
period 390

*t*
targeting correction maneuver 407
Tisserand criterion 401
transfer
bi-elliptic 158
collinear 168
Edelbaum 249, 263, 269, 283
Hohmann 155, 192, 281, 388
impulsive 190
Lambert 387
minimum consumption 207, 243, 263
minimum eccentricity 162
minimum energy 161, 207, 244

transfer (*contd.*)
  minimum impulse 166, 169
  minimum time 243, 253
  semi-tangential 190
  time equation 174
  time equation (Battin) 177, 179
  time equation (Lancaster-Blanchard) 178, 180
transversality conditions 218, 242, 254, 317, 333
trimmed coefficients 418

***u***

universal
  functions 30, 182
  variable 19, 33, 182

***v***

variational equations 74, 117, 124, 132, 208
velocity
  circular 11
  escape 6, 11
  exhaust 241, 304
  infinite 6, 392
  losses 308, 335
  propulsive 308, 335

Printed and bound by CPI Group (UK) Ltd, Croydon, CR0 4YY
17/09/2026
14964336-0001